Green Technology

An A-to-Z Guide

The SAGE Reference Series on
Green Society
Toward a Sustainable Future

DUSTIN MULVANEY, GENERAL EDITOR
University of California, Berkeley

PAUL ROBBINS, SERIES EDITOR
University of Arizona

⑤SAGE | reference

Los Angeles | London | New Delhi
Singapore | Washington DC

Los Angeles | London | New Delhi
Singapore | Washington DC

FOR INFORMATION:

SAGE Publications, Inc.
2455 Teller Road
Thousand Oaks, California 91320
E-mail: order@sagepub.com

SAGE Publications Ltd.
1 Oliver's Yard
55 City Road
London EC1Y 1SP
United Kingdom

SAGE Publications India Pvt. Ltd.
B 1/I 1 Mohan Cooperative Industrial Area
Mathura Road, New Delhi 110 044
India

SAGE Publications Asia-Pacific Pte. Ltd.
33 Pekin Street #02-01
Far East Square
Singapore 048763

Publisher: Rolf A. Janke
Assistant to the Publisher: Michele Thompson
Senior Editor: Jim Brace-Thompson
Production Editors: Kate Schroeder, Tracy Buyan
Reference Systems Manager: Leticia Gutierrez
Reference Systems Coordinator: Laura Notton
Typesetter: C&M Digitals (P) Ltd.
Proofreader: Kristin Bergstad
Indexer: Julie Sherman Grayson
Cover Designer: Gail Buschman
Marketing Manager: Kristi Ward

Golson Media
President and Editor: J. Geoffrey Golson
Author Manager: Ellen Ingber
Editors: Mary Jo Scibetta, Kenneth Heller
Copy Editors: Tricia Lawrence, Holli Fort,
       Barbara Paris

Copyright © 2011 by SAGE Publications, Inc.

Printed in the United States of America

Library of Congress Cataloging-in-Publication Data

Green technology : an A-to-Z guide / Dustin Mulvaney, editor.

p. cm. — (The Sage reference series on green society: toward a sustainable future)
Includes bibliographical references and index.

ISBN 978-1-4129-9692-1 (hardback) — ISBN 978-1-4129-7570-4 (ebc)

1. Green technology. 2. Sustainable engineering. 3. Environmental engineering. 4. Energy conservation. I. Mulvaney, Dustin.

TA170.G75 2011 620—dc22 2011007298

11 12 13 14 15 10 9 8 7 6 5 4 3 2 1

Contents

About the Editors

Green Series Editor: Paul Robbins

Paul Robbins is a professor and the director of the University of Arizona School of Geography and Development. He earned his Ph.D. in Geography in 1996 from Clark University. He is general editor of the *Encyclopedia of Environment and Society* (2007) and author of several books, including *Environment and Society: A Critical Introduction* (2010), *Lawn People: How Grasses, Weeds, and Chemical Make Us Who We Are* (2007), and *Political Ecology: A Critical Introduction* (2004).

Robbins's research centers on the relationships between individuals (homeowners, hunters, professional foresters), environmental actors (lawns, elk, mesquite trees), and the institutions that connect them. He and his students seek to explain human environmental practices and knowledge, the influence nonhumans have on human behavior and organiza tion, and the implications these interactions hold for ecosystem health, local community, and social justice. Past projects have examined chemical use in the suburban United States, elk management in Montana, forest product collection in New England, and wolf conservation in India.

Green Technology General Editor: Dustin Mulvaney

Dustin Mulvaney is a Science, Technology, and Society postdoctoral scholar at the University of California, Berkeley, in the Department of Environmental Science, Policy, and Management. His current research focuses on the construction metrics that characterize the life-cycle impacts of emerging renewable energy technologies. He is interested in how life-cycle assessments focus on material and energy flows and exclude people from the analysis, and how these metrics are used to influence investment, policy, and social resistance. Building off his work with the Silicon Valley Toxics Coalition's "just and sustainable solar industry" campaign, he is looking at how risks from the use of nanotechnology are addressed within the solar photovoltaic industry. Mulvaney also draws on his dissertation research on agricultural biotechnology governance to inform how policies to mitigate risks of genetically engineered biofuels are shaped by investors, policymakers, scientists, and social movements.

Mulvaney holds a Ph.D. in Environmental Studies from the University of California, Santa Cruz, and a Master of Science in Environmental Policy and a Bachelor's Degree in Chemical Engineering, both from the New Jersey Institute of Technology. Mulvaney's previous work experience includes time with a Fortune 500 chemical company working on sulfur dioxide emissions reduction, and with a bioremediation start-up that developed technology to clean groundwater pollutants like benzene and MTBE.

Introduction

What sets humans apart from other living organisms on our planet is the use of technology. *Technology's* etymology derives from the Greek root *techne*, which means craft or art, and the root *-ology* conveys a discipline or field of study. We use technology to grow and prepare food, clothe and house our families, distribute our resources through markets and other financial mechanisms, transport us across the planet and beyond, and to keep us busy and entertain us. Technology has improved the living standards and life expectancy of humans, albeit unevenly. However, technology has given humans the ability to alter and transform Earth in ways previously unimaginable. We can put humans into space, turn mountains into valleys, and transport oil from miles below the sea. Our civilization's rapacious appetite for things and energy has brought considerable disturbance to the Earth's climate and ecosystems, particularly from industrial processes and land use change for agriculture. The evolution of technology has been anything but green.

Yet many argue that it will be green technology that saves human civilization and the planet as it replaces conventional technologies with more environmentally benign ones. As the human–environment relationship evolves, it is possible that technologies can be deployed to make that relationship more sustainable. Renewable energy promises to lessen our impacts to the extent to which it can be deployed. More efficient resource utilization through phenomena like industrial symbiosis and cradle-to-cradle design will lead to materials reuse and recovery, and will lower rates of raw material acquisition. Smart grids, for example, are designed to utilize energy more efficiently and encourage more energy conservation. It is also argued that green markets will drive change and innovation as the environmental externalities created by the economy are internalized, and as market prices reflect the environmental costs of doing business.

This volume's articles explore subjects related to our understanding of the ways that technologies coproduce human civilization and vice versa. There are explicit definitions of particular green technologies: for example, types of solar photovoltaic cells, algae biofuels, and white roofs. But the volume also integrates concepts and frameworks for looking at the interface between technology, society, and the environment. Many of these frameworks are derived from the related fields of the history of technology, science and technology studies, and industrial ecology. These intellectual traditions have deep roots in Marxism and other classical sociological, historical, and anthropological traditions.

Marx argued that technological development in capitalist society is exploitative and alienating. But he remained committed to the idea that working-class struggles can regain control of technological development to suit the purpose of the masses. More contemporary scholars reject the teleology of Marx, and destabilize the notion that technologies are motivated by, and behave in intended ways. Actor-network theory, for example, is a framework that emphasizes contingency and unintended consequences in technological design and deployment.

From what green technology has to offer, it remains difficult to distinguish fact from fiction, and utopian visions from the status quo. Clean coal, for example, is cast as a green technology based on reduced carbon emissions, but how does coal impact the environment through its life cycle? To what extent does solar photovoltaic technology adoption simply let people off the hook for their energy (over)consumption? With the lower carbon emissions involved with nuclear power, does this make it a green technology? Geoengineering likewise promises to bring us out of the climate change quagmire, but could also have uncertain and possibly severe impacts. Nanotechnology might be green for one person's ecological footprint, but it might also create occupational burdens in the manufacturing phase. Which technologies have impacts that are considerable and real? Can they be designed to mitigate these impacts?

Other green technologies are less embroiled in controversy. Green manufacturing, which embraces principles of product stewardship and ecological design, is one important development in green technology deployment. Green chemistry, which looks to substitute toxic chemicals for safer ones, also fits the green technology rubric, as does industrial symbiosis, which looks to employ principles of ecological to industrial systems. Design for recycling practices asks manufacturers to consider the end of life of their products to improve the ease of recycling and to avoid the issues associated with e-waste.

However, truly green technology might be something more transformative. It would change our behaviors and even our needs, instead of simply trading out one technology for one with lower impacts, a notion described as ecological modernization. Taken quite literally, the notion implies that the modernization of industrial society is becoming more and more ecologically minded, even though in practice it more closely resembles technological substitution. Likewise, a truly green technology would be one that is participatory in design and implementation.

Some argue that technologies need to be small, low impact, and decentralized to be green. E. F. Schumacher argued that "small is beautiful." These are echoes from the appropriate technology movement that advocates decentralized solar power, rainwater harvesting systems, biogas, and Earthships. They argue that some centralized technologies like nuclear power have authoritarian tendencies.

There are many questions regarding green technology. What does it mean? How do we assess its impacts? Who gets to define, develop, and benefit from green technology? We hope this volume helps the intrigued reader think through these questions. Answers to these questions are not entirely straightforward as technologies are entangled in politics, culture, and the economy, in addition to the biophysical systems that support human civilization. Yet embracing green technology is possible and even defensible as long as the social and environmental dimensions are carefully evaluated. But even then, technologies can have implications that are not by design.

Dustin Mulvaney
General Editor

Reader's Guide

Biological Processes and Concepts

Adsorption Chiller
Algae Biofuel
Anaerobic Digestion
Bacillus Thuringiensis (Bt)
Biochar
Biochemical Processes
Biogas
Biotechnology
Cellulosic Biofuels
Distillation
Green Chemistry
Maglev
Membrane Technology
Pyrolysis
Thermal Depolymerization
Thermal Heat Recovery
Thermochemical Processes

Cleanup, Recovery, and Maintenance

Desalination Plants
Environmental Remediation
Green Metrics
Green Roofing
Participatory Technology Development
Water Purification
White Rooftops
Zero-Energy Building

Energy Sources and Alternatives

Batteries
Clean Energy
Coal, Clean Technology
Concentrating Solar Technology
Earthships
Light-Emitting Diodes (LEDs)
Offshore Oil Drilling (Gulf Oil Spill)
Passive Solar
Smart Grid
Solar Cells
Solar Hot Water Heaters
Solar Ovens
Waste-to-Energy Technology
Wind Turbine

Environmental Education and Research

Engineering Studies
Environmental Science
Frankfurt School
Information Technology
Science and Technology Studies
University-Industrial Complex

Laws, Government, and Policies

Arms Race
Authoritarianism and Technology
Ecological Modernization
European Union Restriction of
    Hazardous Substances (RoHS)
European Union Waste Electrical and
    Electronic Equipment (WEEE)
    Directive
Green Building Materials
Green Technology Investing

List of Articles

List of Contributors

Badurek, Christopher A.
Independent Scholar

Boslaugh, Sarah
Washington University in St. Louis

Bridgeman, Bruce
University of California, Santa Cruz

Bumpus, Adam G.
Independent Scholar

Das, Kasturi
Research and Information System for Developing Countries

Denault, Jean-Francois
Independent Scholar

Dimpfl, Mike
Independent Scholar

Elmer, Vicki
University of California, Berkeley

Finco, Marcus Vinicius Alves
Federal University of Tocantins

Francis, Sabil
University of Leipzig, Germany, École Normale Supérieure, Paris

Gachechiladze-Bozhesku, Maia
Central European University, Hungary

Gebeshuber, Ille
Universiti Kebangsaan Malaysia

Gibbs, Beverley J.
University of Nottingham

Gill, Gitanjali Nain
University of Delhi

Goodier, Chris
Loughborough University

Gordon, Richard
University of Manitoba

Haik, Yousef
University of North Carolina at Greensboro

Harper, Gavin D. J.
Cardiff University

Harrell, Cassandra R.
Knox College

Helfer, Jason A.
Knox College

Hosansky, David
Independent Scholar

Hostovsky, Charles
University of Toronto

Jarvie, Michelle Edith
Independent Scholar

Jeon, June
Independent Scholar

Johansson, Mikael
University of California, Santa Barbara

Kaldis, Byron
The Hellenic Open University

Kera, Denisa
National University of Singapore

Khetrapal, Neha
Indian Institute of
Information Technology

Kinsella, William J.
North Carolina State University

Kte'pi, Bill
Independent Scholar

Lanfair, Jordan K.
Knox College

Lepsoe, Stephanie
Independent Scholar

Lin, Juintow
California State
Polytechnic University, Pomona

Liu, Jingfang
University of Southern California

Loy, Taylor
Independent Scholar

Macqueen, Mark O.
Aramis Technologies

Maycroft, Neil
University of Lincoln

McKenna, Russell
Karlsruhe Institute of Technology

Moran, Sharon
Independent Scholar

Mullaney, Emma Gaalaas
Pennsylvania State University

Nascimento, Susana
Lisbon University Institute

Nash, Hazel
Cardiff University

Nash, Michael A.
University of Washington, Seattle

Ogale, Swati
Independent Scholar

Ohayon, Jennie Liss
University of California, Santa Cruz

Panda, Sudhanshu Sekhar
Gainesville State College

Paul, Pallab
University of Denver

Purdy, Elizabeth Rholetter
Independent Scholar

Robinson, Robert C.
University of Georgia

Sakellariou, Nicholas
University of California, Berkeley

Salmond, Neil
Independent Scholar

Salsedo, Carl A.
University of Connecticut

Schroth, Stephen T.
Knox College

Smith, Dyanna Innes
Antioch University New England

Surak, Sarah M.
Virginia Tech

Teel, Wayne
James Madison University

Tuters, Marc
University of Amsterdam

Tyman, Shannon
University of Washington

Vadrevu, Krishna Prasad
University of Maryland, College Park

Watson, Derek
Independent Scholar

Whitt, Michael J.
Knox College

Woodworth, A. Vernon
Boston Architectural College

Young, Cory Lynn
Ithaca College

Zehner, Ozzie
University of California, Berkeley

Green Technology Chronology

c. 500,000 B.C.E.: Human beings first use fire.

12,000–6,000 B.C.E.: During the Neolithic Revolution early humans learn to domesticate plants and animals, developing agriculture and the beginnings of settlements in the Fertile Crescent. Previously gathered plants are sowed and harvested, while wild sheep, goats, pigs, and cattle are herded instead of hunted.

c. 6500 B.C.E.: The first known application of metal working with copper begins in the Middle East.

4000–3000 B.C.E.: In a seemingly simultaneous innovation, fledgling civilizations in Europe and the Middle East use oxen to pull sledges and plow fields.

3200 B.C.E.: The wheel is used in ancient Mesopotamia.

c. 3000 B.C.E.: Mules are used as cargo animals in the Middle East, fueling the earliest long-distance trade routes.

c. 3000 B.C.E.: Chinese, Egyptian, Phoenician, Greek, and Roman settlements use heat from the sun to dry crops and to evaporate ocean water, producing salt.

1200 B.C.E.: Ancient Egyptians show knowledge of sailing, but ancient Phoenicians become the first to efficiently harness the power of the wind, using early sailboats to develop an extensive maritime trading empire.

1000 B.C.E.: Egyptians use petroleum-based tar to help preserve the human body during the process of mummification.

1000 B.C.E.: The first known consumption of fossil fuels occurs in China. Coal is unearthed, and likely used to smelt copper in rudimentary blast furnaces.

600 B.C.E.: A rudimentary form of a magnifying glass is used to concentrate the sun's rays on a natural fuel, lighting a fire for light, warmth, and cooking.

200 B.C.E.: Greek scientist Archimedes is said to have used the reflective properties of bronze shields to focus sunlight and set fire to Roman ships, which were besieging Syracuse. In 1973, the modern Greek Navy re-creates the legend, successfully setting fire to wooden boats 50 meters away.

100 C.E.: The Greeks invent the waterwheel.

100–300: Roman architects build glass or mica windows on the south-facing walls of bathhouses and other buildings to keep them warm in the winter.

500: Roman cannon of law, the Justinian Code, establishes "sun rights" to ensure that all buildings have access to the sun's warmth.

500–900: The first known windmills are developed in Persia; uses include pumping water and grinding grain.

700: In Sri Lanka, the wind is used to smelt metal from rock ore.

1088: A water-powered mechanical clock is made by Han Kung-Lien in China.

1300s: The first horizontal axis windmills, shaped like pinwheels, appear in Western Europe.

1306: England's King Edward I unsuccessfully tries to ban open coal fires in England, marking an early attempt at national environmental protection.

1600s: The Dutch master drainage windmills, moving water out of lowlands to make farmland available. During the Protestant Reformation, they use windmill positions to communicate to Catholics, indicating safe places for asylum.

1792: The first electrochemical cell is developed by Alessandro Volta.

1810: Ned Ludd leads a group of weavers to sabotage machines that were set to take the artisan skill and labor out of making textiles, popularizing the term *Luddite*.

1816: Scottish clergyman Robert Stirling receives a patent for the first heat engine using a process that improves thermal efficiency, now called the Stirling Cycle. He calls his invention the Heat Economiser.

1833: English chemist and meteorologist Luke Howard describes the "urban heat island" effect in *The Climate of London*, noting that the city "partakes much of an artificial warmth, induced by its structure, by a crowded population, and the consumption of great quantities of fuel."

1900: Ferdinand Porsche builds the first hybrid electric car.

1920: In response to the perceived failures of existing federal laws to deal with mining of coal and oil resources, the United States passes the Mineral Leasing Act to regulate mining on public lands. The law governs deposits of coal, oil, gas, oil shale, phosphate, potash, sodium, and sulfur.

1947: The International Organization for Standardization (ISO) is formed to coordinate industrial and commercial standards.

1954: American photovoltaic technology makes a giant leap when scientists at Bell Labs develop the world's most efficient solar cell at 6 percent efficiency, enough power to run everyday electrical equipment.

1969: Scottish landscape architect Ian McHarg publishes *Design With Nature,* a landmark work on ecological planning.

1970: The U.S. Occupational Safety and Health Act establishes the Occupational Safety and Health Administration (OSHA).

1970: The U.S. Clean Air Act requires Best Available Control Technology to limit air pollution emissions.

1973: Ernst Friedrich Schumacher's *Small Is Beautiful: Economics as If People Mattered* criticizes the assumption that economic development requires adoption of large-scale Western technologies and a lifestyle based on acquisition of consumer goods.

1976: The U.S. Resource Conservation and Recovery Act (RCRA) is passed to manage solid and hazardous waste.

1980: The Stevenson-Wydler Technology Innovation Act is passed, enabling federal laboratories in the United States to transfer technology to industry, sparking concerns about the university-industrial complex.

1992: U.S. President Clinton's administration develops the U.S. Environmental Protection Agency's Energy Star program to promote energy-efficient devices.

1993: The U.S. Green Building Council is founded as a nonprofit trade organization that promotes self-sustaining building design, construction, and operation. The council develops the Leadership in Energy and Environmental Design (LEED) rating system and organizes Greenbuild, a conference promoting environmentally responsible materials and sustainable architecture techniques.

1996: William Rees and Mathis Wackernagel develop the concept of the "ecological footprint," which signifies all the resources used by a particular population or species in their book *Our Ecological Footprint: Reducing Human Impact on the Earth.*

2001: The U.S. Green Building Council founds the Green Building Certification Institute to certify Leadership in Energy and Environmental Design (LEED) professionals who are qualified to evaluate the sustainability of buildings.

2002: William McDonough and Michael Braungart popularize the term "cradle to cradle," which was introduced by Walter Stahel in the 1970s. "Cradle to cradle" refers to the principle that companies should be responsible for recycling the materials from their products after they are discarded.

2002: The European Union introduces the Restriction of Hazardous Substances (RoHS) and the Waste Electrical and Electronic Equipment Directives regulations that govern the disposal of electrical and electronic equipment, including establishment of collection centers where consumers can deposit discarded goods (rather than putting them in the trash), with the joint purposes of encouraging recycling and reducing the pollution caused by heavy metals and other hazardous materials.

2003: Zipcars, the world's largest car-sharing program, introduces hybrids to its Seattle fleet.

2005: The European Union Emission Trading Scheme, a carbon-trading scheme involving 25 of the then-27 European Union countries, officially begins.

2006: The *New Oxford American Dictionary* selects "carbon neutral" as its word of the year. Ironically, there is no single accepted definition of the term that refers in general to the achievement of net zero greenhouse gas emissions through reducing emissions and purchasing carbon offsets: the question is the scope of the emissions included in the calculations. For instance, should a company include the emissions related to raw materials that they purchase, or is that part of someone else's business?

2008: California Governor Arnold Schwarzenegger signs the two laws that comprise the state's landmark green chemistry initiative.

2008: Honda offers the first fuel cell car, the FCX Clarity, for lease.

2009: First Solar, a photovoltaic (PV) manufacturer, uses cadmium telluride thin-film technologies, and becomes the world's largest PV company based on megawatts (MW) of modules sold.

2009: San Francisco, California, passes the most stringent recycling and composting ordinance in the United States. Several other cities have mandatory recycling but San Francisco is the first to require composting as well.

2010: U.S. PV manufacturer Solyndra is the first PV company to secure a Department of Energy loan guarantee to develop its new CIGS (copper indium gallium diselenide) technology.

Dustin Mulvaney
University of California, Berkeley

ACTOR-NETWORK THEORY

Actor-network theory (ANT) is an integrative approach to science, technology, and society studies (STS) that combines empirical and interpretative methodologies ranging from ethnography to history of science and poststructuralist philosophy. Introduced by French STS scholars Michel Callon and Bruno Latour, and developed by British sociologist John Law and others, it views the relationship of scientific facts to social structures and agency in terms of "material-semiotic" networks. The approach is sometimes described as "situated inquiry" because it acknowledges the unique historically and contextually defined character of these networks. The complex networks are defined by constantly changing relations between heterogeneous agents (actants) that attribute agency not only to humans but also nonhumans. The integrative and processual nature of these networks avoids various forms of reductionism (technological determinism, scientific realism, and social constructivism) and introduces a more performative and normative approach creating opportunities for complex, deliberative, and open policies.

ANT claims that contemporary ecological problems present a technological, social, political, but also philosophical challenge that forces us to rethink some of our fundamental assumptions about the world and our position within it. ANT refuses to see the world in terms of a fragile human habitat in which natural, social, and cultural evolution are succumbing to the effects of foreign technological and scientific transformation. For ANT, there is no world outside the everyday processes of interaction between various actors and no predefined ideal state and equilibrium, or mysterious entities that draw it together (e.g., humanity, Gaia, or liberal democracy). What we call nature, society, and technology form a complex network and assemblages involving various actors from molecules to humans and cities that are often hard to define and follow. In order to understand but also act in such open, processual, and hybrid networks of heterogeneous actors we need a nonreductionist but also nonanthropocentric approach. The ANT approach, methodology, and language aim to help us analyze and influence the complex and emergent relations that humans develop with so-called nonhuman actors to which they have access and with which they negotiate through the experimental apparati of science and technology. It is also important to note that the term *actor* does not refer to the analysis of social relations between individual humans, and the term *network* should not be confused with a technical network, like the Internet, that represents only one possible stabilized state of an actor-network. Rather, networks are nodes that have as many dimensions as they have connections, and actors are any source of action therein.

Society and the Lab

The simple insight about the importance of nonhuman actors and hybrid networks was the core concept of ANT in the early 1980s when a group of thinkers like Michel Callon, Bruno Latour, and John Law decided to challenge the prevailing approach to STS studies. Since the 1970s, STS was dominated by history and philosophy of science and technology and the so-called strong program of the sociology of scientific knowledge. These approaches were radical and provocative but also socially constructionist and reductionist respectively, ignoring core ontological issues and diminishing the importance of material reality to which science and technology were believed to have a privileged access. In response, ANT would offer a conceptual and methodological toolbox with which the entire world suddenly looked like a giant laboratory where history could be seen as sets of experiments involving various actors.

ANT rejected the notion that science and technology merely happened in the laboratories while philosophy happened in books and politics happened in the public sphere. Their approach paradoxically combines empirical modes of inquiry from the social sciences with ontological discussions from continental philosophy in order to outline the problem of a social, political, or cultural sphere existing outside and independent of science and technology, notions that they saw as sources of confusion rather than of solutions. This is not to say that ANT dismissed the existence of society or nature, as such, but rather they draw attention to the often-overlooked work required to bring it into being. While they agreed with the claim by other STS approaches that there can be nothing purely technological or scientific that is not already part of some politics, ANT also asks us to acknowledge that politics are not possible without certain scientific and technological inventions, practices, and transformations creating conditions within which these politics can develop.

Actors and Networks

In order to avoid causality dilemmas between society and technology, ANT developed a processual view of social, political, or technological phenomena in terms of interactions between heterogeneous actors in hybrid networks and processes. ANT's most controversial claim is that the "cosmos" is neither given nor passive, but composed of active agents in which humans' agency matters no more than the agency of things that interact to compose new collectives. These agents gain strength only through their alliances, which are linked through translation, and "nothing is, by itself, either reducible or irreducible to anything else," as Bruno Latour summarizes this approach. Actors do not have an essence outside of these exchanges and networks. They are pure processes that we can study by empirically following various case studies as Michel Callon, Bruno Latour, and John Law did in the case of electric vehicles, bacteria, military projects, industry around the Saint Brieuc scallops, and so forth. Human and nonhuman actors (actants, agents) can be anything and anyone who has a propensity to act, connect, perform, resist, and translate.

In order to understand the networks, we cannot reduce them to social, textual, or objective phenomena and agency or essence but we have to describe the processes of interactions and assembling that form the networks; our methodology must "stick to the actors," which is a famous ANT motto. Rather than "matters of fact" representing an objective reality, ANT agrees with the strong program of STS that there should be a general symmetry between various actors in our explanations of scientific successes as well as scientific failures. With ANT, this principle of "symmetry" was simply given an ontological meaning such that we have to follow the specific interaction between various actors to make any useful distinction between phenomena. The goal is thus to understand the connections of heterogeneous actors in hybrid networks across scales.

From Networks of Actors to the Parliament of Things

In the 1990s, the materialist and realist positions of ANT and its ontology based on the importance of networks and "nonhuman" actors developed into policy- and design-oriented projects. While ANT's original method sought to describe the relations between nature, technology, and society (or culture), this new normative approach to ANT, largely associated with Bruno Latour, has a strong influence on design practices. Design contributes to discussions over the issues of ecological, economic, and political crises by offering various scenarios of future interactions between actors. Design for ANT is a set of tentative experiments bringing together natural processes of evolution, social processes of transformation, and technological processes of innovation. This ability of the design to creatively mobilize actors and intervene in the emergence of hybrid networks inspired ANT to use more scenario-based methodologies that can effectively test different versions of the future via prototypes and mock-up techniques. From the ANT perspective, design has a privileged status as a framework for understanding which version of our future we want to take and how to transform the spontaneous and hybrid networks into what Bruno Latour refers to as a "parliament of things," in which different spokespersons can reason about new networks and connections.

Cosmopolitics: Ecology Without Nature

In recent years, Bruno Latour's influence has grown in the humanities and social sciences, blurring the boundaries between them, thanks in part to the concept of "cosmopolitics." Developed as a direct criticism to some essentialist and romantic notions of ecology, the term *cosmopolitics* was in fact appropriated from a fellow philosopher of science, Isabelle Stengers, to describe how technology and sciences domesticate new entities in our world and open them to the heterogeneity and plurality of different actors. What Latour refers to as cosmopolitics can be understood as a normative approach to the ecological crisis that seeks to make the misalliances between politics, science, and technology more transparent and open for discussion, thereby giving a political voice to all the stakeholders and actors in the so-called parliament of things. While in the earlier descriptive phase of ANT it was used to map relations between various social, political, and technological actors and stakeholders, the recent work sees ANT taking on a more normative and ethical dimension. Cosmopolitics integrates ANT's descriptive methodologies with a normative approach in order to intervene in complex systems involving multiple stakeholders across different scales.

Since complex and emergent networks of human and nonhuman are neither purely social nor biological, critique should avoid the concept of nature altogether, for the same reason that it might avoid an unexamined deployment of the concept of god. When subjected to ANT scrutiny, any social phenomenon, scientific fact, or philosophical discourse is revealed to be part of extremely complex sets of processes, as expressed in this canonical quote: "(t)he ozone hole is too social and too narrated to be truly natural; the strategy of industrial firms and heads of state is too full of chemical reactions to be reduced to power and interest; the discourse of the ecosphere is too real and too social to boil down to meaning effects. Is it our fault if the networks are simultaneously real, like nature, narrated, like discourse, and collective, like society?"

Cosmopolitics is thus an emerging whole created by the relations between humans and nonhumans that has a great potential to reshape the world. Our understanding and interaction with the world is happening on many scales, and we need both challenging philosophical ideas and radical design practices in order to test different versions of this collective. Design lets us experience the limits and the possibilities of these abstruse notions of materiality and agency by involving us directly in different experiments and equipping us with tools that can affect these systems. By experimenting and creating new equilibria between human and nonhuman actors we create more complex, emergent, and symbiotic relations.

See Also: Appropriate Technology; Boundary Objects; Cradle-to-Cradle Design; Frankfurt School; Futurology; Luddism; Marxism and Technology; Participatory Technology Development; Reflexive Modernization; Science and Technology Policy; Science and Technology Studies; Social Agency; Social Construction of Technology; Sociology of Technology; Sustainable Design; Systems Theory; Technological Autonomy; Technological Determinism; Technological Momentum; Technological Utopias; Technology and Social Change.

Further Readings

Callon, M. "Actor-Network Theory—The Market Test" (draft). Actor Network and After Workshop. Centre for Social Theory and Technology (CSTT), Keele University, UK (1997). http://www.keele.ac.uk/depts/stt/stt/ant/callon.htm (Accessed September 2010).

Callon, M. "Society in the Making: The Study of Technology as a Tool for Sociological Analysis." In *The Social Construction of Technological Systems*, W. E. Bijker, T. P. Hughes, and T. P. Pinch, eds. Cambridge, MA: MIT Press, 1987.

Callon, M. "The Sociology of an Actor-Network: The Case of the Electric Vehicle." In *Mapping the Dynamics of Science and Technology*, M. Callon, J. Law, and A. Rip, eds. London: Macmillan, 1986.

Callon, M. "Some Elements of a Sociology of Translation: Domestication of the Scallops and the Fishermen of St. Brieuc Bay." In *Power, Action and Belief: A New Sociology of Knowledge?* J. Law, ed. London: Routledge & Kegan Paul, 1986.

Latour, B. "On Actor Network Theory: A Few Clarifications" (1997). http://www.nettime .org/Lists-Archives/nettime-l-9801/msg00019.html (Accessed September 2010).

Latour, B. *The Pasteurization of France*. Cambridge, MA: Harvard University Press, 1988.

Latour, B. "The Powers of Association." In *Power, Action and Belief: A New Sociology of Knowledge?* Sociological Review Monograph 32, J. Law, ed. London: Routledge & Kegan Paul, 1986.

Latour, B. "The Prince for Machines as Well as for Machinations." In *Technology and Social Process*, B. Elliott, ed. Edinburgh, UK: Edinburgh University Press, 1988.

Latour, B. *Science in Action: How to Follow Engineers and Scientists Through Society.* Berkshire, UK: Open University Press, Milton Keynes, 1987.

Latour, B. "Technology Is Society Made Durable." In *A Sociology of Monsters: Essays on Power, Technology and Domination*, J. Law, ed. London: Routledge, 1991.

Law, John and John Hassard, eds. *Actor Network Theory and After*. Oxford, UK: Blackwell, 1999.

Stengers, I. *The Invention of Modern Science*, trans. D. W. Smith. Minneapolis: University of Minnesota Press, 2000.

Denisa Kera
National University of Singapore

Marc Tuters
University of Amsterdam

ADSORPTION CHILLER

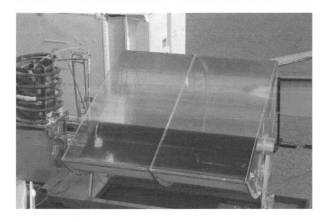

Adsorption and absorption chillers, such as this absorption icemaker, have been promoted as low energy, quiet, and environmentally friendly. They are popular in areas where electricity is costly, difficult to obtain, or an absolute necessity, such as in hospitals.

Source: David Menicucci/Sandia National Laboratory, U.S. Department of Energy

Adsorption chillers, which are driven by hot water instead of by large amounts of electricity, provide an energy-efficient alternative to conventional refrigeration and air conditioning. The source of the hot water may be heat that would otherwise be wasted, such as exhaust or steam from industrial processes, or it can be heat directly generated from solar panels or other devices. Since power plants and manufacturing facilities can lose more than one-half of the energy they consume to wasted heat, the use of adsorption chillers and other waste heat recovery systems can significantly reduce electricity use and carbon dioxide emissions.

Both adsorption coolers and more conventional compressor cooling units use a liquid refrigerant with a very low boiling point. In both types, when this refrigerant boils and evaporates, it takes some heat away with it, providing cooling. (The effect is somewhat analogous to a human becoming cool by sweating.) However, the two types differ in how they change a refrigerant from a gas back to a liquid and repeat the cycle. A compressor cooling unit uses a more energy-intensive process that involves an electrically powered compressor to increase the pressure on the gas. But an adsorption chiller warms the gas back to a liquid without using any moving parts. The adsorption chamber of the chiller is filled with silica gel, which creates an extremely low humidity condition that causes the water refrigerant to evaporate at a low temperature. As the water evaporates in the evaporator, it cools the chilled water. Adsorption chillers use a very small amount of electricity because only their pumps require electrical power to operate.

The technology behind adsorption cooling can be traced back to the mid-19th century, when a French scientist named Ferdinand Carré invented a similar system, known as absorption cooling, that used water and sulfuric acid. Other designs followed, including one created by Albert Einstein and his former student, Leo Szilard. The cooling systems have a variety of designs, using such materials as liquid ammonia, lithium bromide sale, and even a plain saltwater solution. Heat can come from virtually any source, including a campfire.

Adsorption and absorption chillers have been increasingly promoted as low-energy, quiet, and environmentally friendly alternatives to compressors. They do not emit greenhouse gases or use a chlorofluorocarbon or hydrochlorofluorocarbon refrigerant, nor do they consume much electricity or emit heat into the atmosphere or waterways. They are a popular option in locations where electricity is costly or difficult to obtain, where compressor noise can be a distraction, and where there is a readily available heat source. Adsorption chillers are found in a wide variety of settings, from industrial facilities that generate waste heat to recreational vehicles that may use small units to refrigerate food, where they draw heat from burning liquefied petroleum gas. Where there is an abundance of heat or little available electricity, industrial adsorption chillers can be used for both heating and cooling. Using absorptive refrigeration to air condition buildings by tapping waste heat from a gas turbine or water heater is especially efficient because the turbine also produces electricity and hot water.

Adsorption chillers are typically more costly to purchase than conventional compressors. Without an inexpensive source of heat, they are also more costly to operate. For this reason, companies often steer away from them unless their operations are already generating large amounts of heat. If a plant's waste heat stream of air or water reaches 500 degrees Fahrenheit or more, the facility is considered a good candidate for an adsorption chiller. The heat may come from a variety of sources, including ovens, furnaces, incinerators, kilns, dryers, and thermal oxidizers used for pollution control. A growing source of waste heat comes from combined heat and power installations as more and more industries choose to produce their own electricity. In addition to reducing emissions, adsorption chillers can also significantly reduce long-term energy costs and provide a hedge against unpredictable energy expenses. Some industrial facilities use a hybrid system that combines absorption chillers during peak times for electricity demand and electric chillers at other times. Adsorption chillers may also be used on campuses that have a central steam loop while lacking the electrical power distribution needed to run a network of electric chillers.

Adsorption chillers are an important strategy for waste heat recovery. The Department of Energy has estimated that the United States, out of the world's developed countries, has the lowest energy productivity (a measure of how much energy goes into producing each dollar of gross domestic product). This is partly because higher energy costs in other countries have spurred manufacturers to capture greater amounts of waste heat. Advocates of waste heat recovery contend that the technology can make the U.S. economy more efficient in addition to providing environmental benefits. Some policymakers support an investment tax credit for the installation of waste heat recovery systems in industrial settings.

See Also: Appliances, Energy Efficient; Green Building Materials; Sustainable Design.

Further Readings

Kessler, Eric A. "Does Waste Heat Recovery Make 'Cents'?" *Process Heating.* http://www
.process-heating.com/CDA/Archives/8b6e827d50368010VgnVCM100000f932a8c0
(Accessed September 2010).
Sakraida, Vincent A. " Basics for Absorption Chillers." *Engineered Systems.* http://www
.esmagazine.com/Articles/Feature_Article/BNP_GUID_9-5-2006_A_
10000000000000539128 (Accessed September 2010).

David Hosansky
Independent Scholar

ALGAE BIOFUEL

Algae fuel is advanced biofuel produced by algae, small, mostly photosynthetic organisms that consist of one to a few cells. Some are more closely related to bacteria than to plants. In the process of photosynthesis, algae convert carbon dioxide (CO_2), nutrients, and sunlight into oxygen and biomass, including oil. This is achieved in a highly efficient way, and it is estimated that the production of oil from algae could be 10 to 100 times higher than second-generation seed-oil crops, up to 200 barrels per hectare (ha) per year. Also, heterotrophic (nonphotosynthetic) algae can be utilized for oil production. Fossil algae may have produced our crude oil.

The reason why algae produce lipids is still unknown. The lipid oil droplets in diatoms might counter the weight of the dense silica shell and provide buoyancy. But during nitrate depletion, some species change from neutrally buoyant to sinking in spite of increased oil production. Other hypotheses see the lipid pool as reserve products; but the algae also contain significant reserves of polysaccharides. The compromise might be that oil droplets assist the long-term survival in poor environmental conditions, while polysaccharides cover short-term energy needs.

Most algae store oil droplets inside, so to extract the oil, the algae must be dried and centrifuged. The dry mass factor, that is, percentage of dry biomass in relation to the fresh biomass, is economically important. Some algae species (*Botryococcus sp.*) secrete their lipids (long chain alkenes) outside their cells. It is relatively easy to separate the oil without killing the cells. Unfortunately, oil-secreting species grow much slower than other, fast algae. Selective breeding or genetic modification might overcome this limitation.

Algae react to adverse environmental conditions by producing hydrocarbons. The lipid content increases with age, and is dependent on temperature or salinity conditions, "dark phases" (diatoms kept in the dark produce more oil droplets), nitrogen depletion (increases fat production), and drying or desiccation (increases oil production). Selected additives influence which lipids are produced (e.g., organic mercury and cadmium).

The generation of algae biofuels is achieved in bioreactors, transparent tanks with nutrient-enriched water. CO_2 is added, and the algae are illuminated with sunlight. In this environment an oxygenetic photosynthetic reaction is performed by the chlorophyll containing algae. The equation for photosynthesis is $6CO_2 + 6H_2O => C_6H_{12}O_6 + 6O_2$; energy required: 2870 kilojoules/mole (provided by light). Heterotrophic algae can be grown in high densities in containers without any illumination but need energy provided as sugar (glucose).

The classical bioreactor is the "open pond." It is simple but allows little control of algal population, takes much space, and large-scale production is difficult. High performance bioreactors are closed systems that allow better control/protection of the algae. Three kinds of bioreactors differ slightly in cultivation (intensity of sunlight energy, mass transport, size) and production schemes (batch or continuous).

The "plate photo bioreactor" consists of a series of transparent panels that are arranged to achieve a large illumination-surface area. This system is suitable for outdoor cultures and offers good biomass productivity at relatively low cost. Usually the CO_2 and nutrient fluid are injected at the bottom of the panels. As the system is rather flat, it requires sophisticated support infrastructure; also, temperature control and "wall growth" are issues.

A "tubular photo bioreactor" is a system of connected transparent tubes with the algae suspended in circulating fluid. It is suitable for outdoor production and offers a large illumination-surface area in combination with good, continuous biomass productivity at low cost. However, constant pump circulation, which usually introduces gas and nutrients, leads to deficiency of CO_2 and a high concentration of oxygen at the end of the circulation. Fouling and some degree of wall growth result.

A "bubble column photo bioreactor" is a large transparent vertical column. Gas and nutrients are injected at the bottom. The turbulent stream created by the rising bubbles allows good gas exchange, a high mass transfer, and good mixing with low shear stress. The system is easy to sterilize, readily tempered, and reduces wall growth. The main problem is the small illumination-surface area, especially upon scale-up. Large systems might require internal illumination, limiting outside use.

The supply of nutrients is difficult. As freshwater is precious the use of wastewater might be a very good alternative. But—as the algae are rather sensitive to contamination—the fluids must first be processed by bacteria through anaerobic digestion. This increases the complexity of the system. The provision of CO_2 may become a major problem, as concentrated sources are usually from fossil fuels, and therefore not sustainable (except for CO_2 capture). Solar panels containing diatoms or other algae, utilizing atmospheric CO_2, that secrete gasoline rather than provide electricity or hot water, have been envisaged.

Large-scale algae fuel production facilities are still in the development phase. Engineering challenges remain, especially in scaling and dewatering technology. Power companies have established research facilities with algae photobioreactors. The focus here lies on the scaling of the production systems from laboratory dimensions to mass production. Another goal of the research is to find out how efficiently algae fuel production could reduce CO_2 emissions and how much biomass will be produced. Algae biomass by-product can be sold as fertilizer, animal feed, or for pharmaceuticals to generate additional income. The emission reduction can be certified and converted into emission credits that can be sold to industry.

The potential importance of algae in the generation of oil and hydrocarbons has been best illustrated by an estimate from the U.S. Department of Energy (DOE). It states that if all the petroleum fuel needed in the United States were substituted by algae fuel, it would require only about 40,000 square kilometers of land, less than 15 percent of the area where corn is harvested. Algae fuel has the potential to be the most cost-effective renewable alternative energy source on the planet. However, investment in alternative fuels rises and falls with price changes for crude oil. For example, the United States halted a 15-year algal fuel project in 1995.

See Also: Appliances, Energy Efficient; Appropriate Technology; Biotechnology; Cellulosic Biofuels; Geoengineering; Green Chemistry; Green Nanotechnology; Passive Solar; Sustainable Design; Systems Theory; Wastewater Treatment.

Further Readings

"Algae Biofuels: Algal Fuel Producers, High Lipid Content Microalgae, Chevron Corporation, List of Algal Fuel Producers, *Botryococcus Braunii.*" Books LLC, 2010.

Gordon, Richard. "Quitting Cold Turkey: Rapid Oil Independence for the U.S.A." In *The Science of Algal Fuels: Phycology, Geology, Biophotonics, Genomics and Nanotechnology*, Richard Gordon and Joseph Seckbach, eds. Dordrecht, Netherlands: Springer, 2010.

Hartman, Eviana. "A Promising Oil Alternative: Algae Energy." *Washington Post* (January 6, 2001).

Ramachandra, T. V., Karthick B. Durga Madhab Mahapatra, and Richard Gordon. "Milking Diatoms for Sustainable Energy: Biochemical Engineering Versus Gasoline-Secreting Diatom Solar Panels." *Industrial & Engineering Chemistry Research*, 48 (2009).

Sheehan, John, Terri Dunahay, John Benemann, and Paul Roessler. "A Look Back at the U.S. Department of Energy's Aquatic Species Program: Biodiesel From Algae, Close-Out Report." NREL/TP-580-24190. Golden, CO: National Renewable Energy Laboratory, 1998.

Ugwu, C. U., H. Aoyagi, and H. Uchiyama. "Photobioreactors for Mass Cultivation of Algae." *Bioresource Technology*, 99/10 (2008).

Ille C. Gebeshuber
Universiti Kebangsaan Malaysia

Richard Gordon
University of Manitoba

Mark O. Macqueen
Aramis Technologies

ANAEROBIC DIGESTION

Anaerobic digestion is the breakdown of organic matter via microorganisms in the absence of oxygen, which results in the generation of carbon dioxide (CO_2) and methane (CH_4). Materials high in organic content, such as municipal wastewater, livestock waste, agricultural waste, and food wastes, may all undergo anaerobic digestion. The methane gas produced may be collected and used directly as a fuel for cooking or heat, or it can be used to generate electricity. Unlike the production of methane from gas wells, anaerobic digestion is a renewable source of energy.

Anaerobic Digestion Feedstocks

Several feedstocks exist for the anaerobic digestion process, all of which contain organic matter, including municipal and animal wastewaters and agricultural and food wastes. Anaerobic digestion is frequently used in the treatment of municipal wastewaters, often in a series of process steps that also include aerobic digestion (digestion in the presence of oxygen) and sedimentation. The amount of solids produced from the wastewater treatment can be reduced though anaerobic digestion, which in turn reduces the costs associated with solids disposal. Similar to human waste, animal waste may also provide the feedstock for anaerobic digestion.

Materials such as municipal wastewater, food waste, and livestock waste are broken down through anaerobic digestion, such as through this waste-management system for a 900-head hog farm.

Source: Natural Resources Conservation Service/U.S. Department of Agriculture

Confined feeding operations (CFOs) and concentrated animal feeding operations (CAFOs) are large animal feeding operations, typically containing more than 300 cattle, 600 swine or sheep, or 30,000 fowl. When this many animals exist on one farm, the resultant manure and wastewater can have significant environmental impacts if it is allowed simply to run over land and into storm and surface waters. The waste depletes water of its oxygen as it degrades, which can be detrimental to aquatic wildlife. The containment of the animal waste is often required for the protection of water quality. Anaerobic digestion reduces the volume of the waste, produces methane for use, and provides a digestate that can be used as fertilizer.

In addition to animal waste, plant waste from agriculture can also be processed by anaerobic digestion. In Europe, energy crops are grown for plants dedicated to anaerobic digestion, called biogas plants. If the plant accepts more than one agricultural feedstock, it is termed a co-digestion plant. Crops blighted by disease or insects may also be harvested and used as a feedstock for anaerobic digestion.

Most organics can undergo anaerobic digestion, the exception being woody wastes. Wood contains lignin, which most anaerobic microorganisms are not able to degrade. However, in the 21st century, research in the biofuels industry has focused on anaerobes that can break down cellulose for the purpose of producing ethanol from woody wastes.

Anaerobic Digestion Process

The anaerobic digestion process is used in the treatment of domestic and industrial wastewater. Wastewater treatment typically includes the following steps to remove organic solids from the wastewater. Within the typical wastewater process, both the primary and secondary solids can be anaerobically digested. Although this digestion process does produce methane, its primary intent is to reduce the volume of waste solids that must be disposed. Increasingly, municipal plants are viewing methane as a beneficial by-product of solids processing and capturing the methane to be used on site. The organics within the low-oxygen environment of landfills also undergo anaerobic digestion, producing methane that can be harvested for use.

The organics that feed the anaerobic digestion process are composed of various ratios of carbon, nitrogen, and oxygen (C, N, and O). Microorganisms use these organics as a substrate for growth and combine them with water (H_2O) to form carbon dioxide (CO_2) and methane (CH_4). The actual breakdown of organics to methane is not performed by a single microorganism but occurs in three stages through the teamwork of various microorganisms. The first microorganism converts the organics to a substance that other microorganisms can convert to organic acids. Methanogenic (methane-producing) anaerobic bacteria convert the organic acids to methane.

The amount of methane versus carbon dioxide produced depends on the composition of the original organic substrate being broken down. Sugars, starches, and cellulose produce approximately equal amounts of methane and carbon dioxide. When proteins and fats undergo anaerobic digestion, more methane than carbon dioxide is produced. Digested biogas typically contains a maximum concentration of 70 percent when fats are digested. Digested dairy wastes can produce over 60 percent methane.

Gas production is also very dependent upon temperature. Anaerobic bacteria survive in a broad range of temperatures, but thrive (and produce more methane) at 130 degrees Fahrenheit (thermophilic range); below 125 degrees Fahrenheit, the production of methane significantly decreases with temperature. Gas production can be maximized when the temperature is kept high and the feedstock is constant.

Landfill Methane

Although it is not an intentional treatment technique for municipal solid waste, the decomposition of organic matter in the low-oxygen environment of landfills naturally produces gas that is about 50 percent methane and 50 percent carbon dioxide. According to the U.S. Environmental Protection Agency, approximately one-fifth of human-caused emissions of methane come from landfills. Landfill gas can be extracted using a series of wells and collected. The gas can be flared directly if heat or electricity is needed on site. It can also be processed to increase the methane content, providing a higher-quality gas for pipelines or storage in tanks.

End Uses of Methane

Methane is combusted to release energy, which is typically used to heat homes, heat water, cook, or to produce electricity. Methane combusted at wastewater treatment plants is typically used for on-site heat at the plant. This methane would otherwise be flared or vented directly to the atmosphere. Using the methane for on-site heat reduces the overall operating costs for a plant.

In Europe, biogas use for the production of electricity is on the rise. As in many other environmental efforts, Germany has emerged as the global leader in the effort to develop and use biogas for electricity production.

In developing countries, such as India and China, small-scale anaerobic digesters can provide fuel for cooking and lighting within homes. For small farms, it is estimated that the waste from one cow can provide 0.45 cubic meters of methane per day when digested. The United Nations Development Programme recognizes small-scale home and farm anaerobic digesters as one of the most useful decentralized sources of energy. Small home-based systems allow households to use human, animal, and agricultural wastes to produce their own energy.

Whatever the ultimate use of biogas, it reduces the consumption of other, nonrenewable fuel sources of methane. In addition, biogas has less environmental impact because the use of this methane does not require drilling. Also, the carbon-neutral biogas has less global warming impact than traditional methane, as it releases carbon into the atmosphere that would have been released when the organic matter from which the gas was produced decomposed naturally.

See Also: Biochemical Processes; Composting Toilet; Greywater; Membrane Technology; Waste-to-Energy Technology; Wastewater Treatment.

Further Readings

Burke, D. A. *Dairy Waste Anaerobic Digestion Handbook: Options for Recovering Beneficial Products From Dairy Manure.* Environmental Energy Company, 2001.

Kahn, Jeremy. "Waste Not, Want Not: Plastics Maker Sintex Seeks to Solve India's Energy and Sanitation Problems in One Stroke—With an At-Home Biogas Digester." CNNMoney .com (February 27, 2008). http://money.cnn.com/2008/02/26/news/international/kahn_ biogas.fortune/index.htm (Accessed July 2010).

Sustainable Conservation. "Biomethane From Dairy Waste: A Sourcebook for the Production and Use of Renewable Natural Gas in California." http://www.suscon.org/cowpower/ biomethaneSourcebook/biomethanesourcebook.php (Accessed July 2010).

United Nations Development Programme. http://www.un.org/en/development (Accessed July 2010).

U.S. Environmental Protection Agency: Landfill Methane Outreach Program. http://www.epa .gov/lmop (Accessed July 2010).

Michelle Edith Jarvie
Independent Scholar

Anarchoprimitivism

Anarchoprimitivism combines the political framework of anarchism and the cultural critique of primitivism. In many ways, these two ideological lenses share common ground. Anarchism defies hierarchical power relations, particularly in the political domain, whereas primitivism, in general, challenges the conditions of humanity, our way of life, in the civilized world. Each offers critical perspectives on human institutions and the concomitant institutionalization of humanity and ecological systems. In this light, anarchoprimitivists tend to favor small-scale, decentralized technologies such as hand tools, minimalist housing, and wild food sources and are critical of any large-scale technological system requiring a vast infrastructure for maintenance or functionality, such as power plants, automobiles, and the complex technological webs of cities themselves. This position is as much about resisting centralized authority, whether in the form of governmental or corporate entities, as it is about reflecting the ecological concerns of a large, worldwide human population. In reassessing the technologically fueled catastrophes of human history and contemporary crises, the anarchoprimitivist perspective offers a system-level critique of civilization that draws from a continual reinterpretation of the past through archaeology and other anthropological disciplines. The better we understand the choices early humans made and the consequences of those choices, the better we will be equipped to trace what John Zerzan refers to as the "pathology of civilization."

It is important to note that primitivism itself is rarely expressed as a desire to literally reclaim a neolithic lifestyle. While much emphasis is placed on reconnecting humanity to its past ecological proximity, what is sometimes referred to as "re-wilding," little effort is made to outrightly deny or ignore the past 10,000 years. Instead, primitivists seek to retain a more intuitive relationship with natural ecologies, promoting lifestyles on a more human scale with minimal ecological footprints. However, primitivism is more than simply an extreme form of environmentalism; it is a prehistorical humanism that considers the mental, the physical, and the spiritual or mystical well-being of human individuals and societies with respect to their relative proximity and immersion in the ecological realities that have predated the emergence of humanity by many hundreds of thousands of annual cycles. Operationalizing primitivism involves an intentional asceticism with respect to the luxuries of civilization, but more importantly it calls for a categorical denial of the stubborn technological optimism that has been and continues to be the driving force behind the vast, integrated technological infrastructures of civilization. Instead of upping the ante, seeing if human ingenuity can out-engineer the social, psychological, and ecological casualties of civilization, anarchoprimitivists assert that individuals should be seeking to de-escalate civilization's technological momentum and ultimately disengage from its machinery entirely. Given the extreme countercultural nature of such a program, anarchoprimitivism does not have a wide following.

In order to reestablish the natural and ideal limits of human society, it is important to examine when and how humanity became out of sync, so to speak, with the world. For many anarchoprimitivists, the advent of intensive agriculture and animal husbandry represents some of the earliest footholds of civilization. While not an anarchoprimitivist himself, Jared Diamond thoroughly examined this thesis in his popular and expansive work *Guns, Germs, and Steel: The Fates of Human Societies*. Diamond effectively dismissed Thomas Hobbes's assertion in *Leviathan* that human life in the state of nature is "solitary, poor, nasty, brutish, and short." In fact, anarchoprimitivism can be seen as an inversion of Hobbes in that civilization—not nature—acts as the prime engine of alienation and pathological self-annihilation. Worldwide archaeological data concerning the transition from hunting-gathering societies to sedentary agricultural settlements shows a clear pattern: life expectancies drop, diseases flourish and transfer from domesticated animals to humans, labor roles become much more demarcated between the sexes, and the once small, decentralized bands and tribes give way to the centralized authority of chiefdoms and states. These new political structures are in large part a response to two key exigencies: (1) because famine and drought are unpredictable, agricultural societies have to rely on large-scale regional infrastructures for irrigation and for grain and seed storage, and (2) because of rising populations and land pressures, a professional class of war-fighters is required to control and capture resources and choice plots. For these reasons, anarchoprimitivists do not support any intensive agriculture, organic or otherwise.

The practical realities of abandoning the agricultural industry and other technological infrastructures that have enabled the world population to reach over 6 billion people are bleak. The ecological holding capacity for a world constituted solely of hunter-gatherer societies is far below 6 billion. Thus, any anarchoprimitivist program would require the world population to be downsized substantially, either through mass death or mass decisions to forgo reproduction. Since the latter would both necessitate a communication infrastructure to disseminate antireproduction propaganda and require that billions of individuals have universal access to some sort of reliable and technologized birth control to intervene in the reproductive process, the mass death scenario may be the only disengagement from the machinery of civilization that is internally consistent with anarchoprimitivism. This pressing population question coupled with radical anarchoprimitivist activists who seek to further destabilize civilization through violence and terrorism, such as Ted Kaczynski, the infamous "Unabomber," suggests many morally dubious paths for establishing an ideal anarchoprimitivist world. However, just as it would be incorrect to generalize all anarchists as bomb-chucking madmen, so too would be allowing a high-profile activist like Kaczynski to typify other members of the movement. For other anarchoprimitivists who do not actively engage in acts of violence and destruction, the downfall of civilization is seen as an inevitability. In other words, the pathology of civilization is terminal, and civilization will necessarily destroy itself. Some anarchists, such as Andrew Flood, contend that the essentialist claim of anarchoprimitivism—that mass society is inherently unsustainable—is incompatible with the anarchist paradigm, which aims to reform and organize mass society in such a way as to make hierarchical state-level governance obsolete. Coordinating and organizing a mass human society of anarchists requires sophisticated and decentralized social and material technologies; anarchoprimitivism disallows such conveniences.

Some anarchoprimitivists go so far as to position the corrupting force of civilization in the development of symbolic thought and human languages. Before humanity ever exerted its dominion over nature by cultivating crops and selectively breeding animals, it named these things that set each thing apart from the other. Thus, the wholeness of the planetary ecosystem and human consciousness itself became fragmented and malleable, open to intervention and cultivation. Now, symbological knowledge systems and institutions such as science, religion, and technology crowd out, in a totalizing superimposition, the brute experience of existence as a contiguous ecological entity. In this respect, anarchoprimitivism suggests a dismissal of objective or rational modes of being and intervening in the world and promotes a primal mysticism, a Zen-like, unmediated connection with the brute reality and immanence of human experience. For this formulation of the anarchoprimitivist critique the only territories that remain uncorrupted by civilization are an individual's personal experience prior to language acquisition and the archaeological record of societies prior to the advent of symbolic thought.

The diagnosis offered by anarchoprimitivism of civilization's ills is not far removed from other green technology critiques, such as is seen from the appropriate technologies movement. The notion that human civilization and day-to-day life rely too much upon unsustainable and heavily centralized infrastructures at the cost of ecological diversity, third world exploitation, and human health in general is becoming less and less radical every day. However, the idea that the pathology of these ills can be deterministically mapped back to the first symbolic utterance or the establishment of human dominion via the domestication of plants and animals remains an extreme proposition. In admitting such evidence, the prescriptive call to action must go beyond recycling programs, micro-power generation, organic farming, water conservation, and other technological retoolings and convenient cost-saving measures; anarchoprimitivism demands a much more essential paradigm shift: only by completely resisting the centralization of human institutions, large-scale technologies, and all the other concomitant evils of civilization can humanity hope to return to its fundamental ecological orientation.

See Also: Appropriate Technology; Authoritarianism and Technology; Technological Momentum; Unintended Consequences.

Further Readings

Bookchin, Murray. "Ecology and Revolutionary Thought." http://dwardmac.pitzer.edu/Anarchist_Archives/bookchin/ecologyandrev.html (Accessed May 2010).

Diamond, Jared. *Guns, Germs, and Steel: The Fates of Human Societies*. New York: W. W. Norton, 1999.

Flood, Andrew. "Is Primitivism Realistic? An Anarchist Reply to John Zerzan and Others." http://www.anarkismo.net/newswire.php?story_id=1890 (Accessed May 2010).

Goodman, Paul. *New Reformation: Notes of a Neolithic Conservative*. Oakland, CA: PM Press, 2010.

Zerzan, John, ed. *Against Civilization: Readings and Reflections*, Enlarged Edition. Port Townsend, WA: Feral House, 2005.

Taylor Loy
Independent Scholar

APPLIANCES, ENERGY EFFICIENT

The energy crisis of the 1970s served as a wake-up call for many appliance manufacturers, resulting in appliances that consumed less energy than ever before. Over the following decades, it became clear to large segments of the scientific community and to the public that there was a significant connection between irresponsible human energy use and global warming. In response to these demands for decreased energy use, appliances have become increasingly more energy efficient, particularly since the 1990s. While energy-efficient appliances may be more costly than others types of appliances, large savings in utility bills make them more cost-effective in the long run. International energy experts are constantly considering numerous ways in which both industrialized and developing nations can reduce their carbon footprints. Swiss economist Eberhard K. Jochem, for instance, suggests that one of the most effective ways to save energy in member countries of the Organization for Economic Cooperation and Development and in the large cities evolving in emerging countries is through making homes, businesses, and public buildings more energy efficient. That effort includes seeking out the most energy-efficient appliances that are appropriate for individual purposes and budgets. The Leadership in Energy and Environmental Design (LEED) rating system for both domestic and professional buildings promotes the use of such appliances by stipulating the use of energy-efficient lighting and heating and cooling systems as standards for LEED certification.

There are a number of simple and inexpensive measures that most consumers at any economic level can take to increase energy savings even further. These steps include switching from traditional light bulbs to compact fluorescent light bulbs that can last for several years at great energy savings, turning off lights when not in use, changing heating and cooling filters regularly, adding additional insulation, and installing energy-saving window treatments. Replacing old showerheads that can consume some 42 gallons a minute with low-flow heads can save both water and energy. In the United States, the U.S. Environmental Protection Agency aids consumers by issuing Energy Star labels for most appliances and WaterSense labels to identify products that it deems water efficient. Consumers who are interested in determining whether or not older appliances are

Shopping for appliances means more than just searching for an Energy Star symbol; consumers should conduct their own research before purchasing appliances.

Source: iStockphoto

energy efficient can purchase the Kill a Watt, a handheld device that will provide assessments of energy use. However, there has been considerable criticism of the program because standards are considered so lax that most appliance manufacturers have been able to obtain the Energy Star seal of approval.

Consumers should consider a number of factors when purchasing appliances. New models are more energy efficient and cost saving than older models. With new models, certain styles of appliances are more energy efficient than others. A front-load washing machine uses less water than a top-load model. A freezer-on-top refrigerator expends less energy than a side-by-side model, and freezer-on-the-bottom models expend even less. Certain products used with energy-saving appliances may make them even more energy efficient. Low-foaming, high-efficiency laundry detergents are more energy efficient than others. Using cold or warm water saves more energy than using hot water, and some new detergents are designed to work expressly with cold water. Gas-operated washers are more energy efficient than electric-operated models. Models with a moisture sensor inside the drum are more efficient than the more common one in which the sensor is found inside the exhaust vent.

Energy-Efficient Appliances in Practice

Buildings that have received recognition for their use of energy-efficient appliances can be found around the world. The Swiss Re Tower in London has done this partly through using natural ventilation and lighting systems and through passive solar heating. The Menara Mesiniaga building in Subang Jaya, Malaysia, uses natural ventilation, but it also employs louvers on hot external walls and unshielded windows on sides with cool external walls. The Edificio Malecon in Buenos Aires, Argentina, has operable windows that allow for the entry of breezes from a river that flows nearby. In Amsterdam, the world headquarters of ABN-AMRO uses automated blinds and digital climate regulators to allow occupants to respond to exterior temperatures. The Szencorp Building in Melbourne, Australia, has installed dehumidification units for drying and cooling offices simultaneously and has a ceramic fuel cell that provides electricity and heats the hot water tank.

In Hamburg, Germany, one home that has demonstrated outstanding use of energy-efficient appliances uses double-paned windows that cut the flow of heat in half and has silicone foam for edge seals. It also employs floor lamps that reduce energy use by one-fifth to one-fourth and operate on a sensor that can sense when no one is in the room. A booster is used to heat water to high temperatures for dishwashing, allowing for the hot water heater, which is solar and has an insulated storage tank, to be set at a lower temperature. When an old refrigerator was traded in for a more energy-efficient model, it was recycled.

A major emphasis in the fight against global warming is cutting down on amounts of energy expended by business around the world. The fact that energy efficiency does not always equate to cutting production has been demonstrated by a number of companies. For instance, in the years between 2001 and 2005, Procter & Gamble's German-based factory increased production by 45 percent. At the same time, the factory managed to keep energy increases to only 12 percent, and the carbon emission rate remained stable. This was done by using energy-efficient lighting, opting for compressed-air systems, and channeling compressor heat into buildings. Heating and air conditioning systems were redesigned for maximum energy savings, which were meticulously documented. Using natural heating and cooling systems coupled with energy-saving appliances goes a long way in reducing the carbon footprint of most businesses.

Government and Corporate Efforts

In virtually every country in the world, government agencies and the corporate world, often with the assistance of nongovernmental organizations (NGOs), are engaged in promoting efficient energy use for appliances and in establishing and meeting industry standards. In the United States, the creation of the Energy Star program was considered the best way to establish industry standards while providing consumers with information and comparisons on the energy efficiency of appliances. The program, which was established in 1992, was a joint effort of the U.S. Environmental Protection Agency and the U.S. Department of Energy. Energy Star (www.energystar.gov) labels are assigned according to designated energy consumption targets. In general, Energy Star products consume from 10 to 50 percent less energy or water than other products. The American Council for an Energy-Efficient Economy (ACEEE) also regularly ranks products to aid consumers in purchasing decisions.

Prevention Magazine estimates that by using energy-saving appliances, the average American household can save up to $600 a year while reducing the impact of global warming. It advises consumers not to depend solely on the Energy Star label for guidance because there are other factors to be considered. *Prevention* cites the example of a side-by-side refrigerator-freezer that bears the Energy Star label because it is more efficient than other similar models but may actually require more energy to run than a model with a freezer on top that failed to earn the label. Energy-conscious consumers are advised to conduct their own research before purchasing appliances.

Many European countries have also discovered the advantage of energy efficiency in appliances. At least 4,000 homes in Germany, Switzerland, Austria, and the Scandinavian countries have been designed with the intention of increasing energy savings through the extensive use of insulation and windows chosen for this purpose. Energy costs are generally only one-sixth what they would be otherwise.

In many areas, power companies work with consumers to conserve energy. Some now offer options that allow customers to cut down on energy use during the hot summer months by automatically switching off air conditioners for up to 15 minutes at a time. Utility companies also promote energy conservation through the use of rebates to customers who use energy-efficient appliances.

Governments also provide a number of advantages to customers who update old appliances with new energy-saving models. As part of President Barack Obama's stimulus plan, state governments in the United States began offering rebates through the Cash for Appliances program. The Department of Energy website provides information on rebates at www.energysavers.gov.

Governments in many countries also offer tax breaks to energy-conscious consumers. The U.S. Congress passed legislation in 2010 that offered incentives to taxpayers who engaged in greening their homes and businesses. In addition to expensive actions such as installing solar panels and wind turbines, simple measures such as adding insulation and switching to energy-efficient appliances qualified taxpayers for the benefits. Trading in old appliances for Energy Star models could potentially provide a tax credit of up to $1,500. Taxpayers could also claim 30 percent of the purchase price for installing major energy-saving devices such as solar panels, wind turbines, and geothermal heat pumps. There was no cap on major credits. Anyone constructing a new home could also take advantage of the tax credits by taking energy-efficient measures. Certain tax breaks were already available under the Energy Policy Act of 2005, which allowed those constructing new homes to earn a $2,000 credit for constructing a home that saved 50 percent energy as compared to traditional homes and a $1,000 credit for saving 30 percent. Corporations could also receive tax breaks for taking such measures as installing solar water heaters. The Energy Efficient Commercial Buildings Deduction was expressly designed to encourage green building practices.

In the Kitchen

Most energy efficiency experts still consider energy-efficient stoves a work in progress. There is evidence, however, that convection ovens can cut energy use by one-fifth. Efforts to improve other appliances have been more successful. Between 1974 and 2001, refrigerator manufacturers steadily made improvements in the way their products consume energy, cutting energy requirements by approximately one-fourth—equivalent to saving 40 gigawatts of power that would have otherwise been expended by power plants.

In the average home, refrigerators and freezers use more electricity than any other appliances, generally accounting for one-sixth of all energy expended. Smaller units use less energy than larger ones. The most energy-efficient models have the freezer on the bottom of the unit. Freezers that are full require less energy than those that are half full, and more energy is saved when items are placed so as to allow for air circulation around them. Chest-model freezers are more energy efficient than upright ones.

There have been some major advances in research that focuses on improving the energy efficiency of refrigerators. At the Department of Energy's Oak Ridge National Laboratory in Tennessee, researchers have been working with the Appliance Research Consortium to develop a "fridge of the future" that slashes current energy use in half. Researchers switched refrigerants from R-12, which is now banned, to R-134a. They used expensive vacuum insulation only around the freezer, utilizing lower-cost blown-in polyurethane foam elsewhere. Other improvements included switching motors, fans, compressors, and lighting to more energy-efficient types and redesigning the defrosting system. Appliance companies are working with the research team to get this state-of-the art model on the market. Another company has designed an even more sophisticated refrigerator with a 15-inch touch screen for surfing the Web through a broadband connection. From this refrigerator, owners can also send and receive e-mail and video messages, shop for groceries, print out lists, make phone calls, listen to the radio or MP3s, check traffic or the weather, or make appointments using a personal calendar.

As with other appliances, replacing old dishwashers, particularly those manufactured before 1994, with newer models is a big step in becoming more energy efficient. Before purchasing, one should look up particular models on the Energy Star website, aiming for an Energy Efficient number of 0.65 or higher. Multicycle models are more energy efficient than others because they allow consumers to choose the proper cycle for the operation at hand. Opting to dry without using heat can further increase energy savings. Placement of objects can also affect energy use; soiled sides of objects should face the center of the machine while large objects should be placed along the sides or back. Scraping off food rather than rinsing objects beforehand saves water consumption.

In the Laundry Room

Doing laundry generally accounts for between 10 and 15 percent of a household utility bill. In the United States, new energy-efficiency requirements for washing machines were first announced in 1991 by the Department of Energy, but ultimately, implementation was delayed until 2002. After researching energy-saving horizontal-axis models in which the tub rotated around a horizontal axis much like the tumbling action of clothes dryers that were already in use in Europe, one manufacturer released a front-loader that incorporated the new energy requirements while using a third of the water and two-thirds less energy than traditional vertical-axis machines. However, many American housewives objected to the added effort needed to use front-loaders, resulting in a tub that tilted upward and subsequently released higher-load capacities that retained high energy efficiency.

When rating washing machines, the Energy Star label is awarded as a result of both the energy used by the washer and the amount of moisture left on the clothes for the dryer to deal with. The recommended number is 1.72 or above since higher numbers are the most efficient. Clothes dryers are not covered under the Energy Star program, but experts recommend machines that have controls that are able to sense temperature and moisture.

Since 2007, washers that have earned the Energy Star label have had to be 65 percent more efficient than the pre-2007 minimum requirement. According to the Natural Resources Defense Council, making washers and dryers more energy efficient has cut national energy costs by some $200 billion. Now that manufacturers are working toward obtaining the Energy Star label, switching to the new energy-efficient washing machines and clothes dryers can help to reduce pollution and save water while cutting utility bills. Even though horizontal-axis machines are more expensive than vertical-axis machines, they save money over time. Tests of the horizontal washing machines suggest that they can reduce water consumption in a normal household by 25 to 50 percent and cut down on drying time, resulting in a savings of $40 to $60 a year. Additionally, they use less detergent, so consumers have to restock less often. Experts say that there are several other things consumers can do to reduce their carbon footprints, including washing clothes on the lowest appropriate water temperature, keeping the rinse cycle on cold, lowering hot water temperatures to 120 degrees, washing full loads only, and using the fastest spin cycle available to reduce drying time.

When drying clothes, experts recommend using automatic settings rather than timed-dry cycles, regularly cleaning lint filters, and cleaning the moisture sensor, which is usually located inside the front edge of the drum, with a cotton ball and rubbing alcohol. If possible, old washers and dryers should be recycled or donated to charity organizations. Many energy experts suggest that the most efficient clothes dryer of all is the sun. However, some areas have banned the use of outdoor clotheslines. The wave of the future in energy-efficient clothes dryers is the microwave dryer, which is designed to dry up to 65 percent faster than traditional dryers. Since the temperature inside a microwave is cooler, some 110 degrees Fahrenheit as opposed to 350 degrees in traditional models, the process is gentler on items being subjected to the drying process, thereby extending their life.

Traditional hot water heaters generally waste a lot of energy keeping water hot when not in use. Older models can be made more energy efficient by the addition of an insulating blanket, which is relatively inexpensive. Consumers are advised to replace tanks that are more than 12 years old with newer, more energy-efficient models since the average life span is considered to be 10 to 15 years. Consumers are encouraged to consider switching from electric hot water heaters to heat-pump water heaters that can transfer air to the heater from an insulated tank, thus improving energy efficiency. Rebates for energy-efficient upgrades are sometimes available through utility companies.

One of the current trends in green building is the use of tankless or on-demand water heaters. Since much of the energy used by regular water heaters is expanded in keeping stored water at the desired temperature, an inordinate amount of energy is wasted in the average household. In tankless water heaters the water is heated only when it is ready for use. In addition to energy savings, the tanks offer consumers considerable savings on utility bills. For instance, in 2001, the Seisco system manufactured by Microtherm was touted as saving customers from $69 to $89 a year if they switched from a regular gas water heater to the Seisco tankless version. Takagi and Envirotech also offered highly ranked tankless water heaters.

Heating and Cooling

In general, heating a building accounts for about half of its utility bill. One of the most efficient ways to reduce energy use throughout any structure is to install programmable thermostats that control the amount of electricity or other fuel being consumed. The chief advantage to converting to eco-friendly methods of heating is that they emit little or zero amounts of carbon dioxide at the same time they cut utility bills.

Since utility bills tend to skyrocket during the summer months in areas with high temperatures, some experts are now suggesting that lower-impact technology used alone or as a supplement to air conditioning is the solution to both energy efficiency and lowering costs. They recommend that consumers install ceiling fans, attic fans, or whole-house fans. These experts say that unlike air conditioners, fans cool by moving air around, purging interior heat, and drawing in cooler outside air. Ceiling fans, which are generally restricted to home use, are not intended to cool structures; instead they function by cooling a home's occupants by moving the air around them. Although internal temperatures remain the same, they appear to have dropped by approximately four degrees. At the beginning and end of the hot season, ceiling fans may provide adequate cooling for many homes.

There is a great variety in the appearance and cost of ceiling fans, many of which include lighting fixtures. They are also relatively easy to install. Attic fans operate by cooling the hot air that settles around a structure's roof and circulating it though a system of vents. A higher-tech model of the attic fan is the solar version. Some attic fans have exhaust systems that channel air through a gable-end vent. They are operated by a photovoltaic panel installed on the roof. The effectiveness of the attic fan is still being debated, but some experts contend that it is the most energy-efficient tool available to the average consumer. Whole-house fans are a system of fans placed at strategic locations that allow them to draw in cooler air through opened windows and channel it outward through attic vents. They are used in climates that range from moderate to hot and humid and are most effective early in the morning and in the evening.

The need for a range of energy-efficient appliances designed to meet various needs and budgets is abundantly clear. Consumers need to be aware of both government standards and available research when purchasing new appliances. In the face of incontrovertible evidence on global warming and climate change, consumers should also take simple, inexpensive steps to decrease individual, national, and global carbon footprints.

See Also: Earthships; Eco-Electronics; Green Building Materials; Sustainable Design.

Further Readings

Allenby, Braden R. *Reconstructing Earth: Technology and Environment in the Age of Humans.* Washington, DC: Island Press, 2005.

Ashley, Steven. "Energy-Efficient Appliances." *Mechanical Engineering,* 120/3 (March 1998).

Davis, Sam. "New Devices Embrace Energy Efficiency." *Electronic Design,* 58/3 (March 11, 2010).

DeGregori, Thomas R. *The Environment, Our Natural Resources, and Modern Technology.* Ames: Iowa State Press, 2002.

"Efficient Use of Energy." *Physics Today,* 28/8 (August 1975).

"Energy-Efficient Appliances." *Mother Earth News,* 188 (October/November 2001).

Foster, John Bellamy. *The Ecological Revolution: Making Peace With the Planet.* New York: Monthly Review Press, 2009.

Jochem, Eberhard K. "An Efficient Solution." *Scientific American,* 295/3 (September 2006).

"Look Into Energy-Efficient Washers and Dryers." *Mother Earth News,* 219 (December 2006/January 2007).

Mahoney, Patrick. "If You Can't Conserve Energy, Stay Out of the Kitchen." *Machine Design,* 78/19 (October 12, 2006).

Schor, Juliet B. and Betsy Taylor. *Sustainable Planet: Solutions for the Twenty-First Century.* Boston, MA: Beacon Press, 2002.

Torgerson, Douglas. *The Promise of Green Politics: Environmentalism and the Public Sphere.* Durham, NC: Duke University Press, 1999.

"Utility Bill: Free 'Earth Ship' Homes Likely for Brighton." *Professional Engineering,* 19/5 (March 8, 2006).

Elizabeth Rholetter Purdy
Independent Scholar

Appropriate Technology

Technology is the principal means by which a species alters its environment. *Appropriate technology* is a term intended to encompass those technologies that accomplish their ends with a minimum of "inappropriate" consequences. Such consequences could include environmental degradation, erosion of social fabric, exploitation, and excessively hierarchical organization of the workforce. Appropriate technology suggests a scale that is complementary to human activities in small decentralized groups, encouraging autonomy even at the expense of sacrificing maximum efficiency.

Technology is the defining aspect of human potential by which we extend our abilities and influence. The 20th century saw an exponential expansion in the realm of technology as we transitioned from the Industrial Revolution to the nuclear and digital ages. Central to this technological expansion has been a drive toward ever-greater size, scale, and scope. Large automobiles and the freeways that allow them to cover maximum ground at dizzying speeds are preferred over pedestrian streets and bicycle paths. Heating, ventilating, and air conditioning (HVAC) systems for buildings are routinely oversized to account for potential extreme conditions, adding to manufacturing, maintenance, and operation costs. Enormous power plants are designed and built to provide energy to entire regions despite the inefficiencies of transmission losses and environmental consequences. "Bigger is better" acts as an underlying assumption in marketing strategies and corporate policy, despite abundant evidence to the contrary. The potential of this approach to technology to cause significant harm has been demonstrated repeatedly in industrial disasters and is becoming abundantly evident in the pervasive environmental degradation of the planet's biosphere by current patterns of development.

Another downside of technology is its dehumanizing potential, whereby humans become subservient to the machines that were built to serve them. "Appropriate technology" intends to reverse this relationship by promoting technology that draws its modus operandi from the cultural and environmental context, using a minimum of energy to accomplish discrete and finite goals while enhancing the quality of life. The term was coined in the early 20th century, perhaps to describe a central tenet of the Swadeshi (self-improvement) movement endorsed by Mohandas Gandhi. This movement recognized that British imperialism relied in large part on the economic exploitation of the Indian people. By restoring the production of homemade fabrics and encouraging Indians to forgo store-bought clothing, the Swadeshi movement provided employment and loosened the British commercial stranglehold on the populace. In this example, the home-based production of fabric for clothing was "appropriate" in a political and cultural sense for the goals of the independence movement. The dimensions of self-reliance and simplicity are central to the concept of "appropriate technology." Gandhi's followers made a point of spinning thread for the cloth for their own clothing on a *charkha*, an ancient version of the spinning wheel. The importance placed on the significance of self-reliance and appropriate technology to the Indian independence movement is demonstrated by the depiction of a *charkha* on early versions of the flag of India. By Indian law, to this day the national flag is required to be made of *khadi*, hand-spun and hand-woven cotton cloth.

Another essential value of the Swadeshi movement as it relates to appropriate technology is the concept of decentralization. The social structures required for centralized production of mass consumer goods and transportation systems result in inequality and exploitation that deprive individuals of their dignity and self-sufficiency. Labor unrest, class struggle, and violence are the inevitable result. To choose two of Gandhi's favorite examples, the sewing machine provides a means for producing clothing independent of the commercial marketplace, and the bicycle provides maximum flexibility and range of movement at minimum expense. Both examples also principally enable the individual users by extending their creative potential, rather than robbing them of their self-determination in the interests of a corporate or governmental agenda. The evolution of industrialization in the West has been a process of centralization that parallels and reinforces the imperialist impulse. Appropriate technology is a deliberate attempt to find an alternative path to economic development whereby the benefits of technology are maximally distributed while the cultural and environmental costs are minimized.

The economist Ernst Friedrich Schumacher advanced the principles of "appropriate technology" as an author (*Small Is Beautiful* and *A Guide for the Perplexed*), consultant, and maverick, moralist thinker. Schumacher's ideas regarding technology have been given the title *intermediate technology*; however, the principles do not vary widely from those of appropriate technology. Like appropriate technology, intermediate technology does not seek to minimize labor as part of industrial production and does not attempt to justify investment in expensive industrial infrastructure by maximizing market penetration. The emphasis that Schumacher placed on scale and region sought both to keep production local to markets and to guarantee full employment for the residents of the region. Intermediate technology seeks to protect social structures as well as local resources as a higher priority than economic efficiencies. Schumacher saw the replacement of jobs by machines as the most visible symptom of the ills of modern industrialization. His views on this were quasi-religious in nature. Developing a set of principles that he called "Buddhist economics" following a trip to the Far East in 1955, Schumacher sought to promote the idea that "good works" are central to moral development. But intermediate technology is an environmental tool as well as a social and religious one, due to its respect for available resources and technologies and its concern for maintaining these.

Schumacher was a peripatetic advocate for the adoption of appropriate technology as an avenue to sustainable, self-sufficient development in third world countries. His organization, the Intermediate Technology Development Group, originally intended to provide consulting services only, has evolved into a hands-on charity that continues its work today under the name Practical Action, developing appropriate technologies in the areas of food production, renewable energy, water, and sanitation. The efforts of Practical Action have taken on an enhanced meaning as the impacts of peak oil and global climate change become more widely recognized. In this respect, Schumacher must be acknowledged as prophetic in the arena of sustainable economic development. In *Small Is Beautiful*, first published in 1973, Schumacher declared the economic paradigm of the times as unsustainable and provided a road map to a sustainable future based on social and environmental limits that remains valid today.

Practical Action has estimated that for 3 billion people, residents for the most part of third world countries, the primary source of energy for cooking and other activities is "biomass" in the form of either wood, charcoal, or organic waste. Provision of gas pipelines or an electrical grid is not economically practical; however, improved cooking stoves, biogas plants, small-scale wind power generators, and micro-hydro plants can improve the quality of life for this population while reducing waste and emissions and improving indoor air quality. These appropriate technologies also increase free time and income, while providing power for electric lights, computers, and other media.

Practical Action has developed a low-cost solar lantern (the "Glowstar Lantern") that incorporates a radio and can hold up to a six-hour charge. When a school in Sri Lanka received a grant for computers but could not afford the electricity required to operate them, Practical Action provided a small "pico" hydro scheme that runs the equipment without emissions using a nearby stream. In Africa, an enhanced design for construction of clay stoves increases the efficiency of the fuel source (typically wood) while providing an extremely economical cooking tool. Where vehicle batteries are routinely used in households as the sole energy source, such as in Sri Lanka, Practical Action has promoted the construction of small-scale wind generators to reduce the need for travel to commercial recharging stations.

Another organization devoted to the development and distribution of appropriate technologies is the Appropriate Technology Collaborative (ATC). Just as a sewing machine can be operated by foot power, so can the ATC's "Treadle Pump," which looks like a "stair master" exercise machine with the added feature of a water pump and hose. Such a device would be welcomed by anyone facing a flooded basement in a power outage; however, as a means for facilitating irrigation, the Treadle Pump could prove indispensable to anyone living off the grid. ATC reports that 1.4 million of these devices have been sold for $20 apiece in Bangladesh. Other innovations championed by ATC include solar refrigerators, wind belts, and a thermoacoustic engine that uses heat to generate sound waves that are then harnessed to create electricity. ATC's founder, architect John Barrie, sees endless opportunity in his chosen field of "designing for poor people."

Perhaps the most fundamental application of appropriate technology is in the practice of agriculture. While developed countries have evolved an agribusiness approach to food production—reliant on chemical fertilizers, heavy equipment, and transportation infrastructure—the third world requires a less ambitious approach that acknowledges social and environmental constraints. A growing movement toward sustainable agriculture in the developed world has embraced appropriate technology as well, acknowledging that the consequences of corporate agriculture include depletion of topsoil, groundwater contamination, increasing production costs, and negative social consequences. Adapted and diverse crop species requiring minimum irrigation and mulched to provide soil management is appropriate agriculture in comparison to the introduction of water-dependent monoculture crops. The implications of manual (or pedal) pumps for agriculture have already been discussed. Solar voltaic water pumps have also been developed and are in limited use, and wind power has been used to move irrigation water for centuries. Hand- or foot-operated threshers, winnowers, and corn huskers are simple to build and to use. The production of compost, whereby organic material containing nitrogen is mixed at a proper ratio with carbon materials, involves only the application of labor and the technology of a pitchfork to create rich, microorganism-laden soil. Composting does in a few days or weeks what it takes nature years to accomplish, and the process actually binds carbon dioxide rather than releasing it, along with methane, which would contribute to the accumulation of greenhouse gases, as occurs if plant wastes are sent to a landfill.

Soil supplementation technology was known and practiced by the inhabitants of the Amazon Basin as long ago as 450 B.C.E. By adding charcoal, bone, and manure over long periods to the otherwise relatively infertile soils of the region, they created a richer planting medium dubbed "terra preta," a Portuguese phrase meaning "black earth." Vegetative waste, broken pottery, and animal feces were also mixed in to create a nutrient-rich soil with high concentrations of nitrogen (N), phosphorus (P), calcium (Ca), zinc (Zn), and manganese (Mn), as well as significant microorganic activity. Long before the invention of chemical fertilizers, the technology of soil supplementation combined the principles of composting, recycling, and organic gardening to create fertile soils of enduring value. Rich deposits of this soil continue to expand at the rate of 1 centimeter a year due to the activity of the microorganisms it harbors.

A critical aspect of any technology involves the implications for maintenance and repair. If parts or expertise are not readily available, then the costs in terms of time and money for repairs increase, and the effectiveness of the technology decreases. Appropriate technology is repairable locally, ideally by the user. This maintenance aspect is especially critical where the technology in question provides a vital ongoing benefit and in agricultural work where both planting and harvesting have narrow windows of opportunity. A list of prerequisites for appropriate agricultural tools has been provided in the United

Nations Food and Agriculture Organization (FAO) publication *Farm Implements for Arid and Tropical Regions*:

Recommended tools should be as follows:

- Adapted to allow efficient and speedy work with the minimum of fatigue
- Not injurious to man or animal
- Of simple design, so that they can be made locally
- Light in weight, for easy transportation (there are also considerable advantages when threshers, winnowers, and machines such as coffee hullers can be easily moved to where they are needed)
- Ready for immediate use without loss of time for preparatory adjustments
- Made of easily available materials

The principles of appropriate technology have also been applied to design and construction. Construction materials that are locally harvested or manufactured and largely renewable play a part, as does involvement of the end user in the design and construction process. Earth is the most widely used construction material worldwide, with excellent thermal properties for insulation and passive solar storage. Soil-cement and adobe are essentially variants on the principles of earth construction. Soil-cement can be manufactured by means of an easily assembled hand-operated press. The ultimate earth house is wholly or partially underground, whereby the maximum thermal benefits of the material are employed. In terms of vegetative renewable materials, bamboo stands out for its rapid growth and potential for structural applications, including its use as reinforcing material embedded in plaster, cement, and stucco. Rethinking all aspects of construction in terms of the principles of appropriate technology involves the inclusion of the social and cultural sphere, as well as physical and emotional well-being. If design is collaborative rather than imposed, the community is empowered. If labor, materials, and services are local and not imported, the economy is strengthened. If the construction process carefully incorporates principles of sustainability with indigenous techniques, the environment is protected. If the well-being of the worker is made a priority (alternative designs for wheelbarrows increase efficiency while reducing the risk of injury), both productivity and welfare are enhanced.

In a manner analogous to the development of corporate agriculture, the dissemination of uniform architectural detailing and standardized construction techniques has had unforeseen consequences. The unique expression of culture and place in indigenous buildings reinforces identity, gives voice to cultural values, and often responds with maximum effectiveness to climatic conditions. Architecture adapted to climate is attentive to wind and solar impacts, mitigating negative consequences while seeking to harness the energy of these natural forces for the comfort of inhabitants. Passive ventilation techniques can be employed in tropical climates by relating building fenestration to prevailing breezes and encouraging interior air movement by allowing the escape of warmer air at the top of a structure or flue. This is an aspect of the built environment the world over in tropical environs, except where modern construction has introduced sealed windows and HVAC systems. Building roofs, awnings, and canopies can provide enough overhang to protect window openings from excessive exposure to the hot summer sun while allowing the lower solar trajectory of winter to provide welcome solar heat gain. Again, only a building that relies on artificial means of heating and cooling can afford to ignore these common and commonsensical principles of architectural form. The Industrial Revolution combined with the accumulation of unprecedented wealth by first world countries has provided development in the third world with a choice: expensive construction relying on imported materials and equipment with high maintenance and life-cycle costs or refinement of the indigenous principles that yield comfort and economy with minimal operational and environmental costs.

Displaced camp dwellers were employed to weave mats from grass to be used to build water-resistant temporary shelters in Darfur, Sudan, during the recent civil war. During the Iran–Iraq war, Iranian architect Nader Khalili was experimenting with "firing" earth buildings, an in-place ceramic transformation of an indigenous construction method. After Desert Storm, his system for "Super Adobe" homes was used for refugee housing in Iran. Today perfecting his earth dome shelters in California for use in such humanitarian crises as the 2005 earthquake in Pakistan, Khalili has developed a system of flexible earth-filled tubes made from plastic or burlap and laid in courses with barbed wire serving as reinforcing material. After the Indian Ocean tsunami of 2004, a combination of indigenous materials and impact-resistant concrete was used in the design and construction of the Safe(R) house in Sri Lanka, replacing the previous, less-resilient building palette of concrete block and tin or tiled roofs. The new designs cost approximately the same as the previous construction method, with salvaged materials from destroyed buildings reused in the new structures. Straw-bale construction was used by the Hopi Nation for housing and for other buildings. In India, unfired blocks made from soil and sand with a cement or lime binder provided a durable and economical material for a new community center in an impoverished rural village. All of these examples and more are described and illustrated in Architecture for Humanity's 2006 publication *Design Like You Give a Damn*. Focused mostly on shelter for humanitarian response to crisis situations, the work is a textbook of adapted appropriate building technology as well as innovative and higher-tech solutions.

Village Earth is a nonprofit focused on bottom-up decision making and the dissemination of appropriate technology in impoverished communities. Its library of over 1,000 volumes may be the single largest research resource on the topic of appropriate technology. Its mission is "to help reconnect communities to the resources that promote human well-being by enhancing social and political empowerment, community self-reliance and self-determination." Village Earth regards poverty as a symptom of alienation from an appropriate and empowering relationship between a human community and its environmental context. By empowering communities to make meaningful decisions and introducing appropriate technology, Village Earth promotes a healing of this relationship. In the Amazon region of Peru, Village Earth has facilitated local responses to resource exploitation while assisting women's craft cooperatives to connect with capital, training, and markets. Working with a group called Engineers Without Borders in Purulia, India, Village Earth has conducted a "hydraulic assessment" and developed water security plans for individual villages intended to protect and enhance this limited resource.

Village Earth has a particularly succinct definition of appropriate technology:

Appropriate Technologies are small-scale, more labor intensive than capital intensive, culturally relevant, and sustainable technologies appropriate for particular environmental conditions, social contexts, and economic situations. It also includes the recognition that technologies can embody cultural biases and sometimes have political and distributional effects that go far beyond a strictly economic evaluation. Appropriate technologies are not appropriate for all situations, but are more about local people using local knowledge and resources to meet their needs without compromising the ability of future generations to do the same. Therefore it involves a search for technologies that have, for example, beneficial effects on income distribution, human development, environmental quality, and the distribution of political power in the context of particular communities and nations.

This definition acknowledges that while technology extends our influence and abilities, its effect is equally powerful in altering our sense of self and community. The design, selection, and use of technology that seeks only maximum economic results and ignores social consequences results in unforeseen problems and hidden costs. Technology can reinforce community and enhance human dignity or it can destroy these. While appropriate technology may seem backward looking, it is not an exercise in nostalgia. The technology of the automobile and the freeway imagined liberation of human movement, without thought of environmental and social consequences. When economic growth is fed by fossil fuel combustion without consideration of the impact of emissions, a fearful day of reckoning becomes inevitable.

Appropriate technology looks at the holistic impact of a tool or technology and rejects approaches that emphasize profit over people or the well-being of the planet. A technology based on the extension of power and an ethos of domination is doomed to fail. Appropriate technology emphasizes our relationships with one another and with the natural resources on which we all rely. The movement toward a wider use of appropriate technology parallels movements such as holistic medicine and organic farming, which respect and seek to employ natural forces for healing and growth.

The technology of the Industrial Revolution is the foundation of modern capitalism, whereby labor is considered a resource and all resources are considered appropriate for exploitation. This is the thesis of Paul Hawken, Amory Lovins, and Hunter Lovins in their book *Natural Capitalism,* in which they hold that "capitalism, as practiced, is a financially profitable, nonsustainable aberration in human development. What might be called 'industrial capitalism' does not fully conform to its own accounting principles. It liquidates its capital and calls it income. It neglects to assign any value to the largest stocks of capital it employs—the natural resources and living systems, as well as the social and cultural systems that are the basis of human capital." Hawken, Lovins, and Lovins seek to expose the illogic of the current economic model, summarized as the following:

- Economic progress can best occur in free-market systems of production and distribution where reinvested profits make labor and capital increasingly productive.
- Competitive advantage is gained when bigger, more efficient plants manufacture more products for sale to expanding markets.
- Growth in total output (GDP) maximizes human well-being.
- Free enterprise and market forces will allocate people and resources to their highest and best uses.

This model assumes a world of infinite resources and does not account for the waste products either of the manufacturing process or of a consumer economy. Hawken and the Lovinses respond with the simple observation that "The planet is not growing, so the somewheres and elsewheres are always with us. The increasing removal of resources, their transport and use, and their replacement with waste steadily erodes our stock of natural capital." The authors emphasize that our prevailing economic model is enormously inequitable and inefficient, asserting that a system of "natural capitalism" would emphasize the following:

- The environment is not a minor factor of production but rather is "an envelope containing, provisioning, and sustaining the entire economy."
- Misconceived or badly designed business systems, population growth, and wasteful patterns of consumption are the primary causes of the loss of natural capital, and all three must be addressed to achieve a sustainable economy.

- Future economic progress can best take place in democratic, market-based systems of production and distribution in which all forms of capital are fully valued, including human, manufactured, financial, and natural capital.
- Human welfare is best served by improving the quality and flow of desired services delivered, rather than by merely increasing the total dollar flow.
- Economic and environmental sustainability depends on redressing global inequities of income and material well-being.

Technology, like language, appears to be a defining aspect of the human race. Other species communicate verbally and use physical objects for limited purposes, but no others employ such a range of vocabulary and techniques in their communication and physical endeavors. Language stems from perception and then shapes it, and technology similarly expresses and reinforces fundamental aspects of our belief systems. A transformation of our relationship to technology implies an upheaval in our worldview, a paradigm shift that reimagines our sense of self and our place in the scheme of things. The application of appropriate technology has its roots in political dissent and countercultural ideology, but has proven to hold a staying power stemming from an important central tenet: any definition of wealth cannot disregard the health of the community or the natural world. Appropriate technology is wholly in line with the vision for a natural capitalism espoused by Hawken and the Lovinses and may prove to be a central ingredient in the development of a new and sustainable economic paradigm. This potential role of appropriate technology in bringing about an equitable and sustainable economic model for the postindustrial age would vindicate the social activism of Gandhi and the economic insights of Schumacher, while restoring dignity to the worker and balance to the relationship of human development and nature. Signs that such a development could be in the works are found in the increasing prevalence of renewable energy sources, the democratization of knowledge access through the World Wide Web, alternative transportation planning that emphasizes access and storage for bicycle use, and restriction of development in fragile ecosystems. The term *appropriate technology* implies the existence of its opposite, and consensus is growing that waste, pollution, dehumanization, and centralization of resources is "inappropriate" and cannot be condoned.

See Also: Best Available Technology; Intermediate Technology; Technology Transfer.

Further Readings

Architecture for Humanity, Kate Stohr and Cameron Sinclair, eds. *Design Like You Give a Damn: Architectural Responses to Humanitarian Crises.* Los Angeles: Metropolis Books, 2006.

Hawken, Paul, Amory Lovins, and L. Hunter Lovins. *Natural Capitalism.* New York: Little, Brown, 1999.

Hopfen, H. J. *Farm Implements for Arid and Tropical Regions.* Rome: Food and Agriculture Organization of the United Nations, 1969.

Village Earth, The Consortium for Sustainable Village-Based Development. http://www .villageearth.org (Accessed October 2010).

A. Vernon Woodworth
Boston Architectural College

ARMS RACE

As late as the 1950s, people were invited to sit outside and watch nuclear tests, like this Operation Teapot test, for the aesthetic pleasure of the mushroom clouds.

Source: Photo courtesy of National Nuclear Security Administration/Nevada Site Office via Wikimedia Commons

An arms race is any competition among parties for military supremacy through the buildup and development of superior military technology, and especially the phenomenon of escalating sophistication and efficacy of military technology as the result of such competition. When used figuratively, it refers to competitions in which there is no apparent end point, only the relative and transient victory of staying ahead of the other competitors. The weapons used in an arms race need not necessarily ever be used: the term without disclaimer or other context typically refers to the nuclear arms race between the United States and the Soviet Union from the 1940s to the 1990s, in which the only nuclear arms used in war were those used by the United States against Japan to conclude World War II. In a sense, and perhaps in the eventual view of history, the mere fact that 45 years of weapons development occupied the governments and industries of two nations without ever putting those weapons to deadly use is the unique characteristic of the nuclear arms race. However, it is in the postnuclear age that green concerns and military arms races have converged; when the emphasis of weapon development was on the nuclear, there was no way to minimize or negate its environmental impact, by the very nature of nuclear arms. In today's world, where nuclear weapons still exist but are no longer the focus of arms buildup for most militaries, military scientists can speak with a straight face about less harmful bullets, and more environmentally friendly explosives: weapons that kill the enemy without killing the environment.

Even without being used in warfare, nuclear weapons have had and continue to have significant environmental impact. The equivalent of 16,250 Hiroshima-sized nuclear bombs were exploded aboveground before the 1963 Nuclear Test Ban Treaty went into effect. The full effects of the fallout on human, animal, and plant life are still unknown. Proximity to the Nevada nuclear test site (137 miles upwind of the film's set) is thought to have contributed to the statistically significant number of deaths by cancer among the cast and crew of the Howard Hughes film *The Conqueror*: 91 of 220 cast and crew members developed cancer, 46 of them dying, including the six leading actors.

There have been at least 2,000 nuclear tests conducted worldwide since 1945, including 504 atmospheric tests conducted by the United States, the Soviet Union, the United Kingdom, China, and France. The United States alone has conducted more nuclear weapons tests than all other known nuclear states combined. Atmospheric tests produce mushroom clouds of irradiated particles that move downwind and disperse, with smaller particles dispersing great distances. Some explosions kick radioactive material up into the stratosphere, where it takes years before making its way back down to Earth, by which time it has been uniformly distributed. Thus, while any life downwind of a nuclear explosion is exposed to the most radioactive material, all life on Earth has been exposed to some quantity of irradiated material in the form of global fallout created by these tests. Early health risks were expected to be primarily genetic in nature, but time has shown that cancers, which can develop more than a decade after exposure, are by far the more common danger. As late as the 1950s, small-town Nevadans were invited to sit outside to watch the nuclear tests for the aesthetic pleasure of watching mushroom clouds form.

There are also health effects from the production of nuclear weapons. Hanford, Washington, was a nuclear weapons production site, and residents of Hanford brought suit against the federal government when the Department of Energy released 19,000 pages of documentation revealing the extent to which the Columbia River and 75,000 square miles of land had been contaminated with radioactive material, as well as groundwater systems hundreds of miles downstream. One of the significant sources of danger at the Hanford site was the separation and purification of plutonium for use in nuclear weapons, which resulted in releasing radioactive plutonium particles into the air, polluting not only Washington state but Oregon, Idaho, Montana, and parts of western Canada. It was estimated that 2 million people were exposed to radioactive contaminants in their food or water as a result.

The Radiation Exposure Compensation Act of 1990 guarantees that plaintiffs who can show a correlation between a disease they have suffered and their exposure to radioactive material because of federal government activity are eligible to compensation from the federal government, depending on the nature of the exposure: $50,000 for those exposed to fallout, $75,000 to those exposed to radioactive material on-site, and $100,000 for uranium miners. A 2000 amendment further awarded eligibility to uranium mill and ore workers, granting the same compensation as miners. As of October 2009, a total of $1,444,082,096 in compensation had been disbursed for 21,629 approved claims (a further 8,736 claims were denied).

The radioactive waste produced by nuclear programs remains dangerous for thousands of years and must be stored securely for that entire period. Though there have been few confirmed major accidents in military nuclear waste disposal, the 60-some years in which they have had an opportunity to occur represents a small fraction of the lifetime of this waste. Even the task of properly quarantining nuclear waste is one that will cost hundreds of billions of dollars even if further nuclear waste production were to cease immediately. Furthermore, the long-term consequences of nuclear activity may not all have been discovered yet: in particular, the underground tests that replaced aboveground tests after 1963 are believed to be—and most likely are—far less environmentally detrimental than atmospheric tests, but they are also much greater in number. The cumulative effect could be significant and could be as yet unrevealed. The same is true for the release of radioactive materials into the oceans by underwater tests and nuclear submarines.

Today, nuclear weapon states include the United States, Russia (other former Soviet states have transferred their warheads to Russia), France, the United Kingdom, China, India, Pakistan, and North Korea. It is widely considered an "open secret" that Israel possesses nuclear weapons, which it refuses to either confirm or deny. South Africa has the unique status of having formerly been a nuclear weapon state, before voluntarily dismantling its program. In addition, the following states (as well as all of the aforementioned except Israel and North Korea) possess nuclear power: Argentina, Armenia, Belgium, Brazil, Bulgaria, Canada, the Czech Republic, Finland, Germany, Hungary, Japan, South Korea, Mexico, the Netherlands, Romania, Slovakia, Slovenia, Spain, Sweden, Switzerland, Taiwan, and Ukraine.

In the 21st century, the arms race the United States is engaged in encompasses a wider range of competitors, as the world is no longer dominated by the Cold War, and the competition is not always military. The need to wean the country from dependency on foreign oil is expressed in terms of national security, as it makes the United States vulnerable to unstable states—and to Somali pirates, in the case of their 2008 hijacking of the crude oil carrier the *Sirius Star* and its 2 million barrels of oil. By 2020, the U.S. Navy intends to obtain at least half of its energy from alternative sources, and will launch the Great Green Fleet, a strike group powered without fossil fuels, in 2016. The Navy's Green Hornet, an F 18, became the first biofuel-powered aircraft to break the sound barrier.

The war in Afghanistan has underscored the link between energy efficiency and operational effectiveness, as resupply convoys idle in lines eight miles long, providing easy targets for insurgents. More than half of the vehicles in the convoy are supplying fuel and water, resources that can be used more efficiently and resupplied less often.

A "green arms race" may be in its early stages between the United States and China, the state-run economy of which has succeeded in adopting clean energy more quickly. Environmental attorney Robert F. Kennedy, Jr., who also coined the term *carbon incumbents* to refer to the entities in power that benefit from continued reliance on fossil fuels, has explicitly referred to the competition as an arms race, notably in a 2009 editorial for the *Huffington Post*. The Chinese advances in green technology, and the government's massive amounts of spending (at least equal to its military spending), have catalyzed fears of a future in which the United States is dependent on Chinese energy, even more than the country has been dependent on Middle Eastern oil. By 2010, the Chinese share of the solar energy equipment market had already exceeded the American share. Thirty-eight percent of the Chinese economic stimulus plan was spent on green technology (compared to 12 percent of the American stimulus spending), including a plan to increase wind power production twelvefold by 2020 and solar power production by a staggering 20,000 percent. Cheap Chinese solar panels threatened to put American solar manufacturers out of business; even Nellis Air Force Base in Nevada, which has the country's largest solar panel array, responsible for a million dollars in annual energy cost savings, was equipped with Chinese-manufactured solar panels. The United States continued to lead the world in wind installations, with the wind industry overtaking the coal industry in providing jobs, but China led in wind turbine manufacture.

In part because of this perceived threat and in part in response to the Chinese military's push to become more fuel efficient and develop clean energy resources for its own use, the American military has begun work on green military technology. Investments have been made in solar technology for military use as well as a major ocean-energy projects. Military bases throughout the United States installed solar panels all through 2010, and in the previous year, Utah's Hill Air Force Base installed the state's largest solar panel array. Solar-powered battlefield radios and tents equipped with lightweight solar panels are also in development.

The U.S. naval base at Diego Garcia, an Indian Ocean atoll, is engaged in the development of ocean thermal energy conversion (OTEC), a method of generating electricity by operating a heat engine using the temperature difference between deep and shallow seawater. Cold deep water chills the ammonia in a closed system, turning it into a thick liquid that is piped into a turbine. Warm surface water heats the ammonia, vaporizing it into a gas that turns the turbine blades before being cooled down again. There is more ocean energy of this sort available than wave power, but the process has low overall efficiency. On the other hand, the energy source itself is free, placing most of the cost on the initial setup and equipment maintenance, as with solar or wind power. If conducted on a large scale, OTEC could conceivably be cost competitive with other energy sources and could be used to produce sufficient hydrogen to meet the country's fuel needs. Cold water (usable to air-condition buildings) and chilled soil (which is of agricultural benefit in the subtropics where OTEC is best conducted) are both produced as OTEC by-products, and OTEC can easily be adjusted to simultaneously produce desalinated water suitable for a drinking supply. Furthermore, because the process draws up large amounts of deep seawater rich with nutrients that have been depleted at the surface, the process could also be combined with an aquaculture project to raise seafood. OTEC work originally began under the Carter administration, during the energy crisis of the 1970s; the Reagan administration canceled it, but in the 21st century it has been revived. There have also been suggestions that OTEC systems installed on the Arctic coast could exploit the temperature difference between the seawater and the frigid air at a cheaper cost than that of deep/surface OTEC systems in the tropics.

An uneasy alliance has developed between ecological activists, traditionally liberals and progressives, and the American military. Though many activists would rather see no weapons used, for instance, it is difficult for them to argue that the military should not develop weapons with lower ecological impact and collateral damage. Both the British and American militaries are at work on lead-free bullets to avoid poisoning the soil of battlegrounds, as well as explosives that release fewer toxins into the air, soil, and water when they are detonated.

The term *arms race* is also used analogously in ecology, to describe various phenomena. For instance, the coevolution of predators and prey is sometimes described as an arms race: as the prey adapts to better survive attempts at predation, the predator adapts to become a more able predator. Relationships of this sort are not limited to those between living things. Many have warned against the overuse of antibiotics because of a similar phenomenon: as better antibacterial technology is employed, only the strongest bacteria survive, increasing the strength of the bacteria populace and the demand for stronger antibiotics. Similarly, the term *pesticide treadmill* refers to the need to develop different and better pesticides as the targets adapt to previous ones.

See Also: Bioethics; Science and Technology Policy; Social Construction of Technology.

Further Readings

Austin, Jay E. and Carl E. Bruch, eds. *The Environmental Consequences of War.* New York: Cambridge University Press, 2007.

Endicott, Stephen and Edward Hagerman. *The United States and Biological Warfare.* Bloomington: Indiana University Press, 1998.

Mangold, Tom and Jeff Goldberg. *Plague Wars.* New York: St. Martin's Griffin, 2001.

Tucker, Jonathan B. *War of Nerves: Chemical Warfare From World War I to Al-Qaeda.* New York: Anchor, 2007.

Bill Kte'pi
Independent Scholar

AUTHORITARIANISM AND TECHNOLOGY

Technological sophistication does not necessarily entail social progress or augmented individual liberty. In fact, managing technological development is a social and political problem of how decisions about the nature of humankind and its environment are to be made and by whom. Moreover, technology operates like a function of power, which some use at the expense of others. Hence, the vexation regarding the potentialities of technology implies another dilemma, that between democracy and authoritarianism.

This entry addresses "technology" as a power arrangement enacted through material (i.e., artifactual) or social techniques, such as bureaucracy or the nation-state. Speaking, then, to the context of the volume at hand, the genesis of "green" technologies involves both a consideration and a challenge of human and/or environmental values. One way to describe the ways by which technologies support or defy authoritarian rule is by identifying historically specific characteristics that enable or promote their use in supportive or defiant capacities.

On the first consideration, technology is thought to be the primary condition of enhanced autonomy and self-realization; it seems to inherently involve the exact opposite of authoritarianism or intellectual submissiveness.

The Enlightenment-based conception of progress, which has it that advances in science and preoccupation with perfecting technological efficiency unavoidably entail social prosperity, was taken for granted by even the most radical critics of industrial society: the Anarchist Prince, Peter Kropotkin, for example, while calling for the relinquishment of all forms of social organization, left technology outside the scope of his poignant critique. If nothing else, Karl Marx highlighted the connection between freedom and technology in the framework of the early industrial factory proletariat. Therefore, the major technical breakthroughs of the early and mid-19th century in, say, railroad building or textile manufacture, cultivated a compelling promise for utopian and socialist thinking.

Nevertheless, as the experience of the 19th century attests, various critical voices have been at times audible, articulating an evaluation of industrial life as having dehumanizing and alienating effects.

In our own time, writers such as Herbert Marcuse and Jacques Ellul have shown in enlightening detail how the technological mind-set has the ability to become as totalitarian as the religious or despotic ideologies that it sought to replace. Besides, the reappearance in history of conceptions of progress that challenge the viability of "one-dimensional men" has inspired contemporary historiographical genres like Science and Technology Studies (STS) to envision the reconstruction of technologies in harmony with democratic and environmental principles.

In sum, the historical observation that technology has had liberating effects in the 17th and 18th centuries should not be taken to mean that complex technology is always socially beneficial or, for that matter, essentially humanizing. In fact, several instances attest to the dominant rationale's irrationality: national socialism, for example, is characteristic of how a technologically rationalized mode of governance and an efficiency-oriented economy can serve the interests of totalitarianism.

More recently, the proliferation of nuclear power, apart from the vital issue of safety, has raised concerns regarding the oppression of civil liberties in light of the surveillance regimes and the elite decision making that dominate the nuclear industry. In parallel fashion, with uncertainties being endemic to the technological society as a whole, the modern technological nation-state has fostered nationalism and the autocratic ordering of both society and nature.

At least in English, the first explicit remark that technologies must embrace the needs of their respective communities belongs to American writer and public intellectual Lewis Mumford, most remembered for his studies of the American city and urban architecture. In a series of books and articles that he began developing in the 1930s, Mumford described how the historical emergence of large-scale technological systems has involved the association of power, productivity, and authoritarian destruction. Long before STS debated the political qualities of technologies, Mumford was able to distinguish between "democratic" (i.e., person-centered) and "authoritarian" (i.e., power- and systems-centered) techniques, originally claiming that political reconstruction must not overlook the workings of technology itself.

For one thing, the question of whether there are grounds for celebrating the triumph of a liberatory technology has been answered in the affirmative by anarchist thinkers like Murray Bookchin. In addition, optimistic accounts have been advocated by less radical commentators: the examples of the early Mumford and early Marcuse works are quite indicative. Ellul, on the contrary, has consistently held the position that it is not technique that will liberate man, but rather man must liberate himself from the confines of a totalitarian technique. Ellul views the development of an ethics of pluralism as the solution to the problem of authoritarian technology. Interesting middle-ground accounts have been elaborated by the Spanish philosopher José Ortega y Gasset in the late 1930s and more recently by the American philosopher Andrew Feenberg.

But to conclude, we may say that the technological possibility is enigmatic. Machines and social techniques can lead to the simultaneous growth of authoritarianism and freedom. They can contribute to the satisfaction of human needs as well as to generating profit for the few and toil for the many. Information technologies, for instance, might realize the potential to nurture democratization as well as imbalance of power in favor of social elites or major copyright holding companies.

The ethical and political dilemma of whether and under which circumstances humanity must say "enough" to further technological progress becomes pressing, considering that being human increasingly means making technological choices. This fundamental debate on the prerequisites of modern democracies aside, one thing stands clear: until humanity had engineered its inner self, engineering its circumstance could only be superficial, in both senses of the word.

See Also: Anarchoprimitivism; Engineering Studies; Science and Technology Studies.

Further Readings

Bookchin, Murray. *Post-Scarcity Anarchism.* Oakland, CA: AK Press, 2004.

Ellul, Jacques. *The Technological Society.* New York: Vintage Books, 1964.

Marcuse, Herbert. *One-Dimensional Man.* Boston, MA: Beacon Press, 1964.

Mumford, Lewis. "Authoritarian and Democratic Technics." *Technology and Culture,* 5/1 (1964).

Ortega y Gasset, José. *Meditación de la Técnica y otros ensayos sobre Ciencia y Filosofía.* Madrid: Alianza, 1982.

Winner, Langdon. *The Whale and the Reactor: A Search for Limits in an Age of High Technology.* Chicago: University of Chicago Press, 1986.

Nicholas Sakellariou
University of California, Berkeley

B

BACILLUS THURINGIENSIS (BT)

Bacillus thuringiensis (Bt), a bacterium that lives in soil around the world, naturally produces a toxin that is fatal to certain insects. Bt has been used as an insecticide spray since the 1920s, and the genetic engineering of crops to produce Bt toxin has been approved by the U.S. Food and Drug Administration (FDA), in coordination with the U.S. Department of Agriculture and the U.S. Environmental Protection Agency (EPA), since 1995. The genetic material taken from Bt and inserted into Bt crops expresses a protein that is toxic to several orders of insects, including Lepidoptera (butterflies/moths/skippers) and Colcoptera (beetles). The widespread adoption of Bt crops poses the risk that insects will evolve to become resistant to some of the most pervasive pests threatening commercial crops.

Application

In 1901, a solution of Bt toxin crystals was discovered to be highly effective against certain crop pests, including the corn borer, corn rootworm, tobacco budworm, bollworm, and pink bollworm, and was first used commercially as an insecticide spray in 1958. According to industry reports, these insecticidal properties constituted over 95 percent of the biopesticide market as of 2010. Different strains of this bacterium are currently used to control for several insects and their larvae, which feed on fruits, vegetables, and other cash crops, including potatoes, corn, and cotton. Bt toxin can be applied to crops as a spray or, less commonly, in liquid or granular form.

Bt proteins can also be introduced into the crops themselves through genetic modification. Bt crop varieties are engineered to produce a protein toxic to specific insects and used in areas with high levels of infestations of targeted pests. Since 1995, when the EPA first approved use of the technology, commercial production of Bt corn, cotton, potatoes, and rice has increased dramatically in many countries, though plantings have fluctuated significantly, rising and falling depending on pest infestation levels. New research continues on introducing Bt technology to new crops, such as eggplant.

Properties

Susceptible insects must ingest Bt toxin crystals in order to be affected. In contrast to poisonous insecticides that target the nervous system, Bt acts by producing a protein that blocks the digestive system of the insect, effectively starving it. Bt is a fast-acting insecticide: an infected insect will stop feeding within hours of ingestion, and will die, generally from starvation, within days.

Whether applied in spray form or through genetic engineering, each Bt strain is effective against a narrow range of insects. The most commonly used strain of Bt (*kurstaki*) targets only certain species of caterpillars. Since the late 1990s, Bt strains (*israelensis*, or Bti) have been developed that control certain types of fly larvae, including those of mosquitoes, black flies, and fungus gnats. More recently developed are strains (*san diego, tenebrionis*) effective against some leaf beetles, such as the Colorado potato beetle and elm leaf beetle.

Pests Controlled by Various Strains of Bt Toxin

Kurstaki strain:

- Vegetable insects
  - Cabbage worm (Diamondback moth)
  - Tomato and tobacco hornworm

- Field and forage crop insects
  - European corn borer
  - Alfalfa caterpillar

- Fruit crop insects
  - Leaf roller
  - Achemon sphinx

- Tree and shrub insects
  - Tent caterpillar
  - Fall webworm
  - Red-humped caterpillar
  - Spiny elm caterpillar
  - Western spruce budworm
  - Pine budworm
  - Pine butterfly

Israelensis strains:

- Mosquito
- Black fly
- Fungus gnat

San diego/tenebrionis strains:

- Colorado potato beetle
- Elm leaf beetle
- Cottonwood leaf beetle

Advantages

Unlike most insecticides, which target a broad spectrum of species, including both pests and beneficial insects, Bt is toxic to a narrow range of insects. Existing research strongly suggests that Bt does not harm natural enemies of insects (predators or parasites), nor does it impair pollinators, such as honeybees, which are critical to agroecological systems. Bt integrates well with other natural controls and is used for integrated pest management by many organic farmers.

The use of insect-resistant Bt plants can potentially reduce use of chemical insecticide sprays, which are extremely toxic and expensive. Applications of conventional pesticides recommended for control of the European corn borer, for example, have dropped by about one-third since Bt corn was introduced, according to EPA estimates.

Bt proteins are considered safe for human consumption. Although highly toxic to certain insect species, Bt is thought to be nontoxic to humans because a human body lacks the digestive enzymes needed to dissolve Bt protein crystals into their active form. However, any introduction of new genetic material is potentially a source for allergens, and for this reason, certain strains of Bt are not approved for human consumption.

Disadvantages

Bt, when applied in spray or liquid form, is susceptible to degradation by sunlight. Most formulations persist on foliage less than a week following application. Some of the newer strains developed for leaf beetle control become ineffective in about 24 hours. Manufacturers are experimenting with several techniques to increase its persistence. Additives (sticking or wetting agents) often are useful in a Bt application to improve performance, allowing it to cover and to resist washing.

Bt crops are likewise resistant to a narrow spectrum of the insects that prey upon them, which means that, in many cases, additional spray insecticides are required to protect plants from damage.

There is a great degree of uncertainty about the risk that Bt toxin may pose to human health and the impact of Bt proteins on the environment, and existing regulations and safety protections are not always enforced successfully. According to the EPA, more research is needed to study Bt protein accumulation in soils, the risks posed to nontarget organisms, and the likelihood of gene flow from Bt crops, particularly Bt cotton, to wild relatives.

Of great concern in the widespread cultivation of Bt crops is the potential for insects to develop a resistance to the toxin as a result of repeated exposure, which would render useless one of the most environmentally benign insecticides in use today. Already, certain moth and cotton pest populations are reported to have acquired resistance. EPA-mandated risk management strategies include the planting of refuges, such as a plot of non-Bt corn near a field planting in Bt corn, in order to maintain a local population of insects that remain susceptible to Bt toxin.

Ecological Impacts

According to a 1999 Cornell University study by J. Losey et al., eating pollen from corn plants genetically engineered to make their own pesticide can kill the larvae of monarch butterflies. In laboratory tests, about half of the monarch caterpillars that ate milkweed dusted with Bt-corn pollen died after four days, whereas all of the caterpillars that ate regular corn pollen survived. Though previous research had raised questions about the safety of Bt technology, the Losey study helped to raise awareness of these concerns. The EPA itself acknowledged the influence of the study when issuing new rules governing the planting of Bt corn that specifically mentioned the need to protect nontarget species.

A 2003 study from Ohio State University by A. Snow et al., in which wild sunflowers were cross-pollinated with genetically modified Bt sunflowers, raised additional concerns about the ecological impact of modified genes. Wild sunflowers that acquired the Bt gene were found to produce 50 percent more seeds than regular wild sunflowers and gained the ability to fend off predatory insects. These results suggest that modified genes in cultivated crops may drift into closely related weed populations and increase the hardiness of these weeds. Pioneer Hi-Bred International, a company owned by DuPont and the second-largest U.S. producer of hybrid seeds, holds the patent on Bt sunflowers and refused to allow Snow to conduct follow-up research, even with university funding.

See Also: Biotechnology; Intellectual Property Rights; Science and Technology Policy; Technology and Social Change; Unintended Consequences.

Further Readings

Entwistle, Philip Frank. *Bacillus Thuringiensis: An Environmental Biopesticide—Theory and Practice*. Hoboken, NJ: Wiley, 1993.
Losey, John, et al. "Transgenic Pollen Harms Monarch Larvae." *Nature*, 399/214 (1999).
Metz, Matthew. *Bacillus Thuringiensis: A Cornerstone of Modern Agriculture*. Binghamton, NY: Food Products Press, 2003.
Snow, Allison, et al. "A Bt Transgene Reduces Herbivory and Enhances Fecundity in Wild Sunflowers." *Ecological Applications*, 13/2 (2003).
U.S. Department of Agriculture Economic Research Service. "Adoption of Genetically Engineered Crops in the U.S." http://www.ers.usda.gov/Data/biotechcrops (Accessed August 2010).
U.S. Environmental Protection Agency. "EPA's Regulation of *Bacillus thuringiensis* (Bt) Crops." http://www.epa.gov/oppbppd1/biopesticides/pips/regofbtcrops.htm (Accessed August 2010).

Emma Gaalaas Mullaney
Pennsylvania State University

BATTERIES

In their application to green technology, batteries are systems for storing electrical power for use in vehicles or to provide temporary power to compensate for intermittency of generation. With limited ability to store large amounts of energy, they are the weak link in all-electric vehicle applications but are needed to replace disappearing energy supplies. Batteries can also be applied to load leveling of intermittent but renewable electric power sources such as wind and solar and to future hybrid power sources, mixing on-board

Lithium-ion batteries like these are under intensive research after languishing for decades. They offer the highest power density, allowing an all-electric range of about 100 miles, while lead-acid batteries were limited to a reliable range of about 60 miles.

Source: Argonne National Laboratory, U.S. Department of Energy

energy storage with power picked up by a moving vehicle.

There are two classes of batteries: single-use batteries, such as those in flashlights that are discarded after they are exhausted, and rechargeable storage batteries that convert chemical to electrical energy and back again for many cycles. In this article, only high-capacity storage batteries are considered because these are the designs that will contribute to future fossil-fuel–free energy management systems.

Principles of Operation

All modern batteries are based on the same general principle: two metals termed *electrodes* with different electronegativity (affinity for electrons, the carriers of electrical charge) are immersed in a fluid that conducts electricity. Electrons flow from one metal to the other when the circuit is completed with an external load that connects the two metals outside the battery. This is the basic design of a single cell (a vessel containing chemicals that produce electricity when reactions take place between these chemicals); as the term *battery* implies, several cells are usually connected in series to yield practical amounts of power. In his work, Benjamin Franklin realized that a single cell could not provide a significant amount of power. He combined cells in series, creating a battery of storage devices analogous to a battery of cannons.

To recharge a battery, the current flow is reversed; an external voltage is applied with positive charge through the negative terminal, and negative charge through the positive terminal. The flow of ions within the battery is reversed, restoring the potential energy that is stored in the battery's electrodes. There are always some losses, however; not all of the current stored in the battery can be released, because some of it is lost as heat or in incomplete or secondary chemical reactions. Similarly, not all of the charging current is available in the recharged battery. In theory, a storage battery can be charged and discharged for an infinite number of cycles, because the chemical reactions are reversible. In practice, though, a recharge never restores a battery to exactly its original condition. Discharge involves degrading one of the electrodes and depositing material on the other. At recharge, the material returns to the original electrode, but the structure of the restored electrode will be less regular than it was in its previous charged condition. Crystal structures will leave gaps and clumps of material on the electrode, and some of the material will fail to attach to the electrode and be lost. Over many charge/discharge cycles, this reduces the capacity of the battery. Deep discharges usually reduce the life of a battery because of these losses.

Each type of battery has a characteristic voltage, or difference in charge at its positive and negative terminals, determined by the chemistry of its cells. The familiar "dry cell" battery of the sort used in flashlights has a cell voltage of 1.5 volts. The electrodes are a carbon rod as the anode (positive terminal) and the zinc casing as the cathode (negative terminal). The cell is not really dry; the electrolyte solution saturates a paper matrix or is stabilized as a gel, so that no fluid flows as the battery is used in various orientations.

The most commonly known and widely distributed storage battery, the lead-acid battery found in all motor vehicles, golf carts, and so forth, has a cell voltage of a bit more than 2 volts. Like most storage batteries, it is a "wet cell," in this case using dilute sulfuric acid as the electrolyte fluid. To create the standard 12-volt automobile battery, six cells are linked in series; the negative terminal of one cell is connected to the positive terminal of the next one, so that the voltages add. This scheme reflects Benjamin Franklin's original concept—cells connected together in a battery of cells. For high-current applications, several such batteries can be connected in parallel to keep voltage the same while increasing the volume of current (amperage) available.

A problem with all storage battery types is the available energy density, the amount of energy that the battery can store for each kilogram of weight or for each liter of volume. The lead-acid battery is a mature technology, and the materials that make it up are plentiful and inexpensive, but its energy density is low. It takes nearly a ton of lead-acid batteries to provide the energy stored in one gallon of gasoline, weighing about four kilograms. Earlier electric vehicles using these batteries were limited to a reliable range of about 60 miles (95 km) and achieved that range only because electric motors are so much more efficient than gasoline or diesel engines.

Newer battery chemistries offer higher energy densities for electric automobiles, but none reliably exceed the lead-acid standard by much more than an order of magnitude. Nickel metal hydride batteries in the Toyota Prius manage power from a gasoline engine, equivalent to an all-electric range of a few miles. Nickel-cadmium batteries have good recharge characteristics but are more expensive, and cadmium, like lead, is a toxic heavy metal. At present, lithium-ion batteries offer the highest power density, allowing an all-electric range of about 100 miles (160 kilometers). The third-lightest element in the universe, lithium makes much lighter batteries possible, and is relatively plentiful, but most of it is imported. Battery technology is under intensive research, however, after languishing for decades.

Applications

Interest in battery technology has exploded in recent years with the realization that practical electric energy storage will be essential when societies are forced to abandon fossil fuels, only a few years in the future. Alternatives are needed because the world consumes about 1,000 barrels of oil every second, an unsustainable rate. More than half of the recoverable oil that was originally in the ground has already been extracted, a resource that accumulated over millions of years. Oil has enabled Earth's population to quadruple during the 20th century, something that has never happened before and will never happen again. Those people depend on oil for modern farming, transportation, and industry. But Earth has been rather well explored for oil already (deep ocean beds are not geologically suited to oil deposits). The coming decline will be faster than the rise because the population, and therefore the demand, is so much larger. Global warming demands reduced fossil fuel use in any case.

Wind electricity is already cost-competitive with fossil fuels in many locations, and is growing rapidly. European experience has shown that the intermittence of wind can be compensated by fossil fuel or hydroelectric sources up to about 20 percent of the power mix; after that, the power becomes difficult to manage. It becomes necessary to store energy for later use, smoothing out the uneven production from wind and solar. Giant flywheels can store power for a short time, running generators when other sources fail. Scaling this source up to commercial power generation is problematic and has not yet been attempted. Solar installations in deserts, which heat a fluid to run a generator, can become more stable if excess heat acquired in the middle of the day heats molten salts that run the generators a few hours later during evening peaks. The salts can be heated to far above the boiling point of water, resulting in a high energy density. Another solution uses excess wind or solar power to fill a reservoir that can be emptied into hydroelectric turbines when more power is needed. As this nonbattery solution is not an option at most sites, an alternative is large-capacity, stationary storage batteries in the bases of the wind turbines themselves. For this application, weight and volume do not limit battery design—cost per kilowatt (kw) stored is the only criterion.

All of these alternatives smooth out power but incur losses from storing, holding, and releasing their energy. If the electrical power is then applied in a function such as charging an electric vehicle, it is stored and released twice before doing useful work.

One future vision is electric vehicles that plug into a smart grid for long periods of time; when extra power is available, for instance from wind and solar, the batteries would be charged. When things are in equilibrium, charging would continue only for an additional fee, but when more power is needed than the generation sources can provide, power from the storage batteries in millions of vehicles supplements the main supply. The technique has never been tried, however, and compensation of vehicle owners for wear and tear on their batteries is an unresolved issue.

There are two existing visions of how users of electric vehicles can recharge their batteries, a critical issue because of the relatively short range of battery powered vehicles and the long recharging times. First is the conventional approach: recharging at the owners' homes. Overnight charging would be preferred because of the time needed to recharge. Faster charges are possible for some battery types, but they require specialized equipment and might still take several hours.

Home charging would be supplemented by charging stations in parking garages, businesses, and even parking meters to extend the practical range of a daily commute to more than half of the vehicle's battery range. This option imitates the model that has developed around internal combustion engines using fossil fuel. Fixed stations store fuel to fill the tanks of vehicles in a few minutes. With battery-powered vehicles, however, the model works poorly because of long charging times. Drivers still would be limited to the distance they can travel on batteries charged at home, making long continuous trips impossible.

The second vision is to "refuel" quickly by swapping exhausted batteries for fresh ones at a specialized exchange facility. The approach requires vehicles designed from the start to allow swap of a particular battery configuration and enough exchange facilities to maintain the flexibility to which drivers are accustomed. The approach is being tested first in Israel, which has the advantageous characteristics (for battery swapping) of small size, seldom-crossed borders, and sensitivity to dependency on fossil fuels. Another small country with a progressive energy policy, Denmark, will provide a European test bed, and several other small demonstration projects are proposed.

The battery swap strategy promises to overcome the two disadvantages of battery-powered vehicles—short range and long recharging times. No new technology is required, and indeed demonstration battery exchange facilities are already in operation. Robots pull out a 600-pound (270 kilogram) battery from beneath a vehicle and replace it in three minutes. A disadvantage is that a large and expensive infrastructure is required, and the battery format of the vehicles must match that of the exchange facility and the replacement battery pack. Battery geometry and format are still under development, though. If the format becomes fixed before the technology is mature, the exchange technology could be permanently handicapped, as happened with the QWERTY keyboard, the VHS video tape, and even the PC.

A third alternative, not yet in development, is based on the energy model used in all successful long-range electric vehicles. These vehicles obtain their power while in motion, from overhead wires or "third rails." Most are rail vehicles that complete an electrical circuit by using the rails to ground the vehicle. Electric buses, also called "trackless trolleys," use two overhead wires to complete their electrical circuit because their tires insulate them from grounding. These vehicles are completely dependent on power obtained while in motion; they have no batteries. To use such a scheme for private vehicles would require a power and a ground connection that could take the form of a third rail with two sides, and a grip-like device extending out from the vehicle.

Alternatively, vehicles can pick up alternating current from an overhead catenary system or from coils in the road. Such one-wire systems require alternating current and some current management equipment in the vehicle; essentially, the wire provides positive and negative sources alternating 60 times per second. Lanes in freeways along the median could incorporate such power sources; small battery packs would propel vehicles to and from the freeway, which in normal use is not more than a few miles. Drivers would never have to stop for refueling; the need to stop to load up on energy would become a thing of the past.

Other models may emerge for battery applications in electric vehicles; research and development in these areas is just beginning.

See Also: Appropriate Technology; Hybrid/Electric Automobiles; Social Construction of Technology; Sustainable Design.

Further Readings

Hammer, Joshua. "Charging Ahead." *Smithsonian* (July-August 2010).
Lerner, Michael M. "How Do Rechargeable (That Is, Zinc-Alkaline or Nickel-Cadmium) Batteries Work and What Makes the Reactions Reversible in Some Batteries, but Not in Others?" *Scientific American* (October 21, 1999).
Linden, David and Thomas B. Reddy, eds. *Handbook of Batteries*, 3rd ed. New York: McGraw-Hill, 2002.
Rauch, Jonathan. "Electro-Shock Therapy." *Atlantic Magazine* (July/August 2008).

Bruce Bridgeman
University of California, Santa Cruz

BEST AVAILABLE TECHNOLOGY

Best available technology (BAT) is a technical term used in conjunction with environmental standards and regulations. The specific meaning of BAT depends on context but in general it means that the most effective pollution control (or nonpolluting) technology available at the time must be implemented by sources such as factories or municipal sewage plants that emit pollutants. The concept of BAT dates back to at least the 19th century, when the phrase *best practicable means* was used in the United Kingdom (U.K.) in the Salmon Fishery Act of 1861, and remains current today in legislation such as the U.S. Clean Air Act of 1963 and the U.S. Clean Water Act of 1972. Using the term *BAT* rather than specifying a particular technology allows for the fact that what constitutes the best available technology will constantly be changing as improved methods of pollution control are developed. However, this terminology adds a layer of administrative complexity to environmental regulations as specific requirements to meet the BAT standard are not consistent over time and with regard to specific pollutants. Regulatory agencies such as the U.S. Environmental Protection Agency (EPA) have established searchable databases to make it easier to discover what is considered BAT for a particular pollutant at a particular point in time.

Several related terms are also in use. Best available control technology (BACT) is the standard used in the United States for air pollution control under the 1963 Clean Air Act in areas that currently meet national standards for ambient air quality. Under BACT, a major stationary new source of air pollution such as a factory is required to have a permit under the New Source Review (NSR) process, which sets permissible emission levels; these limits are based on levels for each pollutant that can be achieved through the application of BACT. BACT is generally required for each pollutant unless the polluter can demonstrate that it is not feasible due to specific local factors.

RACT refers to reasonably available control technology and is required on existing sources that do not currently meet ambient air quality standards. LAER refers to lowest achievable emission rate and is required on new or modified sources in areas where ambient air quality standards are not currently being met. In Europe and Canada, the phrase *best available techniques* is more common but carries a similar meaning of requiring sources of air or other pollutants to apply the best current techniques to control emissions. The concept of BAT is also used in international contexts, for instance, in documents offering guidance to select water quality technology in different countries.

The New Source Review (NSR) Program

In the United States, a company planning to build a new plant that has been classified as a major source of pollutants (the definition is based on the potential of the source to emit pollutants, with the cutoff generally being either 100 or 250 tons per year) or to make major modifications to an existing plant is subject to the regulations of the new source review (NSR) program. The company must obtain an NSR construction permit, a legal document stating what construction is allowed and the emissions limits that must be met. The NSR program was established in 1977 as part of the Clean Air Act Amendments in order to meet several goals: to protect existing air quality, to see that air quality is not degraded significantly by the construction of new or modified plants and factories, and to assure the public that any such new or modified plants will operate as cleanly as possible and will adopt the best technologies available to control pollution.

The NSR process classifies geographic areas as either "attainment areas," which have ambient air quality as good or better than the national standard as defined in the Clean Air Act, or "nonattainment areas," which do not meet these standards. Because attainment is determined separately for each type of pollutant, a given geographic area could be an attainment area for one pollutant but a nonattainment area for a different pollutant.

For attainment areas, any major new source of pollutants or major modifications of existing sources requires a prevention of significant deterioration (PSD) permit. A PSD permit is not intended to prevent air pollution entirely but to limit it. Specific goals of the PSD permit process include to preserve the air quality in areas such as national parks and wilderness areas, to protect public health and welfare, to ensure that air quality will not be damaged by economic growth, and to allow public participation in the decision-making process by which these permits are granted.

The PSD process has four requirements: use of BACT, an air quality analysis, an additional impacts analysis, and public involvement. The first requirement states that BACT must be used, although the determination of what constitutes BACT varies from case to case and takes into consideration energy, environmental, and economic factors. The second requirement is that an air quality analysis be performed to demonstrate that emissions from the new or modified source (e.g., an industrial plant) will not violate or degrade air quality standards as specified by the National Ambient Air Quality Standards (NAAQS), which set standards for six principal pollutants, also called criteria pollutants: carbon monoxide (CO), lead (Pb), nitrogen dioxide (NO_2), ozone (O_3), particulate matter (PM), and sulfur dioxide (SO_2). The air quality analysis generally involves an assessment of existing quality and predictions based on dispersion modeling of the ambient concentrations expected to result from the new or modified source, taking into account future growth.

The third requirement is an additional impacts analysis, which assesses the expected impact on soil, vegetation, and visibility of increased emissions of any regulated pollutant from the source under review. The additional impacts analysis also considers the expected impact from associated changes such as industrial, commercial, or residential growth in the area expected due to the source being evaluated. The fourth requirement is public involvement: usually 30 days are allowed for public comment on a proposed new permit, which may include a public hearing. In addition, members of the public may appeal permits that have already been issued and bring enforcement actions against sources that do not comply with their permits.

BAT and the Clean Water Act

The U.S. Federal Water Pollution Control Act, also known as the Clean Water Act, is the primary federal law governing pollution of surface waters. It was the first comprehensive statement of federal interest in clean water programs (previously considered primarily a state or local problem) and was first enacted in 1948. Most references to the Clean Water Act today refer to the revisions made in 1972 or later that provide federal funding for technical assistance to improve water quality and establish ambitious goals for water quality standards. The goals of the 1972 revisions goals include zero discharge of pollutants by 1985 and the intermediate goal of achieving "fishable" and "swimmable" water by 1983; neither standard has been entirely met to date.

The Clean Water Act has been called a "technology forcing statute" because of the technical requirements it places on polluters. Initially, industries and municipal wastewater treatment facilities were required to install the best practicable control technology (BPT) to clean up waste discharges by 1977. The BPT standard was defined by category (e.g., by type of plant or industry) as the average of the best existing performance of well-operated plants within a category and focused on controlling discharges of conventional pollutants such as suspended solids and fecal coliform and bacteria. An estimated 90 percent of industrial polluters made the required changes by 1977, but municipal sewage treatment plants lagged behind: only three states had met the requirements by 1977.

A second and higher standard for the Clean Water Act included in the 1977 amendments was the requirement that sources of pollution would have to adopt BAT to control pollution. In these amendments, BAT was defined as best available technology for pollution control that was economically feasible. Several new regulations were also made to the requirements for industries in these amendments, beginning with the definition of three categories of pollutants: conventional, nonconventional, and toxic. A new standard called "best conventional technology" (BCT) was created to deal with conventional pollutants, while the BAT standards were applied to nonconventional and toxic pollutants. The BCT standard is between the BPT and BAT standards and was established because Congress determined that requiring BAT might require an unreasonable degree of treatment for conventional pollutants.

At the same time, Congress recognized the threat posed by toxic water pollutants and required industry to use BAT for existing toxic pollutants by 1984. An additional requirement was that when new toxics were added to the list of pollutants, industries would have to meet BAT standards for them within three years. The category of nonconventional pollutants was created to include pollutants that were judged to be neither toxic nor conventional. Industries are required to use BAT to control nonconventional pollutants with the exception that if an industry could demonstrate that discharge of nonconventional pollutants would not interfere with water quality, it may be granted exemption from the BAT standard.

BATs in the European Union

Concern about pollution and the means to remedy it were reflected in the laws of Great Britain by the mid-19th century. For instance, the Salmon Fishery Act of 1861 prescribes penalties for "every person who causes or knowing permits to flow, or puts or knowing permits to be put, into any water containing salmon, or into any tributaries thereof, any liquid or solid matter to such as an extent as to cause the waters to poison or kill fish," with the proviso that such an individual will not be punished if he can demonstrate to the court "that he has used the best practicable means, within a reasonable cost, to render harmless the liquid or solid matter so permitted to flow or to be put into waters" (quoted in Higgins, 1877).

In the European Union (EU), use of the BAT concept dates back at least to the Air Framework Directive of 1984, which specified BATNEEC (best available techniques not entailing excessive costs) must be applied to control industrial air pollution. Currently, requirements for pollution prevention and control in the European Union are stated in the Integrated Pollution Prevention and Control (IPPC) Directive, passed in 2008. The IPPC Directive applies to industrial and agricultural activities that have a high potential for pollution and specifies a procedure for authorizing such activities and establishes minimum requirements (in particular, regarding the pollutants released to the environment) for such activities. The goal of the IPPC Directive is to reduce the quantities of waste arising for industrial and agricultural applications; to reduce air, water, and soil pollution; and to ensure a high level of environmental protection.

This directive defines BAT as "the cost effective and advanced stage in the development of activities and their methods of operation which indicate the practical suitability of particular techniques for providing in principle the basis for emission limit values designed to prevent and, where that is not practical, generally to reduce emissions and the impact on the environment as a whole" (European Commission, 2008, Article 2, definition 12). The directive specifies that BAT should be applied in setting emissions limits and other regulations that apply to both new and modified installations but allow each member state (i.e., each individual country) leeway to set specific standards with regard to local environmental and geographical conditions. However, it also specifies that the goal must be a high level of environmental protection and that transfer of long-distance pollution must be minimized. The directive also states that it is understood that the specific techniques that constitute BAT will change over time and thus that standards must constantly be revised and updated, and that in the event that the BAT does not result in sufficient environmental quality, supplemental techniques may be required to remove or prevent environmental harm.

Although individual member states set their own regulations, the directive establishes mandatory conditions that must be met for a permit to be issued. These conditions apply to new and existing agricultural and industrial activities considered to hold high pollution potential such as livestock farming, waste management, the chemical industry, and the production and processing of metals. The general obligations of an installation are that it use BAT for pollution prevention and control; prevent all large-scale pollution; handle waste in the least polluting way possible; use energy efficiently; ensure prevention of accidents and limitation of damage; and return sites to their original state at the conclusion of activity. Decisions to grant permits must also contain specific information such as emission limit values (with the possible exception of greenhouse gases if an emission trading scheme is in place); required protection for soil, water, and air; measures for waste management; plans and measures in place for exceptional circumstances such as malfunctions or leaks; release monitoring; and measures to minimize long-distance and transboundary pollution.

Member states make their own decisions regarding whether to accept or reject permit requests but the decision (including the reasons for it) is made public and sent to the other member states as well. Each member state is required to have procedures by which interested parties may challenge permit decisions in court and is also responsible for inspecting industrial installations to ensure that they are complying with the guidelines of the directive.

The IPPC Bureau was established to facilitate the exchange of information among EU states concerning BAT monitoring and development. An important facet of this process is the publication and revision of reference documents called BREFs, which contain information about BAT in different sectors (examples currently available on the IPPC website include the ceramic manufacturing, food, drink, and milk industries; the pulp and paper industry; and the waste treatment industries). BREFs are created by Technical Working Groups that gather and assess relevant information about pollution control technologies currently available or in development for each particular industry. A BREF is fundamentally a technical rather than a policy document whose purpose is to offer information to industrial operators, authorities of member states, the European Commission, and the general public. This information is intended to help guide them in establishing rules and conditions that ultimately should help to improve environmental performance.

The definition of BAT is specific to each sector and requires that the technique designated as "best available" is already developed on a scale such that it could feasibly be implemented within the sector. The costs of techniques as well as their environmental performance are both given consideration, and the BREF does not set emissions standards or levels but contains information about the levels of emissions associated with the proposed BAT. BREFs may also contain information about emerging techniques, that is, novel or under-development techniques that are not currently feasible on a large scale but may be preferable in terms of cost or environmental benefits to current techniques. This section may include preliminary cost estimates and an indication of when the techniques described are expected to become commercially available.

See Also: Best Practicable Technology; Innovation; Intermediate Technology; Participatory Technology Development; Science and Technology Policy; Technological Momentum; Technology and Social Change.

Further Readings

Copeland, Claudia. "Clean Water Act: A Summary of the Law." Congressional Research Service Report for Congress, Order Code RL 30030 (March 17, 2008). http://www.nationalaglawcenter.org/assets/crs/RL30030.pdf (Accessed September 2010).

Europa: Summaries of EU Legislation. "Integrated Pollution Prevention and Control: IPPC Directive." http://europa.eu/legislation_summaries/environment/waste_management/l28045_en.htm (Accessed September 2010).

European Parliament. "Directive 2008/1/EC of the European Parliament and of the Council of 15 January 2008 Concerning Integrated Pollution Control and Prevention (Codified version)." http://eur-lex.europa.eu/LexUriServ/LexUriServ.do?uri=OJ:L:2008:024:0008:01: EN:HTML (Accessed September 2010).

Higgins, Clement. *A Treatise on the Law Relating to the Pollution and Obstruction of Water Courses; Together With a Brief Summary of the Various Sources of Rivers Pollution.* London: Stevens and Haynes, 1877.

Institute for Prospective Technological Studies, European Commission Joint Research Center. "IPPC Brief Outline and Guide—Updated 2005." ftp://ftp.jrc.es/users/eippcb/public/doc/BREF_outline_and_guide_2005.pdf (Accessed September 2010).

Institute for Prospective Technological Studies, European Commission Joint Research Center. "Reference Documents." http://eippcb.jrc.es/reference (Accessed September 2010).

Muskie, Edmund S. "The Meaning of the 1977 Clean Water Act." *EPA Journal* (July/August 1978). http://www.epa.gov/history/topics/cwa/04.htm (Accessed September 2010).

U.S. Environmental Protection Agency. "New Source Review (NSR)." http://www.epa.gov/nsr (Accessed September 2010).

U.S. Environmental Protection Agency. "The Plain English Guide to the Clean Air Act." http://www.epa.gov/air/caa/peg (Accessed August 2010).

U.S. Environmental Protection Agency, Technology Transfer Network. "RACT/BACT/LAER Clearinghouse (RBLC)." http://cfpub.epa.gov/RBLC (Accessed August 2010).

Veenstra, S., G. J. Alaerts, and M. Bijlsma. "Chapter 3—Technology Selection." In *Water Pollution Control: A Guide to the Use of Water Quality Management Principles*, Richard Helmer and Ivanildo Hespanhol, eds. London: Thomson Professional, 1997.

Sarah Boslaugh
Washington University in St. Louis

BEST PRACTICABLE TECHNOLOGY

Best practicable technology is a legal term used in environmental rulemaking that requires polluters to use techniques or tools to reduce their impact on the environment. It is often used interchangeably with terms such as *best available technology, best practicable means,* or *best available control technology.* The term *best practicable technology* came to the fore with the 1972 Clean Water Act, although its derivation dates back to British laws of the 19th century that dealt with such issues as the management of salmon fisheries.

Federal and state regulators have some discretion in defining what is meant by best practicable technology, taking into account factors such as energy consumption, total discharge, environmental impacts, and economic costs. An air flue scrubber or activated sludge wastewater treatment process that substantially adds to a company's bottom line may or may not be regarded as "practicable," and protracted legal battles have been waged over the extent to which a company must implement antipollution measures in order to comply with the law. The law's requirement may encompass both specific antipollution devices and the way in which the overall facility is built or operated. These approaches can significantly change over time as environmental technology becomes more sophisticated.

In general, the best practicable technology requirement may be regarded as a middle course between business interests that prefer a more flexible approach to pollution prevention and environmentalists who want to eliminate pollution almost entirely. On the scale of U.S. Environmental Protection Agency (EPA) standards, best practicable technology and its associated terms are typically treated as more stringent than the "reasonably available control technology" requirement, which is typically used for existing facilities that need to upgrade their systems to meet regulatory requirements. However, it is less stringent than "lowest achievable control technology," a standard that requires polluters to use the best commercially available technology, or sometimes even experimental technology, to reduce emissions to the greatest extent possible. But any of these terms can be applied on a case-by-case basis.

The best available technology standards are contained within the United States' leading environmental laws, including the Clean Air Act, Clean Water Act, Safe Drinking Water Act, and the Comprehensive Environmental Response, Compensation, and Liability Act (better known as Superfund). They are also widely used in state laws and overseas by environmental agencies in the European Union and elsewhere. As the laws are amended or as regulators consider new threats to the environment, the application of these terms can have significant political and legal consequences. With EPA moving toward regulating emissions of carbon dioxide, for example, regulators will determine how to define the best available control technology to minimize emissions of the greenhouse gas—and those regulatory decisions may set off battles in Congress and the courts.

Companies regulated under the Clean Water Act's requirement for best practicable technology use some of the following techniques to treat their waste.

Cooling Ponds and Cooling Towers

Heated water released by power plants or manufacturing facilities directly into lakes and rivers can have significant localized environmental effects. Facilities can mitigate this by pumping the water into artificial ponds, giving it time to cool down, and then potentially drawing it back again for use inside the facility. The construction of the ponds needs to be undertaken without creating additional environmental impacts. They often have the side benefit of providing recreational uses to the community or even being used as fish hatcheries. Cooling ponds may not be a good option in arid areas because too much water may be lost to evaporation. Alternatives include cooling towers, which dissipate heat to the atmosphere instead of to waterways, and cogeneration, a process by which the heat can be harnessed for power or heating.

Animal Waste Treatments

Large-scale farms with livestock and poultry generate enormous amounts of animal waste and are subject to scrutiny because of mounting concerns about their environmental impacts. Manure can be diverted into lagoons or constructed wetlands and, over time, eventually pumped into spreaders and applied to croplands or grasslands. In warmer areas, anaerobic lagoons use organisms naturally present in the manure or environment to decompose the waste. The process produces biogas that can be used for energy or heating. Animal waste can also be treated with a mix of straw and composted, producing a bacteriologically sterile nutrient that can be applied to soil.

Erosion Controls

Sediment that gets washed into waterways from construction sites, farm fields, and other sources can have deleterious effects on wildlife and plants. Construction sites often minimize sediment impacts by building temporary ponds, known as sediment basins, that capture disturbed soil washed away during rainstorms. The sediments are allowed to settle to the bottom of the pond before the water is discharged. Such ponds can, in certain cases, serve as the basis for capturing storm water after construction is complete. Construction sites can also use a silt fence, constructed of a synthetic fiber and often reinforced with a chain link fence, that filters out sediments from runoff after rain.

Farmers seeking to minimize sediment loss use a number of techniques, such as contour plowing, crop mulching, and the installation of riparian buffers that preserve their soil as well as safeguard the environment.

Oil and Grease Treatments

To recover oil released onto the surface of open water, companies typically use skimming devices. Relatively cost effective and reliable, skimmers can sometimes remove large amounts of oil on their own; at other times, their use can be augmented with membrane filters and the application of certain chemicals. Recovering oil that has degraded or descended deeper into water is more challenging, requiring solvents that can cause additional water quality problems. To treat wastewater with large amounts of oil and suspended solids that has been discharged from factories, industries often use an oil–water separator. The suspended solids settle to the bottom of the separator and the oil rises to the top, with the cleaned wastewater in between.

Treatment of Acids and Toxics

Certain industrial processes, such as final processing of iron and steel products, involves the use of powerful acids to remove rust and prepare the surface for plating or other treatments.

Acids, as well as alkalis, can usually be neutralized. The process usually involves a chemical reaction that produces a solid, known as a precipitate, which may also be toxic and require specified disposal techniques. Industrial processes may also produce trace toxic metals and other substances that make their way into cooling water and cannot be discharged into waterways. The toxins can sometimes be treated through chemical processes but they often must be disposed of in a landfill.

Mining Waste

Mines and quarries can generate slurries of rock particles that are washed into waterways from rock crushing and washing processes, as well as from haul roads. The waste can be contaminated by fine particulates that may contain arsenic, lead, and other poisonous elements, as well as oils and other industrial materials. This mining waste can be extraordinarily difficult to treat. The construction of a subsurface limestone system, known as a limestone drain, can sometimes act to remove contaminants from wastewater. Other systems may include constructed wetlands, ceramic filters, or chemical treatments.

Sewage

Sewage is typically treated in a series of steps, beginning with a holding area or basin that allows heavier solids to sink to the bottom while oil, grease, and lighter solids float to the top. The remaining water is treated by microorganisms and then disinfected.

Suspended Solids

Wastewater that contains solids may be held in detention areas to allow the solids to sink, or it can be run through specially designed filters. A chemical process known as flocculation is sometimes used to remove particularly small particles. The solids remain in the form of a slurry, which can be recycled for other applications if toxins are not present.

Solvents, Pharmaceuticals, Pesticides, and Other Materials

Solvents and other synthetic organic materials can be difficult to remove from wastewater. The materials sometimes can be separated out through physical processes that may involve the application of heat or chemical treatments involving oxidation. In other cases, the waste may need special disposal in a landfill.

Urban Runoff Systems

Controlling oil and other pollutants from city streets that wash into waterways can be more challenging than minimizing the pollution impacts of an individual facility. In addition to designing developments to minimize runoff, city planners also use systems to trap and filter water. Where soil conditions are suitable, engineers can design shallow ponds known as infiltration basins that capture water during a storm and infiltrate it through permeable soils to an underground aquifer. Another process, known as bioretention, removes sediments and other contaminants from storm water runoff by slowing the water flow with a sand bed and then distributing it through a ponding area that includes grass buffer strips, plants, and rich soils. The stored water gradually infiltrates into the soil, leaving contaminants behind.

Antipollution systems are also applied to air pollution and toxic waste cleanup, although they are typically known as best applicable control technologies rather than best practicable technologies. Air emissions at a power plant or industrial facility are often treated by a series of devices, including an electrostatic precipitator, a denitrification unit, and a scrubber to remove particulates, nitrogen oxides, sulfur dioxide, and mercury. Engineers are developing systems to remove additional mercury as well as to capture carbon dioxide. Cleaning up Superfund and other hazardous waste sites can involve a wide range of technologies, including physical, chemical, and biological processes, to separate and safely dispose of the many waste products that may have been left at the site.

See Also: Anaerobic Digestion; Best Available Technology; Carbon Capture Technology; Coal, Clean Technology; Electrostatic Precipitator; Flue Gas Treatment; Wastewater Treatment.

Further Readings

Copeland, Claudia. "Clean Water Act." Congressional Research Service. http://ncseonline.org/nle/crsreports/briefingbooks/laws/e.cfm (Accessed September 2010).

Hanna, John. "Analysis: Helplessness Fuels Kansas Anger Over EPA." *Business Week*. http://www.businessweek.com/ap/financialnews/D9I73A080.htm (Accessed September 2010).

Koncelik, Joe. "What Would BACT Be for CO_2?" Ohio Environmental Law Blog. http://www.ohioenvironmentallawblog.com/2009/01/articles/climate-change/what-would-bact-be-for-co2 (Accessed September 2010).

David Hosansky
Independent Scholar

BIOCHAR

Biochar is pyrolized woody biomass that has a long life span as a soil amendment. It has similarities to the hardwood charcoal that you can buy in a store, but it is made at higher temperatures than charcoal and it is not compressed. The key word in the manufacture of both charcoal and biochar is *pyrolysis*: the process of burning in the absence of oxygen. The charcoal-making process is known in nearly every country and goes back millennia. The simplest method involves burying a pile of wood under a mound of soil with a small opening to the wood at each end of the pile, and then igniting the wood. At first the smoke from this air-limited fire will be white. After hours or days, the smoke turns blue, a sign that the now-carbonized wood is burning. The fire is then smothered completely and the pile allowed to cool. The wood is now charcoal and can be used for fuel or other purposes, such as biochar.

Biochar as soil amendment seems to have originated in the Amazon. There, perhaps in the past 5,000 years, native peoples discovered that adding charcoal to soil changed its properties in positive ways. Most Amazonian soils are highly leached, infertile red clays. Most of the nutrients in the region are captured and held in the above-ground vegetation. The addition of charcoal changes the characteristics so that the soil becomes black, is able to capture nutrients and hold them even when dowsed by the frequent rains, enabling use for continuous agriculture. People in the Amazon call these soils "Terra Preta dos Indios," or "black earth of the Indians" in Portuguese. The value or the existence of these soils was not realized until recently because the people who made them were wiped out by European diseases in the early years of exploration and conquest.

This biochar, created from manure pellets, is used to replenish agricultural soil and increase its fertility while also sequestering carbon from the air in an effort to reduce atmospheric carbon dioxide (a contributor to climate change).

Source: U.S. Department of Agriculture/ARS of Prosser, Washington, via Pacific Northwest National Laboratory

Biochar has become a major topic because of its potential in replenishing agricultural soils and as a means of sequestering carbon. As we humans have extracted and burned fossil fuels to power a nearly global consumer society, we have raised the level of carbon dioxide in the atmosphere, affecting global climate. At the beginning of the Industrial Revolution, carbon dioxide concentrations in the atmosphere hovered just under 280 parts per million (ppm). Now, over 200 years later, we are rapidly climbing above 390 ppm. Computer models predict further rises in temperature beyond the 1 degree Celsius rise already experienced, and the fear is that this will lead to catastrophic changes in sea levels, ecosystems, and our ability to survive. The question then becomes how do we effectively and quickly reduce this concentration of carbon dioxide back to an acceptable level?

The carbon cycle is a simple concept, though complex in pathways. Carbon is pulled from the air in the form of carbon dioxide, captured by plants through photosynthesis, and then proceeds through the complex web of life to be released through respiration by all animal organisms back to the atmosphere as carbon dioxide again. This process can happen quickly, as it does in the Amazon, or very slowly, as it does in the Arctic and Antarctic. The carbon made into biochar and used as a soil amendment does not progress through the carbon cycle quickly. If we imitate the Amazonian idea of putting biochar into the soil as an amendment, we could sequester this carbon and improve soil fertility at the same time, thus helping reduce atmospheric carbon dioxide in the near future.

Scientists have grabbed hold of this with increasing enthusiasm since the early 1990s. The questions they are trying to answer include: How much biochar can a given field and soil type hold? What is the best way to make biochar at a scale needed to sequester carbon and improve soil? What are the best types of biochar? Are all biochars the same? Can we do this without causing unintended harm to the environment?

It is clear that biochar was made on a small scale over a long time span in Amazonian terra preta soils. Some argue that this is the best way to proceed now. Thousands, or even millions, of small producers can make biochar, incorporate it into local soils, and reap the benefit of increased agricultural production. Others argue vehemently that this actually leads to emissions of greenhouse gases like methane because traditional methods of making charcoal are inherently inefficient and wasteful, doing more harm than good. They advocate instead more centralized and larger-scale biochar production processes. These larger production systems do work, says the counterargument, but they involve greater transport costs of raw materials to and biochar from the processing plant. The compromise probably lies in the ability to make biochar locally without releasing greenhouse gases beyond low levels of carbon dioxide during pyrolysis.

One aspect of biochar production attracting attention of both small and large producers is the large amount of energy liberated in the process of making biochar. Some have devised stoves that cook food using pyrolysis gases and produce biochar as a by-product. Others have devised much larger systems that either condense pyrolysis gases into liquid fuels or use them to make electricity. In all cases, use of the heat from the manufacture of biochar is a win-win proposition.

As studies continue, more evidence of the positive properties of biochar for soil enhancement will appear. A project by the Biochar Fund in Cameroon, in conditions similar to the Amazon, shows markedly improved growth in crops after incorporating biochar enhanced with manure-based compost as compared to their normal practices where fertilizer is unavailable. This idea has spread to the Congo as well. Other projects in North America have found that synthetic fertilizer needs of crops are reduced with the introduction of biochar because of reduced nutrient losses from leaching. At the same time, inexpensive biochar-making technologies have emerged that allow people to make biochar while they cook. Such multi-use approaches to biochar production effectively reduce overall pollutants.

Biochar will make an increasingly significant contribution to both agriculture and carbon sequestration. While probably not the entire answer to halt climate change, which will require major reductions in fossil fuel use and conversion to alternative energy sources, it has the potential to become a major player in the reduction of carbon dioxide in the atmosphere.

See Also: Carbon Capture Technology; Pyrolysis; Thermochemical Processes.

Further Readings

Biochar Fund. http://www.biocharfund.org (Accessed August 2010).

International Biochar Initiative. http://www.biochar-international.org (Accessed August 2010).

Lehmann, J. and S. Joseph. *Biochar for Environmental Management*. Sterling, VA: Earthscan, 2009.

Mann, Charles. *1491. New Revelations of the Americas Before Columbus*. New York: Vintage, 2006.

Wayne Teel
James Madison University

BIOCHEMICAL PROCESSES

Biochemical processes are the chemical processes that occur in living organisms, involving biomolecules. Biomolecules are organic molecules produced by organisms and include both polymers—large, complex molecules such as proteins, peptides, cellulose, hemoglobin, nucleic acids, and polysaccharides—and monomers like amino acids, monosaccharides, and nucleotides, as well as small molecules like carbohydrates, metabolites, hormones, vitamins (originally shortened from "vital amines," referring to compounds vital to an organism's survival but only in trace quantities), and neurotransmitters. One of the first biochemical processes studied was alcoholic fermentation: the conversion of sugars into cellular energy with ethanol (alcohol) and carbon dioxide as by-products. Fermentation often leads to a second process, that of autolysis, in which cells are destroyed by their own enzymes—in this case, the yeast cells that converted the sugar to ethanol die and sink to the bottom of the liquid (either when nutrients and sugar run out or when the alcohol level creates an environment toxic to yeast cells); if these dead cells, which in winemaking are called lees, are left in contact with the wine, enzymes break the cells down and produce polysaccharides and mannoproteins. In winemaking this is usually considered undesirable because it can lead to hydrogen sulfide aromas; however, when autolysis occurs in wine that has been sealed in wine bottles, a secondary fermentation occurs that produces the bubbles of "sparkling" wines such as champagne, while the mannoproteins produced bind with the wine's tannins and reduce the sensation of bitterness.

One of the biochemical processes best known to the layman is photosynthesis, which occurs in plants, algae, and cyanobacteria: the use of solar energy to convert carbon dioxide into organic compounds. Through photosynthesis, these organisms create their own food, releasing small amounts of oxygen as a waste product. With the exception of chemoautotrophs, which derive energy from chemical reactions, all other forms of life on Earth are dependent on photosynthesis—either they engage in it themselves, or their food sources do, or their food sources' food sources do. Plants typically have a photosynthetic efficiency, depending on light wavelength and irradiance, temperature, and carbon dioxide concentration, ranging from less than 1 percent to as much as 8 percent, with the highest efficiency being equivalent to the least efficient solar panels.

Chemosynthesis is similar to photosynthesis, converting carbon dioxide or methane into organic material, but uses the oxidation of methane or inorganic molecules as its energy source instead of sunlight. Chemoautotrophs, the organisms that rely on chemosynthesis, live in dark but chemically rich environments like hydrothermal vents and methane clathrates. Chemoautotrophs are only recently discovered: the first were discovered in the deep sea in the 1970s, and it seems likely that most species have not yet been discovered. Chemosynthesis suggests the possibility of life that could survive on other planets in the solar system, as well as applications in nanotechnology.

Mineralization is the process of converting organic material into inorganic material, such as the way birds form eggshells from a buildup of calcium carbonate (about 95 percent) stabilized by protein, as well as the formation of the baby chick's egg tooth that lets it break through the shell in order to hatch.

Autophagy is one of the fundamental processes of catabolism, the breakdown of molecules into smaller units, and release of energy and cellular waste like lactic acid and carbon dioxide as by-products. The various catabolic processes provide the chemical energy necessary to maintain and grow cells. There are a number of processes that perform autophagy, in which the cell's own components are degraded through the machinery of lysosomes (cube-shaped organelles that contain enzymes that break down waste materials and cellular debris). Autophagy is key to living organisms because it reduces the amount of resources used on dying or unhealthy cells—the cell helps to recycle itself to free its resources up for other use. Furthermore, infected cells can be destroyed by autophagy, helping to stop the spread of the infection.

Human Applications of Biochemical Processes

Observing naturally occurring biochemical processes suggests a number of technological possibilities. Chemosynthesis may someday inspire a method of powering nanomachines, for instance. In 2010, the energy firm Krebs and Sisler announced a faster method of purifying water, using photosynthesis and photocatalysis (a chemical process that can be used to destroy organic contaminants in water by accelerating their degradation). The process works faster and more cheaply than reverse osmosis, the standard water purification process for seawater, and creates as a by-product a 50-percent carbon algae biomass that is high in protein and suitable for being dried to use as livestock feed, human nutrients, or fuel. The process has the additional environmental advantage of consuming carbon dioxide.

The process of autophagy is one that has fascinated scientists for decades. Because autophagy removes damaged cells and cell components, the effects of aging are blamed on an accumulation of cell damage as autophagous processes fail; by extension, those effects could be postponed by somehow assisting the autophagous processes and keeping them going. Furthermore, induced autophagy seems like a promising remedy for Alzheimer's disease and other degenerative neurological disorders.

The Atkins Nutritional Approach ("Atkins diet") introduced by cardiologist Robert Atkins relies on autophagy, as body fat is reduced by ketogenesis, the production of ketone bodies by breaking down fatty acids.

Though photosynthesis is much less efficient than man-made solar cells, even low-efficiency methods of converting solar energy to human-usable energy would be invaluable if the operating cost were low; that operating cost is the main obstacle to solar cells currently on the market. Genetic engineering has been suggested as a way to bridge the gap between photosynthetic organisms and the photoelectric cells used for man-made solar-energy collection; if new organic "solar cells" could be grown in bulk, solar energy could be much cheaper, with potentially less carbon impact compared to the manufacture of inorganic solar cells. In recent years, there have been interesting developments in "artificial photosynthesis," including the development of dye-sensitized solar cells, which replicate the first stage of photosynthesis and release hydrogen, which can be used as a fuel.

See Also: Anaerobic Digestion; Biotechnology; Composting Toilet; Environmental Remediation; Green Chemistry; Thermochemical Processes.

Further Readings

Campbell, Neil A., Brad Williamson, and Robin J. Heyden. *Biology*. Boston, MA: Pearson, 2006.

Hunter, Graeme K. *Vital Forces: The Discovery of the Molecular Basis of Life*. San Diego, CA: Academic Press, 2000.

Michal, Gerhard and Dietmar Schomburg. *Biochemical Pathways: An Atlas of Biochemistry and Molecular Biology*, 2nd ed. Hoboken, NJ: Wiley-Interscience, 2010.

Bill Kte'pi
Independent Scholar

BIOETHICS

Bioethics is a relatively new field of applied ethics devoted to moral and social questions concerning human and nonhuman biological life and death. The social and ethical impact of advanced technology related to life sciences has been momentous. Genetic-engineering technology, effecting the transfer of genes between organisms of different species and even, in some cases, the insertion of an artificially synthesized gene into a natural organism, has added new moral problems. Modifying the genetic material of natural beings, thus affecting the natural environment, has also important implications for green technology.

From its inception, biotechnology has been a highly controversial technology, even when it may appear that there are only unalloyed good results to be expected. One cannot fail to realize the enormity of the moral dilemmas or the uncomfortably irresolvable moral deadlocks involved when considering transgenic organisms that may be detrimental to ecosystems but advantageous to alleviating human hunger, or, by contrast, of "designer bacteria" used for oil-spill cleaning; of ever advancing medical technology that can maintain life longer; or perhaps using organs from a malformed and possibly dying newborn for life-saving transplants on the one hand, but from a newborn whose handicap must not be seen as a license to withdraw medical care on the other.

Not unlike other types of technology that have affected people's stance vis-à-vis moral attitudes, for example with "green" or environmentally sensitive viewpoints, biomedical technologies have also raised moral sensibilities, all the more so given their direct impact on human nature, either in terms of alleviating pain or holding out the promise of curing debilitating or life-threatening/life-shortening maladies. These possibilities rest primarily on further extending medicine or conventional biotechnology into genetic engineering allowed by recent advances in molecular biology, genetics, and by genomics (copying DNA tracts and controlling their sequences to see how genes function in new bioenvironments).

Bioethics must be distinguished from popular hype, misinformation, or superficial journalistic coverage of hypothetical ghoulish biotechnological eventualities; for example, transgenic organisms "running amok," Frankenfoods, or fear-inducing dangers especially associated with human cloning. This does not mean that bioethics should only be of arid academic interest and that people's fears should not be heeded, or that protesting against genetic modification (GM) in general has no value.

One crucial element in the vehemence of bioethical confrontations has also to do with language: surreptitiously value-laden descriptions are used to buttress arguments against opponents. Bioethics is also very much linked to religion, either fueling religious passions or forming the topic of considered church positions. Moreover, the phenomenon of a booming bio-industry profiting in GM, and arguably influencing by funding the direction of "free" academic research, calls for moral evaluation and legal regulation.

Ethical Theories

Ethical theories may roughly be distinguished into those that morally assess an action as opposed to those that focus on the ethical status of the agent: "what ought one to do" versus "what kind of person ought one to be." When judging the moral value of an action from the perspective of its good or bad consequences, theories are said to be "consequentialist," emphasizing the maximization of possible foreseeable good (or minimizing evil), whereas theories that privilege not the goodness of an action but its rightness or wrongness are usually called "deontological."

The German philosopher Immanuel Kant (1724–1804) is the paradigm deontological moral philosopher, having developed an intricate moral system based on moral rules that must be absolute. These moral rules specifying what makes an action right are the universally holding and necessary demands of reason, not contingent empirical calculations of good or bad outcomes, and must be followed by all rational beings motivated only out of respect for them as a duty—not because of their good consequences or out of benevolent sentiment of sympathy. When it comes to human beings, Kant's categorical imperative encapsulates three formulations issuing in maxims, one of which is, "always act so that you treat humanity, in your own person or in another, as an end and never as a means only." Many bioethical positions have drawn on this, as in respect for autonomy, or of not exploiting human beings whatever the beneficial effects of a genetic or other medical experiment.

Of the consequentialist theories, utilitarianism is the most prominent, and was developed principally by the English philosophers Jeremy Bentham (1748–1832) and J. S. Mill (1806–1873). The principle of utility requires that between two or more courses of action open to an agent at a time, one should perform the action maximizing the overall utility for the greatest possible number of people (or minimizing overall disutility or cost). Utility measured in terms of pleasure (or absence of pain) can be equated to happiness or well-being, being the ultimate end for which everything else should be done. Mill believed adult human beings should be free to pursue what they freely decide—provided no harm to others results from it—without being prevented by society.

While deontological theories suffer from possible conflict between rules, utilitarianism suffers from difficulties amounting sometimes to the impossibility of calculating costs and benefits of a future outcome and distributing it fairly, as well as from sometimes disregarding rights, autonomy, promises, or justice (e.g., sacrificing an innocent person) if the outcome to be achieved is significantly far more superior than not engaging at all in those acts as means.

A third type of ethical theory emphasizes instead that morality is the product of a (hypothetical) "social contract" for the purposes of mutual cooperation and survival. It is not difficult to see how such theories can be used to assess the moral responsibilities of bioindustry for indigenous peoples or the public at large if one assumes the plausibility that in cases of plants' genome-exploitation, an underlying contract has been violated. Similarly, in cases when GM pharmaceutical-patenting leads to social inequality of accrued benefits between rich and poor patients, or to double standards in experimental-research protocols respected in Western affluent societies, but not, say, when poverty-stricken African women are involved.

In contrast to theories centering on the action to be performed and on a certain foundational moral principle operating as a "law," an older theory looks instead to the side of the actor. Virtue ethics, going back to the Greek philosopher Aristotle (4th century B.C.E.), looks at the goodness of human character, which at its most excellent exhibits a firm condition of unwavering virtue. Virtues are prerequisites for a flourishing human life or character traits that humans find admirable. Virtue is a character trait to be achieved by hard effort. Aristotle did not expect everyone to achieve a virtuous state of being—being virtuous was to large extent open only to some and involved a considered standpoint based on reasoned judgment of a specific case, differing from one occasion to another.

Style of Ethical Reasoning

By contrast, the modern theories encountered above view morality as the action of an "impartial spectator," whose standpoint is universal, gender-neutral, and disinterested. These are "principle-based" theories, whereby principles operate as rules guiding action. For instance, a classic recent formulation of "principled bioethics" puts forward four main principles: respect for autonomy, nonmaleficence, beneficence, and justice. There have also been proposals for rules of procedure and of ranking to decide on each occasion which principle(s) cover(s) the case in question and whether there is a conflict among them when a particular case in bioethics does not entirely fit the criteria. Obviously, such approaches, despite their criticized theoretical validity, may be useful in policymaking and legislation. Yet such a version of principlism is dismissed by the paradigm approach of absolute rules calling for the absolute exceptionless of normative "precepts" to action based on natural law. According to this theory, such an infrangible precept prohibits an action such as abortion on the grounds that it is a case of "intentionally killing innocent humans."

Against this moral absolutism, some contemporary positions emphasize the "cultural relativity" of ethics, and in particular of mainstream bioethics, accusing them of promoting falsely a set of universal values that is in effect only Western ethical beliefs; they even go so far as to challenge the universality of rational principles as such. They open up to other kinds of voices, like "feminist ethics" or "care ethics," distancing themselves from abstract theorizing like rigorous deontology, duty-bound morality, contractually bounded responsibility, or impartial utility-maximization. Care ethics, promoting interhuman caring relationships and awareness of sympathy as opposed to "doing the right thing," has been quite prominent in medical ethics. Feminist bioethics, too, challenge the theoretical presuppositions embedded in mainstream Western bioethics and insist on underlining, on the one hand, the different ethical perspectives between men and women in evaluating issues of life and death, but also, on the other, the fact that biomedical practice has been a site of women's exploitation and oppression.

Issues: GM Biotechnology

Sophisticated GM has been applied to crops and food as well as in recombinant products used as drugs. GM has raised fears of high risks of irreversible bad effects and/or noncontained GM growth, of invading and altering the wilderness, of adversely affecting biodiversity, and fears of misuse demanding legal control. GM microorganisms, growth hormones, viruses beneficial to crops, and insecticide- and herbicide-resilient GM plants may have some environmentally beneficial effects making them suitable for green technology, but at the same time pose high risks of adversely invading and altering nature. Risk avoidance and the precautionary principle are particularly pertinent critiques, as well as critiques resting on intrinsic wrongness using arguments from respect or unnecessary suffering, or violating the intrinsic value or essence of a species or animal rights. In the case of GM, animals used or produced for human organ transplants or human proteins of GM farm animals engineered for human consumption (allegedly saving human lives from starvation), it is crucial to point out that these cases are different in terms of kinds of moral questioning, thus requiring a different ethical assessment.

In terms of effects, standard criticisms against all GM, whether in human or nonhuman contexts, are based on the high and unknown risks that may be involved (potentially affecting wider segments of humanity), unfair distribution of costs and benefits, and transgressing species boundaries. Here, Mill's principle of liberty and his no-harm-to-others principle may be in conflict. Moreover, from a consequentialist perspective, a so-called precautionary principle has been widely asserted, especially in the context of green technology, whereby advances with statistically high risk of harm must be stopped even prior to full-proof demonstration of such harmful effects. Kantian autonomy and informed consent as well as respecting all humanity's rights may be accommodated by a GM bioindustry heeding them. Arguments by Catholic Natural Law ethicists are used against GM to prohibit altering the essential *telos,* or substantial end defining each biological species, as decreed in God's plan. In virtue ethics, it would hardly be sufficient for a virtuous person to exercise virtues alone in judging GM applications and risks without, in some cases, an in-depth knowledge of biology. But a "virtuous" person can help show or criticize how technology-inspired values may have entered into new character traits now considered desirable.

A new type of issue is appearing on the ethical–legal horizon of genetic engineering having to do with gene patenting and gene piracy. While the former refers to the contentious practice of securing identified gene sequences under utility patents as if they were inventions (not simply discoveries of what already exists in nature), the latter has a more obviously socioeconomic impact. The use of animal or plant genes or genetic material without the consent of those from whom they were taken (or without sharing the profits accrued) raises a number of issues. These issues have to do with proprietary rights of humans to their organs or to tissue after these have been removed and with rights of countries or indigenous peoples to their flora or environmental infrastructure.

Combined with criticisms of the expense of green technology being impossible to be borne by poor farmers in underdeveloped countries, and with the controversy over the desirability of a "Second Green Revolution," genetic ownership in general raises questions on the beneficial outcome (distributed justly or not), and on the presumed or explicit contract between bioindustry and indigenous peoples. It also raises political questions about the role of governments, and what "consent" means and how it is secured, as well as issues regarding how far one's labor is a substantive ingredient that transforms an otherwise common good into a private one to be owned by the person or company working on it. As in the case of modern eugenics and enhancement, both in the case of gene patenting and gene piracy, the erstwhile divide between what is "natural" and what is "artificial," or between "discovery" and "invention," is questioned.

Issues: Human Life Technology

Before Birth or Reproductive Technology

Reproductive technology allows people not only to choose to procreate without sex, but also to avoid certain disabilities or to choose certain traits and other vital characteristics of their offspring by means of in vitro fertilization (IVF) techniques and intracytoplasmic sperm injection (ICSI) subsequent to donor insemination (DI), as well as by assisted reproduction techniques ranging from surrogate motherhood to fertility drugs and other technological aids. Thanks to genetics, parents who are aware that they suffer from a hereditary disorder can secure a disability-free future for their children by receiving gametes from unaffected donors.

All such technological facilitating of human reproduction has been the subject of state legislation in certain countries in the West, but also the target of religious criticism, arguing that it is impermissible to separate sex from procreation or to use technology or other means to block the natural law embedded in God's design, in accordance with which only a procreating end or natural *telos* for sperm and ovum is acknowledged as divinely set. IVF, DI, or surrogate motherhood raise a number of ethical concerns, including whether children so conceived should have a "right to know the father," whether donors should be paid as vendors, whether it involves exploitation of the poor, or whether technology should be put into the service of people with no partners wishing to procreate.

The ethical dilemmas surrounding abortion and reproduction converge on one of the styles of argumentation. The central question is: "When does human life begin?" Given different starting points of distinct approaches—theological, metaphysical, biological— certain conclusions are drawn regarding whether the embryo after syngamy or the early fetus can be said to be merely human material, organelle, or a person—actual or potential. Decisions are then made as to whether aborting or experimenting on the fetus is morally permissible.

But even avoiding establishing definitive criteria of personhood, there are serious doubts about how to maintain a position that regards even a fertilized egg as the beginning of human personhood, given the absence of genetic uniqueness at that point.

Similar to abortion, which can be debated without mentioning the question of person-hood, in genetic testing and in assisted reproduction, a rights-based approach can be adopted. Within this argument framework, it is often claimed that the rights of an existing human being override those of a nonexisting being. Suing for having been born is a real case, while the right not to be born is an issue that further depends on a wider philosophical problem regarding personal identity of "possible": human beings whose existence depends on current actions or inactions. Can it be said that a child has been harmed if brought to life with a mild handicap if the alternative would have been not to be born at all?

Preimplantation and Prenatal Genetic Testing

Deciding what kind of children to have at this stage of technological development primarily involves either testing in vitro embryos before they get implanted, or fetus diagnosis carried out to determine whether there is a risk of a child being born with a genetic disease or a disability. Of the several embryos produced in IVF, some are discarded if found to be carriers of the mutant gene. Quite a few gene tests have been devised to diagnose inheritable abnormalities connected to mutant genes, and the issue links up directly with abortion. Genetic screening involves checking for gene-induced disabilities, and some countries allow abortions. In clear-cut specific cases of painful and debilitating genetic conditions, testing is implemented by governments.

Apart from these cases, what constitutes a disease or disability is contentious. Given reliable genetic testing, coupled with the denial of nonexistent people possessing a right to exist, it is maintained that it is a duty of parents not to bring to the world children at predictable risk of a degenerative or lethal malady with no prospects of a satisfying life. By contrast, disability viewed as a plight only because it is caused by adverse social attitudes and negative discrimination would make prenatal diagnosis and selective abortion impermissible.

Abortion

One of the usual starting points in abortion debates (apart from those adhering to a notion of the "sanctity of life") is the potential personhood of the human embryo and fetus over and above its being genetically human. The same issue is also raised about the status of infants and nonhuman animals, depending on which criterion is being proposed as essential for conferring *personhood* to the entity in question. Having the potential to become a person is taken as, barring adverse extraneous effects, the entity will achieve that state thanks to its inherent disposition toward it. Therefore, preventing it from becoming a person is regarded as a morally blameworthy action, for it robs its potential possessor of morally significant status. In addition, the fact that that status, when achieved, confers rights preventing its removal from the class of living persons, licenses the same right being conferred to any entity at the initial stage.

For others, regardless of whether the human conceptus, embryo, or fetus is a not-yet-fully developed human organism or whether it is a potential person or not, it still has a valuable future and for others even a set of "intrinsic interests"—an argument in support of prohibiting the production of GM animals with humanly compatible organs for the use of human patients. Terminating the life of a fetus, as such, amounts to depriving it of its future, something that is not permissible. Whether a fetus has a valuable future is also an issue that figures in deciding what to do with severely handicapped newborns, or late-in-life severely handicapped adults, or comatose patients in terminal vegetative states.

Others point to the rights of women against the unborn, either claiming that the unborn have no correlative rights or even if they do have, those of the mothers-to-be/women bearing them trump them since the rights of actual humans must override those of the not yet existing. Here it is pertinent to contrast such a rights-based approach or the consequentialist one with that of modern virtue-bioethicists with a completely different take on abortion: some claim that it is not a matter of rights, but whether rights are exercised virtuously or not, and whether a decision is virtuous as a result of having deeply reflected on the values involved.

Modern Eugenics: Gene Selecting and Human Enhancement

Future human enhancement by various means (GM, medical prostheses, or computer interphases) refers to drastically altering bodily, mental, and even emotional nature to such an extent that one transcends the limited human nature, reaching a "transhuman" state. Gene selecting is an important ingredient, and techniques, such as alternative ribonucleic acid (RNA) splicing related to therapies of neurological conditions, already point to future applications. Sex selection is a troubling possible choice with wider social implications, such as upsetting sex ratios in societies. Those on one extreme favor a "genetic supermarket"; bioconservatives or bio-Luddites on the other extreme fight for a total ban on this, with some pointing to the irreparable detriment in losing the invaluable "ethic of giftedness" people should be embracing life with.

Ethical, social, and theological issues (beyond gut reactions against "playing God") are raised with regard to the rights of future human beings, the sanctity of life, and the extent of parents' right to reproductive freedom. Though the latter sounds like "eugenics," contemporary forms of so-called liberal eugenics should be distinguished from disreputable eugenics whereby a state imposes its own control of "racial hygiene" by forceful sterilization or compulsory breeding. Modern eugenics advances the view that parents are free to exercise their right to procreation in accordance with their own wishes without harming future persons, to the extent that reproductive technology enables them to engage freely in procreation choices regarding desirable mental and biological characteristics of their offspring.

Proponents of liberal eugenics also claim that if states not only forbid parents to harm their children, but also place a parental responsibility upon them, then they should have the freedom to ensure that their children are biologically and mentally nondisabled. However, the choice for genetically modified offspring is not always about ordinarily preferred traits: there are cases where such characteristics desired by the parents-to-be may be what is normally categorized as a disability, for example, bringing deaf children into life on the grounds that there is a valuable cultural identity involved in this that parents wish their offspring to experience.

Ethical debates about human enhancement via genetic engineering of future children decided upon by parents pivot around the distinction between harmful disabilities prevented versus positively enhancing traits conferred upon offspring beyond what is required for normal health. In terms of ethical reasoning, there may well be a conflict of principles between the autonomy of the future individual to shape its own life as opposed to the principle of beneficence that a parent ought to follow when exercising parental duties.

One camp considers foreseeable disabling conditions to be a clear disadvantage that parents have a moral reason to prevent. Furthermore, proponents of such a view take preventing disability to be morally symmetrical with positive enhancement. Hence, they argue for enhancement as a moral obligation.

Equally controversial is the issue of knowingly bringing to life severely disabled children; advocates of "disability rights" and some severely disabled people themselves have argued against the easy dismissal of their view, claiming that far from being an automatically miserable existence, there is room for a possible worthwhile life or satisfying life to be lived by people with severe bodily disabilities.

Finally, another contentious case, eventually leading to the topic of cloning, has to do with the decision to bring to life children for the purpose of assisting their dying or severely suffering existing siblings as donors of vital tissue. While this practice has existed for some time, ethical arguments against it usually center on the Kantian maxim never to use human beings solely as means but always as ends in themselves. Proponents of the practice are quick to point out that no child is ever brought into the world for a Kantian reason unalloyed by many other desires parents happen to have. What is worrisome for some, though, is that sidestepping the chancy method of just giving birth to another child who might turn out to be a suitable donor, preimplantation testing that determines which embryo is indeed a bearer of the compatible tissue material may lead to killing nonsuitable embryos until the right one is hit upon.

Thus, in cases like these, medical–genetic technology is used to choose between fetuses—for some, not so different from objectionable human cloning. Cloning humans is thus regarded as the further, undesirable, step in this slippery-slope argument.

Cell Technology: Cloning, Stem Cells, and "Parthenotes"

Reproductive cloning refers to the biotechnology of gene regulating by nuclear transfer, whereby the nucleus of a specialized cell of an adult is substituted for the removed nucleus of an ovum. The enucleated egg is thus activated to go through a reproductive process such that a new individual is born genetically identical to—a clone of—the adult whose cell provided the nucleus containing the reproduced genetic material. Human reproductive cloning is the production by such a nuclear transfer of human clones, whereas the term *human therapeutic cloning* refers to the production of human clones for therapeutic purposes. Despite the associated frantic hype surrounding cloning, one thing is clearly ruled out: that the genetically identical human clone is ever identical to the original person. A person's phenotype is clearly more than its genetic barcode (genotype), given differences in social environment and upbringing, while even the physical conditions enveloping the two different genotypes when they were being formed are clearly different.

Though nowhere near realization as far as humans are concerned, significant benefits are already pointed out by proponents, such as clones having ahead of them their genetic copies to gain from, and promoting parents' reproductive freedom. Its detractors point to myriads of negatives, including high risk, disrespecting sanctity of life, procreative selfishness, ignoring possible psychological problems cloned people may face knowing they are clones or that they have no natural-looking father, and violating the right to be a genetically unique individual.

Tissue and organ repair and transplant is a critical issue, given the high numbers of humans plagued by degenerative diseases caused by malfunctioning of vital tissue that could be replaced. The most promising source for such tissue replacement is stem cells: embryonic, fetal, or adult.

A fertilized egg at the 14-day stage comprises an outer wall that will morph into the placenta and a hanging inner mass of cells containing the DNA. It is obvious that stem cells have a vital role to play. In particular, embryonic stem cell (ES) lines can be used to replenish lost tissue in an adult organism. Besides somatic ESs, use can be made of postabortion fetal germ cells (EGs). Recent ES research on embryos from which ESs where harvested has shown that ESs can be induced to morph into a gamut of specialized cells in an adult human body. But it is equally obvious, nevertheless, that whether ES material should be extracted from induced or spontaneous abortions, or whether embryos should be created by IVF solely for such a use (as "spare parts") is a serious moral issue. Though no one disputes the enormous value of these applications of stem cell research, critics claim that this sort will lead to a slippery slope of making reproductive cloning admissible.

One application of styles of ethical reasoning is "principlism," which assesses the merits of stem-cell research, grounding it on the four principles while trying to determine what each one would involve and whether possible conflicts among them could be accommodated. Similarly, enthusiastic proponents of stem-cell research, reminding critics that accepted practices like natural sexual intercourse and IVF also lead to embryo destruction, put forward a "principle of waste avoidance." Virtue ethics may demand that though no utilitarian-type of arguments based on felt pain are applicable, a virtuous mode of dealing with stem-cell research and application is possible by taking a thoughtful, responsible, and respectful attitude toward it by forming appropriate moral habits. If an embryo cannot be guaranteed a positive right to a definite life, the agent who makes decisions about it may at least show a virtuous disposition toward dealing with this issue.

Moral Status: Blastocyst to Parthenote

Finally, concerns regarding the commodified use of oocytes and women's rights and risks can become more complicated if recent research into human parthenogenesis develops. Virgin Birth (ova behaving reproductively without being fertilized by sperms produce an embryo though never developing into a fetus) is encountered in certain parts of the animal kingdom and can also be induced chemically in some animals. It has been claimed that it can now be done with humans producing embryos, supplying stem cells to the possessor of the ovum. A embryo could escape all adverse consequentialist arguments and those regarding personhood-with-potentiality or those about being a subject of rights, but could still be considered an entity with interests of its own.

Generally, some rest moral hope on screening, germ-line intervention, cloning, and stem-cell research as avoiding the social uneasiness for the immorality of induced selective abortions, though those methods introduce their own ethical worries.

After Birth: Postnatal Gene Testing

Carried out on newborn babies, children, and adults, postnatal genetic testing may be helpful in initiating therapy. Parental informed consent, the ability to cope with knowledge of genetic diseases without cure, and whether mandatory newborn screening should be legally implemented are issues. There is an argument, though, that every newborn acquiring a "genetic report card," as some have called it, may be a valuable thing in future. Genetic screening is also a postnatal option, but it may also involve a biological profiling of the whole of genetic conditions that, if kept in the form of public records, would jeopardize individual rights to privacy, including the right not to know one's own genetic disability, or even the autonomy of screened children whose consent has not been given. On the other hand, knowing the incidence of affected carriers may spur societies into favorable health programs, affected individuals into choosing therapies, and even science into advancing better cures. But the controversial status of such profiling is shown by the fact that it may well lead to social discrimination, loss of employment, inability to ensure insurance, and increased feelings of pity, even for people whose "disability" is a matter of statistical probability and may set in only later in life.

Severely Disabled Newborns

Philosophers have strongly disagreed as to whether there is an equivalent moral significance between an action of killing and not doing anything to prevent a sure death. Whether the latter involves intentionally letting die and thus "causing death" is directly relevant to cases when doctors' inaction terminates—by not providing life-sustaining treatment—the lives of newborns with severe disabilities.

Gene Therapy

Transplanting a correctly functioning gene to replace the malfunctioning one responsible for certain diseases has been used for human therapy. Those somatic cells carrying the gene malfunction are singled out as sites for the insertion. Unlike such somatic gene modification where only the affected individual is involved, germ-line modification involves future generations—the new gene is inserted in the ovum, thus passing it on to offspring. Both techniques can be ethically acceptable in cases of harmful genetic disabilities, though in the latter case, it can only be of any use when it is certain that all offspring will be carriers. Apart from risk and precautionary arguments with regard to all sorts of GM, or slippery-slope ones predicting widespread modification if legalized, and barring severely debilitating diseases of outright suffering, deontologists point out that the autonomy of future human beings and their "open future" are compromised in germ-line therapies. The line dividing therapy from enhancement is fuzzy and its morality is unclear. According to so-called transgeneration ethics, there is a duty not to leave a worse gene pool for descendants—but this brings into conflict the principle of beneficence against that of nonmaleficence.

See Also: Futurology; Intellectual Property Rights; Luddism; Social Construction of Technology; Technology and Social Change; Unintended Consequences.

Further Readings

Beauchamp, T. and J. Childress, eds. *Principles of Biomedical Ethics*. Oxford, UK: Oxford University Press, 2008.

Berry, R. *The Ethics of Genetic Engineering*. London: Routledge, 2007.

Burley, J. and J. Harris, eds. *A Companion to Genethics*. Oxford, UK: Blackwell, 2002.

Dyson, A. and J. Harris, eds. *Ethics and Biotechnology*. London: Routledge, 1994.

Finegold, D., et al. *Bioindustry Ethics*. Oxford, UK: Elsevier, 2005.

Glover, J. *Choosing Children: Genes, Disability, and Design*. New York: Oxford University Press, 2008.

Glover, J. *What Sort of People Should There Be?* New York: Penguin, 1984.

Gordijn, B. and R. Chadwick, eds. *Medical Enhancement and Posthumanity*. New York: Springer, 2008.

Harris, J. *Clones, Genes and Immortality*. Oxford, UK: Oxford University Press, 1998.

Harris, J. *Enhancing Evolution*. Princeton, NJ: Princeton University Press, 2007.

Hodge, R. *Genetic Engineering*. New York: Facts on File, 2009.

Hubbard, R. and E. Wald. *Exploding the Gene Myth*. Boston, MA: Beacon Press, 1994.

Kaldis, B. "Could the Ethics of Institutionalized Health Care Be Anything but Kantian? Collecting Building Blocks for a Unifying Metaethics." *Medicine, Health Care and Philosophy*, 8/1 (2005).

Kuhse, H. and P. Singer, eds. *Bioethics: An Anthology*. Oxford, UK: Blackwell, 2006.

Nussbaum, M. and C. Sunstein, eds. *Clones and Clones*. New York: W. W. Norton, 1998.

Ruse, M. and C. Pynes, eds. *The Stem Cell Controversy*. New York: Prometheus Books, 2003.

Sandel, M. *The Case Against Perfection*. Cambridge, MA: Harvard University Press, 2007.

Byron Kaldis
The Hellenic Open University

Biogas

Biogas refers to gas used in energy production produced from biological processes rather than chemical processes. Biogas is generated from anaerobic digestion (the breakdown of organic matter by bacteria without oxygen), resulting primarily in methane gas, carbon dioxide (CO_2), and trace amounts of nitrogen, hydrogen, and carbon monoxide. It differs from natural gas—commonly used as a fuel source—in that it has not had all other gases (aside from methane) removed through chemical processing. Biogas occurs naturally in compost heaps, as swamp gas, and as a result of enteric fermentation in cattle and other ruminants, which are the estimated source of 3 percent of greenhouse gas emissions related to global warming. The biogas produced from anaerobic digestion can be burned to produce heat, or used in combustion engines to produce electricity. Source organic material used to produce biogas includes animal waste, such as manure and sewage, and municipal solid waste (MSW), such as that harnessed from landfills. The natural process of digestion is used in harnessing the biogas from MSW contained in landfills, sometimes referred to as landfill gas (LFG).

Biogas, like that being produced in this anaerobic digester, is created from biological processes such as animal waste and other municipal solid waste rather than chemical processes.

Source: Wikimedia

The natural decomposition of organic matter in landfills occurs over many years, and the gas produced can be collected from a series of interconnected pipes located at various depths across the landfill. The composition of this gas changes over the life span of the landfills. Generally after one year, the gas is composed of 60 percent methane and 40 percent carbon dioxide. It is estimated that large, older landfills will produce 150–300 cubic meters of gas per ton of waste with a total energy potential of 5–6 GJ (gigajoules) per ton of waste. A landfill with a million tons of waste is estimated to have a capacity of about 2 MW over a 15–20 year period. However, most landfill collection is variable according to percentage of waste composition and age of the facility, with average energy potential within range of 2 GJ per ton of waste.

The collected gas can be burned at or near the site in furnaces or boilers or used to generate electricity. LFG is increasingly being used in internal combustion engines or gas turbines to create electricity since there is often limited need for the heat produced at most remote landfill locations. There is an energy loss in the conversion of gas to electricity, and generally, 10 percent of energy is actually used. Despite the lack of efficiency, LFG systems are increasingly being implemented as the gas needs to be burnt off to prevent explosions from methane accumulation inside the landfill or to prevent the loss of methane, a greenhouse gas, into the atmosphere. It is estimated there were 1,000 landfills collecting biogas as of 2001 with 325 in the United States and over 130 in Germany as well as in the United Kingdom. The largest landfill in the United States, the Puente Hills landfill near Los Angeles, California, also contains the largest biogas plant in the world, producing 50 MW. Estimates from 2007 indicate U.S. landfills captured 2.6 million tons of methane each year with 70 percent used for generating heat or electricity. The U.S. Department of Energy (DOE) estimates the United States consumed 147 trillion Btu from LFG in 2003, or 0.6 percent of U.S. natural gas consumption.

Animal and plant wastes can also be processed in an anaerobic digester to produce biogas as a liquid or by converting them to a slurry composed primarily of water. Anaerobic digesters are composed of a feedstock source holder, digestion tank, biogas recovery unit, and heat exchangers in cooler climates to maintain ideal temperature for bacteria to digest the feedstock. Small-scale household digesters containing as little as 200 gallons can be used to provide cooking fuel or electric lighting in rural homes. It is estimated that millions of homes in less developed countries, including China, use household digesters as a renewable energy source. Large-scale farm digesters store liquid or slurried manure from farm animals, such as cattle or pigs, or sewage waste, continuously, or for shorter time periods ranging from days to many weeks. The primary types of farm digesters are covered lagoon digesters (collect biogas from within the unheated covered extent of manure lagoons), complete mix digesters (heated tanks for slurry manure), and plug-flow digesters (heated underground tanks used for dairy manure). Heat is usually required in digesters to maintain a constant temperature of about 35 degrees Celsius for bacteria to decompose the organic material into gas. An efficient digester may produce 200–400 cubic meters of biogas containing 50–75 percent methane for a total output of 8 GJ per dry ton of input waste. The gas produced by the digester can then be burned to produce heat or in internal combustion engines to generate electricity.

The use of biogas as a green technology has environmental benefits in addition to its use as a renewable energy source. Biogas technology enables the effective use of accumulated animal wastes from food production as well as municipal solid waste from urbanization. The conversion of animal waste into biogas also assists in the reduction of the greenhouse gas methane. Overall, methane (CH_4) is nearly 23 times more effective in trapping heat than CO_2, and efficient combustion replaces CH_4 molecules with CO_2 resulting in a substantial net reduction in greenhouse gas emissions. Estimates from the United Kingdom indicate that greenhouse gas emissions were 10 percent less in 2002 because of the combustion of LFG. In addition, the CO_2 and hydrogen sulfide in biogas can be removed to make renewable natural gas and used for fuel in gas-powered vehicles. The DOE reports biogas is increasingly being used as an alternative fuel in natural gas vehicles in the United States and in Europe.

The U.S. Environmental Protection Agency (EPA) reports 8 percent of methane emissions in the United States are from animal waste. Burning this waste reduces methane concentrations as a greenhouse gas and also offsets energy that would be generated from fossil fuels. The conversion of animal waste to biogas also disposes of animal waste, keeping it from affecting groundwater and reducing odors and ammonia produced from manure from large-scale livestock operations. EPA has enacted its AgSTAR program to encourage livestock producers to use anaerobic digesters as a means of handling waste. Since 1994, the number of anaerobic digesters has increased to 151 across the United States with a total electricity generation of 374,000 MWh in 2009. In addition, biogas holds significant potential as a reliable energy source in developing countries. The United Nations Development Programme (UNDP) estimates 2.8 million biogas plants were installed in India as of 1998 with a future potential of 12 million plants. The use of biogas has clear environmental and economic benefits and strong potential as a green energy source derived from the primary sources of waste from human activities, food production, and solid waste management.

See Also: Anaerobic Digestion; Solid Waste Treatment; Waste-to-Energy Technology.

Further Readings

Alternative Fuels and Advanced Vehicles Data Center. "What Is Biogas?" http://www.afdc .energy.gov/afdc/fuels/emerging_biogas_what_is.html (Accessed September 2010).

Andrews, J. and N. Jelley. *Energy Science: Principles, Technologies, and Impacts.* Oxford, UK: Oxford University Press, 2007.

Boyle, G., ed. *Renewable Energy: Power for a Sustainable Future*, 2nd ed. Oxford, UK: Oxford University Press, 2004.

Themelis, N. J. and P. A. Ulloa. "Methane Generation in Landfills." *Renewable Energy*, 32 (2007).

United Nations Development Programme (UNDP). *World Energy Assessment: Energy and the Challenge of Sustainability.* New York: United Nations and UNDP, 2001.

U.S. Environmental Protection Agency. "The AgSTAR Program." http://www.epa.gov/agstar (Accessed September 2010).

U.S. Environmental Protection Agency. "Landfill Methane Outreach Program." http://www .epa.gov/lmop (Accessed September 2010).

Christopher A. Badurek
Independent Scholar

BIOTECHNOLOGY

Biotechnology has been defined as the use of techniques to modify deoxyribonucleic acid (DNA) or the genetic material of a microorganism, plant, or animal in order to achieve a desired trait. The word *biotechnology* was coined by Karl Ereky, a Hungarian engineer, in 1919. He used *biotechnology* as an umbrella term to describe methods and techniques that allow the production of substances from raw materials with the aid of living organisms. This emerging field of applied biology has tremendous scope in engineering, technology, and medicine, and includes techniques such as genetic engineering, and cell and tissue culture technologies. The 1993 United Nations (UN) Convention on Biological Diversity defines biotechnology as "any technological application that uses biological systems, living organisms, or derivatives thereof, to make or modify products or processes for specific use." In essence, it can be defined as the use of living organisms to create products.

While some form of biotechnology, such as baking bread or making wine, has always existed, it is the possibility to manipulate matter at the genetic level that is of prime importance in green technology, especially in the field of agriculture. In the case of foods, genetically engineered plant foods are produced from crops whose genetic makeup has been altered through a process called recombinant DNA, or gene splicing, to give the plant desired traits. Genetically engineered foods are also known as biotech, bioengineered, and genetically modified (GM, GMO), although "genetically modified" can also refer to foods from plants altered through methods such as conventional breeding. While in a broad sense biotechnology refers to technological applications of biology, common use in the United States has narrowed the definition to foods produced using recombinant DNA.

Biotechnology, which has applications in a wide range of life sciences, is one of the most controversial of all technologies. Those who support it argue that it is the only way to feed a world population that might reach 9.1 billion by 2050 according to UN estimates; those who oppose it argue that it at best can only buy temporary relief at enormous costs, some of them unknown. Biotechnology, especially in confluence with other emerging technologies like nanotechnology, has the potential to change life—a new industrial revolution. An increasing understanding and manipulation of living systems, optimists argue, will lead to a world free of disease, with abundant food for all, and clean energy based on biotechnology, while pessimists contend that it would leave large parts of the world infertile and could even create Frankenfoods or new organisms that could either wipe out life on Earth or seriously damage it.

In agriculture and animal husbandry, GM technology falls into three main categories: (1) the use of hormones such as Bovine somatotropin (rBST, also known as rBGH), a synthetic hormone that when injected into a cow's bloodstream increases the production of milk by stimulating the IGF-1 hormone, which occurs in both cow and human milk, in cows; (2) "terminator seeds" that are genetically engineered to make the second generation sterile; and (3) the modification of seeds so that they have special properties, as, for example, genetically modified canola or soybean seeds.

Biotechnology: A Panacea?

One of the appeals of biotechnology is that it can be a panacea for several problems. In the face of an expanding population, organic farming may not be able to sustain humankind, especially in Africa where slash-and-burn agriculture is still widely prevalent. Habitat destruction is one consequence of such agriculture. The savannahs and tropical forests of Central and South America, Asia, and Africa are treasures of biodiversity, and they are immense carbon sinks. Since 1972, about 200,000 square miles of Amazon rainforest have been cleared for crops and pasture; from 1966 to 1994, all but three of the Central American countries cleared more forest than they left standing. Rainforests once covered 14 percent of Earth's land surface; now they cover a mere 6 percent, and if present rates of clearance continue, it is projected that the last remaining rainforests could be consumed in less than 40 years. It is estimated that one and one-half acres of rainforest are lost every second with tragic consequences for both developing and industrial countries, destroying the planet's best defense against global warming. While such clearance cannot be ascribed to agriculture alone, at the core is demographic pressure and the increasing paucity of arable land. Advocates of biotechnology argue that the only way to prevent catastrophe is to use biotechnology to produce enough to feed a growing population.

Biotechnology draws its inspiration from one of the early successes of 20th-century agriculture, when the use of pesticides, fertilizer, and high-yielding dwarf wheat, first introduced in Mexico and then later in India, resulted in an enormous expansion of the food supply, especially in famine-ridden India. By 1974, India, a nation that throughout its history was haunted by the specter of famine, became self-sufficient in the production of all cereals. In both India and Pakistan, successor states of British India, food production since the 1960s has increased faster than the rate of population growth. In 1950, the world produced 692 million tons of grain for 2.2 billion people; by 1992, production was 1.9 billion tons for 5.6 billion people—2.8 times the grain for 2.2 times the population. Global grain yields rose from 0.45 tons per acre to 1.1 tons; yields of corn, rice, and other foodstuffs improved similarly. From 1965 to 1990, the globe's daily per capita intake grew from 2,063 calories to 2,495, with an increased proportion as protein. Together, the changes that this early application of biotechnology brought were so dramatic that they came to be known as the Green Revolution. However, questions have been raised by environmentalists regarding the long-term feasibility of such pesticide-intensive agriculture; and even if it could be continued, the amounts of pesticides, fertilizer, and other chemicals required to make it effective would be prohibitive.

GMO foods, created using biotechnology, are seen as a panacea by its advocates. They argue that modifying plants at the gene level (as, for example, Monsanto's Roundup Ready crops that have been genetically modified to permit direct application of the Monsanto herbicide glyphosate) so as to kill nearby weeds without killing the crops would save a fortune on pesticides. MON 810, the GM corn seed, is genetically engineered to produce a toxin to fight the voracious larvae of the corn borer moth. Anti-GM activists argue that such plants could kill neighboring plants, and that given the monopoly power of seed companies such as Monsanto, GM crops would amplify input costs, reduce the right of farmers to choose their seeds, and force poorer farmers out of agriculture. Uniform corporate–capitalist agriculture would dominate and even transfer control over the world's food supply to the private sector. Countries such as Brazil, India, and South Africa, where the new technology has been introduced, have seen massive protest by farmers.

Unintended Consequences?

One of the greatest fears regarding GMO foods echoes earlier beliefs in the power of science for human good that later had unintended consequences. For example, the synthetic pesticide dichlorodiphenyltrichloroethane (DDT), a key factor in the Green Revolution, which Monsanto produced along with several other companies, was so harmful that it was banned by the United States in 1977, and worldwide by the Stockholm Convention on Persistent Organic Pollutants (2004), which restricted the use of DDT to malaria control.

With biotechnology, similar products could have extremely deleterious effects. One example is the inadvertent creation of evil pigweed, a superweed that is resistant to Roundup. Genetically modified crops have failed, too, as, for example, in 2009 when South African farmers were hit by millions of dollars in lost income when 82,000 hectares of genetically manipulated corn (maize) failed to produce seeds despite an outwardly green and lush appearance. The company that produced the seeds, Monsanto, claimed that "under-fertilization processes in the laboratory" were responsible for this, and has offered to compensate the farmers. Similar problems of crop failure have affected Monsanto seeds in India.

However, the greatest concern with regard to biotechnology is the possible risk that GMO foods would impact humans at a genetic level, since the future impact of GMO foods is still unknown and consumers may be at risk. Genetic manipulation has awakened fears in the public, especially in Europe. Hungary, Austria, Greece, France, and Luxembourg have banned genetically modified (GM) corn, claiming that MON 810 is harmful for the environment, though it had been approved for cultivation in Europe by the European Union in 1998. Moreover, in Europe, unlike in the United States, genetically modified food has to be clearly labeled.

Biotechnology is also a key area for intellectual property rights (IPR) conflict. Environmental activists argue that it gives power to big companies and often victimizes even those who have not deliberately violated a company's IPR. Thus, for example, from 1997 to July 2009, Monsanto filed 138 lawsuits in the United States, from a customer base of 250,000 customers and others worldwide. In one of the most prominent cases of this nature, Monsanto sued Canadian farmer Percy Schmeiser in 1997 after Monsanto agents found the company's patented gene in canola plants on his farm near Saskatoon, Saskatchewan. After extensive legal proceedings, Canadian courts ruled that Schmeiser was not obliged to pay Monsanto damages or the company's legal costs, agreeing with Schmeiser that the plant was a higher life form and could not be patented, but upholding Monsanto's right to the patented gene.

Biotechnology has also raised concerns regarding patterns of corporate ownership, especially in areas such as the patenting of genes. In an October 2005 paper in *Science*, Fiona Murray and Kyle Jensen plotted patent activity on the human genome and discovered that 20 percent of genes are explicitly claimed as U.S. IP, which amounts to 78 percent of total gene ownership. Roughly 63 percent of patented genes are assigned to private firms. Of the top 10 gene patent assignees, nine are U.S.-based, including the University of California, Isis Pharmaceuticals, the former SmithKline Beecham, and Human Genome Sciences. The top patent assignee is Incyte Pharmaceuticals/Incyte Genomics, whose IP rights cover 2,000 human genes, mainly for use as probes on DNA microarrays.

Biotechnology and Climate Change

Climate change has also opened up new areas for bio prospecting in hitherto untapped areas, such as the Arctic. For example, Norwegian scientists are prospecting for new drug candidates by isolating the secondary metabolites from bacteria, traditionally isolated from the soil, living in the sediments of one of Norway's largest fjords and within cold-water sea sponges that global warming has made accessible. Pharmaceutical and biotech industries have aggressively scouted untouched areas, such as coral reefs and tropical rainforests, to identify flora and fauna that may yield new medicines and processes, such as a cure for cancer. This is because nature remains the largest source of new products. Between 1981 and 2002, almost half of the small molecule new chemical entities introduced were natural products or inspired by them.

Investment in biotechnology has also increased dramatically. According to the respected journal *Nature,* over the past few years, research spending has increased by nearly $3 billion annually and moved up from $22.8 billion in 2007 to $25.5 billion in 2008. It was one of the few sectors that remained unaffected by the financial crisis of that year, and even turned a profit of $3.8 billion, a marked increase over the barely $1 billion in profit Nature Biotechnology reported for the sector in 2007. The United States remains the leader in biotechnology funding, approving an additional input of $2 billion in 2010. However, since the 1980s, the share of private sector investment in this area has increased dramatically. The private sector is also active in the funding of controversial research such as embryonic stem cell research, which is deemed too controversial to be funded by the federal government. Finally, in contrast to earlier technological initiatives, at least part of the funding is being devoted to projects that analyze the political and social impact of new technologies such as biotechnology.

The United States has the largest number of public biotech companies, followed by the European Union, Canada, Australia, and the rest of the world. There are also joint projects for the development of biotechnology such as the EC-US Task Force on Biotechnology Research that has, since 1990, been coordinating transatlantic efforts to promote research on biotechnology and its applications for the benefit of society. The task force was established in June 1990 by the European Commission and the U.S. White House Office of

Science and Technology. In the European Union, industrial specialization has a large role to play in the size and direction of private investment, with the Netherlands, France, and the United Kingdom seeing a majority of investments going to agro food, while in Germany, chemical and pharmaceutical companies are at the forefront.

Biotechnology is one of the areas where the new collaboration among universities, academe, and the state is very visible. Biotechnology in its present form exploded only when genetic technology had risen to a level where copies of human DNA genes could be inserted into bacteria, a fact that revealed the immense commercial potential of this new life science. This meant that once the codes of growth hormone, insulin, and the like, were inserted and expressed in bacteria, pharmaceutical companies could reap millions by sponsoring research into biotechnology.

The first biotechnology company, Genentech, was a spin-off from Stanford University and the University of California at San Francisco and went public in 1980. The foundation of the company rested on two new technologies: recombinant DNA and monoclonal antibodies. Moreover, the nature of the knowledge that made this innovation possible was a new way of making certain substances or producing certain antibodies in quantity. Crucially, this knowledge could not be easily written down, and ongoing collaboration was essential to any commercialization of it. This continuous collaboration within laboratory conditions, between the inventor and the learner, create newer and closer links between university researchers and new biotechnology firms. One key element in such partnerships is the creation of spin-offs, where one technology or its derivatives are used to set up a new company.

See Also: Green Nanotechnology; Intellectual Property Rights; University-Industrial Complex.

Further Readings

"A Blueprint for Biotech's Blues." Editorial. *Nature*, 27/8 (2009).

Borras, Saturnino M., Marc Edelman, and Cristóbal Kay. *Transnational Agrarian Movements Confronting Globalization*. Hoboken, NJ: Wiley-Blackwell, 2008.

Easterbrooke, Gregg. "Norman Borlaug: Forgotten Benefactor of Humanity." *Atlantic Monthly* (January 1997). http://www.theatlantic.com/past/docs/issues/97jan/borlaug/borlaug.htm (Accessed July 2010).

Pew Initiative on Food and Biotechnology. "Feeding the World: A Look at Biotechnology and World Hunger." Washington, DC: Pew Charitable Trust, 2004.

Rauch, Jonathan. "Will Frankenfood Save the Planet?" *Atlantic Monthly* (October 2003). http://www.theatlantic.com/past/docs/issues/2003/10/rauch.htm (Accessed August 2009).

Renault, Catherine Searle. "Academic Capitalism and University Incentives for Faculty Entrepreneurship." *Journal of Technology Transfer*, 31 (2006).

Shiva, Vandana. *Stolen Harvest: The Hijacking of the Global Food Supply*. Cambridge, MA: South End Press, 2000.

U.S. Food and Drug Administration. "Biotechnology." http://www.fda.gov/Food/FoodIngredientsPackaging/ucm064228.htm (Accessed August 2010).

Sabil Francis
University of Leipzig, Germany,
École Normale Supérieure, Paris

BOUNDARY OBJECTS

In sociology, a boundary object is something—not necessarily a physical object—that acts as an interface between different communities. A boundary object may be used differently by different communities (indeed, a community need not understand or even be concerned with how the object is used by any community but their own) but must also be sufficiently robust to retain its identity despite the different ways it is used. For instance, a nongovernmental organization (NGO) that accepts donations from a manufacturer and distributes them to a community in need is acting as a boundary object between the manufacturer and community. The goods distributed could also be considered boundary objects. The concept of the boundary object is useful in social theory because it offers a way to explain how one object can be viewed and used in multiple ways by multiple communities without conflict arising—in fact, boundary objects can facilitate cooperation among communities without requiring that they agree on the purpose or function of the boundary object. This understanding may play a critical role when addressing environmental concerns because this process may require enlisting the cooperation of communities that have quite different interests regarding the object in question.

The Health Effects Institute is a boundary object between the Environmental Protection Agency, which is focused primarily on the health effects of pollution, and the automobile industry, which is focused primarily on the financial health of automobile companies.

Source: Wikimedia

This concept may be clarified by considering an example. S. L. Star and J. R. Griesemer coined the term *boundary object* in a 1989 study in which the boundary object in question was the Museum of Vertebrate Zoology (MVZ) at the University of California at Berkeley. According to their theory, the MVZ was used in several different ways by different communities, yet retained its identity as the Museum of Vertebrate Zoology for all of them. Another way to say this is that the MVZ had different meaning for the different communities (according to how it coincided with their interests), yet all would recognize it as the same museum. The different communities were able to collaborate on the museum and further each other's goals without necessarily endorsing any interests but their own.

Joseph Grinnell, the museum curator, as a member of the scientific community, was interested in advancing Darwinian theory, in particular, studying how changes in the environment can drive natural selection. To this end he required access to large quantities of information about the environment and the fauna that lived in it. The community of conservationists, including the philanthropist Annie Montague Alexander, saw the preservation of California flora and fauna as a good in and of itself (i.e., they were not necessarily motivated by scientific goals) and saw the MVZ as a means to further this end.

Alexander partially financed the museum and she, along with others in the conservation community, provided some of the labor necessary to collect specimens for the museum. The university administration was willing to provide some of the funding and space for the MVZ because they saw it as a means to enhance the university's prestige (putting it on a footing similar to eastern universities that had their own museums) and also to fulfill their mandate to serve the people of California. The result of the efforts of these three communities was a world-class museum that fulfilled in some way the desires of each community without ever requiring them to agree on the goals of the museum. Therefore, the museum served as a boundary object among people with radically different goals—scientific support for the theory of natural selection, preservation of local flora and fauna, and pursuit of academic prestige—and facilitated their working together, each making contributions that served their own desires but also benefited those of the others involved.

David Guston and colleagues have extended the notion of the boundary object to include boundary organizations, that is, organizations that must reconcile the interests of different communities regarding a common object. For instance, in dealing with environmental issues an organization may be required to serve the demands of both the scientific and public policy communities, drawing incentives from and producing output for both. Boundary organizations require participation from all the communities involved and are accountable to each. Boundary organizations are successful when they produce results that satisfy all the communities involved.

The Health Effects Institute (HEI) created in the United States in 1980 is an example of a boundary organization. It is an independent research organization chartered as a nonprofit corporation that receives about half its core funds from the U.S. Environmental Protection Agency (EPA) and the other half from the motor vehicle industry, with additional funds for special projects or research programs occasionally provided by other public and private organizations. The HEI is governed by an independent board of directors (including both scientific and policy personnel) while separate bodies oversee its operations. The research committee decides what research should be funded by HEI, the review committee reviews and critiques the scientific quality and regulatory implications of HEI research projects and publishes their findings, and special expert panels are appointed as needed to undertake scientific reviews and other special projects.

The HEI was founded in part to deal with the antagonism that developed in the late 1970s between EPA and the automobile industry. EPA was concerned with protecting public health and the environment from damage by (in this case) vehicular air pollution, while the automobile industry was concerned, as any commercial industry would be, with the financial health of companies involved in manufacturing automobiles and related lines of work. The 1977 Clean Air Act Amendments included a provision requiring EPA to ban any new vehicle or vehicle component that was a threat to public health and placed the burden of proof on the automobile industry to demonstrate that any new product did not provide such a threat. EPA was understandably concerned about the quality of research that might be provided by the automobile industry in order to meet this requirement.

The HEI was created as a boundary organization that would incorporate the interests of both EPA and the automobile industry by providing the best possible scientific information about the health effects of pollution created by motor vehicles. In pursuit of this goal, HEI conducts two primary activities: funding independent research on topics related to vehicular air pollution and reviewing existing literature about the current science in the field. HEI regularly publishes requests for research proposals in areas related to motor vehicle–related air pollution and health, and grants are awarded competitively after scientific review. HEI itself is also subject to external review to be sure it is fulfilling the needs of all the communities involved. For instance, in 1993 a review by the National Academy of Science reported that HEI was funding research that was of high scientific quality but was lacking in relevance and timeliness and recommended several internal reforms, including a broadening of leadership and staff, seeking sponsorship outside the automobile industry, and diversifying the research agenda. In response, HEI expanded its mission, broadened its activities to include re-analysis of existing data, expanded its board of directors and its definition of its constituency, and encouraged its staff to strengthen ties with other researchers.

See Also: Science and Technology Policy; Social Construction of Technology; Technology Transfer.

Further Readings

Guston, David H. "Boundary Organizations in Environmental Policy and Science: An Introduction." *Science, Technology & Human Values*, 26/4 (2001).

Guston, David H., William Clark, Terry Keating, David Cash, Susanne Moser, Clark Miller, and Charles Powers. "Report of the Workshop on Boundary Organizations in Environmental Policy and Science" (April 2000). http://www.hks.harvard.edu/gea/pubs/huru1.pdf (Accessed September 2010).

Health Effects Institute. http://www.healtheffects.org (Accessed September 2010).

Star, S. L. and J. R. Griesemer. "Institutional Ecology, 'Translations' and Boundary Objects: Amateurs and Professionals in Berkeley's Museum of Vertebrate Zoology, 1907–39." *Social Studies of Science*, 19/4 (1989).

Sarah Boslaugh
Washington University in St. Louis

CADMIUM TELLURIDE PHOTOVOLTAIC THIN FILM

Cadmium telluride photovoltaics, also known as cadmium telluride thin film or CdTe thin film, are a type of photovoltaic device. Photovoltaic (PV) devices produce direct current electricity upon exposure to light. They are a member of a class of technologies known as "thin film solar cells"—that is to say, devices that produce electricity from light from a "thin film" of chemical compounds, rather than a wafer of crystalline silicon.

Cadmium telluride is a compound that can be used in the manufacture of photovoltaic devices. They differ from "crystalline silicon" photovoltaic technologies in that they use a smaller amount of semiconductor—a thin film—to generate the electricity. This can be cheaper to produce; however, there is a trade-off in that the efficiency of thin film devices is less than that of crystalline silicon devices. Cadmium telluride thin film technology is particularly interesting, as it has the potential to surpass silicon in terms of cost per kilowatt (kW) of installed capacity.

Market Share

Thin film technologies account for a small share of the market in photovoltaic devices; however, this segment is expected to rapidly grow with crystalline silicon technologies comprising around 76.6 percent of sales in 2009, while thin film devices occupied the remaining 23.4 percent of sales in 2010. However, it is significant when you consider that this share has nearly doubled from 12.9 percent in 2008. CdTe thin film sales made up around 43 percent of the market in 2010 for thin film photovoltaic devices. There is also much interest in the development of novel manufacturing methods for thin film devices that have the potential to unlock economies of scale, which could increase the market share of thin film devices.

Thin Film Solar Technologies

Thin film solar technologies differ from crystalline photovoltaic technologies in that rather than using a crystalline solid to form the semiconductor junction (e.g., mono- or polycrystalline silicon), they instead use thinner semiconductor layers to absorb sunlight and convert the light energy into electrons.

The first thin film technology to be developed was amorphous silicon, wherein silicon is randomly deposited onto a substrate (as opposed to the regular crystal lattice seen in wafer crystals). There were a number of problems with this technology: the process of depositing the silicon onto the substrate was time consuming, driving up the cost of these cells, and furthermore, the efficiency of the technology was relatively poor.

Cadmium telluride is another thin film technology that comprises a junction of n-doped cadmium sulfide, known as the "window layer," on top of a p-doped layer of cadmium telluride, known as the "absorber." A transparent conductive front contact covers the cadmium sulfide, while the cadmium telluride is in contact with a conductive rear surface substrate. One of the advantages of a cadmium telluride junction is that its band gap is 1.4 electron volts, which matches the solar spectrum very well.

Key Technical Points

One of the reasons that cadmium telluride technology is seen as showing lots of promise, is that its band gap—a physical property of the material—is well matched to the way that photons are distributed in the spectrum of the sun's light output.

In terms of the technical challenges of transferring the concept from the laboratory to production, cadmium telluride thin film technology has a number of favorable characteristics: it can be deposited onto the substrate quickly and is a high-throughput technology, lending itself to production.

Concerns About Cadmium Telluride Cells

The electronics industry has moved to try to remove elemental cadmium from personal electronics because cadmium is a cumulative poison. In Europe, the Restriction of Hazardous Substances (RoHS) legislation has been powerful in eliminating cadmium from electronic devices due to health effects. cadmium represents a health risk—for miners during extraction of the raw materials, for workers processing the material, and at end of life during disposal.

To put the risks into perspective, a square meter of cadmium telluride thin film solar cell contains around the same amount of cadmium as a single "C" cell nickel-cadmium battery; however, in the form of a thin film solar cell, the cadmium is more stable and less soluble.

Researchers posit that if CdTe PV modules were to be used on a large scale, there would be little risk to health and the environment as the alloys are encapsulated within the modules; however, there have been concerns regarding cadmium leaching from broken CdTe modules expressed by researchers at the Norwegian Geotechnical Institute and Wuppertal Institute.

While it has been promoted that closed-loop recycling would address any concerns over end-of-life disposal, it is important to note that even closed-loop recycling systems do not recover everything.

Clearly there is a need to balance the risks associated with producing and handling cadmium with the urgent need to address issues of energy supply and carbon emissions reduction.

See Also: Clean Energy; Copper Indium (Gallium) Selenide (CIGS or CIS) Solar Photovoltaic Thin Film; Crystalline Silicon Solar Photovoltaic Cell; European Union Restriction of Hazardous Substances (RoHS) Directive; Passive Solar; Solar Cells.

Further Readings

Bonnet, D. and P. Meyers. "Cadmium-Telluride—Material for Thin Film Solar Cells." *Journal of Materials Research* (October 1998).

Chu, T. L. and S. S. Chu. "Recent Progress in Thin-Film Cadmium Telluride Solar Cells." *Progress in Photovoltaics: Research and Applications*, 1/1 (1993).

German Solar Energy Society. *Planning and Installing Photovoltaic Systems: A Guide for Installers, Architects and Engineers*, 2nd ed. London: Earthscan, 2008.

Jha, A. R. *Solar Cell Technology and Applications*. Boca Raton, FL: Auerbach Publications, 2009.

Gavin D. J. Harper
Cardiff University

CARBON CAPTURE TECHNOLOGY

Carbon capture refers to the means to mitigate the contribution of carbon dioxide (CO_2) emissions to global warming. Cumulative emissions of CO_2 (representing 62.5 percent of all greenhouse gases) have been identified as a major contributor to global warming. Due to the greenhouse effect, the average surface temperature of the Earth is 35 degrees Celsius higher than its radiation temperature. With higher CO_2 concentrations, the greenhouse effect is enhanced. It was reported that the mean surface temperature would be 2–3 degrees Celsius higher for doubled CO_2 levels with a strong amplification of 8–10 degrees Celsius warming in the polar areas. The fate of CO_2 produced as a result of burning fossil fuel depends on the exchange rates of carbon between the atmosphere and oceans, shallow water sediments, and terrestrial biosphere.

Capturing CO_2 is primarily applied to large emission producing facilities, such as fossil fuel power plants. Power plants contribute to more than 30 percent of the CO_2 emissions worldwide. Industrial processes, such as ammonia manufacturing, fermentation, and hydrogen production, significantly contribute to CO_2 emissions. Scrubbing CO_2 from the atmosphere would essentially require substantial amounts of energy, hence increasing the CO_2 emission as a result of producing the required energy from fossil fuel power plants.

Storage of the captured CO_2, often referred to as "sequestering of CO_2," is a complementary process to reduce the escape of CO_2 emissions to the atmosphere. Due to the impact of reducing the volume of greenhouse gases from the atmosphere, a significant amount of scientific work on the sequestering and utilization of CO_2 has been reported. Once the CO_2 is captured, it needs to be prevented from reaching the atmosphere for a period of sufficient time to allow climate stabilization. A number of studies have been carried out to understand the behavior and long-term fate of CO_2 when large volumes of CO_2 are injected in the subsurface. The three main options for subsurface storage of CO_2 are saline aquifers, exiting oil and gas fields, and un-mineable coal seams. There is an economic benefit for this injection, in addition to lowering the CO_2 tax, in which CO_2 is utilized to enhance oil recovery or coal-bed methane production. Alternative subsurface storage options exist, but are still at an early stage of technology development. These options include deep ocean sequestering, carbonate production, and injection in deep carbonate sediments. The major drawback of the injection processes is the need to have high concentrations of CO_2, which have a high potential to escape to the atmosphere.

CO_2 Capture From Combustion

The technology to capture CO_2 generated from fuel combustion can be classified into three categories; precombustion, postcombustion, and oxyfuel combustion. With the precombustion technology, the CO_2 is removed before the combustion. This technology is applied to gaseous fuel, such as methane, propane, and hydrogen, used in power generation. Oxygen is used in a gasifier to oxidize the gaseous fuel. Carbon monoxide and hydrogen will result from the oxidation, which can be shifted to carbon dioxide and hydrogen. The resulting CO_2 is then captured and stored while the hydrogen can be used as fuel. The advantage of this technology is that the CO_2 is captured from a pure CO_2 exhaust stream prior to mixing with other gases that result from the combustion stream.

With the postcombustion technology, the CO_2 is removed from the stream of exhaust gases that results from the burning of fossil fuels. Several gases and unburned fuel gases are generally mixed in the exhaust stream. CO_2 is captured before it escapes to the atmosphere. Technologies to selectively remove CO_2 from a gas stream and store it in a solid medium have also been proposed. A microporous gas diffusion membrane to facilitate the separation of CO_2 from among the emission gases has been reported. The captured CO_2 gases are then further processed to form stable by-products, such as carbonic acid. Zeolitic imidazolate frameworks (ZIFs) have been synthesized with complex cages that contain up to 264 vertices. These ZIFs have shown selectivity to capture CO_2 from different gas mixtures at room temperature. Some classes are able to store up to 28 liters of CO_2 per liter of the material at standard room temperature and pressure. Ammonium carbonate has been utilized to absorb CO_2. The resulting ammonium bicarbonate is converted back to ammonium carbonate in a regenerator, and is reutilized to repeat the process; the CO_2 is captured and stored.

With oxyfuel combustion, the fossil fuel is burned in oxygen, instead of air, limiting the formation of other combustion gases. The flue gases, primarily CO_2 and water vapor, are circulated in a cooling coil to condense the water, allowing the CO_2 to be captured at the end of the circulation loop. Power plants using this technology are often termed *zero emission power plants* because most of the CO_2 generated is captured and stored. A recent technique to utilize oxygen combustion is to use metal oxide particles as carriers for the oxygen to the combustion chamber. Following the combustion, the CO_2 and water vapor are cooled to condense the vapor and capture the CO_2. The metal particles are pushed to a fluidized bed, where they react with air to oxidize, and are utilized again in the combustion chamber.

CO_2 Capture by Microalgae

Microalgae are photosynthesizing microorganisms, considered the most primitive form of plants. Photosynthesis in microalgae is generally more efficient than photosynthesis in higher plants because of their simple cellular structure. These microorganisms grow in aqueous suspension; they have more efficient access to water, carbon source, and other nutrients. Oil-rich microalgae have recently gained much attention because of their ability to consume CO_2 emissions and produce oil, which can be further utilized as biodiesel. Placing oil-rich microalgae reactors in the vicinity of CO_2 point sources has been proposed, whereby the CO_2 produced from the fossil fuel combustion is channeled to the microalgae reactors. The economic and environmental impacts of the application are considerable. Oil-rich microalgae are capable of producing 30–100 times the amount of oil per unit area of land, compared to terrestrial oilseed crops.

CO_2 Reuse

Captured CO_2 can be utilized in the chemical industry for producing vulcanized rubber, polyurethane foam, and polycarbonates. Several other proposals exist, including reformation of methane gas using CO_2, dehydrogenation of ethylbenzene to styrene, and conversion of CO_2 into methanol using enzyme-like catalytic activity.

Economics

Carbon capture is composed of four different cost processes: separation, compression, transport, and storage. There are many factors that affect the cost, including the technology utilized in the processes, the carbon source, the energy required to pass the CO_2 through the absorption column, the energy required to separate the CO_2, the carbon compression and storage characteristics, and the distance between the source and storage.

Based on economical studies for different types of power plants, the average cost increase in electricity in the United States will be 1–4¢/kWh if capture technologies are adopted.

See Also: Ecological Modernization; Technology Transfer; Waste-to-Energy Technology.

Further Readings

Chisti, Y. "Biodiesel From Microalgae." *Biotechnology Advances* (2007).
Laboratory for Energy and Environment, MIT. "Carbon Capture and Storage From Fossil Fuel Use." http://sequestration.mit.edu/pdf/enclyclopedia_of_energy_article.pdf (Accessed April 2011).
U.S. Department of Energy (DOE). "The Capture, Utilization and Disposal of Carbon Dioxide From Fossil Fuel–Fired Power Plants." DOE/ER-30194. Washington, DC: DOE, 1993.

Yousef Haik
University of North Carolina at Greensboro

CARBON FINANCE

Carbon finance is the commonly used term to refer to specific investment in technologies or processes that reduce greenhouse gas (GHG) emissions and result in the creation of carbon credits. Carbon finance (CF) is therefore directly associated with carbon offsets as a mechanism that allows a company, organization, or individual to reduce its environmental impact on the atmosphere in one area by investing in projects that reduce GHGs in another. As a result, carbon finance helps deploy green technology into new areas. However, it is a complex mechanism and does not always create wide-scale green benefits in creating low-carbon economies. This article outlines different forms of CF, its relationship to waste and energy management, and possibilities and pitfalls of CF use for green technology.

The Carbon Finance Market

Carbon finance can be seen as an investment tool that allows the deployment of green technology in the places where it is most economically efficient to do so. This often means carbon finance flows from developed to developing countries, where technologies may be older or processes less efficient. In this way, the same investment dollars can create larger amounts of net emissions reductions. Carbon finance principally reduces emissions through projects or programs of activity that are known as carbon offsets. These offsets use investment to generate carbon credits, and link people, places, and environments across space, transcending traditional economic geographies as finance flows one way and carbon credits flow the other.

The biggest carbon finance mechanism in the world is the Kyoto Protocol's Clean Development Mechanism (CDM), which generates credits that can be used for compliance under Kyoto and in the European Union Emissions Trading Scheme. In addition to the compliance market, a fast-growing parallel Voluntary Carbon Offset (VCO) market exists to assist companies that want to invest in low-emissions projects outside formal regulation. Regional initiatives, such as the Western Climate Initiative in North America, also help spur carbon finance activities because they use carbon offsets as a cost containment mechanism. The first carbon offsets came from large emitters in the United States investing in forestry projects in Latin America, but the wheels of the modern-day carbon markets were oiled by public institutions such as the World Bank and national governments in the early 2000s. Given the profits to be made in emissions trading, however, since 2005 primary investment has come from the private sector. In 2008, CF creating carbon offsets totaled about $7.2 billion, the bulk of which was between developed and developing countries through the CDM. (The amount of carbon finance fluctuates annually because of factors that include prices and availability in connected emissions trading schemes such as the European Union's [EU] Emissions Trading Scheme, climate policy developments at multiple scales, and overall economic performance. Because of the global economic downturn, carbon finance in 2009 amounted to only $3.3 billion.) By creating a policy framework that linked project activities to emissions reductions and emissions reduction targets, the CDM aimed to reduce emissions as cheaply as possible and help developing countries "leapfrog" dirty industrial development. Although criticized for its geographical skew toward the most industrialized developing countries, the CDM has channeled finance to technologies that reduce emissions primarily in industrial, waste, and energy processes.

Creating Emissions Reductions Through Carbon Finance

Carbon finance has the ability to tip the scale in clean technology investments by improving the attractiveness of investment in low-carbon activities. The role of carbon finance is to provide the capital to overcome barriers and ensure the project creates additional environmental benefits that go beyond "business as usual." As such, and in order to be credible, carbon finance must flow to projects that are "additional," use conservative calculations and specific methodologies, and create reductions that are real, measurable, reportable, and verifiable. Carbon finance implies a relationship between an emitter in one place and a reducer in another. Therefore, although projects may produce local benefits, the total net global emissions reductions associated with a project are important to understand its overall net environmental impact.

Carbon finance has been used to fund a wide variety of projects. Nearly half a billion tons of carbon dioxide (CO_2) equivalent have been reduced by the CDM, with over three-quarters of a billion in the CDM pipeline predicted to be reduced. The United Nations (UN) optimistically expects 1.8 billion tons to be reduced by 2012. Almost half of these projects in the CDM financing pipeline are for hydro and wind renewable electricity generation, but a further third potentially relate to waste management, biomass energy, methane avoidance, and landfill gas capture (see Table 1).

Table 1   The Rankings of Number of Projects, Amount of Carbon Credits Produced, and the Ranking in Amount of CERs Produced in Total for Each Carbon-Financed Technology Type in the CDM Pipeline

	Technology	Projects	1,000 CERs	CER Rank
1	Hydro	1483	162,868	1
2	Wind	1029	107,543	2
3*	Biomass energy	703	45,665	8
4*	Methane avoidance	601	28,158	11
5	EE own generation	452	57,757	4
6*	Landfill gas	321	46,081	7
7	EE industry	142	5,195	15
8	Fossil fuel switch	127	49,977	6
9	EE supply side	80	32,398	10
10	N₂O	71	50,000	5

Notes: Waste-related technologies are marked with an asterisk. CERs = certified emissions reductions; EE = energy efficiency. Data correct as of September 1, 2010.

Waste projects are important to reducing GHG emissions because organic waste trapped in an oxygen-scarce environment in landfill decomposes anaerobically and emits methane (normal aerobic decomposition in air emits biogenic carbon dioxide; by volume, landfill gas typically contains 45 to 60 percent methane and 40 to 60 percent carbon dioxide). Methane has a global warming potential of 21, meaning that it is 21 more times as potent as a GHG than CO_2 over a 100-year time scale. This means carbon financiers can claim 21 tons of CO_2 equivalent (tCO_2e) for every one ton of methane reduced. Carbon financed projects either flare this methane to turn it into CO_2 or use it in energy generation. The largest landfill gas capture project to date is in Bandeirantes near São Paulo in Brazil: the project generated 8 million tons' worth of CERs and used the methane to fuel a 23 MW electricity plant.

Although carbon financing can provide developing countries with an effective mechanism to retrofit modern waste management technologies, its primary environmental purpose was to help set economies on a pathway to low carbon growth, specifically by incentivizing low-carbon infrastructure investments over traditional fossil fuel–intensive infrastructure. Now that the "low-hanging fruit" of industrial emissions (e.g., HFC and N_2O destruction and, to some extent, methane capture and landfill) has been taken up, we should expect to see waste-related projects drop down the CDM pipeline list in the future.

The Limits to Carbon Finance

Proponents of carbon finance have consistently pointed out that billions of dollars of investment have flowed to finding the cheapest emissions reductions possible, and in some cases assisted in sustainable development. They therefore point to the use of market mechanisms as a powerful tool in channeling investment toward a low-carbon economy. Others have claimed that as much as two-thirds of the CDM projects are nonadditional and, therefore, do not create the environmental benefits they promised. As noted above, this is an important consideration in creating "green" technology through carbon finance, given the link between emitters and reducers. Carbon finance may help implement green technology, but it depends on the complexities of creating the carbon credit and the lens of scale used to examine the project (i.e., there are differences when looking locally versus globally) to understand its total contribution to greening the environment.

Carbon finance should also only be seen as a stepping-stone to low-carbon economies. Especially in the case of waste, where in many jurisdictions methane emissions are reported and regulated, carbon finance should provide a mechanism for cleaning up existing projects and providing best practices that will become the future minimum requirements. Green technology financed through carbon finance mechanisms should continue to focus on fundamentally shifting energy infrastructure to lower-carbon options.

Carbon finance also relies heavily on a steady climate change policy framework that links the generation of emissions reductions to credits that can be used in an emissions reductions compliance market. Strong policy signals allow capital investment in projects, whereas uncertainty in future climate policy freezes investment and paralyzes innovation in green technology that relies on carbon finance. Carbon finance, therefore, is not applicable to all technology types in all contexts: complementary investment vehicles and policies are needed to support or replace carbon finance in certain green technology developments.

Carbon finance also does not deal directly with other environmental issues associated with waste such as, for example, water pollution or local non-GHG pollutants. Wider environmental policy and investment channels are needed to apply green technologies at a scale beyond carbon. Similarly, carbon finance is not intended to spur innovation, but to implement proven technology in new geographical areas.

Conclusion

Carbon finance is an investment tool that has helped finance the retrofitting of greenhouse gas–polluting technologies and overcome barriers to investments in renewable energy principally in developing countries through UN mechanisms. The waste sector constitutes a significant amount of projects in the global carbon market to date because methane emissions have a high global warming potential and are, therefore, economically attractive to carbon finance investors. Carbon finance, however, is precarious because it relies on steady and secure climate policy, and is only able to incentivize certain technology shifts. It also implies a relationship between emitters and reducers across global space, which means that although local green benefits may exist, if carbon finance is not implemented with effective environmental integrity, global-scale emissions reductions may not occur. Carbon finance is a stepping-stone to lower-carbon futures that ultimately should be dealt with by broad geographical and sectoral low-carbon policies.

See Also: Carbon Footprint Calculator; Carbon Market; Science and Technology Policy.

Further Readings

Arcadis. "Bandeirantes Landfill Gas to Energy Project—Monitoring Report." São Paulo (2008). http://cdm.unfccc.int/UserManagement/TileStorage/ EBA9WO5SEM8Y5RRUGSHJ4LGHTBHHCM (Accessed September 2010).

Bumpus, A. G. "The Matter of Carbon: Understanding the Materiality of tCO_2e in Carbon Offsets." The Governance of Clean Development Working Paper Series, School of International Development, University of East Anglia UK, no. 008. The Governance of Clean Development Working Paper Series, 2010.

Bumpus, A. G. and D. M. Liverman. "Accumulation by Decarbonization and the Governance of Carbon Offsets." *Economic Geography*, 84/2 (2008).

Kossoy, A. and P. Ambrosi. "State and Trends of the Carbon Market 2010." Washington, DC: World Bank, 2010.

United Nations Environment Programme (UNEP). Risoe CDM/JI Pipeline Analysis and Database (2008). http://cdmpipeline.org (Accessed September 2010).

United Nations Framework Convention on Climate Change (UNFCCC). "CDM: CDM-Home" (2010). http://cdm.unfccc.int/index.html (Accessed September 2010).

Wara, M. "Is the Global Carbon Market Working?" *Nature*, 445/8 (2007).

Wara, M. W. and D. G. Victor. "A Realistic Policy on International Carbon Offsets." Program on Energy and Sustainable Development Working Paper 74, Stanford University.

Adam G. Bumpus
Neil Salmond
Independent Scholars

CARBON FOOTPRINT CALCULATOR

The carbon footprint is a measure of the total amount of carbon dioxide (CO_2) and other greenhouse gas emissions that are directly or indirectly caused by an activity, or that are accumulated over the life span of a product, person, an organization, or even a city or state. Carbon footprinting is a measure by which a company or individual can calculate how much carbon emissions have been produced during a project or time period. Usually, there are two major reasons for wanting to determine a carbon footprint—to manage the footprint and reduce emissions over time, and/or to report the footprint accurately to a third party.

An organization's carbon footprint is a measurement of its three main types of carbon emissions: those from activities the organization controls (i.e., gas to heat the building), indirect emissions from products or services used by the organization, and electricity, as shown here, for lighting and equipment.

Source: iStockphoto

Knowing the carbon footprint of an organization can be an effective tool for ongoing energy and environmental management. If this is the main reason for calculating the carbon footprint, it is often enough just to understand and quantify the key emissions sources through a basic process, typically including gas, electricity, and transport, which is relatively quick and straightforward to do. Having quantified these emissions, potential opportunities for carbon reduction can be identified.

Companies and organizations also increasingly want to calculate their carbon footprint in detail for public disclosure, for example, for marketing purposes, to fulfill requests from customers or investors, or to determine what quantity of emissions they need to offset for them to become "carbon neutral." This, therefore, requires a more robust approach, covering the full range of emissions for which the organization might be responsible.

The basic carbon footprint can be calculated fairly quickly for most organizations. A basic footprint is likely to cover direct emissions and emissions from electricity as these are the simplest to manage, but exclude some of the indirect emissions. Major emissions sources that must be quantified include on-site fuel and electricity usage and the use of transport that you own. Data must be collected from all utility meters, and the distances traveled by the organization's vehicles must be recorded. The fuel, electricity, and transport consumption figures are then converted to CO_2 by using standard emissions factors, which in the United Kingdom are available from Defra and the Carbon Trust.

A methodology that is popular is the Green House Gases (GHG) Protocol produced by the World Resources Institute (WRI) and the World Business Council for Sustainable Development (WBCSD). This methodology provides detailed guidance on corporate emissions reporting and is available at www.ghgprotocol.org. A more recent standard from the International Organization for Standardization, ISO 14064, also provides guidance on corporate carbon footprint calculation and is available at the ISO website (www.iso.org). The United Kingdom (U.K.) has introduced a new measure of CO_2 emissions for goods and services to standardize carbon footprint calculations. PAS 2050:2008, the Specification for the Assessment of the Life Cycle Greenhouse Gas Emissions of Goods and Services produced by BSI British Standards, is designed to show customers how much CO_2 has been emitted during production, consumption, and disposal of a range of products. The new calculation, available at the BSI Group website (www.bsigroup.com/pas2050), will give the first "life cycle" CO_2 measure that is standardized across industry. The measure is the first standardized calculation of greenhouse gas emissions for goods and services that will enable consumers to compare products easily. A variety of carbon footprint calculators exist, many of them online, from organizations such as the following:

- U.K. government (http://actonco2.direct.gov.uk/index.html)
- U.S. government (www.epa.gov/climatechange/emissions/ind_calculator.html)
- WWF (http://footprint.wwf.org.uk)

Main Types of Emissions

An organization's full carbon footprint encompasses a wide range of emissions sources from direct use of fuels to indirect impacts such as employee travel or emissions from other organizations up and down the supply chain. When calculating an organization's footprint it is important to try to quantify as full a range of emissions sources as possible in order to provide a complete picture of the organization's impact. In order to produce a reliable footprint, it is important to follow a structured process and to classify all the possible sources of emissions thoroughly. A common classification is to group and report on emissions by the level of control that an organization has over them. A standard classification is defined by the Greenhouse Gas Protocol, a widely utilized standard for corporate emissions reporting produced by the World Business Council for Sustainable Development and the World Resources Institute. Three main types of emissions exist:

- Emissions that result from activities that the organization controls. The majority of direct emissions will result from combustion of fuels that produce CO_2 emissions, for example, the gas used to provide heating for a building.
- Emissions from the use of electricity, for example, electricity for lighting and equipment, and electricity generation. In the United Kingdom, around 75 percent is produced through the combustion of fossil fuels such as coal and gas.
- Indirect emissions from products and services. Each product or service purchased by an organization contributes toward emissions. The way the organization uses products and services therefore affects its carbon footprint.

Incorporating all three types of emissions in a carbon footprint calculation can therefore be a complex task. In addition, published carbon footprints are rarely comparable for several reasons:

- Despite emerging international standards, not all organizations follow the same methodology to calculate their carbon footprint.
- Some carbon footprints are expressed on a time-period basis, such as annually, and some are measured on a unit basis, such as per product produced.
- Carbon footprints are usually calculated to include all greenhouse gases and are expressed in tons of CO_2 equivalent (tCO_2e). Some, however, calculate the carbon footprint to include CO_2 only and express the footprint in tCO_2 (tons of CO_2).

Carbon Neutral

The term *carbon neutral* is commonly used for something having net zero emissions (e.g., an organization or product). As the organization or product will normally have caused some greenhouse gas emissions, it is usually required to use carbon offsets in order to achieve neutrality; carbon offsets are emissions reductions that have been made elsewhere and that are then sold to the organization that seeks to reduce its impact. Carbon neutral can be anything from a person, to a building, to an organization, or even a city or state—the Vatican in Rome is quickly moving to become the first "carbon neutral state" with the installation of solar panels and the planting of a 37-acre forest in Europe, which is hoped to offset up to 80 tons of CO_2 a year.

How to Reduce the Carbon Footprint

A variety of methods exist for reducing an individual's or an organization's carbon footprint. Planting trees is one of the most common and simplest forms of carbon footprint reduction because trees absorb CO_2 from the atmosphere. Recycling waste materials such as household, industrial, and construction waste is also beneficial, as the carbon content of the new materials that would have otherwise been used can be offset. Many energy-saving technologies also exist that can contribute toward carbon footprint reduction, from cheap and simple measures such as installing low-energy light bulbs to more expensive measures such as using electric vehicles. Renewable energy generation can also be used for offsetting a carbon footprint, such as wind turbines and solar panels.

See Also: Carbon Market; Clean Energy; Recycling; Sustainable Design.

Further Readings

British Standards Institute (BSI). PAS 2050:2008. "Specification for the Assessment of the Life Cycle Greenhouse Gas Emissions of Goods and Services." http://shop.bsigroup.com/en/Browse-by-Sector/Energy--Utilities/PAS-2050 (Accessed August 2010).

Brown, M. A., F. Southworth, and A. Sarzynski. *Shrinking the Carbon Footprint of Metropolitan America* (2008). http://www.brookings.edu/reports/2008/05_carbon_footprint_sarzynski.aspx (Accessed January 2011).

Global Footprint Network. http://www.footprintnetwork.org (Accessed August 2010).

Greenhouse Gas Protocol (GHG Protocol). http://www.ghgprotocol.org (Accessed August 2010).

International Organization for Standardization. ISO 14064-1:2006. "Greenhouse Gases Part 1: Specification With Guidance at the Organization Level for Quantification and Reporting of Greenhouse Gas Emissions and Removals." http://www.iso.org/iso/catalogue_detail?csnumber=38381 (Accessed August 2010).

Chris Goodier
Loughborough University

CARBON MARKET

Economic and industrial activities and changes in land use, such as deforestation, have resulted in a constant increase in the emission of greenhouse gases (GHG) to the atmosphere since the Industrial Revolution. A high concentration of GHGs (e.g., carbon dioxide, methane, nitrous oxide, hydrofluorocarbons) may in turn increase the average temperature due to the so-called greenhouse effect. On this basis, the subject has been analytically discussed at international conferences, especially at Conferences of the Parties (COP), which occur once a year in different regions of the world. A result of these discussions has been the proposition of market-based instruments to assist industrialized countries (developed countries)—hitherto the most responsible for the greatest percentage of GHGs—in reducing their GHG emissions. Another reason for the emergence of these market instruments is the variation in GHG abatement costs that exist among countries. There are differences in costs, but economic incentives exist for countries to begin providing this service, thereby generating a carbon market. The idea is that the reduction, stabilization, and/or elimination of a given pollutant can be achieved through trading carbon credits, since this trade lends greater flexibility for countries to reduce GHG emissions.

In this context, the actions proposed during the COPs emphasize that the use of market mechanisms aims not only to reduce the costs to developed countries in mitigating global warming, but also to foster sustainable development in developing countries. In particular, the COP held in Kyoto, Japan, in 1997 stands out as one of the most important. This COP successfully reached an agreement that set targets for reducing GHG emissions for developed countries (Annex B countries), as well as criteria and guidelines for the use of market mechanisms. This agreement became known as the Kyoto Protocol and states that developed countries must reduce their emissions by an average of 5.2 percent between 2008 and 2012 (the first commitment period) compared to levels observed in 1990. The Kyoto Protocol also highlights the potential ability for carbon markets to help reduce GHG emissions through the creation of a tradable value for these reductions. On this basis flexible mechanisms were established that seek to reduce GHG emissions, such as the Emission Trade, the Joint Implementation, and the Clean Development Mechanism.

The Emission Trade is defined in Article 17 of the Kyoto Protocol and stresses that any Annex B country can sell emissions reductions that exceed their committed targets to another Annex B country that needs to diminish its own GHG emissions. However, in this case, carbon trading should happen as supplement to domestic actions, that is, internal abatement of emissions should take precedence before a country has access to carbon markets. On the other hand, the Joint Implementation (Article 6 of the Kyoto Protocol) is a tool for negotiating bilateral joint implementation projects to reduce GHG emissions among Annex B countries. Through Joint Implementation a developed country can offset its emissions by participating in projects in another Annex B country. It implies, therefore, a transfer of carbon credits from the country where the project is being implemented to the issuing country, enabling the latter to buy these carbon credits, known as Emission Reduction Units (ERUs).

Clean Development Mechanism

The Clean Development Mechanism (CDM; Article 12 of the Kyoto Protocol) is the only flexible mechanism that allows the participation of developing countries in the carbon market. Each ton of carbon left to be issued or that is withdrawn from the atmosphere by a particular developing country may be traded on the world carbon market, giving an incentive for reducing global GHG emissions. In this sense, the Annex B countries that cannot reach their assigned amounts, based on preset targets, can buy carbon credits from developing countries—in this case, defined as Certified Emission Reductions (CERs)—and use them to fulfill their target obligations. Developing countries that handle the licenses should in turn use the resources to promote their sustainable development, applying the resources in environmental conservation and poverty alleviation, for instance.

In order for flexible mechanisms and carbon markets to work properly, it is essential that certain steps be followed. In the case of CDM, for example, it is imperative that an executive board be established. Also, the COP should nominate a Designate Operational Entity (DOE), whose functions include certifying the entire process surrounding the project. On this basis, the life cycle of a project proposed by the executive board involves (1) the project validation and registration, (2) monitoring and verification, and (3) certification. The project should also generate additional reductions, that is, the emissions of GHG must be smaller than those that occur in the absence of the project, and/or the carbon sequestration must be greater than in the absence of the project. For this, a baseline must be drawn based on a scenario that reasonably represents the anthropogenic emissions of GHG that would occur in the absence of the project (so-called business-as-usual scenario). The executive board will manifest itself in favor of or against the baseline and the monitoring methodology proposed in the project. Once accepted, the project can be registered and the carbon credits generated can be traded at carbon markets through stock exchanges, funds, or even through brokers.

Due to the high costs of the projects that reduce emissions and/or sequester carbon, there is also the possibility of the approval and registration of small-scale projects, which incur more modest costs. These projects have slightly different regulations when compared to large-scale projects, allowing, for example, that traditional communities in developing countries can sell the carbon credits generated by a small-scale project and implement the action in promoting their sustainable development. In this case, local nongovernmental organizations (NGOs) and regional governments can assist in the design of the project and act as brokers in the sale of carbon credits to private companies, for instance.

The value of afforestation (planting forests) was internationally recognized in 1997, when it was included in the Kyoto Protocol agreement on global action to reduce the risk of human-induced climate change. However, historically speaking, decreasing deforestation and forest degradation has been absent from international negotiations, mainly because of difficulties in monitoring. Projects involving activities of land use, land use change, and forestry (well known as LULUCF) are intensively debated within the negotiation process of the COPs. Carbon sequestration is also still controversial due to the lack of robust methodologies in understanding the role of forests as carbon sinks. Based on this, apart from the flexible mechanisms promoted by the Kyoto Protocol, there are also voluntary (parallel) carbon markets that trade carbon credits generated by carbon sequestration and other forest activities. The Prototype Carbon Fund (PCF) and the BioCarbon Fund (BF) of the World Bank illustrate this. For these, independent consultants should approve the baseline as well as the methodology for verification/certification of the project. BF also has the objective of financing agro-forestry projects for carbon sequestration focusing on biodiversity conservation, combating desertification, and socioeconomic development. In addition, there are also regional markets for carbon. The Chicago Climate Exchange (United States), the Certified Emission Reduction Unit Procurement Tender (CERUPT) from the Dutch government, the Emission Trade Scheme (ETS) of the United Kingdom, and the European Climate Exchange (ECX) are only some examples of where carbon credits can be traded. In official markets (i.e., markets that follow Kyoto Protocol guidelines) as well as voluntary markets, each carbon credit refers to one metric ton of carbon dioxide (CO_2). Therefore, GHGs should be measured in carbon dioxide equivalent (CO_2 equivalent); in other words, the credit generated by an emission reduction and/or carbon sequestration project is turned into a CO_2 equivalent and thereafter is traded on the carbon market. The monetary value for this credit varies according the type of project (CDM, Joint Implementation) as well as the market where it is being negotiated (official, voluntary).

REDD-plus

Since deforestation is responsible for part of GHG emissions and despite the previous shortfall in climate change policies in considering LULUCF activities, REDD-plus (Reducing Emissions from Deforestation and Forest Degradation) was created, in which developing countries are given incentive to reduce deforestation and forest degradation. In this way, carbon stored in forests is given a monetary value and thus supports developing countries in investing in low-carbon paths to sustainable development. Moreover, REDD-plus includes some activities that might have serious implications for indigenous people, local communities, and forests. It comprises policy approaches and positive incentives for issues relating to reducing emissions from deforestation and forest degradation in developing countries as well as the role of conservation, sustainable management of forests, and the enhancement of forest carbon stocks in developing countries. Based on this, if cost-efficient carbon benefits can be achieved through REDD-plus, CO_2 concentration increase could be slowed, effectively buying much-needed time for countries to move to low-emission technologies. Support for efforts to reduce emissions from deforestation and forest degradation have been expressed at the highest political levels and are included in the Bali Action Plan of the United Nations Framework Convention on Climate Change in 2007. So, although REDD-plus is not yet formally established in the United Nations Framework Convention on Climate Change (UNFCCC), some REDD-plus credits are already being sold in voluntary carbon markets, and some initial finance is provided for pilot projects. The World Bank's Forest Carbon Partnership Facility, for instance, includes a readiness mechanism to help governments participate in REDD-plus. In particular, it helps developing countries estimate their forest carbon stocks, establish national reference scenarios, calculate opportunity costs, and design monitoring, reporting, and verification systems.

Based on the discussion above, since the "forest standing" usually has no economic value to local communities, their opportunity costs might lead toward deforestation. A win-win approach could be achieved by paying these communities to sequester carbon (like a small-scale CDM project), which sets up a situation where: CO_2 is removed from the atmosphere; high soil organic matter increases agro ecosystem resilience; and improved soil fertility leads to better yields. If a carbon market successfully allows the trade of sequestered carbon on the international market, deforestation of native forests in tropical countries might decrease. If deforestation is business as usual (or the baseline scenario), then the conservation of native forests would implement change and generate additional positive externalities. Thus, the REDD-plus project would create a stimulus for the conservation of native forests and increase the amount of carbon sequestered, thereby decreasing the amount of carbon that is emitted to the atmosphere. Nevertheless, the REDD-plus project that is implemented will pay for environmental services and should adhere to certain prerequisites such as guaranteeing a reduction in GHG emissions compared to the baseline scenario, like a CDM project, for instance. Once the REDD-plus project is implemented, additional positive externalities are generated and credits can be traded at carbon markets.

See Also: Carbon Finance; Carbon Footprint Calculator; Zero-Energy Building.

Further Readings

United Nations Framework Convention on Climate Change. "Compliance Under the Kyoto Protocol." http://unfccc.int/kyoto_protocol/compliance/items/2875.php (Accessed June 2010).

United Nations Framework Convention on Climate Change. "Kyoto Protocol Reference Manual on Accounting of Emissions and Assigned Amount." http://unfccc.int/resource/docs/publications/08_unfccc_kp_ref_manual.pdf (Accessed June 2010).

United Nations Framework Convention on Climate Change. "Reducing Emission From Deforestation and Forest Degradation (REDD)." http://unfccc.int/methods_science/redd/items/4531.php and http://unfccc.int/documentation/documents/advanced_search/items/3594.php?such=j&symbol="FCCC/AWGLCA/2008/16/Rev.1"#beg (Accessed May 2010).

Marcus Vinicius Alves Finco
Federal University of Tocantins

CELLULOSIC BIOFUELS

Cellulosic biofuels are created with nonedible vegetation, such as this switch grass, and are considered second-generation biofuels (the first generation being those from edible plants).

Source: Wikimedia

Cellulosic ethanol is a second-generation biofuel that is manufactured by transforming vegetation unsuitable for human consumption. As such, it can be produced by using raw materials such as wood, grass, or nonedible parts of plants. It is an alternative to first-generation biofuel, which uses edible feedstock (such as seeds and grains). Hence, one of its main advantages is that its production lessens the strain on the food chain while also being renewable. While it is true that when cellulosic crops are grown, less land is available for edible crops, some of the crops (such as switch grass) are farmed on lands that are only marginally useful for regular farming. One of the main challenges related to manufacturing cellulosic ethanol is that the conversion rate of raw materials to final product is lower than for first-generation biofuels. Also, all biofuels are heavily subsidized by government stakeholders, and without subsidies, it is uncertain that they could continue to be manufactured. The future of cellulosic ethanol might be in blended fuels, rather than focusing on completely replacing petroleum.

Background on Cellulosic Ethanol

Cellulosic ethanol is a biofuel produced from lignocelluloses, which constitute a large part of all plants. It is manufactured from nonedible plant components (such as the stem or leaves of edible plants) or biomass waste (such as woody waste). It can also be manufactured by processing nonedible plants such as switch grass. It is often referred to as a second-generation biofuel, in contrast to biofuels that are manufactured from edible plants (referred as first-generation biofuels). It is touted as a viable alternative to petroleum-based fuel, mitigating petroleum scarcity, reducing petroleum dependence, and reducing the environmental impact of consumers and industries that use petroleum as a primary energy source since biofuels produce less green emissions than their fossil counterparts. It also qualifies as a renewable energy since the crops can be grown cyclically, without worrying about exhausting raw materials. Conversion technologies are somewhat immature, and the technology is the subject of research and development by both government and private corporations. Research is also being dedicated to developing genetically modified enzymes as they are a key component of cellulosic fermentation, improving yield from some feedstock (such as sugar cane) as well as experimenting on the different types of vegetation that can be used as raw components.

Emergence of Cellulosic Ethanol

The first biofuels used regular feedstock as a primary component (such as cornstarch and rapeseed oil). The use of regular feedstock led to some controversy, which is now referred to as the "food versus fuel" debate. This debate was brought on by the massive use of feedstock for purposes not related to food, diverting it from the human food chain. This had a direct impact on the price of food for consumers. Early biofuels used limited amounts of feedstock, but as capacity grew, the need for raw materials (such as wheat, sunflower seed, or corn/maize) increased rapidly. Since the companies manufacturing first-generation biofuels needed these raw materials, they were often willing to pay more than food manufacturers, encouraging farmers to sell their crops to biofuel manufacturers instead of to their traditional purchasers. This created food shortages and rapid price increases, leading to remarks that first-generation biofuels were not as environmentally friendly as first believed. Furthermore, authors like Mark Delucchi suggested that the use of first-generation biofuels increased stress on water supplies, quality, and land use, especially when compared with traditional petroleum fuels, further impacting arguments against the green nature of the technology.

During this debate, cellulosic ethanol emerged as an interesting alternative for many stakeholders (such as politicians and producers) because it uses raw materials that were either outside the human/animal food chain (such as switch grass), uses lands unviable for agriculture, or uses unused by-products of agriculture/forestry. It also consumes less fertilizer to generate, increasing its positive green output. As such, cellulosic ethanol rapidly emerged as a viable and popular alternative. It solved some of the problems that first-generation biofuels presented (such as reducing the impact on the food chain and increasing food prices) while retaining most of its advantages (renewable in nature and reducing dependence on petroleum).

Challenges and Criticisms Related to Cellulosic Ethanol

One of the many challenges related to manufacturing cellulosic ethanol is that the conversion rate is lower than with first-generation biofuels. The conversation rate is the quantity of raw material necessary to produce a set quantity of biofuel. The conversion process for cellulosic ethanol is quite expensive, especially at a larger scale, and the enzymes necessary for this process continue to command a high market price. Therefore, it is more costly to manufacture cellulosic ethanol. Since it is more costly, it is much more difficult to justify financially. Some have suggested that many cellulosic ethanol manufacturing plants in operation require a petroleum market price of $120 per barrel for the technology to be financially feasible. (As a reference for the reader, prices for a petroleum barrel varied between $70 and $87 for the first half of 2010, and the U.S. Department of Energy does not expect prices to reach $120 per barrel until 2027). As such, a lot of research is necessary before cellulosic ethanol can viably compete with existing energy resources.

There are also scale issues confronting cellulosic ethanol. Petroleum consumption around the world is imposing, and it is impossible to displace current consumption to a different product. Hence, the quantity of cellulosic ethanol necessary for complete substitution is massive, and very few countries have the geography and climate to produce enough cellulosic ethanol to make the transition. As such, many countries favor blended fuel, where cellulosic ethanol is blended into regular petroleum, enabling cleaner burning and less pollution. Most blends are in the 5 percent to 10 percent range so they can be used in current vehicles without modifying the engine.

This brings us to the last critical issue, a topic that applies to biofuels in general. Biofuel is a heavily subsidized energy source, and it is unable to compete in the current market. Without subsidies, it is doubtful that cellulosic ethanol could continue to be manufactured. Legislation requiring fuel manufacturers to have a set percentage of biofuel blended into their products is one of the main market drivers for market growth. An economic slump and legislative hesitation could severely impact the future of this technology.

The Future of Cellulosic Biofuels

Cellulosic ethanol continues to be a popular political solution, being touted as a possible solution to energy independence, as well as being a method of interest for reusing some waste materials. There are many initiatives in the United States that were adopted to stimulate ethanol production, such as the Renewable Fuel Standard (RFS1), which called for the production of 7.5 billion gallons of renewable fuel to be blended into gasoline by 2012. This was followed by the second Renewable Fuel Standard (RFS2), which further encouraged growth in biofuel production as well as requiring greenhouse gas reductions.

Although it is possible that production costs of cellulosic biofuels will one day be lower than production costs for petroleum, it is conditional on the sharp rise of fossil fuel prices as much as on optimizing production costs (such as lowering the costs of enzymes). Cellulosic ethanol continues to be a popular political solution, being touted as a possible solution to energy independence, as well as being an interesting method to reuse some waste materials. Also, the sheer amount of cellulosic fuel necessary to viably replace fossil fuel makes it unlikely that it will completely replace it. Blended fuels are more likely to be a feasible long-term solution. The emergence of third-generation biofuels (such as algae-based biofuel) could also have a significant impact on future production of cellulosic ethanol.

See Also: Algae Biofuel; Biogas; Hybrid/Electric Automobiles.

Further Readings

Delucchi, Mark. "Impacts of Biofuels on Climate Change, Water Use and Land Use." *Annals of the New York Academy of Science* (2010).

Lange, Jean-Paul, et al. "Cellulosic Biofuels: A Sustainable Option for Transportation." In *Sustainable Development in the Process Industries: Cases and Impacts*, J. Harmsen and Joseph B. Powell, eds. Hoboken, NJ: Wiley, 2010.

Nigam, Poonam and Anoop Singh. "Production of Liquid Biofuels From Renewable Resources." *Progress in Energy and Combustion Science* (May 2010).

U.S. Department of Energy. "Annual Energy Outlook 2010 (With Projections to 2035)." http://www.eia.doe.gov/oiaf/aeo/pdf/0383(2010).pdf (Accessed April 2011).

Jean-Francois Denault
Independent Scholar

CLEAN ENERGY

Clean energy is an informal term used to refer to low- or zero-emission energy sources, and is generally synonymous with renewable energy (as used in this article). Renewable energy resources include wind and solar energy, and are distinguished from nonrenewable resources by the time frame upon which they are available. Thus, renewable energy resources are continuously available when viewed from the perspective of human time scales of thousands of years. On the other hand, nonrenewable energy sources, such as fossil and nuclear fuels, are finite because the time scales required for them to form, namely millions of years, are much greater than human dimensions. Another way of considering this distinction is to look at the respective rates of supply and demand: at current rates of consumption, nonrenewable energy resources would be exhausted within a few hundred years, but renewable resources would not likely be depleted or exhausted within human time scales (because they mainly stem from the sun). There are, however, limits on the exploitation of renewable energy resources, such as technical, economic, and social constraints on the available potential at any given time. As will be discussed in this entry, the main constraints for a stronger exploitation of renewables are presently the sometimes-restrictive economic cost and technical difficulties associated with integrating large amounts of intermittent renewable electricity generation into electricity systems that were originally designed to deal with large centralized generating plant operating at relatively constant outputs.

First, this entry will introduce the different types of renewable or clean energy resources, followed by a definition of the different types of potential. The individual technologies and the current state of the art will then be presented in more detail, whereby not all renewable technologies are considered here for the sake of brevity. The general focus in this entry is on developments and the state of the art in Europe, whereby a wider perspective is taken where it is deemed necessary to illustrate a point or to demonstrate the state of the art for specific technologies that have a low penetration in Europe.

Types of Renewable Energy Resources

The vast majority of renewable energy resources derive, either directly or indirectly, from the sun; hence, the importance of the sun as a energy source cannot be understated, and is

The United States is the largest producer of geothermal energy, which is one of the five types of clean energy. Shown here is the steam pipe to the Geysers geothermal plant in Sonoma County, California.

Source: Don Follows/Yellowstone National Park, U.S. Department of the Interior via Lawrence Livermore National Security, LLC, for the Department of Energy's National Nuclear Security Administration

the reason why the lifetime of most renewable resources is closely linked to the estimated lifetime of the sun—on the order of billions of years. The most direct form of solar energy is in the form of radiation, which streams through the atmosphere, and can be directly exploited for heating and to generate electricity. Indirectly, this solar irradiation leads to the unequal heating of the Earth's and the ocean's surface, which leads to the ocean currents on the one hand, and means that the temperature of the Earth is generally a few degrees higher than that of the atmosphere—a difference that can be exploited with a heat pump, for example, as described below. Indirectly, the sun is also responsible for the existence of several other forms of renewable energy, including wind, water (hydro energy), and biomass. The unequal heating of the Earth's surface and the resulting pressure differences result in local and regional movements of air that manifest themselves in the form of wind. The sun also drives the water cycle, in that it evaporates water from seas and rivers (as well as on leaves, etc.), which then falls in other areas as precipitation. When this water falls to Earth and collects in streams and rivers, it has a higher gravitational potential than the water in the sea; for example, the gravitational potential energy relates to the distance of a mass of water above sea level and is converted into kinetic energy as water flows toward the sea. This potential can be utilized to generate electricity, as in hydroelectric power plants. Finally, the sun also drives the process of photosynthesis, which occurs in plants on the Earth, and in which energy (in the form of sunlight) converts water and carbon dioxide (CO_2) into oxygen and energy (sugar). The energy in the sunlight is thus "locked up" in plants and can be used if the plant material is later harvested. The CO_2 released when plant matter is burned corresponds to the CO_2 stored in plants during their lifetimes, so that one speaks of sustainably managed biomass resources as being "carbon neutral." Of note is that biomass is, strictly speaking, only renewable if the resources are sustainably managed; for example, clearing a forest for firewood without replanting does not constitute a renewable resource.

There are other renewable energy sources on Earth that do not derive from solar energy, however, including tidal energy and geothermal energy. Tidal energy, which results from the difference in gravitational potential energy between two bodies of water at different heights above the Earth's surface, results from the movement of the Earth and the moon relative to one another—in particular, the movement of the moon around the Earth and the rotation of the Earth. Geothermal energy, on the other hand, refers to the relatively high temperatures to be found below the Earth's surface in rocks and aquifers (underground lakes), which results from thermal processes occurring within rocks; for example, as they are transformed under great pressure and at high temperature into metamorphic rocks.

All of these renewable resources can be used to generate different forms of energy, depending upon the availability of the resource, its form, and its energy density. Generally, the higher the energy density of a resource, the better it is suited to generating higher-quality energy forms such as electricity, but energy uses such as generating heat, and the cogeneration of heat and power is also often an attractive option for biomass-based products. Almost all of the above resources require conversion into a form that is useful, often in terms of being directly usable, as in the case of electricity. The exception here is the generation of electricity in photovoltaic cells, in which case the energy form (electricity) is exactly the one that is required and ultimately used. It is important to note at this stage that, with the exception of large-scale pumped storage hydroelectric power plants, electricity is not generally storable on a large scale. This means that integrating large amounts of electricity generation from renewables is challenging because the energy either needs to be stored and used at times when it is required, or the demand needs to be shifted so that the demand and the supply coincide. In this respect, biomass has the great advantage of being able to be stored as fuel, and used as needed; in this sense, it is a renewable energy resource, offering significant flexibility over other resources.

Potential Definitions

In the field of renewable energy, it is common to distinguish between different potentials, which means the realizable amount of a renewable resource according to specific criteria. Potentials are generally measured in terms of the resource per unit of land area and unit of time; for example, solar irradiation per square kilometer per year. Typical potentials include theoretical, technical, economic, and realizable. The theoretical potential refers to the absolute maximum resource that is available, but that could never be practically exploited, such as the total amount of solar irradiation incident on a particular land area within a year. This theoretical potential is an average over the time period being considered, often a year, such that seasonal variations in availability (e.g., rainfall) are generally not considered. The technical potential is the fraction of the theoretical potential that could be realized with current technology; it therefore takes into account factors such as conversion efficiencies and the fraction of land and building/roof areas that could feasibly be covered with solar cells, for example. Once again, the technical potential is rarely, if ever, realizable in practice because of the restrictively high costs relating to renewable energy technologies, as well as other obstacles, such as social opposition.

Therefore, the fraction of the technical potential that is economic at current market conditions (i.e., capital interest rates and expected amortization periods) is known as the economic potential. This is typically only a small fraction of the technical potential, and requires an investment appraisal based on approximating assumptions in order to be determined. Depending on the perspective taken, namely that of businesses or individuals, the economic potential can strongly differ, which is due to the different financial expectations of these two groups. Finally, the realizable potential is the potential that is currently realizable under current market conditions in the region under consideration.

The realizable potential accounts for the fact that markets for renewable energy technologies are not necessarily functioning perfectly, and therefore what appears to be economical on paper may not necessarily be realized in practice. The realizable potential can be less than, equal to, or even more than the economic potential. The latter is the case in countries like Germany with strong support policies for renewables, in which the fractions of renewable potential are made realizable, which would be considered uneconomical if left to the market.

Hydroelectric/Hydropower

Within the water cycle, water falls to the Earth as precipitation and then flows through rivers to the sea, during which time its potential energy is transformed into kinetic energy. Hydroelectric power plants exploit the height difference along a flowing watercourse, and the fact that, as the water flows, this potential energy is converted into kinetic energy. This kinetic energy is used to drive a turbine, which in turn drives a generator and generates electricity. While the principle of operation is essentially the same throughout, there are two general types of hydroelectric power plants. The first type are run-of-river plants, in which a constant flow of water (e.g., in a river) drives turbines to generate electricity. In dammed plants, a river is dammed and the water is allowed to turn the turbine in a controlled manner, depending on the current demand. The second type, dammed plants, can further be broken down into conventional and pumped-storage plants, in that conventional plants are unidirectional and pumped-storage plants are bidirectional: water is pumped to a higher reservoir during the night when demand for electricity is low and therefore prices are also low, and the water falls into the lower reservoir at times of high demand, when a correspondingly high price for the generated electricity can be demanded. Thus, pumped-storage plants are used largely for load management as they can be started very quickly (in a number of minutes) during times of peak load and used to store electricity that is not currently required. Strictly speaking, these plants do not generate any renewable energy on a net basis; instead, they are storage devices.

The electrical power of a hydroelectric plant is directly proportional to the amount of water flowing through it (the flow rate in m³/s, or cubic meters per second), as well as the height difference between the water surface in the dam and that below (the head in meters). For different combinations of these two key parameters, there are several types of turbines available for electricity generation, some examples of which include the Pelton, Kaplan, and Francis turbines. The former operates on the same principle as a water wheel, in that water flows into buckets and turns the wheel. The latter two models are known as overpressure turbines because in contrast to the Francis turbine, they operate at a pressure higher than the surroundings. In other words, the head or pressure of the water above the turbine is extracted, or converted, into mechanical energy as the water flows through the turbine. In general, the Pelton turbine is most suited to applications with large fall heights but low to moderate flow rates, whereas the Francis and Kaplan turbines are suited to more moderate fall heights and moderate to high flow rates. Only by employing the correct turbine can the optimal efficiency of around 80 to 90 percent in hydropower plants be achieved.

Hydroelectric plants are characterized by relatively high initial investments and very low running costs: as with most renewable technologies, the variable or "fuel" costs are negligible, and the fixed costs of operation consist mainly of service and maintenance costs. The lifetime of these plants is around 50 years, and with a complete overhaul of the turbine(s), they can potentially run for even longer periods.

About one-third of all countries generate more than half of their electricity from hydro-electric power plants (European examples include Norway, Switzerland, and Austria). Around three-quarters of the economic potential for hydroelectric power is currently exploited within Europe, but on the world scale, this figure is around one-third. Some of the main problems with hydropower plants include the fact that most dams are impassable and disrupt migration of fish and other aquatic life. In general, there are significant negative influences on the local ecology and population, as has been seen in the recently completed Three Gorges Dam project in China, where over 1 million people were displaced.

Wind Energy

Wind resources are varied as they depend on several regional factors, including the latitude (i.e., the effects of atmospheric circulation) and the distance from the nearest coastline: the smooth surface of water means that wind velocities are on average higher over or near the sea. Local factors such as the altitude, the topographical shape of the landscape, and the presence or absence of obstacles (such as vegetation or buildings) also play a significant role.

The power from wind, and therefore the power of a wind turbine, is proportional to the cube of the wind velocity. This means that higher wind speeds have a disproportional effect on the power output of a turbine, and also that higher wind speeds are favorable. There is a maximum amount of energy or power that can be extracted from the wind, which is around 60 percent, which is referred to as the Betz constant after the person who discovered it, Albert Betz. This figure relates to the conversion of kinetic energy in the wind to the kinetic energy in the turbine, and does not account for further losses within the gear box and generator, and so forth.

There are two general types of wind turbine, depending upon how the axis of rotation of the turbine blades is aligned. This axis may be horizontally or vertically aligned, and may use the flat or aerofoil-type blades, the latter being more efficient at extracting energy from the wind. Vertical axis wind turbines (VAWT) are generally simple in their construction and therefore relatively inexpensive. They tend to operate on the resistance principle, although some models such as the Darrieus and Savonius rotors also employ aerofoils as blades (i.e., to generate lift as on an aircraft wing). Nevertheless, the efficiency of this type of turbine is generally quite low, and they are better suited to onetime applications, perhaps where the erection of a horizontal axis turbine would not be feasible. Their simplicity in design, in particular the lack of moving parts and control systems for wind direction tracking, presents the advantage that they are robust and can therefore withstand large fluctuations in wind speeds. Their proximity to the ground means that the wind speeds are significantly lower than in the case of horizontal axis wind turbines (HAWTs).

HAWTs, on the other hand, can achieve higher conversion efficiencies; for example, through the use of aerofoil blades, blade pitching, and wind tracking systems, but these advantages are all associated with additional costs. A typical HAWT consists of several blades connected through a drive shaft to a generator, often through a gearbox. The whole head of the turbine may be automatically controlled to track the wind, and the pitch of the turbine blades may be automatically adjusted in order to optimize the generation efficiency in the current prevailing wind conditions. Overall efficiencies of approximately 40 percent are achievable with HAWTs, and capacity factors (i.e., the fraction of the time during which the turbine is operating) in the range of 10 to 30 percent are typical. Trends over the past few decades in the development of HAWTs include the tendency toward ever-larger turbines, from around 200 kW in the early 1990s to turbines rated in excess of 5 MW today. While larger turbines have significant technical (particularly structural) challenges to overcome, the advantage with larger turbines is that the economies-of-scale effect is exploited (although the investment is higher, the cost per generated unit of electricity is lower) and the turbine is better able to exploit the higher wind speeds in any given region.

General issues with wind energy include the perception that turbines and large wind parks are an eyesore in the landscape, but this problem is often exaggerated and not properly understood in the context of the wider debate about renewable energy sources and a shift away from fossil fuels. Bird strikes are another major problem associated with large wind farms, whereby migrating birds inadvertently fly into the moving turbines. This, as well as the problem of wind parks interfering with radar systems, does not appear to have been fully quantified, and it is therefore fair to say that these problems are perceived issues more than real issues.

Solar Energy

The direct exploitation of solar irradiation presents two main opportunities for energy conversion: first, the generation of electricity through photovoltaic cells, and second, the generation of heat through solar thermal systems. The useful fractions of solar irradiation include direct and indirect components: often the indirect part, which is reflected or refracted from surfaces and through clouds, is as large as the direct fraction, which means that even on cloudy days, energy can be captured with solar-based energy converters. In addition, the indirect possibilities for energy generation include via heat pumps.

Photovoltaic (PV) cells utilize the photoelectric effect, in which light falling onto a semiconductor material (containing a proton and a neutron layer) can give the electrons in one layer of the material enough energy to "jump" into a higher energy level, and thus generate a potential difference across the cell. If the two sides of the cell are connected, the current flowing between the poles can then be used locally or fed into the national grid. The overall efficiency of PV cells is currently generally under about 20 percent, but several recent developments look promising for the near future, including the use of materials other than silicon, the employment of multilayer cells (with several layers that each absorb a specific wavelength of radiation), and the use of concentrator technology, which focuses incident radiation and thereby increases the power per unit area. Laboratory tests of such technologies have shown that efficiencies in excess of 40 percent are possible. In general, the yield from PV systems can also be increased by integrating sun-tracking systems that tilt the plane of the PV cells in one or more directions, depending upon the location of the sun. The cells can also be integrated into building facades and windows, although they generally cannot be used to replace windows, given their low transmission of light.

With solar thermal forms of energy conversion, one tends to distinguish between active and passive methods, with active referring to the capture and conversion of the solar energy into heat energy, and passive referring to measures such as building design through which one can utilize the sun's energy to efficiently heat a building. Active solar thermal devices employ collectors that transfer the energy in the sunlight to a working medium through an absorber. This working medium can then be used to provide supplementary heating and/or hot water, but will rarely provide enough hot water to completely replace a conventional heating system, especially at times of peak load. In combination with a heat pump and PV cell generating renewable electricity, a completely renewable heating system may be feasible. For hot water generation alone, two water circuits are required, along with an intermediate heat exchanger, so that the water in the collector does not contaminate the warm water system. In the case of central heating, an additional circuit and heat storage device is required. Typically, about 60 percent of energy in the incoming solar radiation may be converted into heat in the working medium.

Larger, so-called concentrating solar thermal power plants concentrate the sunlight with large mirrors, either in a line or onto a point, and thereby achieve much higher temperatures than roof-mounted collectors. In this case, the working fluid is usually oil, which reaches temperatures above 500 degrees Celsius, and can thus be used to produce steam, which drives a turbine and generates electricity through a generator. Large heat stores are required in these concentrating plants in order to balance out the peak in supply from the sun, at approximately midday, with the peak in demand in the early evening. In most cases, the large concentrating mirrors used in such plants are tracked in one or two axes in order to maximize the energy conversion efficiency. The overall efficiency of such a plant, if related to the total area of the mirrors, is on the order of 20 percent.

Heat pumps upgrade heat energy from a generally low temperature, diffuse source (heat source) to a higher temperature, more concentrated source (heat sink) by using mechanical energy. They do this by employing a working medium that absorbs low temperature heat (e.g., from the ground, in the case of ground source heat pump) and transfers this heat to another working fluid by running a thermal compression and expansion cycle similar to a refrigerator (although in reverse). By employing one unit of electricity to provide the mechanical power, about three units of heat energy can be produced, and this ratio is referred to as the coefficient of performance, or COP.

Bioenergy

The use of biomass has a long history: wood, in particular, has been exploited for many centuries as a source of energy for heating applications. Only relatively recently has the use of other biomass-based energy sources significantly increased. *Bioenergy* is a collective term for all plant material and residues, dead (but not yet fossilized) plant matter, and all other animal and organic waste. The diversity of this energy resource is also reflected in the possibilities for processing it and thereby utilizing the energy contained in it. With bioenergy in general, there are three potential uses of the material: either as foodstuff, as material (i.e., for industrial process, such as to make paper), or for energetic purposes. Sometimes these uses conflict with one another, such that it is a complex process to decide what is the most "optimal" means of using any given biomass resource.

Conclusion

Renewable energy resources and technologies are characterized by a globally and regionally unequal distribution, fluctuating temporal availability, relatively high initial investments but low or negligible fixed and variable costs, and low conversion efficiencies. Their high investment costs are currently the main reason for their constrained development, although in some countries, such as Germany, support policies are putting these technologies on an equal playing field with conventional fossil fuel technologies. The further development of renewable energies and whether or not national or international targets for their energy and electricity generation are met will depend on the suitable implementation of further support policies and technical developments, which will further reduce the costs of these technologies. In recent years, learning effects have led to significant reductions in the generation costs from renewable technologies, but there is still a way to go before these technologies are competitive (without subsidies) with fossil fuel–based energy forms. The integration of large amounts of renewable energy generation capacity into conventional energy systems requires significant adaptations of the system in order to be able to deal with the fluctuating supply.

See Also: Algae Biofuel; Anaerobic Digestion; Biochemical Processes; Biogas; Cogeneration; Concentrating Solar Technology; Crystalline Silicon Solar Photovoltaic Cell; Green Markets; Intermediate Technology; Microgeneration; Passive Solar; Pyrolysis; Smart Grid; Solar Hot Water Heaters; Solar Ovens; Wind Turbine.

Further Readings

Ayres, Robert U. and Edward H. Ayres. *Crossing the Energy Divide: Moving From Fossil Fuel Dependence to a Clean-Energy Future*. Upper Saddle River, NJ: Pearson Prentice Hall, 2009.

Boyle, Godfrey. *Renewable Energy: Power for a Sustainable Future*, 2nd ed. Oxford, UK: Oxford University Press, 2004.

Fox-Penner, Peter. *Smart Power: Climate Change, the Smart Grid, and the Future of Electric Utilities*. Washington, DC: Island Press, 2010.

Lior, N. "Energy Resources and Use: The Present Situation and Possible Paths to the Future." *Energy*, 33 (2008).

Quaschning, Volker. *Understanding Renewable Energy Systems*. London: Earthscan, 2005.

Russell McKenna
Karlsruhe Institute of Technology

COAL, CLEAN TECHNOLOGY

Coal is one of the most popular fossil fuels around the world since it easy to find, harvest, and manufacture. It is used mainly to generate electricity. Most countries with heavily developed industrial sectors (such as China and the United States) continue to use and invest in coal-based power production. Since it has been extensively used, it is an efficient technology when one looks at it from a "raw material to electricity generated" ratio.

Nonetheless, burning coal generates negative by-products in the environment, as coal is very carbon intensive. Burning coal to generate electricity is one of the major carbon dioxide (CO_2) sources today. Since CO_2 is released directly in the atmosphere, its treatment is a constant challenge. Some direct environmental impacts include acid rain, terrain contamination, groundwater pollution, and greenhouse gases.

Currently, private and government stakeholders are privileging the development of new technologies under the header of clean coal. These technologies include integrated gasification combined cycle power generation (IGCC) and carbon capture and storage (CCS) as well as coal transformation technologies. Clean coal is of interest since it presents three main strong points: it reduces the environmental harm of regular coal, it can be implemented in existing facilities (limiting required investments), and it ensures the generation of consistent energy. The main weak points related to clean coal are the impact of clean coal technology on plant efficiency, the relative cost of some proposed solutions, the continued harm to the environment while harvesting coal, and the unknowns surrounding long-term carbon storage. On the whole, with coal currently occupying an important part of the current world energy mix, with the absence of any comparable technologies, with the current political support clean coal enjoys and with dependence on coal increasing (rather than decreasing), development of clean coal technologies is expected to continue in the foreseeable future.

Overview of Clean Coal

Coal is one of the main energy sources used around the world. It is available in abundant quantities worldwide, and it is relatively inexpensive to harvest and to transform. It is estimated that there are sufficient reserves to continue using it at the current rate for hundreds of years. As such, despite its documented negative impact on the environment, its popularity has not waned. And because coal has a long history of use, many countries and energy producers have developed an expertise in the building of coal energy facilities. Energy conversion technologies have been optimized, and the energy efficiency is quite elevated. Producing energy with coal is currently the benchmark against which newer technologies are evaluated. Most countries with heavily developed industrial sectors (such as China and the United States) continue to use and invest in coal-based power production.

In an age where environmental concerns are omnipresent, the development of coal-based power facilities has become controversial. Building new facilities (or just expanding existing facilities) is challenging as stakeholders push for the development of cleaner and renewable technologies (such as solar and wind power). These pressures from stakeholders, combined with environmental concerns, have contributed to the emergence of multiple alternatives to traditional coal, which are called "clean coal" technologies.

Understanding the Technology: Coal and Clean Coal

Coal is used to generate electricity in a variety of methods. In an energy generation model, coal is brought to a power facility, where it is burned. The heat released by the burning coal boils water in a reservoir, which generates steam. This steam then makes the steam turbines spin, which feeds a connected electrical generator. Overall, the design is quite efficient in terms of energy transfer but generates a lot of polluting elements.

Clean coal refers to an ensemble of technologies that reduce the overall environmental impact of coal by reducing or storing the polluting elements. Some technologies capture and store carbon emissions, referred to as carbon capture and storage (CCS) technologies (which some refer to as postcombustion capture technology). Another popular technology development alternative is called Integrated Gasification Combined Cycle (IGCC) power generation wherein the technology reduces greenhouse gas emissions by the gasification of coal (some refer to this technology as precombustion capture technology). Other technologies include chemically cleaning the coal to remove impurities or steam treatment of raw materials to reduce sulfur dioxide.

Strengths of Clean Coal

Clean coal is an interesting technology that presents three main strong points: (1) it reduces the production of harmful by-products that are generated when burning coal, (2) some technologies can be added to existing facilities (which limits required investments), and (3) it ensures the generation of consistent energy.

The use of clean coal technologies can eliminate much of the pollution produced by coal. Some of the technologies being developed can reduce the particulates that are emitted during burning, such as the sulfur oxides as well as nitrogen oxides. They can also reduce the CO_2 emissions (sometimes by as much as 80 percent). This is important since the main disadvantage related to coal is the polluting impact.

Another strength that clean coal presents is that some of the technologies being developed optimize existing facilities, rather than focusing on building new structures. This is quite important since companies building and maintaining coal-power facilities have invested greatly in their current infrastructure, and few are inclined to close their facilities if they are still efficiently producing power. Additionally, the use of coal is unlikely to diminish due to its abundance and technology know-how, so any technology that is built around the concept of "cleaning it up" will require less investment, making it more attractive to current producers. Hence, some technologies (especially CCS technologies) that favor the retrofit of existing facilities can be very cost effective (when compared with building completely new coal power plants). As many government subsidies and programs exist that target these technologies, there are a number of incentives that further encourage their growth.

A third advantage that clean coal presents is related to coal itself, which is the constancy of energy produced. Many renewable technologies like solar, wind, and hydro technologies are touted as replacements for fossil fuels, but they are all more costly, less consistent in terms of power production, and all have geographical restrictions in terms of implementation. Solar technology works best during peak sunny periods, wind technology only works during optimal windy conditions and needs a location with regular windy conditions, while hydro technologies need the appropriate hydro geography to be implemented.

Weaknesses of Clean Coal

Clean coal does also have a number of weaknesses that merit further review. The main weaknesses are the impact of clean coal technology on plant efficiency, the relative costs and economic feasibility of some proposed solutions, the continued harm to the environment while harvesting coal, and the unknowns surrounding long-term carbon storage.

One of the main impacts related to clean coal is that the implementation of this type of technology directly impacts the efficiency of coal-based power production. Some studies have found that the thermal efficiencies of an IGCC power station are reduced when the plant is configured for CCS; some studies point to a reduction between 30 and 40 percent. This is a huge drop in power production for a facility and might incur the perverse effect of encouraging the construction of new facilities to compensate for that drop in power production. It could also encourage increased mining since (ironically) more coal is needed to produce comparable levels of electricity.

It is inevitable that clean coal is an expensive proposition. Some researchers believe that clean coal could double the cost of the electricity generated by the facility. Other studies have found that the cost of CCS deployment is close to $80–$120 per ton of CO_2 stored, and while there is much excitement about optimizing existing plants, critics believe that some existing plant designs cannot cost effectively capture carbon dioxide; some studies estimate that adding the technology to a conventional coal plant would dramatically increase cost and reduce energy output. It is becoming clear that it might be extremely expensive to add clean coal technology, particularly as a retrofit to some facilities. As a counterargument, some proponents have presented new designs for clean coal production facilities, but then these facilities have to compete with other clean technologies (like nuclear facilities) to get the necessary investment. As an indication, the U.S. Department of Energy estimates that a clean coal facility with carbon storage could cost up to 30 percent more than a conventional facility.

Another strong criticism of clean coal is that it does not address one of the main issues of coal, which is the mining of raw material. Some will agree that while the output of clean coal is cleaner, the coal mining process remains the same. This is a process that has been heavily criticized for its impact on the local environment. Coal mining has been demonstrated to damage the environment, and some contend that even cleaning the output of coal does not deal with the pollution that occurs during mining.

A fourth concern exists related to the storage of the carbon. Some recent incidents have brought to light that carbon capture and storage is vulnerable to leakage, since once carbon is stored underground, there is a structural risk that the carbon can leak back out into the surrounding environment. For example, geological shifts can damage storage areas, which could create a slow carbon leak, which could seep into underground water reservoirs, harming the health of surrounding communities as well as the local ecological system. It is still undemonstrated that carbon can be stored indefinitely, and undetermined what type of underground geological formations are best suited for this type of storage. Finally, carbon storage can require monitoring to ensure the carbon is safely stored. This monitoring incurs extra costs that cannot be neglected when calculating financial viability.

Finally, a series of other criticisms and concerns have emerged that merit mentioning. These include concerns that CCS is the only technology worthy of the term *clean coal*, since other technologies (such as coal cleaning or filters) do not really reduce emission but only displace the problem from emission to waste by-products, which present challenges for disposal. Others criticize the heavy investments being made in clean coal technologies, believing they would be better invested in new renewable energies.

The Future of Clean Coal

Current dependence on coal around the world makes it clear that development of clean coal will have to continue. No other renewable or clean power technologies can effectively replace coal as a primary energy source, so most governments are invested in finding ways to optimize their current facilities. Major coal users are even predicted to increase their usage of coal in the next 20 years, even if they are committed to investing in renewable energies. Also, even if industrialized countries were to reduce their use of coal, it is quite likely countries demonstrating strong growth (such as India and China) will be unwilling to reduce their usage of coal. Hence, clean coal enjoys strong political support as a great short-term solution to developing cleaner energies. It must be stressed that renewable energies cannot replace coal in the next few years due to challenges linked to technology maturity or consistency issues: the only short-term alternative to coal is nuclear power, which also has its own advantages and disadvantages.

Therefore, while there are few examples of functional facilities using clean coal technologies, investment in research and development as well in new installations are being subsidized by governmental agencies to favor their expansion. Furthermore, different storage options for captured CO_2 are being examined. Some alternatives include underwater storage and exportation abroad so it can be used in mining. Finally, with many traditional coal facilities reaching the end of their life cycle, some proponents have suggested that they be replaced by IGCC facilities instead, but since very few such facilities exist at the moment, they are not very comparable.

Like many clean technologies, it remains to be seen if clean coal can be financially feasible without government support.

See Also: Carbon Capture Technology; Clean Energy; Thermochemical Processes.

Further Readings

Burgelam, Robert and Andrew Grove. "Toward Electric Cars and Clean Coal: A Comparative Analysis of Strategies and Strategy-Making in the U.S. and China, Chapter 6: Carbon Capture and Storage in the U.S.: Fact or Fiction—Two Paths to 2030." *Stanford Graduate School Papers* (February 2010).

Hoadley, Andrew. "The Sustainability of Clean Coal Technology: IGCC With/Without CCS." In *Process Safety and Environmental Protection*, David Edwards and Tom Stephenson, eds. New York: Elsevier, 2010.

Sen, S. "An Overview of Clean Coal Technologies I: Pre-Combustion and Post-Combustion Emission Control." *Energy Sources, Part B: Economics, Planning, and Policy*, 5/1 (2010).

Wang, Hao and Toshihiko Nakata. "Analysis of the Market Penetration of Clean Coal Technologies and Its Impacts in China's Electricity Sector." *Energy Policy*, 37/1 (2009).

Jean-Francois Denault
Independent Scholar

Cogeneration

Cogeneration or combined heat and power (CHP) refers to the coupled generation of heat and electricity (power) by one plant. CHP generally offers overall energy-efficiency benefits compared to the decoupled generation of heat (e.g., in boilers) and power (in centralized stations), but is not always appropriate for applications with greatly or quickly varying heat and/or electricity loads. Hence, CHP is not always an appropriate technology to employ, despite the potential energy and costs savings it may offer, and the calculation of such energy savings should be based on system-specific comparisons. There are a wide variety of technologies available for cogeneration that are employed depending on the application specifics, such as the overall heat and power demands, load fluctuations, and mode of operation (i.e., heat or power led). Despite the advantages of CHP, it is not always suitable because of the inflexibility associated with cogenerating and often unfavorable economics.

CHP is nevertheless employed in a wide variety of economic sectors, including providing space and process heating alongside electricity in the industrial sector. In this case the heat output may need to be processed and/or upgraded in order to meet the demand requirements, such as temperature and pressure, for example. CHP is also employed in several other sectors, mainly for providing space and water heating, including the services (e.g., hotels, airports, public buildings) and the domestic sectors.

Due to the fact that heat and electricity demands are generally not constant over time, but rather have seasonal, daily, and hourly variations, along with the fact that CHP units are most energy efficient (and therefore economical) when operating at their design output, they tend to be employed alongside a peak load plant. The sizing of a CHP plant for a particular application therefore involves choosing heat outputs that can meet the heat load that occurs throughout most of the year, known as the "base load." The peaking plant is then employed to meet the heat demand when it exceeds this value, such as in winter.

The heat-to-power ratio refers to the proportion in which heat and electricity are generated, and this parameter is an important characteristic for CHP plants. There are two main types of CHP technologies, namely those in which the heat-to-power ratio is fixed, and those in which this ratio can be altered to match individual demands. The former have only one degree of freedom, namely the total output, whereby heat and power are generated in the same proportions, and the latter have two degrees of freedom because the total output as well as the heat-to-power ratio can be independently varied. The devices used for CHP plants are generally referred to as prime movers; the most suitable technology depends in particular on the size of the plant, the required heat-to-power ratio, and the load factor (i.e., the fraction of the time, on average, for which the plant is required to operate), among other things. Load factors can vary from about 40 percent for a small internal combustion engine micro-CHP plant up to about 90 percent for an industrial combined cycle gas turbine, although this very much depends upon the application. The heat and power output from a CHP unit is usually measured in watts, with ratings referring to the electrical and thermal capacity of a plant, usually declared in We and Wh, respectively. The overall efficiency of a CHP unit is expressed as the sum of the electrical and thermal efficiency. It depends on the technology and precise application but tends to be on the order of 80 percent.

Technology Options

Several technology options are available and in widespread use, depending on the specifics of the application. In general, one can distinguish between the very small, or micro, scale; an intermediate, or mini, scale; and large-scale plants. The boundaries between these categories are somewhat vague and there is no universal consensus, but micro-CHP plants tend to be suited to individual households, having a power rating up to around 5 kWe. Plants for providing heat and power to larger buildings such as hotels, universities, and hospitals might range in size from this figure up to around 200 kWe or even 500 kWe. Plants larger than this, operating in the megawatt (MW) region, can be considered large scale, and generally include those plants supplying industrial facilities and/or feeding into district heating networks. In general, CHP plants have heat-to-power ratios in the range of 1:3, and larger plants tend to have lower ratios, although there are obviously exceptions.

There are several different types of prime movers for CHP plants, where *prime mover* is the term given to the device that actually converts the fuel into heat and power. At the lower end of the scale (i.e., smaller than about 5 kWe, about the size for an individual house), internal combustion engines operating on petrol, diesel, or gas tend to dominate because they are reliable, require relatively little maintenance, and are flexible in their operation, that is, they can respond well to load changes. At larger sizes than this, gas turbines, steam turbines, and combined cycle (gas and steam) turbines tend to dominate due to their high overall efficiencies and reliabilities. It is likely that fuel cells, in which heat and power are produced from a chemical reaction between a fuel and an oxidant in the presence of an electrolyte, will become more economical in the future, and will therefore be widely exploited due to their numerous advantages, including no noise, no vibrations, no moving parts (hence low maintenance), and low heat-to-power ratios. The efficiency of CHP plants is strongly affected by economies of scale, such that larger plants are much more efficient and the cost per installed unit of heat or electricity output is less.

The electricity generation sector in many countries is currently dominated by large, centralized, fossil fuel–based power plants that produce electricity by combustion of the fuel and using the released heat energy to drive a turbine. This is either direct, as in the case of a gas turbine, or indirect, in which case steam is raised and used to drive a steam turbine. The overall electrical generation efficiency of these large plants is around 50 percent for coal or oil systems and almost 60 percent for combined cycle gas turbine–based plants. The remainder of the input energy is not utilized but instead is released from the process as waste heat.

In several countries, for example, Denmark, this centralized generation model is based not around only power plants, but also CHP plants that feed into district heating networks. Hence the heat that would be otherwise wasted is transported to local houses and other buildings through insulated pipes to meet the demand for space heating and hot water. In these countries, the heat distribution infrastructure has generally evolved in tandem with the power plant, which is an important factor when one considers replacing centralized power-only plants in other countries with CHP units, because they are often situated large distances from the nearest heat demand (known as a "heat sink").

There are also other reasons, some quite complex, why CHP along with district heating has evolved more in some countries than in others. If prevalent climatic conditions do not consistently lead to extremely cold winters, the proper sizing of district heating networks and their prime movers is complicated, not least because it is difficult to design for the expected heat demand. There are also challenges associated with centrally and individually metering the heat distribution through the network, and there tends to be resistance to inflexible centralized supply from customers who are used to exercising their choice between suppliers, fuels, and technologies. Finally there are a several social factors that go some way to explaining the drastically different levels of uptake in CHP around the world.

The main problem with CHP is the inflexibility, which is mainly due to power grid regulations in individual regions or countries. The price received for selling electricity back to the grid also plays a crucial role in the economics of CHP. If the price of fuel (e.g., gas) rises relative to the cost of electricity, then there is less economic incentive for cogeneration. There is also no general consensus on evaluating the environmental benefits of CHP compared to conventional, separate generation of heat and power. This means that CHP is still relatively a niche technology in some countries, while it is the bedrock of the energy system in others.

See Also: Best Available Technology; Clean Energy; Microgeneration; Smart Grid; Thermal Heat Recovery.

Further Readings

Horlock, J. H. *Cogeneration: Combined Heat and Power Systems: Thermodynamics and Economics.* Oxford, UK: Pergamon Press, 1987.

International Energy Agency (IEA). "Cogeneration and District Energy." Paris: IEA, 2009.

Rogers, G. and Y. Mayhew. *Thermodynamics: Work and Heat Transfer*, 4th ed. Harlow, UK: Longmann Scientific, 1992.

Russell McKenna
Karlsruhe Institute of Technology

COMPACT FLUORESCENT LIGHT BULBS (CFLs)

Compact fluorescent light bulbs (CFLs) are an energy-efficient alternative to traditional incandescent light bulbs. CFLs are designed to fit into sockets that are also used for incandescents. Compared to incandescents that give off the same amount of light, CFLs use less power and have a longer rated life. As a result, they can provide cost savings over the lifetime of the bulb, even though the initial purchase price of a CFL is generally higher than for an incandescent. Governments in the United States and several other countries are strongly encouraging the widespread adoption of CFLs, even to the point of phasing out incandescents. CFLs do have certain disadvantages: they generally are not recommended for outdoor lighting, or situations in which they would be rapidly turned on and off; their light quality is not always comparable to incandescents; and they contain mercury, which can create complications for disposal.

Fluorescent light bulbs, including CFLs, are more energy efficient than incandescent bulbs because they use an alternative method for creating light. Incandescents heat a filament inside the bulb, making the filament white-hot and producing light. A lot of the energy used to create the heat that lights an incandescent bulb is wasted. A fluorescent bulb, on the other hand, contains a gas that produces invisible ultraviolet (UV) light when electricity excites the gas. The UV light strikes the white coating that lines the inside of the fluorescent bulb. The coating changes the UV into visible light. Because no heat is used in the process, fluorescent bulbs are far more energy efficient than standard incandescent bulbs.

A CFL uses about 75 percent less energy than a standard incandescent bulb and can last 8 to 15 times longer. It can save an estimated $30 or more in electricity costs over its lifetime. Since lighting accounted for approximately 9 percent of household electricity usage in the United States in 2001, widespread use of CFLs could reduce the nation's household electrical consumption by as much as 7 percent. The U.S. Department of Energy has estimated that if every American home replaced just one light with an Energy Star–rated CFL, the nation would save enough energy to light 3 million homes for a year, save about $600 million in annual energy costs, and prevent 9 billion pounds of greenhouse gas emissions per year, equivalent to those from about 800,000 cars.

In addition to drawing less energy than incandescents to produce a comparable amount of light, CFLs also produce about 75 percent less heat. For this reason, they can reduce air conditioning costs. While CFLs require more energy in manufacturing than incandescent bulbs, this is more than offset by the fact that they last longer and use less energy than equivalent incandescent bulbs during their lifespan.

Due to the potential to reduce electric consumption and pollution, government agencies, utilities, and other organizations have encouraged the adoption of CFLs and other efficient lighting. At times, electric utilities and local governments have subsidized CFLs or provided them free to customers in order to reduce demand and defer the need to construct more power plants.

Some governments have gone further, using taxes, production bans, or other policies to phase out incandescent light bulbs that do not meet energy efficiency requirements. The U.S. Congress passed a 2007 law that set new energy efficiency standards for all common household lamps, effectively phasing out traditional incandescent bulbs by 2020. The European Union, Australia, and Canada have also put policies in place to require more efficient lighting and end the use of traditional incandescents.

This does not mean, however, that CFLs will have a monopoly on future lighting. Solid-state lighting, which uses light-emitting diodes (LEDs), is used for specialty applications such as traffic lights and may compete with CFLs for household applications if the price comes down. Halogen incandescents, which use about 70 percent of the energy of traditional incandescents, may provide consumers with another option.

History

The technology behind fluorescent bulbs dates back to Cooper Hewitt lamps in the 1890s, which were used in photographic studios and for certain industry applications. General Electric patented the first fluorescent lamp designed for widespread use, displaying fluorescent lighting at the 1939 World's Fair in New York. During the energy crisis of the 1970s, an engineer with General Electric, Edward E. Hammer, worked to make fluorescent bulbs smaller, creating the spiral design now associated with CFLs. However, General Electric decided against large-scale production because of the costs required to build new factories to produce them. Following additional technological innovations, including the addition of an electronic ballast to reduce flickering and enable the bulbs to reach full brightness more quickly, China began manufacturing CFLs in the 1990s. Sales of the energy-efficient bulbs quickly climbed. Fluorescent lights radiate a different light spectrum from that of incandescent lamps. Traditional fluorescents as well as early CFLs have been faulted for their harsh, cold light quality. Manufacturers, however, have worked to provide a warmer color by improving phosphor formulations. Every extra phosphor added to the coating mix improves the light quality, but at a cost of reduced efficiency and higher price. Good-quality consumer CFLs use three or four phosphors to achieve a white light that can be similar in color to that of standard incandescents.

Although they represent a major advance in energy efficiency, CFLs can create challenges when it comes to disposal. CFLs, like all fluorescent lamps, contain small amounts of mercury as vapor inside the glass tubing. Most CFLs contain 3–5 milligrams per bulb, with some brands containing as little as 1 milligram. Because mercury is poisonous, even these small amounts are a concern for landfills and waste incinerators where the mercury from lamps may be released. Many jurisdictions regulate the disposal of CFLs, requiring them to be properly disposed or recycled instead of being sent to a landfill. Consumers are urged to follow certain safety procedures if a CFL breaks in the home in order to minimize the danger of any mercury contamination. Workers in CFL manufacturing plants also are at risk. Hundreds of workers at CFL manufacturing plants in China have been exposed to unhealthy mercury levels, with some requiring hospitalization.

However, CFLs also can lower the threat of mercury emissions by reducing demand for electricity—especially electricity from coal-fired power plants, which are a major source of mercury emissions. A 2008 report by the U.S. Environmental Protection Agency concluded that the use of CFL lighting led to a net decrease in mercury emission, based on rates of mercury emissions from power plants and the average estimated release of mercury from CFLs dumped in landfills. The EPA estimated that if all 270 million compact fluorescent lamps sold in 2007 were sent to landfill sites, the amount of mercury would represent around 0.13 metric tons—the equivalent of about 0.1 percent of all U.S. emissions of mercury, which amounted to about 104 metric tons that year.

Potential Problems

The use of CFLs may be restricted by several factors, including the following:

- *On/off cycle*: A CFL typically has a rated lifespan of between 6,000 and 15,000 hours, far greater than an incandescent's life span of about 750 hours or 1,000 hours. However, this is dependent on a CFL's use. The life of a CFL becomes significantly shorter if it is turned on and off frequently. A CFL on a five-minute on/off cycle, for example, may have a lifespan close to that of incandescent light bulbs. For that reason, the Department of Energy recommends that fluorescent lamps be left on when leaving a room for less than 15 minutes.
- *Full illumination*: Incandescents reach full brightness a fraction of a second after being switched on. CFLs can turn on within a second, but many still take time to warm up to full brightness. During that warm-up period, the light color may be slightly different. Some CFLs are marketed as "instant on" and have no noticeable warm-up period. The warm up period, coupled with the shorter life of CFLs when turned on and off for short periods, may make CFLs less suitable for applications such as motion-activated lighting.
- *Dimming*: Standard CFLs are not suitable for use with a dimmer, although specially manufactured CFLs can be used with a dimmer. If a standard CFL is put on a dimmer switch, the dimming may be ineffective and the CFL's effective life can be shortened. A dimmed CFL may also emit an unpleasant bluish color.
- *Infrared signals*: The infrared light emitted by a CFL can interfere with the use of remote controls used to operate televisions and other devices, as well as cell phones. The electronic devices can intercept a CFL's light and interpret it as a signal, causing such problems as a television set suddenly changing channels. For this reason it is sometimes necessary to limit the use of CFLs near household appliances.
- *Outdoor use*: Regular CFLs are not designed for outdoor use and some will not start in cold weather. However, there are special CFLs available with cold-weather ballasts that can be used in subfreezing temperatures.
- *Ceiling fan and other vibrating environments*: CFLs generally are not suitable for use with ceiling fans, garage door openers, and other environments where there are a lot of vibrations. The vibrations can cause the electronics in the CFL to fail. Specially manufactured CFLs may be used in such environments. Fortunately, this only happens when light is produced at the same wavelength as the electronic device signals, which is rare.
- *UV emissions*: CFLs and other fluorescent bulbs can damage paintings and textiles that have light-sensitive dyes and pigments. This happens because strong colors can fade after repeated exposure to UV light.

See Also: Appliances, Energy Efficient; Clean Energy; Green Building Materials; Light-Emitting Diodes (LEDs).

Further Readings

Kanellos, Michael. "Father of the Compact Fluorescent Bulb Looks Back." CNET News. http://news.cnet.com/Father-of-the-compact-fluorescent-bulb-looks-back/2100-11392_3-6202996.html (Accessed September 2010).

Shogren, Elizabeth. "CFL Bulbs Have One Hitch: Toxic Mercury." NPR. http://www.npr.org/templates/story/story.php?storyId=7431198 (Accessed September 2010).

David Hosansky
Independent Scholar

COMPOSTING TOILET

A composting (dry or biological) toilet is a waterless sewage treatment system that decomposes human excreta into an inert, nitrogen-rich material similar to humus. Because they eliminate the water use associated with typical toilets, composting toilets circumvent the costs associated with traditional sewage treatment. By holding and processing waste material on-site, they capture nutrients in human waste for local reuse. They are well suited to rural areas and water-scarce regions, but they are increasingly used in institutional and urban settings. This includes environmentally conscious users who are seeking to decrease their impact on water resources as well as cases where the water and sewer infrastructure is at capacity. Composting toilet technology provides a means to decrease daily water consumption and capture the nitrogen and phosphorous in human waste.

Composting toilets usually require the addition of a bulking agent like sawdust after each use and have a wide range of design complexity. They can decrease household water use by one-third or more.

Source: Way of Nature Corporation/Wikimedia Commons

In traditional household systems, the dirty water from sinks, showers, and washing machines (greywater) is combined with wastewater from toilets (blackwater) and discharged to a sewer or on-site septic system. As waterless devices, composting toilets circumvent the production of blackwater or discharge wastewater. They do not use water to move waste from the toilet to the next stage of waste treatment. In a privy or outhouse, waste is typically deposited on-site, covered with lye and buried, or removed for traditional sewage treatment elsewhere. Composting toilets differ from these primitive systems in that waste material is biologically processed on-site, allowing it to be used as a soil nutrient.

Composting toilets vary in terms of complexity of design, energy requirements for optimal operation, and capacity. The simplest form of composting toilet is a "humanure" system, which can be built with a five-gallon bucket, a few two-by-fours, and a pile of hay. More complex systems can be constructed with 55-gallon drums and minimal additional parts. The common goal is to ensure aerobic conditions for bacterial decomposition in the compost (the addition of a bulking agent like sawdust or coir is usually required after each use), remove exhaust from the compost reactor or catchment basin (often using a small fan), manage leachate (with gravity or a heating element), and provide a means to easily remove the finished product. Site-built and single-chamber systems can be built with few moving parts. In remote areas, for example, a solar-powered fan connected to an aeration chimney is all that is needed to ensure effective year-round processing of waste. For self-contained units within households or other indoor settings, toilets can have mechanical batch stirrers, electrically powered rotating chambers, and heating elements to drive off excess moisture.

Commercially built composting toilets can be grouped in two types by size and intended use. Small, all-in-one systems process waste in a reactor (a small, partially sealed container) directly attached to and below the toilet bowl. Models resemble a flush toilet and are popular in residential applications because they require little modification to existing bathrooms. Larger, centralized (or remote) systems use gravity or a small amount of water (a micro-flush) to direct waste to the compost reactor. Centralized systems are ideal in high-use settings as well as off-the-grid applications where solar energy may be the only available source of power. They can be multistory and often require subfloor or basement space to accommodate their larger compost reactors.

Both the smaller and larger systems can have single or multiple chambers. Single-chamber (or continuous) systems rely on gravity and little additional energy to operate. New material is added at the top of the reactor pile and processed as it moves downward. Finished compost is removed from a small opening at the base of the chamber. Urine introduced to the system maintains moisture, and the weight of the material helps ensure the right temperature for bacterial activity. By contrast, multiple-chamber (or batch) systems have rotating or removable chambers that produce individual batches of compost. These systems often feature heating elements to evaporate moisture and are better suited to intermittent use. By dividing waste into smaller batches, it is easier to ensure that the finished compost is fully processed, as no new waste is added to the chamber once it is full.

While regulations continue to develop to reflect increased use of composting toilets, there remains an education and information gap. Overlapping and sometimes confusing regulatory policies complicate decisions on installation and use. Regulations must protect both public and ecosystem health: these policies remain precautionary, particularly concerning the use or disposal of finished compost. For example, some policies allow direct application for agriculture, some require burial of compost, and some mandate removal by licensed sanitation workers.

One drawback to composting toilet technology is that it disrupts the easy convenience of the flush-and-forget systems that use water to convey waste to treatment. However, this disruption is a part of what makes composting toilets a green technology. By eliminating the need for centralized treatment, composting toilets can save energy, material, and infrastructure costs associated with traditional sewage systems. In daily use, composting toilets have been shown to decrease household water use by one-third or more; institutional applications can save up to 60 percent of water consumed. Though they have clear environmental benefits, they present different risks than traditional flush toilets. For example, if a composting toilet is not fully processing waste, pathogenic bacteria and viruses may remain present in compost material.

Commercial manufacturers include BioLet, Envirolet, Sun-Mar, and Clivus Multrum. The first three focus on residential-scale models. Clivus Multrum has recently focused on larger systems frequently used in institutional settings. Composting toilets are used widely by state and national park systems at remote sites. Their primary application is practical, to preserve and protect natural resources and fragile ecosystems, particularly in the Southwest and other dry or remote regions. In addition, they can serve as educational tools, to help inform users about water management challenges. One institutional application that has showcased composting toilets for the general public is located in the Bronx Zoo, and the facility managers have placed informative signage right in the bathrooms, next to the composting toilets. Other institutional installations are found in the C. K. Choi Building at the University of British Columbia, where the building's wastewater system is not connected to the municipal sewer system, and at Warren Wilson College in a residence hall. An example of a larger-scale residential application is the Skaneateles Lake community in New York state, where 75 households have been using Sun-Mar toilets for the past decade to protect an adjacent, unfiltered drinking water source.

See Also: Solid Waste Treatment; Sustainable Design; Waste-to-Energy Technology; Wastewater Treatment.

Further Readings

Del Porto, D. A. and C. J. Steinfeld. *The Composting Toilet Book*. Whiteriver Junction, VT: Chelsea Green Publishing, 1998.

Jenkins, Joseph. *The Humanure Handbook*. Whiteriver Junction, VT: Chelsea Green Publishing, 1999.

Lazarova, V., S. Hills, and R. Birks. "Using Recycled Water for Non-Potable, Urban Uses: A Review With Particular Reference to Toilet Flushing." *Water Science and Technology: Water Supply*, 3/4 (2003).

National Sanitation Foundation. "Composting Toilets." http://nsf.org/consumer/wastewater_treatment_systems/wastewater_nonliquid.asp (Accessed June 2010).

National Small Flows Clearinghouse. "Wastewater and Onsite Systems." http://www.nesc.wvu.edu/wastewater.cfm (Accessed June 2010).

ReSource Institute for Low Entropy Systems. "ReSource Composting Toilets: Integrating Design, Beauty, and Sustainable Maintenance." http://www.riles.org/tech.htm (Accessed June 2010).

U.S. Environmental Protection Agency. "Water Efficiency Technology Fact Sheet: Composting Toilets." http://www.epa.gov/owm/mtb/comp.pdf (Accessed June 2010).

Mike Dimpfl
Sharon Moran
Independent Scholars

Concentrating Solar Technology

Concentrating solar technology refers to a family of different solar technologies that employ either a mirrored or lensed surface to reflect or focus the sun's beams to a central element, from where the solar energy is extracted. This technology can be deployed on a variety of scales, from small, roof-mounted units where both concentrator and receiver are integrated into the device to massive arrays of mirrors focusing light onto a heliostat (central collector).

The basic technology of concentrating solar power can be deployed in a variety of ways using different technology pathways that are explained below.

Economics of Concentrating Solar Power

Concentrating solar power (CSP) may be one technology pathway that can reduce the cost of solar power and ease its deployment on a much larger scale. As "mirror" and "lens" can be manufactured using existing optics capabilities on a large scale, more effective use can be made of the material or technology that converts the solar energy to useful power. This is based on the principle that concentrating material, which harnesses the sun's energy from over a wide area, is relatively cheap to manufacture. Mirrors (for reflecting the sun's light) and lenses (for focusing it) can be produced readily and cheaply for fractions of the cost of the technologies that actually convert the solar energy into electricity. Where solar photovoltaic technologies are used, a smaller amount of material can be used relative to the area over which light is being collected. Additionally, there are some solar thermal technologies that rely on concentrating the solar energy from over a very wide area to raise heat—these technologies require solar concentration in order to generate the necessary temperatures that make this technology viable.

History of Concentrating Solar Power

It is said that Archimedes used a "burning glass" that would focus the sunlight over a large area to a point, where it reached a sufficient intensity that it could be used to burn the boats of the invading Roman fleet in 212 B.C.E. Although this is contested by some historians, a number of experiments have been conducted that give the idea validity. As early as 1866, Auguste Mouchot used a parabolic surface to raise steam to drive an engine, believing coal would eventually run out—all at the height of the Industrial Revolution. The first patent for a solar collector was not granted until 1886, when Alessandro Battaglia obtained one in Genoa, Italy. By 1913, solar concentrating technology was being used by Frank Shuman to power an irrigation plant in Meadi, Egypt. By 1968, the technology was being deployed to produce energy at a site near Genoa, Italy, with a plant that was capable of generating 1 MW; light was reflected off an array of mirrors and focused to a central point. Later, in 1981, Solar One, a 10 MW CSP plant was built; this also used the same technology of focusing light to a central point. In 1984, the Solar Energy Generating System facility was unveiled in California's Mojave Desert, taking until 1991 to fully commission the site; the plants that make up the facility have a capacity factor of 354 MW.

Solar Photovoltaic Concentrating Technologies

Photovoltaic devices are expensive to manufacture. Photovoltaic concentrating technologies try to reduce the amount of semiconductor content required by deploying a large lens or reflector in order to focus the sun's rays from a large area onto a smaller cell.

This also opens the door to using more complex, expensive photovoltaic devices, such as multilayer devices, which can be more expensive to manufacture but have the advantage of capturing light from several different areas of the light spectrum.

Some approaches that have been used on this small scale include parabolic dishes and linear trough mirrors, focusing light to a point and a line, respectively, where a photovoltaic device is positioned. Some other novel approaches have employed lightweight reflectors made from an inflated balloon of Mylar and transparent plastic. While this lacks the optical precision of a true parabolic mirror, it is very cheap to manufacture. Some other approaches employ lenses to focus the sun's light. Often, variations of the Fresnel lens can be used as these are both cheap to manufacture and require a minimum of material.

There has also been some investigation of developing hybrid solar photovoltaic/thermal devices using this approach—a concentrator focusing a large area onto a small photovoltaic device, which in turn is cooled using a matrix of pipes from which low-level heat is extracted.

Stirling Dish Concentrators

A medium-scale solar concentrating technology uses an array of mirrors that form a "dish" to focus solar energy onto a Stirling engine. A Stirling engine is a mechanical device that produces movement from a temperature differential. By alternately compressing and expanding a working fluid, mechanical energy can be extracted from a temperature differential. A Stirling engine has an "expansion" cylinder, which must be kept at a constant high temperature, and an expansion cylinder that must be kept cooler. Solar Stirling dish concentrators employ concentrated solar energy to maintain the temperature of the expansion cylinder. A conventional generator can then convert the rotary motion of the Stirling engine into electricity.

Thermal Solar Concentrating Technologies

Thermal solar concentrating technologies rely on generating a large quantity of heat using the sun's focused rays. This heat can then be used to generate electricity by employing a conventional "thermal" generating plant. Here, water is turned into steam, and the steam is in turn used to drive a steam turbine, which is coupled to a generating set. This technology has some advantages; there is the potential for the "heat" generated by the sun to be readily stored in the form of molten salt or thermal stores. This means that when energy is actually generated can be a different time from when the sun is actually shining—introducing the ability for solar thermal plants to adjust their outputs in the manner of a conventional power station.

There are several different implementations of solar thermal concentrating technologies. Some use a large array of mirrors that rotate to face the sun (heliostats) in order to focus light onto a central "tower" where a solar furnace receives the heat. This approach is known as a solar power tower.

An alternative approach is known as the concentrating linear Fresnel reflector, or CLFR. This method employs long mirrors mounted on an assembly that allows them to rotate parallel to the Earth's surface to track the sun, and focus its rays on an absorber mounted at the focal point of where the focused beams from the mirror array converge. Heat transfer oil flows through this pipe, and through a heat exchanger: this can be used to raise steam to power a generator.

Challenges of Solar Concentrating Technology

One of the most important challenges with concentrating solar technology is reliability; as with any technology in early stages of deployment, there are technical challenges that have yet to be comprehensively addressed. As many designs of solar concentrator must "track the sun" in some way, there is always a mechanical element required to provide movement—and this mechanism is vulnerable. The technology has proven its worth in a variety of demonstration projects, and many commentators believe this technology's time has come.

There remain a number of challenges in bringing this technology to market; fossil fuel and nuclear technologies have a history of government support and intervention—however, by comparison, the market for electricity from solar concentrating plants is relatively immature—therefore, the technology will need a variety of incentives and support mechanisms in order to help bring it to a broader marketplace. Some sources believe that in the right location, concentrating solar power can already offer a cost advantage over some other "dirty" technologies.

Finding the right location, however, is not always easy. While CSP technologies require a location with a good solar resource, they also require water for cooling: unsurprisingly, those areas with clear skies and little clouds also tend to be areas that do not receive a great deal of rain, and therefore are vulnerable to drought. This is a problem that has been witnessed to be particularly acute with CSP plants in the U.S. Southwest, where it has been important to find a balance between water conservation and energy vulnerability. Therefore, some simple, cheap solutions that employ evaporative cooling may not be deemed appropriate for such regions. To get around this problem, developers are often encouraged to either seek water rights from neighboring states or to employ alternative technologies that use dry or hybrid cooling. It can be seen that water challenges do not preclude CSP developments; however, they do have bearing on technology selection.

As with all energy technologies, it is often the case that the resource is located some distance away from the point of use. Concentrating solar technologies, for example, are well suited to areas of clear, flat land with good access to the solar resource. Therefore, one of the main challenges in developing this technology is deploying the infrastructure that allows the energy to be transported from concentrating solar plants to the points of use.

Throughout the 1980s, Luz International built a total of 354 MW of CSP plants in the Mojave Desert—nine plants in total. The plants have proven to be robust and durable over their 20 years of operation. Unfortunately, the business model of the company that built them proved less so, with Luz International going bankrupt in 1991. While this perhaps discouraged investment in the technology throughout the 1990s, since the turn of the millennium a resurgent CSP industry has attracted renewed enthusiasm from the investment community.

While alternating current is more commonly seen in electricity transmission and distribution networks, for moving "bulk electricity" over long distances it seems probable that high voltage direct current (HVDC) links may be used. HVDC links employ a converter at each end that transforms the alternating current into high voltage DC for transmission; this results in low losses. DC is not used for short-range power distribution as the cost of equipment would be prohibitive, but over longer distances, DC distribution at high voltages makes economic sense.

Conclusion

With increasing pressure on the global community to transition to lower-carbon energy sources, the array of concentrating solar technologies has the potential to help in the transition to clean green energy sources as the technology and political challenges associated with the technology are solved.

See Also: Cadmium Telluride Photovoltaic Thin Film; Clean Energy; Copper Indium (Gallium) Selenide (CIGS or CIS) Solar Photovoltaic Thin Film; Crystalline Silicon Solar Photovoltaic Cell; Solar Cells.

Further Readings

Bingham, C., A. Lewandowski, K. Stone, R. Sherif, U. Ortabasi, and S. Kusek. "Concentrating Photovoltaic Module Testing at NREL's Concentrating Solar Radiation Users Facility." Paper presented at the National Center for Photovoltaics and Solar Program Review Meeting, 2003. http://www.nrel.gov/docs/fy03osti/33617.pdf (Accessed June 2010).

German Aerospace Center. "Trans-Mediterranean Interconnection for Concentrating Solar Power." Stuttgart: German Aerospace Center, 2006. http://www.trec-uk.org.uk/reports/TRANS-CSP_Full_Report_Final.pdf (Accessed June 2010).

International Energy Agency (IEA). "Technology Road Map: Concentrating Solar Power" (2010). http://www.iea.org/papers/2010/csp_roadmap.pdf (Accessed June 2010).

Swanson, R. "The Promise of Concentrators." *Progress in Photovoltaics: Research and Applications*, 8 (2000).

Trieb, Franz, Christoph Schillings, Marlene O'Sullivan, Thomas Pregger, and Carsten Hoyer-Klick. "Global Potential of Concentrating Solar Power." SolarPaces Conference, Berlin, 2009. http://www.solarthermalworld.org/files/global%20potential%20csp.pdf?download (Accessed June 2010).

U.S. Department of Energy. "PV FAQs: What's New in Concentrating PV?" (2005). http://www.oilcrisis.com/apollo2/photovoltaics/WhatsNewInConcentratorsNREL.pdf (Accessed June 2010).

Gavin D. J. Harper
Cardiff University

COPPER INDIUM (GALLIUM) SELENIDE (CIGS OR CIS) SOLAR PHOTOVOLTAIC THIN FILM

CIGS is an acronym that stands for copper indium gallium selenide. This is a type of thin-film photovoltaic device that can be produced using a process that has the potential to reduce the cost of producing photovoltaic devices by combining a low cost of manufacturing with a high production yield. While they currently only occupy a small market share,

Copper indium gallium selenide solar cells are a member of the thin film family of photovoltaic (PV) panels, shown here, and absorb light energy better than silicon PV cells.

Source: Fieldsken/Ken Fields/Wikimedia Commons

CIGS cells have the potential to be a "disruptive" technology in the field of solar photovoltaics, as their manufacturing processes could be configured toward large-volume production.

Thin Film Solar Technologies

Thin film solar technologies differ from crystalline photovoltaic technologies in that rather than using a crystalline solid to form the semiconductor junction (e.g., mono- or polycrystalline silicon), they instead use thinner semiconductor layers to absorb sunlight and convert the light energy into electrons. CIGS solar cells are a member of the thin film family of photovoltaic devices.

The thin film PV relies on a combination of materials that form a "solid solution" of copper indium selenide and copper gallium selenide. A trace amount of sodium is also required. The CIGS film forms what is known as a direct band gap semiconductor; what this means in principle is that due to its electronic properties, it absorbs light very well, and a very thin layer can be used. This differs from silicon, which is an indirect band gap semiconductor. This does not absorb light energy so easily, necessitating a thicker active layer. The material forms a "heterojunction," as the band gaps of the two different materials are unequal. This is more complex than the "homojunction" of a silicon solar cell.

The cell is deposited onto a substrate; this can be soda lime glass or a polymer such as polyamide, or sometimes a metallic foil such as stainless steel—this forms the rear surface contact. Where a nonconductive material is chosen for the substrate, a metal such as molybdenum is used, but if deposited on a stainless steel foil, no additional conductor is required. The front surface contact must be able both to conduct electricity and be transparent to allow light to reach the cell. Material known as a "transparent conductive oxide" such as indium tin oxide, doped zinc oxide, or, more recently, advanced organic films based on nano-engineered carbon are used to provide an ohmic contact.

The cells are designed so that light enters through the transparent front ohmic contact, and it is absorbed into the CIGS layer. Here electron-hole pairs are formed. There is a junction of p- and n-type materials at the junction of the cadmium-doped surface of the CIGS cell; here a "depletion region" is formed. This separates the electrons from the holes and allows them to be collected, and generates an electrical current.

There is some concern surrounding the use of cadmium in solar cells, and investigation is under way into alternatives, with zinc sulfide being advanced as a possible alternative.

Manufacturing Process

CIGS films are deposited onto a substrate. Commonly, this is done by a process of deposition in a vacuum, using either an evaporative or a sputtering process. Copper, gallium, and indium are deposited in turn, with this film being annealed with a selenide vapor resulting in the final CIGS structure. One of the challenges is to control the profile of the gallium in the CIGS layer. Other approaches have been developed, including processes that do not require a vacuum. In one process, nanoparticles of the materials are first deposited onto the substrate, and then sintered to form the final CIGS structure. Others have tried using electroplating as an alternative to vacuum-based deposition approaches.

In looking at non–vacuum-based approaches, the goal is to investigate lower-cost methods of CIGS cell production that can be produced quickly, cheaply, or in a continuous process in order to reduce the cost per watt. Novel approaches are being developed that are closer in nature to "printing" technologies than traditional silicon solar cell fabrication. In one process, a printer lays droplets of semiconducting ink onto an aluminum foil. A subsequent printing process deposits additional layers and the front contact on top of this layer; the foil is then cut into sheets.

Efficiency and Market Status

In laboratory experiments at the National Renewable Energy Laboratory, an efficiency of 19.5 percent was produced by a CIGS cell with a modified surface structure. Module efficiencies for CIGS cells are lower, with current best modules attaining 13 percent conversion efficiencies.

CIGS photovoltaics can be considered to be in the early stages of large-scale commercialization. In 2010, thin film solar cells comprised only about 23.4 percent of the world's photovoltaic cell market, with the majority of the thin film market currently dominated by cadmium telluride cells, which take up around 13 percent of market share. However, as the performance, uniformity, and reliability of thin film CIGS products improves, the technology has the potential to expand its market share significantly in the medium term.

The production cost of solar devices is partly dependent on the cost of raw materials. In times of silicon shortage (caused by lack of processing capacity and demand from the semiconductor industry), the cost of silicon can rise. Furthermore, silicon solar cells are expensive to produce because they must be made using "batch" production techniques, which limit the throughput of plants. By contrast, many believe that CIGS solar cells could be produced on a continuous process, driving down the cost.

Applications of CIGS Solar Cells

CIGS solar cells have the possibility of being manufactured on flexible substrates. This would make them suited for a variety of applications for which current crystalline photovoltaics and other rigid products are not suitable.

For building integrated applications, CIGS solar cells on a flexible substrate have the advantage of giving the architect a greater range of possibilities in styling and design, making the cells attractive and better integrated with the built environment.

For consumer products, the cells have the advantage of being resistant to shattering as they contain no glass, and they are a fraction of the weight of silicon cells. There are also possibilities of integrating these cells into transport applications as their low profile lends them to applications where air resistance could be an issue. Furthermore, on lightweight substrates, they would not add significantly to the weight of the vehicle.

Conclusion

There are a variety of technical challenges that will be resolved in due course concerning the manufacturing of CIGS cells, which will improve their attractiveness as a technology option. The potential to manufacture these devices at low cost in a continuous process offers promising indications for this technology. Furthermore, there is less environmental concern with the materials used in CIGS cell manufacture compared to cadmium telluride thin film cells.

See Also: Cadmium Telluride Photovoltaic Thin Film; Clean Energy; Crystalline Silicon Solar Photovoltaic Cell; Passive Solar; Solar Cells.

Further Readings

German Solar Energy Society. *Planning and Installing Photovoltaic Systems: A Guide for Installers, Architects and Engineers,* 2nd ed. London: Earthscan, 2008.

Jha, A. R. *Solar Cell Technology and Applications.* Boca Raton, FL: Auerbach Publications, 2009.

Ramanathan, K., J. Keane, and R. Noufi. "Properties of High-Efficiency CIGS Thin-Film Solar Cells." 31st IEEE Photovoltaics Specialists Conference and Exhibition, 2005, National Renewable Energy Laboratory. http://www.nrel.gov/docs/fy05osti/37404.pdf (Accessed June 2010).

Gavin D. J. Harper
Cardiff University

CRADLE-TO-CRADLE DESIGN

The term *cradle to cradle* refers to a holistic design approach that mimics natural systems. Among its founding principles are the use of fewer materials, less energy, and the creation of little to no waste in the production of the material things that we use every day. It is often abbreviated as C2C. Like biomimicry, cradle to cradle's inspiration is found in nature's biological metabolism, where very little input is required to create complex organisms and diverse materials, and all nutrient cycles are closed. Rather than externalize wastes, which provides a threat to environmental and human health, cradle to cradle turns waste into fuel for the production process. The fate of all materials is considered throughout the life cycle of products, structures, and more. Such closed-loop production is at the heart of sustainable practices and mirrors other system models, such as biodynamic's farm organism. The cradle-to-cradle framework unites industrialists and environmentalists in its inherent efficiency, which saves both costs and resources.

History and Precedents

The term *cradle to cradle* was coined by Walter Stahel and Genevieve Reda in a 1976 research report to the European Commission in Brussels titled "The Potential for

Substituting Manpower for Energy," which presented a vision of a loop or circular economy. Consequently, in the 1990s, *A Technical Framework for Life-Cycle Assessment* by Michael Braungart outlined the concept of life-cycle development. In 1991, Braungart and William McDonough coauthored a document for the World Urban Forum titled "The Hannover Principles," which acknowledges human interdependence with nature and seeks to eliminate waste. Other influential texts included E. F. Schumacher's 1973 book *Small Is Beautiful: Economics as If People Mattered* and the Club of Rome's *The Limits to Growth* and *Beyond the Limits*. These documents culminated in the widespread popularization and elaboration of the term *cradle to cradle* by architect McDonough and chemist Braungart in the 2002 manifesto-like book *Cradle to Cradle: Remaking the Way We Make Things*. Since 2005, Cradle to Cradle Certification® has been a registered trademark of McDonough Braungart Design Chemistry (MBDC).

Cradle to cradle is a response to the current linear industrial model that creates cradle-to-grave waste flows. Instead, cradle to cradle is founded on the recognition that natural resources are limited and builds on previously developed ideas at the intersection of ecology and economy. Herman Daly's *Steady State Economics*, originally published in 1977, argued that total resource-to-waste throughput must be brought to a sustainable level. In 1999, Paul Hawken, Amory Lovins, and L. Hunter Lovins published the influential book *Natural Capitalism: Creating the Next Industrial Revolution*. In it they outline a new economy whose four principles are (1) radically increasing resource productivity, (2) redesigning industry on biological models with closed loops and zero waste, (3) shifting from the sale of goods to the provision of services, and (4) reinvesting in the natural capital that is the basis of future prosperity. Cradle to cradle has been heavily influenced by both of these texts as well as the larger fields of ecological economics and industrial design. It aims to provide an industrial philosophy and technological know-how to successfully achieve an ecologically friendly economy.

Philosophy: Waste Equals Food

Cradle to cradle is not a new technology, but rather a new approach to the use of existing technology. Cradle to cradle is founded on the three E's of sustainability: ecology, economy, and equity. Industrial design, according to cradle-to-cradle philosophy, need not be divorced from environmental intelligence.

Energy and material efficiency are key to cradle-to-cradle design, but MBDC strives to move beyond mitigating negative impacts. Thus, they are designing to optimize ecological benefits. The philosophy underlying this design framework is that waste equals food. The goal is a zero-waste product that can be infinitely recycled. MBDC evaluates materials as either "biological nutrients," organic nutrients, or "technical nutrients," inorganic nutrients. These are then evaluated for their environmental and human health characteristics throughout the entire life cycle of the product, including consideration of the recyclability or compostability of the material.

But McDonough and Braungart maintain that ecoefficiency is not enough to create a true paradigm shift. Indeed, they question the value of efficiency as a goal in itself. It is only as "good" as the larger philosophy in which it resides. Instead, they aim for "ecoeffectiveness," at the intersection of economy and ecology. The term describes, in Braungart's own words, cradle to cradle's strategy for intelligent and healthy materials use, designing human industry that is safe, profitable, and regenerative, producing economic, ecological, and social value. It thus attempts a more comprehensive vision of economy and human life.

Cradle to cradle responds to local circumstances and recognizes interdependent relationships. Ideally, it uses local materials and taps into latent energy flows. Rather than mourn limits, cradle to cradle works with abundance.

McDonough and Braungart introduce the concept "product of service" to describe a unique way of viewing products. The manufacturer maintains ownership of the product's material assets for continual reuse while the customer receives the service of the product. The product is thus described and valued through its human use value. This requires awareness of the use goal of unique consumers and varying individual preferences. Cradle to cradle is not meant to be a one-size-fits-all solution, but a particularized and thoughtful design strategy.

To the traditional design considerations of cost, performance, and aesthetics, McDonough and Braungart add ecological intelligence, justice, and fun. Importantly, cradle to cradle incorporates concern for social justice as well as environmental well-being. One of the central tenets of the design framework is "celebrate diversity." For McDonough and Braungart, this translates to the promotion of prosperity, health, security, community, peace, and culture.

Cradle to cradle can be differentiated from other environmental design philosophies in its comprehensive nature and its celebration of human creativity, culture, and productivity as well as its appreciation of fun. The cradle-to-cradle philosophy is one of cooperation with natural and human systems. It pursues ecological, cultural, and societal growth within an open system that derives its free energy from sunlight.

Cradle to Cradle® Certification

Before C2C principles coalesced into the current-day certifications, the principles were applied by Braungart's Hamburger Umweltinstitut (HUI) and the Environmental Institute or O Instituto Ambiental (OIA) in Brazil. In 1995, William McDonough and Michael Braungart founded MBDC with the self-stated goal of reorienting the design of products, industrial processes, and systems toward financial, environmental, and societal benefits.

According to the MBDC website, "Cradle to Cradle® Certification is a multi-attribute ecolabel that assesses a product's safety to humans and the environment and design for future life cycles." Cradle to Cradle Certification comprehensively evaluates the sustainability of a product and the practices employed in manufacturing the product, focusing on the use of safe materials that can be disassembled and recycled as technical nutrients or composted as biological nutrients. The materials and manufacturing practices of each product are assessed in five categories: Material Health, Material Reutilization, Renewable Energy Use, Water Stewardship, and Social Responsibility. There are four levels of Cradle to Cradle Certification: Basic, Gold, Silver, and Platinum.

Though MBDC is the primary certifier, the company has partnered with the German company Environmental Protection and Encouragement Agency (EPEA), founded by Braungart; the Spanish company EcoIntelligent Growth (EIG); and the U.S.-based company Zimmerman Management Solutions. At the two highest levels of certification, manufacturers are required to obtain third-party social responsibility or Fair Labor certification.

Cradle to Cradle Certification provides the following:

- External verification of a product's recyclability and safety for human and environmental health
- Expert assessment of toxicity hazards of all product ingredients throughout the supply chain down to 100 parts per million (ppm), or 0.01 percent
- Defined trajectory for optimizing product design and manufacturing processes

Cradle to Cradle Certification qualifies manufacturers for Leadership in Energy and Environmental Design (LEED) points and allows them to sell to federal agencies under the U.S. Environmental Protection Agency Environmentally Preferable Purchasing (EPP) Program.

McDonough and Braungart have been involved in many projects and partnerships around the world. They collaborate with the Green Products Innovation Institute and designed Gap headquarters. McDonough designed a prototype ecovillage for the Chinese government. Other cradle-to-cradle clients include the U.S. Postal Service, which has received Silver Cradle to Cradle Certification on its corrugated boxes, paperboard envelopes, and mailing labels and label tape for Priority and Express Mail, among other products.

The Next Industrial Revolution

The principles of cradle-to-cradle design are not limited to industrial products and a building's architecture, but can be applied to the organization of entire metropolitan regions, social systems, and economies. The triple bottom line of sustainability—people, planet, and profit—requires commitment to significant changes. McDonough and others argue that the cradle-to-cradle framework offers just that and thus will lead to the next industrial revolution. This transition will be defined by the transition from petroleum-based, nonrenewable resources to solar-sourced renewable energy, including thermal-generated wind energy. This "next industrial revolution" will see a dramatic reconsideration of the relationship between people and their environment that optimizes nature's resources and actively engages an ecologically friendly vision of society.

Criticisms

Criticisms of C2C focus primarily on the proprietary nature of the trademarked term. The exclusivity that is created by private certification has, it is argued, meant that few companies are able to pay the costs of certification and thus few products have been able to meet the standards of C2C. McDonough and Braungart have been encouraged to join in public/private partnerships, notably with the Dutch government, to address this critique but have not entered into any such agreements. Others encourage the owners to make cradle to cradle open source, or available to all at no cost. Friedrich Schmidt-Bleek of the Wuppertal Institute doubts that the model could be realized on a large scale.

See Also: Green Markets; Product Stewardship; Sustainable Design; Technological Utopias.

Further Readings

McDonough, William and Michael Braungart. *Cradle to Cradle: Remaking the Way We Make Things*. New York: North Point Press, 2002.
McDonough Braungart Design Chemistry (MBDC). "Cradle to Cradle Framework." http://www.mbdc.com/detail.aspx?linkid=1&sublink=6 (Accessed September 2010).
Sacks, Danielle. "Green Guru William McDonough Must Change, Demand His Biggest Fans." *Fast Company* (February 26, 2009). http://www.fastcompany.com/blog/danielle-sacks/ad-verse-effect/william-mcdonough-must-change (Accessed September 2010).

Shannon Tyman
University of Washington

CRYSTALLINE SILICON SOLAR PHOTOVOLTAIC CELL

According to the British Petroleum (BP) "Statistical Review of World Energy" reported in 2008, the entire world consumed approximately 15 TW equivalent of power. As of 2005, 85 percent of the world's total energy was being produced from nonrenewable and environmentally unfriendly sources, such as fossil fuel (37 percent), natural gas (23 percent), and coal (25 percent). The total energy consumed in the United States in 2009, according to the U.S. Energy Information Administration, was 35.3 percent, 23.4 percent, and 19.7 percent from fossil fuel, natural gas, and coal, respectively. However, a completely free and renewable source of energy, the sun, provides about 100,000 TW of power to the Earth, which is approximately 6,700 times greater than the world's present rate of power consumption. Photovoltaic (PV) cells are used to tap into this huge and environmentally friendly resource. Since 1954, modern-era PV cell technology, using a diffused silicon p-n junction, has been developed with the goal of making greater use of this infinite resource.

The basic working theory of PV technology—converting solar energy into electricity—is a simple process. Photons in sunlight are absorbed by silicon in the solar panel. During this process, electrons are knocked loose from the atoms as described below. These freed-up electrons flow in a single direction through the semiconducting material, due to the special composition of the solar cells, to produce electricity. Direct current (DC) electricity is produced by an array of solar cells due to solar energy conversion. In a conventional p-n junction, two PV cell layers are stacked on top of each other. The bottom layer is doped with a p-type material such as aluminum, gallium, or indium to produce holes, and the n-type layer is doped with phosphorous, arsenic, or antimony to create mobile electrons (Figure 1). In light phase, electrons become excited and dislodged when light quanta penetrate in to the bottom p-layer and holes are produced. With the correct photovoltaic circuit setup, the freed-up electrons move to the top n-layer through the external circuit. The charge on the top mobile carrier is negative while the bottom layer is positive. When the two materials make contact, diffusion of electrons occurs from the region of high electron concentration (the n-type) into the region of low electron concentration (p type). This spontaneous transfer of negative and positive charges across the junction produces an excess positive charge on the side of the n-doped silicon (Figure 1), also known as

Shown are several different types of photovoltaic cells. Four types of photovoltaic cells are available for commercial use, all of which absorb photons to create energy.

Source: iStockphoto

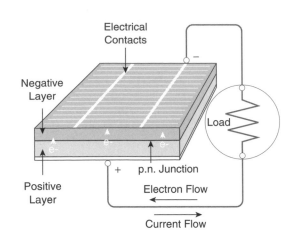

Electrical Contacts

Negative Layer

Load

− +

p.n. Junction

Positive Layer

Electron Flow

Current Flow

Figure 1   Schematic of a Conventional Solid State (p-n Junction PV Cell) Photovoltaic Device Working Process

Source: Polar Power Inc., 2010.

"dopant," and an excess negative charge on the opposite p-doped (Figure 1) side, known as "wafer." Thus, it results in the creation of an electric field. A net photocurrent is generated by the collection of the photo-induced carriers (Figure 1) through the help of the electric field.

In general, four types of PV cells are available commercially. They are single-crystal silicon, polycrystalline silicon (multicrystal silicon), ribbon silicon, and amorphous (or thin film) silicon (aSi). Typical crystalline silicon solar cells are quadruple cells, and typical thin film solar cells are narrow, long cells. Single crystalline silicon PV cells are the most common, in which the crystal lattice of the entire sample is continuous and unbroken with no grain boundaries. They are made by purifying, melting, and then crystallizing the silicon into ingots. The ingots are then sliced into thin wafers to make uniform individual cells, blue or black in color. Most single crystalline silicon solar cells produce about 1.5 volts, regardless of the surface area of the cell, but more current is produced with larger cells.

Multicrystal or polycrystalline silicon (PS) is a material that consists of small silicon crystals or crystallites. PS cells can be recognized by their visible grain or a metal flake effect. Present-day solar panels use PS as a key component because it is cheaper to produce and is more cost effective than single-crystal silicon, although solar power production is less. In 2008, PS cells accounted for 80–85 percent of the total PV cells manufactured worldwide. Intrinsic and doped PS is used on a large scale in the electronics industry. Using PS as the active and/or doped layers in thin film transistors is preferred over plastics. R. W. Birkmire and E. Eser (1997) found that PC thin film solar cells, on copper indium diselenide (CuInSe2) and its alloys and cadmium telluride (CdTe), show potential for large-scale application of solar energy production. This material achieved more than 15 percent laboratory efficiency.

Ribbon silicon (RS) PV cells are a ribbon structure as opposed to an ingot, grown from the molten silicon. These multicrystalline silicon cells are formed by flat thin films of molten silicon. Although the RS cells work in the same way as single silicon and PS cells, they have lower efficiencies. Thin film silicon (TFS) or aSi PV cells do not have the distinct crystal structure that the previous three types have. Very thin layers of vaporized silicon are deposited, while in a vacuum, onto a base of glass, plastic, or metals to produce TFS units. The efficiency of aSi PV modules is poor and often below 50 percent of the previously described three types. However, due to a shortage of the other three types of PV cells, TFS usage has increased to 15–20 percent of world production.

PV cells are often connected electrically in series and summed up as a module for increased voltage. A sheet of glass covers the sun-facing top of the PV module, which allows light to pass through but simultaneously protects the semiconductor wafers from rain, snow, hail, and other destructive elements. Also, many PV modules are connected to one another for increased current. An antireflective coating, typically silicon nitride, is often painted in the solar modules to increase the amount of light absorbed by the cells.

To ascertain the efficiency of PV technology, comparisons are made between production costs and energy payback time. The energy payback period means the number of years it would take producing electricity in this manner for the savings to equal the energy needed to produce the cells/module itself. TFS has proven efficiency in both areas. Yet solar energy generated this way is considerably more expensive than energy from other sources. According to M. Graetzel (2007), the present cost of solar power per kilowatt hour (kWh) is $0.25–$0.65 as opposed to $0.05 per kWh for current wholesale electricity. Again, the shortage of raw materials to produce crystalline silicon PV cells is still a big hurdle to overcome. However, new silicon technology is evolving with the goal of reducing the cost of solar energy production. More PV-grade silicon is being produced to meet demand and make solar power energy competitive. Graetzel (2007) proposed that if just 0.1 percent of the Earth's surface was covered with PV devices, with a conversion efficiency of 10 percent, that there would be no need for electricity from any other source.

See Also: Cadmium Telluride Photovoltaic Thin Film; Clean Energy; Copper Indium (Gallium) Selenide (CIGS or CIS) Solar Photovoltaic Thin Film; Passive Solar; Solar Cells.

Further Readings

Birkmire, R. W. and E. Eser. "Polycrystalline Thin Film Solar Cells: Present Status and Future Potential. Annual Review of Materials." *Science*, 27 (1997).
BP. "Statistical Review of World Energy 2010." http://www.bp.com/productlanding.do?categoryId=6929&contentId=7044622 (Accessed April 2011).
Graetzel, M. "Photovoltaic and Photoelectrochemical Conversion of Solar Energy." *Philosophical Transactions of the Royal Society A*, 365 (2007).
Hutchinson, K. and P. Holland. "Crystalline Silicon Solar Cell Manufacturing Requires Vacuum-based Solutions." *Solid State Technology*, 51/2 (2008).
Polar Power Inc. "Photovoltaic Electricity." http://www.polarpowerinc.com/info/operation20/operation23.htm (Accessed September 2010).
U.S. Department of Energy, Energy Information Administration. "Annual Energy Review 2009." http://www.eia.doe.gov/emeu/aer/pdf/aer.pdf (Accessed September 2010).

Sudhanshu Sekhar Panda
Gainesville State College

DEMOCRATIC RATIONALIZATION

Democratic rationalization is a term coined by Canadian philosophy professor Andrew Feenberg to describe the nondeterministic development of technology according to social factors—a notion he put forth in opposition to technological determinism, a popular model of the development of technology. Feenberg's work has been mainly in the philosophy of technology, his views on which he developed primarily in his Critical Theory of Technology trilogy: his 1991 *Critical Theory of Technology* (later revised and republished as *Transforming Technology*), his 1995 *Alternative Modernity*, and his 1999 *Questioning Technology*.

The critical theory of technology builds on the social critique of technology as articulated by Karl Marx, Herbert Marcuse, and Martin Heidegger (all of whom Feenberg has written books about), amplified with science and technology case studies. Feenberg describes technology as one of the most significant sources of public power, and keeping its development out of the hands of the people as inherently undemocratic and unjust. He calls for a "redesigned industrialism" that serves democracy and democratic institutions better than the worker- and consumer-exploiting industrialism that developed out of the 18th century.

"What does it mean to democratize technology?" Feenberg asks in the first book of the trilogy. He explains the following:

> The problem is not primarily one of legal rights but of initiative and participation. Legal forms may eventually routinize claims that are asserted informally at first, but the forms will remain hollow unless they emerge from the experience and needs of individuals resisting a specifically technological hegemony. That resistance takes many forms, from union struggles over health and safety in nuclear power plants to community struggles over toxic waste disposal to political demands for regulation of reproductive technologies. These movements alert us to the need to take technological externalities into account and demand design changes responsive to the enlarged context revealed in that accounting. Controversy becomes a tool for bringing about technological change that is more democratic.

Feenberg uses as a key example the history of acquired immune deficiency syndrome (AIDS) treatments and the ability of AIDS patients to mobilize movements to increase their access to treatment because "the networks of contagion in which they were caught were paralleled by social networks that were already mobilized around gay rights at the time the disease was first diagnosed." The business and activity of medicine are principally technical, the industry organized technocratically, and patients are treated as individual objects of technical work. However, because no cure for AIDS exists, medicine can best serve AIDS patients through clinical research. Access to experimental treatments was highly restricted because of the industry's concerns for patient safety, but AIDS patients successfully forced the issue and won for themselves a greater right to dictate their own medical futures and choose their risks, expanding access to experimental research as well as speeding up the process of making treatments available. Feenberg refers to this as "hack[ing] the medical system."

Democratic rationalization is not specifically socialist, which Feenberg describes as "an old idea," but shares its concern for workers' lives, dignity, and "the larger contexts modern technology ignores." Ultimately, democratic rationalization is the desired alternative to an authoritarian system of management that chooses technological options according to the whims and methodologies of the powerful few. Further, this is not simply about technology: in Feenberg's view, echoing Heidegger, history demonstrates that a fundamentally different form of modern civilization cannot be brought about without fundamentally different technology, and seizing control of technology is thus not only politically expedient but politically necessary.

See Also: Best Available Technology; Best Practicable Technology; Technological Determinism; Technology and Social Change.

Further Readings

Feenberg, Andrew. *Alternative Modernity.* Berkeley: University of California Press, 1995.
Feenberg, Andrew. *Questioning Technology.* New York: Taylor & Francis, 2002.
Feenberg, Andrew. *Transforming Technology: A Critical Theory Revisited.* New York: Oxford University Press, 2002.
Kaplan, David M., ed. *Readings in the Philosophy of Technology.* Lanham, MD: Rowman & Littlefield, 2004.
Scharff, Robert C. and Val Dusek, eds. *Philosophy of Technology.* Hoboken, NJ: Wiley-Blackwell, 2003.
Veak, Tyler J., ed. *Democratizing Technology: Andrew Feenberg's Critical Theory of Technology.* Albany: State University of New York Press, 2006.

Bill Kte'pi
Independent Scholar

Desalination Plants

Desalination plants remove salts and minerals from seawater, brackish water, or treated wastewater, typically using thermal or membrane technologies. Seawater is the most important source of desalinated water globally. Proponents of desalination plants argue

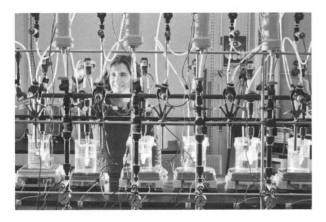

Here, a researcher performs some reverse osmosis (RO) desalination. RO is less energy-intensive than thermal distillation and is the predominant technology for new and planned desalination plants.

Source: Randy Montoya/Sandia National Laboratories, U.S. Department of Energy

that they provide a reliable and secure water supply for areas struggling with water shortages and can alleviate pressure from overstressed and environmentally unsustainable water supplies. Opponents of desalination, on the other hand, cite the negative effects on ecological systems that can result from the intake of seawater and the discharge of brine and by-products by plants, as well as the high energy usage, high costs, and equity issues associated with the technology.

Desalination plants commonly employ thermal distillation or reverse osmosis technology to separate salts from water. Distillation is the oldest method, accounting for the majority of the world's desalination capacity, and consists of evaporating seawater and then condensing the resulting vapors to produce freshwater. The predominant technology for new and planned desalination plants is reverse osmosis (RO). In RO, water is pumped at high pressure through semi permeable membranes that allow freshwater molecules to pass through while barring salt.

Water Security

Desalination of seawater is expanding as an alternative source of freshwater in arid regions of the Middle East, North Africa, Spain, and Australia where the availability of freshwater is limited. Desalinated water supplies less than 1 percent of the total water use in the United States, although interest in seawater desalination has been growing rapidly since the early 1990s, particularly in California and Florida. While seawater is the main source globally for desalination, it accounts for less than 10 percent of U.S. desalination capacity, with most plants being small scale and used to treat brackish water or wastewater. There has been, however, an increase in the construction of new seawater desalination plants in the United States.

Supporters of desalination state that the technology can help ensure reliable water supplies for regions that are struggling with scarcity concerns as a result of population growth, inadequate regulation of water supply, drought conditions, and ecosystem degradation to current water bodies. Opponents, however, argue that expanding seawater desalination deflects attention away from exhausting more environmentally benign options, such as improving water conservation, use efficiency, and recycling practices. Furthermore, concerns have been expressed that desalination will only be used to further increase unsustainable urban, agricultural, and tourism development.

Energy Use and Water Rates

Reverse osmosis desalination is less energy intensive than thermal distillation, and improvements to its design have further enhanced its efficiency in terms of cost and energy consumption. The U.S. Bureau of Reclamation estimates a 4 percent efficiency improvement per year for RO technology. Despite advances, desalination technology continues to be a highly energy-intensive and expensive procedure for providing potable water compared to other supply sources. The high energy consumption of desalination plants can lead to an increase in greenhouse gas emissions and the worsening of climate change. Furthermore, the public can assume some of the high costs associated with designing and operating seawater desalination facilities through subsidies or increased water rates, which can have impacts on equitable access to water. Proponents argue that having a local and drought-proof water supply is worth the higher cost of desalinated water.

To date, using renewable energy in large-scale desalination is still in its early stages, with most operating and planned desalination plants relying on coal-based or nuclear-based energy generation. Australia and Israel have been experimenting with large-scale renewable energy desalination plants.

Ecological Issues

Operating desalination plants can have adverse impacts on coastal and marine ecosystems. This includes the killing or harming of marine species, mainly adult fish, invertebrates, birds, and mammals, when they are trapped against intake screens by the velocity and force of water, a process called impingement. Small fish, larvae, plankton, and eggs are killed when they bypass the intake screens, a process called entrainment. As impinged and entrained organisms are disposed of in the marine environment, decomposition can decrease the oxygen content of the water and create additional stresses.

Technologies can be employed to minimize impingement and entrainment, such as low intake velocities, allowing some organisms to swim away, and newer screening devices. Desalination plants can use open sea intakes or subsurface intakes such as beach wells, which collect water via slow filtration through sand. While surface water intake systems have been traditionally used, subsurface intakes can avoid the environmental impacts associated with impingement and entrainment. Furthermore, as subsurface intakes provide natural filtering, this approach could reduce dependency on chemical pretreatment. Subsurface designs may, however, negatively impact coastal aquifers during construction, limit water intake flows, and are likely not feasible for a large (e.g., over 50 million gallons per day) desalination plant.

Another ecological issue associated with seawater desalination is the discharge of large volumes of highly concentrated brine back into the ocean. Depending on the efficiency of the desalination process, every 100 gallons of seawater produces 15 to 50 gallons of potable water and discharges 50 to 85 gallons of salt-enriched effluent. This waste stream contains up to twice as much salt as the feed water, high concentrations of seawater constituents such as manganese, lead, and iodine, and heavy metals that may accumulate as a result of the corrosion of desalination equipment. Discharge may also contain chemicals that are used during the desalination process to treat impurities in the water and prevent membrane fouling (which results from the growth of microorganisms). The impact of brine discharge on coastal and marine ecosystems can be mitigated somewhat by prohibiting disposal in sensitive habitats. Negative impacts from chemicals can be decreased by treatment before discharge and substitution of hazardous substances. For example, ultraviolet light instead of biocides can be used for pretreatment.

Proponents of desalination argue that a properly designed desalination plant may be environmentally advantageous if it is used to replenish or reduce water demand on other water bodies, such as stressed river ecosystems and groundwater. The effects of desalination on marine ecosystems vary according to local environmental conditions and plant design. Research on the ecological effects of desalination is limited, and without proper assessment, desalination could interact with pollution and overfishing to add substantially to the negative impacts on species biodiversity and water quality.

Privatization of Plants

In the United States, many of the proposed or operating desalination plants are privately owned. Arguments have been made that if desalination facilities remain public, water use could be more efficiently incorporated into government water management plans and thereby be better protected. It is also unclear how current or future versions of international trade agreements, such as the North American Free Trade Agreement, could impact water decisions if international companies have significant involvement in the consumptive use and provision of water. Concerns have been expressed that multinational corporations could use agreements to circumvent local, state, and federal regulations if they act as barriers to trade.

See Also: Environmental Science; Greywater; Membrane Technology; Water Purification.

Further Readings

Cooley, H., P. H. Gleick, and G. Wolff. *Desalination, With a Grain of Salt—A California Perspective*. Oakland, CA: Pacific Institute, 2006.
Pankratz, T. *An Overview of Seawater Intake Facilities for Seawater Desalination: The Future of Desalination in Texas, Vol. 2: Biennial Report on Seawater Desalination*. Austin: Texas Water Development Board, 2004.

Jennie Liss Ohayon
University of California, Santa Cruz

DESIGN FOR RECYCLING

Design for recycling would, at first sight, seem to be an unproblematic idea, even if its specific content were in need of elaboration. It implies that manufactured objects are designed in such a manner that once they finish their period of active use and have "worn out," they can either wholly or in terms of their constituent components be remade into other useful products. However, both of the terms *design* and *recycling* can be defined in multiple ways. Depending on which definition of each is adopted, the idea of what constitutes "design for recycling" becomes a much broader one. Added to this are questions concerning the durability of products, what constitutes their obsolescence, the nature of the social networks into which designed products are embedded, questions of who does the recycling, how and where, and so on. In short, design for recycling is not a neutral technical issue but a complex and contestable social one, albeit one that intricately connects with overlapping technical considerations.

Recycling, however defined, appears in and of itself to be a good thing. It extends the life of durable objects and components, it lessens the impact of production of new objects on scarce natural resources, especially energy, and it reduces the amount of postconsumer waste that has to be addressed at household, neighborhood, national, and global scales. For example, keeping material out of landfill sites has the added benefit of reducing the amount of methane produced via decomposition within those sites. Recycling—narrowly defined in terms of products that may be reconstituted, remanufactured, or reconfigured into new products—can, however, be usefully supplemented by other forms of product revalorizing, including reuse, repair, and functional reassignment. Moreover, many practices and circuits of product "disposition" have been identified, including giving, lending, sharing, and reselling, all of which offer opportunities for extended use of products whether they have or have not been recycled in some manner.

Despite what may be termed these *natural* or *organic* opportunities to extend the life of useful objects and to socially recycle them in various ways, the number of discarded, defunct, obsolete, and broken objects proliferates as does the total amount of postconsumer waste of which they form a significant portion. The success of ideological initiatives aimed at reducing this waste and wasting of useful goods is contestable; more resources are devoted to educational and other forms of persuasion while concomitantly the levels of waste and discarded products rise. This is often despite increasing success in getting households and businesses formally to engage in often and increasingly obligatory recycling practices such as the sorting of domestic waste into identifiable repositories. Consequently, various reformers, activists, and commentators have stressed two things: first, that it is within the political economy of the production, use, and disposal of goods that the most potent opportunities for successful recycling lie, namely, that the same economic incentives that produce the "wasteful" products can be levered to address the problem of recycling. Second, it is argued that "behavior-steering" initiatives, rather than ideological ones, provide a surer footing for the expansion of attitudes, behaviors, and dispositions that will support a significant growth in recycling. Such initiatives include most prominently legal obligations to treat one's discarded, unwanted, or broken products in particular ways, for example, the requirement to pay for the removal of various consumer goods from the home by specialist services. In such case, the convergence of the two approaches is apparent. However, behavior, it is increasingly argued, can be altered through design, that is, that design can instantiate in products various material characteristics or "affordances" such that conscious attention must be paid to issues such as repair, reuse, and recycling as part of the bond of ownership between product and consumer.

Designers have, then, become a vocal presence at least in terms of the discourses concerning recycling, even if many of their ideas have been only partially realized in the world of consumer goods themselves. However, before considering design for recycling in more detail, a couple of general points remain significant. First, not all consumer goods can be designed to be recycled or even reused except in limited ways. There will always be a "material remainder" of waste matter that cannot be recycled, items such as tea bags, uninvited detritus such as junk mail, objects beyond repair, worn-out clothes, and so on. While much is made of processes such as home composting of organic materials, and of the creative scavenging and vernacular remaking of various goods, their overall impact of waste reduction remains small. It is the industrial system of production that produces the goods that may or may not lend themselves to recycling and the effect of ameliorative vernacular practices at the domestic level will doubtless continue to pale in comparison. Indeed, for many, such practices are neither possible nor meaningful.

People have also become accustomed to an economy of consumer goods comprising many that have been specifically designed to be disposed of, such as various forms of packaging, disposable diapers and razors, and so on. The convenience and time-saving afforded the consumer is a factor that must be borne when advocating practices that may be more geared toward recycling from the designer's viewpoint but that represent for the consumer more work, more hassle, more engaged involvement in attending to the materials of consumption, and so on. Second, and related to the last point, when societies are materially poorer in terms of incomes and ownership of consumer goods, yet richer in the availability of free time away from work, they tend to recycle far more. For example, the 19th-century slop pail and grease pots redolent of much of American society gave way in the 20th century to the waste disposal systems and casual binning. In the first instance, such food scraps were seen as valuable components for subsequent meals, as raw materials for new domestically produced goods such as candles, and as feed for animals. During the mid-20th century, the same food waste was widely regarded as the justifiably disposable remnant of an abundant and convenient consumer culture of food production. It is certainly the case that the societies that currently recycle, repair, and reuse most intensively are both population dense and poor by Western, affluent standards.

Design for Disassembly

Recycling, reuse, repair, and so on, depend on goods that can be recycled, reused, and repaired. Consumer societies have developed sophisticated consumer goods economies that work against this in many ways, and it is here that designers have been vociferous both in critique and in the advocating of alternatives. One series of design interventions that holds much promise comes together under the rubric of design for disassembly (DfD). This approach advocates the use of recycled materials and reused components, design that facilitates ease of disassembly and a product's recycling at the end of its useful life. It is argued that reasonable-quality components can be refurbished or reused, that metallic parts can be easily separated, so increasing their recycling value, that disassembled plastic parts can be easily removed and recycled, and that parts made from glass or hazardous materials can be easily separated and reprocessed.

Consequently, proponents of DfD recommend that various features be integral to finished manufactured goods. These include that all parts should be clearly marked and easily removed, that the number of joints, attachments, and fasteners be minimized, that the number of tools needed for disassembly be minimized and standardized, that the highest-value parts be the most easy to locate and remove, and so on. In relation to products that are covered legally by a producer's responsibility for their disposal, design for disassembly is attractive and is often undertaken by either the original manufacturer of the product or by a company having a contractual arrangement with the manufacturer. However, elements of this approach can easily be seen to be adaptable to and adoptable at the household level, an extension of domestic refuse sorting, for example. An extension of this kind of design logic, one that recognizes many consumer goods as assemblies of discrete, and often valuable, components, could be the increasing adoption of the requirement for home assembly through the promotion of kits. It is argued that assembling domestic products from kit form would lessen consumption as our appreciation and familiarity with our products would increase through the act of self-assembly. Overall consumption would also diminish, it is claimed, as familiarity with modular kit-form products would allow minor home repair as well as encouraging safer and more enduring use of these products.

Kit buying is already a common feature of products such as flat-packed furniture, which requires some self-assembly.

In terms of design, such strategies would depend on the adoption of conventions at odds with those that currently predominate. First, widespread standardization of components, fastenings, tools, and such like would need to be adopted. Currently, it is often the marked lack of compatibility between similar objects and their components (often "competing" brands actually produced by the same manufacturer) that is notable. Second, manufacturers would need to move to a system of goods that can be opened, resealed (without loss of functional efficiency), repaired, and serviced by consumers themselves. Currently, this is often prohibited, especially in relation to domestic goods containing electronic components, arguably among those products that most lend themselves to the potential of home assembly and disassembly. Manufacturers benefit from this by offering consumers various warranties and extended guarantees that cover the maintenance, servicing, repair, and sometimes disposal of the goods in question. The monopolizing of these services effectively works against design strategies aiming to increase the recycling potential of products.

Recycling and Durability

This somewhat gets to the heart of the problem in relation to design for recycling: that of durability. Recycling seems by definition an answer to a consumer culture of goods that lack durability. Surely, products would not need recycling if they lasted long enough? Manufacturers would argue that they attempt to extend the durability of goods via the provision of these various services while critics would argue that this is a cover for an increasing lack of material durability in favor of attempting to engineer a lasting relationship between the producer and the consumer. Consequently, designers have themselves offered visions of product-service webs and networks through which consumers would either lease a consumer good such as a washing machine and along with it the various services that would guarantee its durability—repairs, replacement, cleaning, and eventual disposal—or would lease a service, for example in this case, clean clothes, rather than an object at all.

In this way the basic question of durability tends to recede from focus as it does when other design strategies that aim to tie consumers closer to their products are considered. These have generally been of two kinds: first, using design to try to produce extended emotional attachments between consumers and their goods. Variations of form, often organic in nature, color, texture, materials, and other elements of product "semantics" have been suggested as a basis for generating more durable emotional attachments. Second, designers have looked at ways in which required user engagement is necessary in order to make the product function effectively. Necessitating that the user of a good has to engage in some act of care in relation to his or her product is, it is argued, a means of ensuring greater attention, care, and attachment to it. Again, the question of material durability is effaced.

The most visible manifestation of the concept of design for recycling thus remains that of products cleverly designed and manufactured such that they can wholly, or some substantial proportion of their components, be recycled into new recognizable products. One well-known example is that of the recycling of plastic bottles into fleece garments. Admittedly, there is not much "design" here, rather recognition of the properties of the materials in question. This is true also for ideas such as "downcycling" in which, for example, glass can be mixed with other materials to make road surfacing. Designers can focus on the manufacture of goods from materials for which recycling is shown to have a smaller environmental impact, in terms of greenhouse gas emissions, than incineration; for example, paper, cardboard, glass, plastics, aluminum, and steel.

Only so many goods can be manufactured from such materials, however, and in some cases the nature of those very materials may not persuade consumers to return them to the circuits of recycling but to opt instead for other forms of disposition; the home composting of paper and cardboard, the collecting and selling of metals. There has to be a willingness on the part of consumers to engage in the practices of recycling that often seem to have been handed to them by the very same culture that promotes wasteful consumption in the first place. Consequently, incentive schemes have been proposed, including "pay-as-you-throw" or "earn-as-you-recycle," which promise rebates on local taxes for recycling more or fines for not recycling enough.

Designers' ingenuity for proposing recycling innovations remains undiminished: computer printers that use leftover coffee rather than ink cartridges, edible packaging, and so on. However, such solutions tend to appeal to consumers who already have a propensity to reuse or adapt the use of goods to other ends. Consequently, adding "affordances" that allow such adaptability can facilitate such reuse and adapted use. In general, such approaches favor low technological content, and those designers who favor the exploration of new materials and high technology innovations have turned much of their attention to the design of devices that harvest and save energy, monitor and adjust environmental conditions, that is, to a host of new products that need to be manufactured and that will be more or less durable and more or less designed for recycling. Design for recycling is a noble, notable, and worthwhile pursuit but the political economy of production and consumption against which it is proposed erects many barriers to its widespread and successful implementation.

See Also: Cradle-to-Cradle Design; Extended Producer Responsibility; Product Stewardship; Recycling; Sustainable Design.

Further Readings

Maycroft, Neil. "Re-Valorizing Rubbish: Some Critical Reflections on 'Green' Product Strategies." *Capital & Class Special Issue: Environmental Politics: Analyses and Alternatives*, 72 (2000).

Slade, Giles. *Made to Break: Technology and Obsolescence in America*. Cambridge, MA: Harvard University Press, 2006.

Walker, Stewart. *Sustainable by Design: Explorations in Theory and Practice*. London: Earthscan, 2006.

Neil Maycroft
University of Lincoln

DISTILLATION

Distillation is a method used to separate liquids into different fractions that have different boiling points or to separate liquids from dissolved solids. It may be used to identify the materials in a compound and also to purify that compound. There are many commercial and environmental applications for distillation, from distilling crude oil into different petroleum products to creating potable water from seawater by removing the salt. The simplest methods of distillation require minimal technology; distillation has been used to separate and purify liquids for over 2,000 years. Distillation also occurs as a natural process and has been used to explain the high concentration of certain pollutants in areas in which they have never been used.

The distillation process is possible because different materials have different boiling points, that is, they become vapor at different temperatures. If a compound liquid is heated in a closed vessel, when a given substance in the compound reaches its boiling point that substance will become gaseous, while the other substances in the compound (those with higher boiling points) will remain in their liquid state. The gaseous substance can be collected and condensed back into its liquid state, and it is now purified of the other elements. Because boiling points differ among elements, recording the boiling point for an element in a compound can also be used to identify that element.

One of the best-known uses for distillation is to remove salt and other impurities from water, for instance, to make seawater drinkable or to remove minerals that could clog machinery (this is the reason distilled water is used in a steam iron) or interfere with a chemical process (this is the reason distilled water is used in car batteries).

A simple example of using distillation to purify a liquid is the operation of a solar still, which uses the heat of the sun to cause water to evaporate and condense on a surface, then collects the condensation (now a liquid again) in a vessel. This method can be used to purify drinking water from seawater or otherwise contaminated water (although technically speaking, the simplest solar stills use evaporation rather than distillation, and because they do not boil the water, they may not remove all the contaminants). Purifying water through a solar still is a slow process but is still in use in some parts of the world because it requires only the use of solar energy, which is free and inexhaustible. Solar stills are also used to provide drinking water in emergency situations: for instance, people hiking in the desert often carry the components of a simple solar still to produce an emergency water supply should they become stranded.

More advanced technological means of desalination through distillation were developed in the 20th century: the world's first land-based multiple effect distillation (MED) desalination plant was built in 1928 in the Netherlands Antilles. Further developments in the 1950s, applying means developed for industrial uses and for shipboard distillation plants developed during World War II, have resulted in much more efficient methods of using distillation to produce desalinated water. Modern methods of distillation require a source of energy that may be renewable (e.g., wind power or solar-generated electricity) or traditional (e.g., burning coal) and generally also use technologies that reduce the boiling point of water by reducing vapor pressure within the distillation unit, thus limiting the amount of fuel required for boiling.

Because of the large amounts of energy required to produce potable water through the desalination process, it is an expensive way to obtain drinking water and plants. The U.S. Geological Survey estimates that producing drinking water by desalinating seawater costs $1,000–$2,200 per acre-foot as opposed to the $200 per acre-foot cost of obtaining water from conventional sources. As of 2002, there were about 12,500 desalination plants around the world, primarily in the Middle East and North Africa, producing about 14 cubic meters of freshwater per day (less than 1 percent of the total global consumption of freshwater). However, due to improved technologies that can lower costs and reduce environmental effects, spending on desalination plants is expected to increased in the coming years, with substantial investments in Saudi Arabia, the United Arab Emirates, the United States, China, and Israel.

The term *global distillation* (also called "the grasshopper effect") refers to a natural process that results in high concentration of certain chemicals such as PCBs (polychlorinated biphenyls) in cold regions of the Earth, thousands of miles from where they were actually used. The mechanism for global distillation is the same as for human-organized distillation: a substance is heated until it vaporizes and is released into the air, where it is carried by air currents to a colder region, where it condenses. One major area of concern is the global distillation of pesticides and other pollutants that can also evaporate out of the soil where they have been used and travel in the atmosphere to colder climates far from the original point of application. For instance, Inuit people living Arctic regions have some of the highest levels of toxic chemicals in their bodies despite the fact that such chemicals have never been used in the regions where they live.

See Also: Environmental Science; Green Chemistry; Science and Technology Studies; Water Purification.

Further Readings

Organization of American States. "Desalination by Distillation." http://www.oas.org/DSD/publications/Unit/oea59e/ch21.htm (Accessed September 2010).

Stichlmair, Johann and James R. Fair. *Distillation: Principles and Practices*. Weinheim, Germany: Wiley-VCH, 1998.

University of Arizona. Sustainability of Semi-Arid Hydrology and Riparian Areas. "Solar Stills @ UA." http://www.sahra.arizona.edu/education2/solarstill (Accessed September 2010).

University of Guelph. "The Grasshopper Effect and Tracking Hazardous Air Pollutants." http://www.arctic.uoguelph.ca/cpe/arcticnews/articles/Grasshopper/Grasshopper.htm (Accessed September 2010).

U.S. Geological Survey. "Thirsty? How 'Bout a Cool, Refreshing Cup of Seawater?" http://ga.water.usgs.gov/edu/drinkseawater.html (Accessed September 2010).

Yates, Roger, T. Woton, and J. T. Tlhage. *Solar-Powered Desalination: A Case Study From Botswana*. Ottawa, Canada: International Development Research Centre, 1990.

Sarah Boslaugh
Washington University in St. Louis

E

EARTHSHIPS

This Earthship at Stanmer Park in Brighton, United Kingdom, exemplifies several standard features of Earthships: the rear walls are burrowed into the hillside, the roof harvests rainwater, and there are solar panels to create energy.

Source: Wikimedia

Designed to promote sustainability, Earthships were the brainchild of New Mexican architect Michael Reynolds, who during the energy crisis of the 1970s came up with the idea of creating environmentally friendly structures that had no need to draw on nonrenewable resources to support modern living. Instead, Reynolds's buildings used wind turbines, solar panels, and biodiesel generators to generate energy for heating and cooling. Each Earthship roof consisted of a giant cistern designed to catch rainwater and transfer it to a Water Organization Module (WOM) that purified water for drinking and transported wastewater to plants scattered both inside and outside the Earthship. These plants were essential to achieving Reynolds's goal of creating structures that looked as if they grew naturally in their environments. In order to achieve maximum sustainability, the rear walls of Earthships were either covered by dirt, or they burrowed into hillsides to promote the generation of passive energy. The structures generally faced south to absorb maximum heat, and these slanted southern walls were usually made of glass. This material was rarely used elsewhere in the structure. Inside, walls were made of brick-like materials consisting of old tires filled with dirt that had been hammered in with sledge hammers. Open spaces were filled with recycled aluminum cans, and the entire surface was covered by adobe, plaster, or stucco.

Because Earthships are generally built below the frost line, the temperature of the internal mass walls is naturally maintained at 59–60 degrees Fahrenheit, regardless of the weather outside. In Reynolds's home in Taos, New Mexico, with an elevation of 7,000 feet and typical winter temperatures of –30 degrees and summer temperatures of over 100 degrees, internal temperatures of Earthships range from 65 to 75 degrees Fahrenheit without using outside resources. Heat absorbed by the walls during the day is retained for hours after the sun sets, and the design of the walls ensures that they will release heat whenever internal temperatures drop. Skylights and windows add to the self-sufficiency of Earthships. Outside walls are often constructed of two rows of recycled aluminum cans separated by insulated air space and then covered with materials such as adobe.

In the early 1990s, Reynolds received a good deal of attention after the press learned that he had built an Earthship for actor Dennis Weaver, known chiefly for the TV series *Gunsmoke* and *McCloud*. Much was made of the fact that Reynolds's "bizarre" house plan included the use of discarded tires. Undaunted, Reynolds continued to promote sustainability by building Earthships. By the end of the decade, there were 20 contractors in North America building Earthships at a cost of $160,000 to $180,000 each in New Mexico, Montana, Colorado, California, Florida, and Ohio. Do-it-yourselfers were able to erect Earthships for less than $100,000 by studying with Reynolds's firm, Solar Survival Biotecture. Beginning in 1996, Reynolds built three Earthships in Taos: Lemuria, known locally as the "Gravel Pit"; Reach, in the Sangre de Cristo Mountains; and Star, located in the nearby desert.

Also in the 1990s, Canadian Earthship builders and environmental activists Pat and Chuck Potter set out to adapt Reynolds's design for harsh Canadian winters after studying with him in Taos. Some 37 acres near Bancroft, Ontario, were selected for their Earthship, which was to become the model for other North American projects. Because they had witnessed the devastation resulting from massive fire, one advantage to the Earthship, in the Potters' view, was that fire inside an Earthship was impossible because the tires used in the internal walls contained no oxygen. Major selling points included the Earthship's self-sufficiency and its low operational costs. Because Canadian winters are much harsher than those in New Mexico, and due to the fact that the area receives a good deal of rain, the Potters added a vapor barrier between the walls and the floor and used total outer wall insulation. It was also necessary to insulate the roof to prevent heat loss during the cold winter months. Special features included toilets that reduced waste to ash, a solar hot water tank, an insulated cold box that negated the need for a refrigerator, and a wood stove.

By 2006, about 3,000 Earthships existed, but most of these were still located in the United States. There were only two Earthships in Great Britain, one in Scotland and one in Brighton, England. That year, the Low Carbon Network convinced the Brighton and Hove City Council that Earthships could be used to save money as well as to protect the environment, and Earthship Biotecture drew up plans for building the first utility bill–free housing project in Great Britain in Brighton. Initially consisting of 16 two- and three-bedroom units, the Earthships were designed to use wind turbines and solar panels to generate sufficient energy for cooling and heating all homes. Rainwater via the rooftop system was intended to provide water for domestic use, and septic tanks were included to handle waste. Inside the homes, some 15,000 earth-filled scrap tires were used as walls. The project was completed in late 2007.

Many environmentalists argue that Earthships are the answer to sustainable building in the 21st century. Reynolds's designs have been used around the world, ranging from the rebuilding of homes in Asia that had been destroyed by hurricanes to the erection of million-dollar Earthships for the environmental-minded homebuilder.

See Also: Green Building Materials; Green Roofing; Participatory Technology Development; Recycling.

Further Readings

Allenby, Braden R. *Reconstructing Earth: Technology and Environment in the Age of Humans*. Washington, DC: Island Press, 2005.

Buchan, Jane. "Earthship North." *Natural Life*, 51 (September/October 1996).

DeGregori, Thomas R. *The Environment, Our Natural Resources, and Modern Technology*. Ames: Iowa State Press, 2002.

Foster, John Bellamy. *The Ecological Revolution: Making Peace With the Planet*. New York: Monthly Review Press, 2009.

Knipe, Tim. "Mass Appeal." *Mother Earth News*, 128 (October/November 1991).

Levy, Marc and Will Nixon. "At Home in an Earthship." *E: The Environmental Magazine*, 6/1 (January/February 1995).

Reynolds, Michael. *Comfort in Any Climate*. Taos, New Mexico: Solar Survival Press, 2000.

Schor, Juliet B. and Betsy Taylor. *Sustainable Planet: Solutions for the Twenty-First Century*. Boston, MA: Beacon Press, 2002.

Torgerson, Douglas. *The Promise of Green Politics: Environmentalism and the Public Sphere*. Durham, NC: Duke University Press, 1999.

Elizabeth Rholetter Purdy
Independent Scholar

ECO-ELECTRONICS

Eco-electronics, or green electronics, can be defined as machines that are designed to combine low energy consumption with minimal environmental impact. Sustained pressure from environmental nongovernmental organizations (NGOs), the possibility of expensive litigation, and the growing green market have contributed to encouraging innovation in the electronic sector.

Defining Eco-Electronics

Using a narrow technical definition, eco-electronics involves developing new ways to create household power (through solar power and fuel cell systems), storing energy through new lithium-ion battery technologies, smart energy management systems to prevent the waste of clean energy in homes, and through ecocentric manufacturing and product procurement practices, including developing stricter policies on the practices of partners and materials suppliers. Eco-electronics also incorporates designs and features material and manufacturing processes that consume less energy and attempt to use renewable and natural materials.

For example, using the Solio Hybrid 1000 portable solar charger, an hour of sunshine can be converted into 15 minutes of cell phone talk time or 40 minutes use of an MP3 player. In a broader definition, eco-electronics also includes initiatives taken by companies to minimize the impact of electronics on the environment. In this approach, eco-innovation is best seen as a series of initiatives that take place at the design stage and continue across products, processes, organizations, and institutions. Thus, eco-electronics can be seen at three levels: (1) individual design projects that improve upon traditional methods of manufacturing to make them more environmentally friendly, (2) broadening this approach to incorporate an ecodesign involvement of the entire company that is epitomized by a product-oriented environmental management system, and finally, (3) the life-cycle perspective that not only considers the useful life of a product, but also actively addresses how to dispose of it. Thus, in this definition, eco-electronics is not a matter of technology alone, but also includes institutional initiatives. This involves establishing separate environmental divisions in companies and targeted research and development efforts, collaboration with other sectors of the industry, and corroborative research networks.

Eco-Practices in the Electronics Industry

In the electronics sector, this is evident in the focus given to the new state of electronic products, especially in the use phase, where the maximum amount of energy is consumed. Thus, eco-electronics focuses on two aspects: (1) an attempt to achieve low energy consumption through product modification and redesign and (2) to engage in institutional arrangements that enhance product recycling. One example of this is the Managed Print Services (MPS) system introduced by Xerox, an American manufacturer of printing supplies. The focus of this business model is an enterprise-wide management service that tries to cut costs by minimizing the energy consumption of printing devices, providing for the ultimate maintenance of the equipment, and trying to minimize paper consumption and printing costs by the introduction of the pay-per-use scheme.

Japanese manufacturer Sharp shows another example of eco-initiatives in developing recycling technology, both at the design stage and through institutional initiatives. The company took the step of providing for the safe removal of mercury backlights in liquid crystal display (LCD) TVs. In 2009, Sharp launched the operation of a safe and efficient flat-panel TV dissembling and recycling line. The company has also been active in the recovery and recycling of plastics. In 1999, Sharp developed the technology for close looped material recycling, an initiative that put plastics recovered from televisions, air conditioners, refrigerators, and washing machines back into use. In 2008, the use of recycled plastics reached about 1,050 tons, which was a 100 percent increase from 2005. Other initiatives included a proprietary technology for the recovery and recycling of indium, a rare metal that is contained in the transparent electrodes of LCD panels.

Pressure From NGOs, Government Policies, and the Public

Publications, such as Greenpeace's *Guide to Greener Electronics*, first published in 2006, have put pressure on companies to adapt green technology. The guide uses two criteria to rank companies: (1) the absence of hazardous materials used in manufacturing, and (2) the ease of recycling obsolete electronics. Companies are ranked on their efforts to eliminate polyvinyl chloride (PVC) and brominated flame retardants (BFR) from electronic products. Rather than create waste that must be disposed of, eco-electronics tries to foresee and solve problems at the design stage. Current methods to dispose of e-waste leave much to be desired. Old computers and mobile phones are often dumped in landfills or burned in smelters, while many are exported (often illegally) from Europe, the United States, Japan, and other industrialized countries to Asia, where workers at scrap yards (including children) are exposed to a cocktail of toxic chemicals and poisons.

NGOs have been active in raising public awareness of the issue, such as the Silicon Valley Toxics Coalitions (SVTC) report "Toxic Sweatshops," which details the deplorable health and safety conditions within electronics recycling factories run by UNICOR, also known as Federal Prison Industries. The NGO also has partners in Asia and Africa that try to improve conditions in recycling plants, and is a founding member of the Electronics TakeBack Coalition—a national coalition of community/advocacy groups that work with recyclers, policymakers, consumers, and responsible businesses to try to force the electronics industry to take full responsibility for the full life cycle of their products.

For example, the average lifespan of computers in developed countries dropped from six years in 1997 to just two years in 2005, while electronic production has increased dramatically. Electronics impact the environment in two major ways—through their consumption of power, and through problems with their safe disposal, as they contain toxic substances like lead, mercury, cadmium, and chromium that leak into landfills and eventually seep into the environment. According to some studies, although consumer electronics constitute from 1 to 4 percent of municipal waste in Europe and the United States, they account for up to 40 percent of the lead in the system. Smaller items, like mobile phones, can be extremely harmful because the cheaper they are to purchase, the more likely they are to be disposed of.

In the United States, for example, of the 2.25 million tons of TVs, cell phones, and computer products in 2003 ready for end-of-life management, only 18 percent were are collected for recycling, while 82 percent went into landfills. This figure increases by an estimated 15 percent every year. More than 4.6 million tons of e-waste ended up in U.S. landfills in 2000. In Asia, where regulations are not stringent, more e-waste piles up. In Hong Kong, according to Greenpeace figures, it is estimated that 10–20 percent of discarded computers end up in landfills. The other often-used alternative, incineration, is even more dangerous. Incineration releases heavy metals like lead, cadmium, and mercury into the air and in the form of ashes. Mercury released into the atmosphere can bioaccumulate in the food chain, particularly in fish—the principal route of exposure for the general public. If the products contain PVC plastic, highly toxic dioxins and furans are also released. Brominated flame retardants generate brominated dioxins and furans when e-waste is burned.

With growing environmental consciousness and decreasing cost differentials, consumers are more likely to choose eco-electronics over traditional ones. Some companies, like HP, Canon, and Apple, which take back equipment and peripherals for recycling, are seen as innovative in their approach to electronics.

Several other factors have pushed the creation and development of eco-electronics, prominent among them being growing consumer awareness of toxins in electronic products, forceful legislation (especially in Europe) for environmentally sensitive products, and a change in the way the environment has been seen by business, particularly the shift from end-of-pipe approaches that traditionally marked technology to the environmentally sensitive designs that began in the early 1990s. However, in spite of legislation and intensive research programs, ecodesign has not become a natural part of the electronics business. Nevertheless, given the growing market for such products (considering the energy saving options, and more legislation being passed, especially in Europe), this is changing.

Electronics also have a great impact on the environment through their energy consumption. In addition to appliances like washing machines, dishwashers, and refrigerators, a major portion of average household consumption comes from small consumer electronics. For example, according to U.S. government figures, the typical household holds around 24 consumer electronic devices, including stereos, clock radios, flat screen TVs, and DVD players. Energy efficiency in the operation of these devices is not only good for the environment, but can also cut the cost of energy bills. This cost savings advantage is a prime driver in the eco-electronics market.

One of the key areas of energy consumption, especially given the dramatic growth of call centers and other back office operations, is power consumption by corporate computer servers. For example, in 2006, data centers were responsible for 1.5 percent of the entire electricity consumption in the United States, and it is estimated that the average data center consumes 40 times more electricity than a conventional office building. The adoption of energy-efficient methods and technologies has become a pressing concern.

The U.S. Environmental Protection Agency (EPA) estimates that energy efficient methods and technologies of power consumption can reduce power consumption of corporate computer servers by an estimated 56 percent. Some companies have taken a proactive approach to cutting power consumption. For example, IBM's Project Big Green invested $1 billion to design a five-step approach that would sharply reduce the energy consumption of data centers and lower carbon dioxide emissions associated with energy use and costs. An example of this was the consolidation of distributed servers, which consume more power than consolidated ones. Thus, as part of Project Big Green, an estimated 3,900 distributed servers were consolidated into 33 System Z servers IBM data centers worldwide. The company estimated that this measure alone saves as much as 119,000 megawatt hours (MWh) of electricity per year, or enough energy to power about 9,000 average U.S. homes for an year.

Another key reason for the rise of eco-electronics is the changing legal climate in regard to the environment. Several international environmental conventions exist, but the European Union (EU) has taken the lead in promoting environmental protection programs. Taking effect in July 2006, the EU's 2003 Restriction of Hazardous Substances Directive (RoHS Directive) prohibited the import of electronics products containing more than specified amounts of lead, cadmium, mercury, hexavalent chromium, and halogenated fire retardants such as polybrominated biphenyls (PBBs) and polybrominated diphenyl ethers (PBDEs) into the EU. Furthermore, the EU's Waste Electrical and Electronic Equipment Directive (WEEE Directive) mandated that, beginning in December 2006, manufacturers must employ high proportions of recycled and reused materials and assess products' environmental impact and recycling performances. The implementation of these EU directives has compelled the electronics industry to incorporate waste disposal considerations into product design and manufacturing. Such directives have had an impact on global markets. For example, the enforcement of lead-free manufacturing means that there are changes in supply chains, raw materials, manufacturing, distribution, and waste disposal operations in a process that has been termed *green supply chain activities*. Japan has also seen the introduction of institutional initiatives to encourage recycling. The country's home appliance recycling law mandates the recycling of TVs, air conditioners, refrigerators, and washing machines. In pursuit of this goal, a consortium of Japanese companies, including Sharp, operate around 190 sites in Japan for dropping off old appliances. In 2005, approximately 1.5 million home appliances manufactured by Sharp were recycled under this initiative.

Standards, such as the International Organization for Standardization (ISO) 14031 Environmental Performance Evaluation (EPE) standards, are now used to assess the environmental impact of electronics. Labeling standards have emerged to certify the cost advantage of electronic appliances. These include the Energy Star label in the United States, which ensures that electronics are 10 to 25 percent more energy efficient than government standards. Another such label is the Electronic Product Environmental Assessment Tool (EPEAT). The current legal climate and growing consumer awareness means that electronics manufacturers need to provide an eco-rating on products sold in the United States. Moreover, EPEAT ratings are broken down by country to identify products sold under different names or with slightly different configurations around the globe. Pressure has also been mounted by international organizations like Greenpeace both with the *Green Electronics Product Survey,* a report that ranks companies based on a broad selection of company practices ranging from the use of certain chemicals, to manufacturing standards, to each company's policies on waste management. Greenpeace also uses direct pressure on companies to adopt more green practices, as for example, a 2010 campaign aimed at forcing computer manufacturer Dell to honor its 2006 promise to eliminate PVC plastic and brominated flame retardants (BFRs) from their products by 2009, a deadline that the company could not meet.

A key element in eco-electronics is a whole life-cycle approach that addresses concerns regarding the extraction of raw materials, their manufacture, operation, and disposal. One of the key reasons behind intelligent design of eco-electronics is that decisions made at the design stage allow producers to optimize use of the product and garner market share at an early stage, thus becoming competitive in a growing market, and more importantly, drastically reducing costs they might incur if products need to be modified and corrected during manufacturing, or even after a product is introduced into the market. In the design phase, there are many ways in which eco-electronics can be developed, ranging from simple checklists to complex methodologies, such as tools that are oriented to life-cycle analysis of the product. Examples of these systems include the environmental priority strategies (EPS) system, life-cycle planning (LCP), Environmentally Responsible Product Assessment (ERPA), and material cycle, energy use and toxic emissions (MET Matrix) approaches.

See Also: Appliances, Energy Efficient; Best Practicable Technology; Cradle-to-Cradle Design; Design for Recycling; Ecological Modernization; European Union Waste Electrical and Electronic Equipment (WEEE) Directive; E-Waste; Extended Producer Responsibility; Innovation; Materials Recovery Facilities; Participatory Technology Development; Product Stewardship; Sustainable Design.

Further Readings

Chiang, Shih-Yuan, et al. "How can Electronics Industries Become Green Manufacturers in Taiwan and Japan?" *Clean Technologies and Environmental Policy*, 13/1 (March 2010). http://www.springerlink.com/content/qu1134u5u7p94087 (Accessed October 2010).

Electronics TakeBack Coalition. http://www.electronicstakeback.com/home (Accessed October 2010).

Fargnoli, Mario and Fumihiko Kimura. "The Optimization of the Design Process for an Effective Use in Eco-Design." In *Advances in Life Cycle Engineering for Sustainable Manufacturing Businesses*, 59–64 (2007).

Greenpeace International. "Guide to Greener Electronics." http://www.greenpeace.org/
international/en/campaigns/toxics/electronics (Accessed October 2010).
Organisation for Economic Co-operation and Development (OECD). *Eco-Innovation in
Industry Enabling Green Growth*. Paris: OECD Publications, 2009.
Salkever, Alex. "How Green Is Your Laptop? Eco-Electronics Registry Now Global." http://
www.dailyfinance.com/story/how-green-is-your-laptop-eco-electronics-registry-now-
global/19124737 (Accessed October 2010).
Silicon Valley Toxics Coalition. http://svtc.org (Accessed October 2010).

Sabil Francis
University of Leipzig, Germany,
École Normale Supérieure, Paris

ECOLOGICAL MODERNIZATION

Ecological modernization (EM) is the realization that environmental problems are not apocalyptic in nature, but a challenge for social, technical, and economic reform that involves the transformation of core institutions of modernity, especially science and technology, production and consumption, politics and governance, and the market on local, national, and global scales. Thus, it is simultaneously an analytical approach, a policy strategy, and an academic environmental discourse. From a management perspective, ecological modernization can be defined as the movement of an organization or industry toward less environmentally destructive activities. Originating in Europe, ecological modernization makes the environment an actor in modernization theory and practice and ecological considerations as important as economic considerations in forging policy. In this approach, clean technology, including green technology, and innovation are key to economic progress. Rather than see the two as mutually exclusive, EM theorists argue that environmental sustainability, both by reducing consumption and wisely using resources, is key to the maintenance and increase of economic competitiveness.

The scope of ecological modernization, however, remains undefined. While some argue that it would include only tangible actors such as the state, corporations, civil society, and markets, others argue that it incorporates a whole paradigm shift in how humankind relates to the environment. However, one key aspect is the increasingly important role of innovative structural change that includes eco-innovations such as green technology and how these interact with various other actors in society, including scientific, institutional, legal, political, and cultural factors. Ecological modernization literature is also concerned with the role of social movements and the role that civil society plays in bringing about environmental change. As an academic discipline, it has close connections with subjects such as industrial ecology and industrial metabolism.

The term can be seen in four different perspectives. First, there is an identifiable school of ecological modernizationist/sociological thought. Second, it can be employed as a notion for depicting prevailing discourses of environmental policy. Third, especially in policymaking circles, ecological modernization is often used as a synonym for strategic environmental management, industrial ecology, eco-restructuring, and so forth; and finally, some scholars broaden the concept to refer to almost any environmental policy innovation or environmental improvement.

Origins

The idea originated in Germany, a nation that has been at the forefront of the modern environmental movement since the 1960s, and was later developed in the United Kingdom and the Netherlands. Joseph Huber, a German sociologist who has been hailed as the founding father of ecological modernization, contends that modernization has three distinct phases. The first, the early phase or the industrial breakthrough (from the late 1700s to around 1850) witnessed the development of industrial structures, with mechanical spinning and the steam engine as the key technologies a distinct, but not the dominating, feature in a predominantly agrarian society. This was followed by the development of a society (1850–1980) where industry predominated, and technological progress or an aspiration to it was the key defining feature of society. Finally, Huber argues, the late 1980s were marked by an age of superindustrialization where ecological forces became an important actor. In all three stages the driving forces are economy and technology, but the third stage of development is driven by the need to reconcile the impacts of human activity with the environment. Consequently, for Huber, economic actors and entrepreneurs were key to realizing the transformation associated with ecological modernization. In fact, he imagined the spirit of ecological modernization as a solution to environmental problems by saying in his book *Die Regenbogengesellschaft: Okologie und Sozialpolitik* (translated in English as *The Rainbow Society: Ecology and Social Politics*) "the dirty and ugly industrial caterpillar will transform into a[n] ecological butterfly."

The subject came into its own in the early 1980s, centered on the Free University and the Social Science Research Centre in Berlin. In addition to Joseph Huber, Martin Jänicke and Udo E. Simonis were important figures in this new subject. Earlier attempts that did not formalize the subject but touched upon it can be found in the discussions of ecological modernization by social theorists such as David Harvey in *Justice, Nature and the Geography of Difference* (1996) and the British sociologist Anthony Giddens in *The Third Way* (1998).

The subject has influenced both policymaking and academia. For example, the Organisation for Economic Co-operation and Development (OECD) conceptualized a green growth strategy in 2009 emphasizing eco-innovation and exploring the kind of policy instruments that would contribute to a fairer and more competitive and sustainable economy. In academia, for example, the original purpose of studies of technological innovation systems had been to show how technological innovation gives rise to economic growth. However, recent contributions in this area of research have been more concerned with the emergence of clean technologies that address sustainability problems. The emergence of ecology as a powerful agent mirrored the questioning of technological progress as an unmitigated good. In the United States, this was evident in the 1960s when a series of books, including Ralph Nader's *Unsafe at Any Speed* (1965), which accused Detroit automakers of knowingly producing unsafe automobiles, and Rachel Carson's *Silent Spring* (1962), which revealed the ecological devastation that was the result of the widespread use of chemical pesticides. All this brought into question the idea that technological progress was value neutral and not to be questioned.

In Germany, Huber's work was built upon by, among others, Ulrich Beck, who in his book *Risikogesellschaft. Auf dem Weg in eine andere Moderne* (published as *Risk Society: Towards a New Modernity* in English) argued that the risks of modernity are the dominant central mobilizing political and social force in restructuring major institutions and that future politics will be dominated by such concerns. Beck's work, among others, pointed out how ecological concerns, once marginal, have now become an independent, major force that must be included along with the traditional fields such as the socioideological, economic, and political. In other words, the environment is an actor in both academic discourses and policymaking on modernity. Central to this viewpoint is an approach to nature that is accommodative, rather than predatory, and an appeal for a sustainable society and institutional change to support this, thus "ecological modernization."

New Twists

This gave a new twist to standard theories of modernization that had sidelined the power of ecological interests to produce transformation. Ecological modernization sees the environment becoming a central player in the process of transformation, and this, combined with the emergence of the discourse of sustainability, has made the environment important, especially after the 1970s, a change that has been reflected in policymaking. For example, in many countries in the West, environmental nongovernmental organizations (NGOs) such as Greenpeace or Friends of Earth have become serious players in the policymaking arena and play the role of lobbyists and watchdogs in environmental matters. In industries such as chemicals or plastics, their input is as important as that of economists or policy planners, and businesses are awakening to the economic potential of sustainable development. One tangible example of this is the increasing importance of environmental assessment. There are several aspects of ecological modernization that have been translated into policymaking. These include product life-cycle assessment, which considers the complete life cycle of any product, from creation to eventual disposal, rather than the traditional perspective that neglects waste disposal. Thus, as for example, in green nanotechnology, an effort is made even at the design stage to incorporate eventual methods of waste disposal into products. This is in contrast to earlier technologies, such as information technology, that led to the generation of an enormous amount of electronic waste.

However, such concerns are much more evident in the developed and industrialized north rather than in the newly developing south. The north–south divide is evident here, as in many other aspects. In the developed world, environmental politics is a major force. Rather than assuming that environmental degradation is a by-product of modernization, a huge network of actors confronts it. These include intrasocietal ecological modernization processes; developments at transnational levels, such as scientific exchanges; pressure from NGOs both domestic and transnational; international conventions and treaties; the sharing and copying of environmental laws and regulations; and the reactions of local and transnational corporations to increasing pressures from laws and from public scrutiny. In the West, decades of environmental legislation and practice have made the no-holds-barred industrialization that marked the late 19th and early 20th centuries untenable for the present. Ecological modernization has made ecological rationality a legitimate arena of analysis and has forced both corporations and the state to add ecological calculations into their planning and projects.

EM in the Developing World

However, in the developing world, especially in countries such as China and India, EM is not so much a factor. Here, the "treadmill of production" model, which sees environmental degradation as an unfortunate by-product of economic progress, is dominant. This results in a capital-intensive extraction of natural resources and a dumping of waste back into nature. Thus, economic considerations trump environmental considerations. Enormous public support for environmentally damaging projects such as the Three Gorges Dam project in China or the Narmada Dam project in India testify to this. In such countries, environmental legislation is usually weak, and even if it exists in the statute books, is easily bypassed. Often, commitment to the environment is confined to rhetoric.

Ecological modernization, as a social science discipline, treads a middle path between those who argue that globalization has deleterious or contributive effects on the environment. Neoliberal thinkers argue that putting a market value on the environment is the best way to preserve it—this would include steps like privatizing forests and oceans. Ecologists and anti-globalization activists contend that integrating underdeveloped countries under the present capitalist paradigm would accelerate environmental degradation, eclipse the authority of the state and civil society, and subordinate the environment to the economic force of globalization. They see capitalism that is geared to increasing production and respect for the environment as mutually exclusive. Some even argue that the present pattern of ecological exploitation is inherently flawed, and that ideas like ecological modernization create a chimerical notion that present economic systems can be converted into "sustainable" ones.

Ecological modernization sees globalization and its relationship to the environment as a complex and intricate system of opposing and complementary processes of interdependence. EM theorists, however, have been faulted for the emphasis that they give to the market when they insist that businesses in industrialized nations are ecologically restructuring in response to market signals, and that economic actors increasingly perceive a business case for sustainability. This approach has been criticized because it accepts that modernization is compatible with the environment, and that it does not address the inherent inequities of the capitalist system or the fact that the cost of environmental degradation is disproportionately borne by marginalized people such as indigenous people. Moreover, it assumes that once the environment is a key factor in policymaking, ecologically feasible solutions will be found.

Often, environmental consciousness and regulations are the result of a wide range of factors, including the ability of various actors to bring pressure. Contrasting approaches to environmental legislation in the private sector–oriented United States, and the more environmentally conscious societies of Germany and the Netherlands where ecological modernization was first developed make this clear. In the latter, the series of National Environmental Policy Plans and associated instruments developed and implemented by the Dutch government in the 1990s have had a major impact on the ecological modernization debate. The European Union has been very proactive in trying to incorporate ecological modernization perspectives in policymaking. In 2009, Sweden, as president of the European Union, pressed for an ecoefficient economy, arguing that there was a need to create more wealth using less natural resources. On the other hand, in the United States, fossil fuels are seen as key to economic progress, and strong lobbies see environmental legislation as a hindrance to economic progress. At the global level, one of the key markers of the rise of ecological modernization was the United Nations Conference on Environment and Development that took place in Rio de Janeiro, Brazil, June 3–14, 1992, and adopted the principles of the Brundtland Report, which defines sustainable development as "development that meets the needs of the present without compromising the ability of future generations to meet their own needs."

See Also: Green Chemistry; Green Nanotechnology; Technological Utopias.

Further Readings

Beck, Ulrich. *Risk Society: Towards a New Modernity*. London: Sage, 1992.

Carson, Rachel. *Silent Spring*. Boston, MA: Houghton Mifflin, 1962.

Giddens, Anthony. *The Third Way*. Queensland, Australia: Polity Press, 2010 [1998].

Harvey, David. *Justice, Nature and the Geography of Difference*. Oxford, UK: Wiley Blackwell, 1997 [1996].

Huber, Joseph. *Die Regenbogengesellschaft: Okologie und Sozialpolitik* [The Rainbow Society: Ecology and Social Politics]. Berlin: S. Fischer, 1985.

Mol, Arthur P. J. *Globalization and Environmental Reform: The Ecological Modernization of the Global Economy*. Cambridge, MA: MIT Press, 2003.

Nader, Ralph. *Unsafe at Any Speed*. New York: Grossman Publishers, 1965.

Sabil Francis
University of Leipzig, Germany,
École Normale Supérieure, Paris

ELECTRONIC PRODUCT ENVIRONMENTAL ASSESSMENT TOOL (EPEAT)

An electronic product environmental assessment tool (EPEAT) is an evaluation and procurement tool that helps a buyer compare and select certain electronic products, taking into account their environmental characteristics. It sets out environmental criteria for examining desktop computers, laptops, computer monitors, workstations, and thin clients throughout their life cycles. Only those products whose environmental attributes conform to EPEAT's environmental criteria can be registered under it. Purchasing EPEAT-registered products results in various quantifiable benefits, for example, saving electricity, reducing the use of toxic materials, precluding disposal of wastes, and lowering emissions. Awareness about the EPEAT process and its tiers, criteria, and benefits is important for moving toward "greener" technologies and making an informed choice of "greener" products.

The Electronic Product Environmental Assessment Tool (EPEAT) system ranks desktop computers, laptops, computer monitors, and workstations as bronze, silver, or gold. More electronics manufacturers are voluntarily joining the EPEAT system, and around 1,840 products in 40 countries are already EPEAT-registered.

Source: iStockphoto

EPEAT was designed under a grant from the U.S. Environmental Protection Agency in response to purchasers' demand for a tool that would enable them to choose environmentally friendly electronics, as well as to manufacturers' calls for clear procurement standards. Developed through a multiyear, multistakeholder process involving government, academics, nonprofit organizations, electronics recyclers, and other experts, EPEAT became operational in 2006. In the same year, it was registered as Institute of Electrical and Electronics Engineers Standard 1680 for the Environmental Assessment of Personal Computer Products, and one year later all U.S. state and federal agencies were formally required to acquire EPEAT-registered products for at least 95 percent of electronics acquisitions, unless no EPEAT standard existed for the product.

Overall, EPEAT has gained growing recognition in information technology markets, spreading far beyond U.S. borders. This is partly because the EPEAT criteria contain or are consistent with international standards and formal requirements of other countries and jurisdictions, such as the European Union Directive 2002/96/EC on waste electrical and electronic equipment. As of mid-2010, an increasing number of electronics manufacturers such as Apple, Dell, Hewlett-Packard, Samsung, Sony Electronics, and Toshiba were voluntarily joining the EPEAT system, and around 1,840 products in 40 countries are EPEAT-registered.

The EPEAT process consists of two key elements: registration and verification. The condition for registering desktops, laptops, or monitors is the manufacturer's declaration about the products' conformance with the 23 required environmental criteria, such as "identification/removal of components containing hazardous materials," "minimum 65 percent reusable/recyclable materials," "demonstration of corporate environmental policy consistent with ISO 14001." Manufacturers then have to sign a formal Memorandum of Understanding with EPEAT, which requires them to provide accurate product and company information and pay annual fees.

EPEAT additionally contains 28 optional environmental performance criteria, including "elimination of intentionally added cadmium," "availability of replacement parts," or "provision of take-back program for packaging." The required and optional criteria are clustered into eight categories:

- Reduction/elimination of environmentally sensitive materials
- Materials selection
- Design for end of life
- Life-cycle extension
- Energy conservation
- End-of-life management
- Corporate performance
- Packaging

The EPEAT-registered products are ranked in three tiers, depending on the extent to which they meet the following optional criteria:

- Bronze (meets all 23 required criteria)
- Silver (meets all 23 required criteria plus at least 50 percent of the optional criteria)
- Gold (meets all 23 required criteria plus at least 75 percent of the optional criteria)

The EPEAT tiers stimulate technological and operational improvements among the participating manufacturers and allow for flexibility of choice among purchasers.

A verification system has been established to ensure the accuracy and credibility of the EPEAT registry. It is managed by the Green Electronics Council (GEC), which is responsible for the overall operation of EPEAT, including the maintenance of an online product declaration system for manufacturer listing of EPEAT products for buyers; providing guidelines for manufacturers, purchasers, and resellers; and enforcing the conditions of Memoranda of Understanding. Each year, the GEC randomly selects registered products for verification that the declarations are accurate. Any inaccuracies or errors revealed are subject to remedy; otherwise, the product is removed from the registry. The results of verification are published online.

Total environmental benefits arising from the acquisition of all EPEAT-registered products are annually estimated by the GEC. Annual benefits snapshots and full reports are published online. To calculate the specific environmental benefits of an EPEAT-registered computer or monitor, one can use an Electronics Environmental Benefits Calculator (EEBC) developed by the University of Tennessee. The EEBC compares an EPEAT-registered model with a conventional model and shows the savings in energy and finances, reductions in toxic substances, emissions, waste, and so forth.

The EPEAT system clearly demonstrates that "greening" of electronics technologies as well as purchasing greener products can bring about many environmental benefits. Consequently, EPEAT is evolving to include other electronic products, such as printers and televisions, thereby providing a further tangible possibility for individuals and institutions to contribute to sustainable development.

See Also: Design for Recycling; Eco-Electronics; Extended Producer Responsibility; Sustainable Design.

Further Readings

Green Electronic Council. "EPEAT. Green Electronics Made Easy" (2006). http://www.epeat .net (Accessed June 2010).
O'Neill, Mark G. *Green IT for Sustainable Business Practice: An ISEB Foundation Guide.* Swindon, UK: British Informatics Society, 2010.
Webber, Lawrence and Michael Wallace. *Green Tech: How to Plan and Implement Sustainable IT Solutions.* New York: AMACOM, 2009.

Maia Gachechiladze-Bozhesku
Central European University, Hungary

ELECTROSTATIC PRECIPITATOR

An electrostatic precipitator is a device that cleans up particulates, including dust and ashes, from smokestacks and other flues by using an induced electrostatic charge. It is a tool to help companies meet regulatory requirements and minimize their impacts on air pollution. The precipitator, also known as an electrostatic air cleaner, applies energy only to the particulate matter being collected and therefore is relatively energy efficient. It also works effectively without significantly impeding the flow of gases.

Electrostatic precipitators are widely used in numerous industrial applications as well as in some household settings. Highly effective at reducing particle pollution, including those in the micrometer range, they can handle large volumes of gas at various temperatures and flow rates, removing either solid particles or liquid droplets. Some of the more common applications include the following:

- Removing dirt from flue gases in steam plants
- Removing oil mists in machine shops and acid mists in chemical process plants
- Cleaning blast furnace gases
- Cleaning air of bacteria and fungi in medical settings, including operating rooms and facilities that produce antibiotics and other drugs
- Purifying air in ventilation and air conditioning systems
- Recovering potentially valuable materials from the gas flow, including oxides of copper, lead, and tin
- Separating rutile from zirconium sand

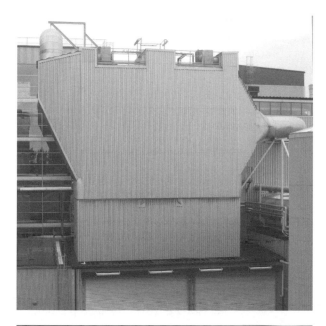

Precipitators, like this heating plant precipitator, can collect 99.9 percent or more of dust from the gas stream; when enough particles have accumulated they are transported away for disposal or recycling.

Source: Wikimedia

In many industrial plants, particulate matter created in the manufacturing process is carried as dust in the hot exhaust gases. If released into the atmosphere, the particulates can lead to serious health problems, including lung damage and bronchitis, as well as reduce visibility and affect climate. Fine particles that are smaller than 2.5 microns in diameter can be especially dangerous because they are drawn deep into the lungs and can trigger inflammatory reactions. The United States and other nations regulate particle pollution, although the regulations have often sparked controversy and lawsuits.

Electrostatic precipitators are available in many different sizes and types, designed for various dust and water droplet characteristics and gas volume flows. Very large power plants may have multiple precipitators for each unit. Precipitators can collect 99.9 percent or more of the dust from the gas stream, depending on the temperature and flow rate of the gas, the size and chemical composition of the particles, and the precipitator design and voltage it applies to the gas.

Although precursors to electrostatic precipitators date back to the early 19th century, a chemistry professor at the University of California, Berkeley, Frederick Cottrell, is credited with inventing the modern electrostatic precipitator in 1907. Early units were used to remove sulfuric acid mist and lead oxide fumes emitted from acid-making and smelting activities. The devices helped protect vineyards in northern California from lead emissions.

Precipitators are an important tool in the process of cleaning up flue gases and are often deployed with denitrification units that remove nitrogen oxides, and scrubbers or other devices that remove sulfur dioxide. Precipitators function by electrostatically charging particles in the gas stream. The charged particles are attracted to and deposited on plates or other collection devices. The treated air then passes out of the precipitator and through a stack to the atmosphere. When enough particles have accumulated on the collection devices, they are shaken off the collectors and deposited into hoppers. A conveyor system transports them away for disposal or recycling.

The most basic precipitator design consists of a row of thin vertical wires and a stack of large, flat, vertical metal plates. The plates are spaced from less than a half-inch to about seven inches apart, depending on the application. The stream flows horizontally between the wires and through the stack of plates. A negative voltage of several thousand volts is applied between the wires and plates.

A wet electrostatic precipitator operates with saturated air streams that have 100 percent relative humidity. Such precipitators are commonly used to remove liquid droplets, including oil, resin, tar, and sulfuric acid mist, from gas streams in industrial settings. They are applied where the gases are highly moist, contain combustible particulates, or have particles that can be sticky.

Although precipitators are effective at removing particles from the gas stream, the subsequent disposal of those particles can create environmental problems. Both wet and dry particles can be harmful to works and the surrounding community if not handled correctly.

Plate precipitators are often marketed to the public as air purifiers or as a permanent replacement for furnace filters. Unlike some other air purification technologies, they typically do not become breeding grounds for harmful forms of bacteria. However, the plates can be difficult to clean and can also produce ozone and nitrogen oxides.

Some consumer precipitation filters are sold with special soak-off cleaners. The entire plate array can be removed from the precipitator and soaked for several hours in a large container overnight, thereby loosening the particulates.

See Also: Best Available Technology; Flue Gas Treatment; Thermal Depolymerization.

Further Readings

Lloyd, D. A. *Electrostatic Precipitator Handbook*. Oxfordshire, UK: Taylor & Francis, 1988.
Neundorfer KnowledgeBase. "Electrostatic Precipitator." http://www.neundorfer.com/knowledge_base/electrostatic_precipitators.aspx (Accessed September 2010).
Parker, K. R. *Electrical Operation of Electrostatic Precipitators (IEE Power and Energy)*. Hertfordshire, UK: The Institution of Engineering and Technology, 2002.

David Hosansky
Independent Scholar

ENFRAMING AND STANDING RESERVE

Enframing and standing reserve are interrelated concepts introduced by the philosopher Martin Heidegger (1889–1976) in his critique of modern science, technology, and metaphysics. Heidegger regarded modern society as existing in a state of deep existential crisis, and his later work emphasized the roles of science and technology in that crisis. His argument has been adopted, applied, and disputed by scholars working in the philosophy of science and technology, environmental philosophy and ethics, media studies, science and technology studies, and other associated fields. Applied to the realm of green technology, his concepts challenge and expand prevailing notions of ecology and sustainability. They also link those notions to larger questions of existential choice and authentic engagement with the nonhuman and human world.

In his earlier work, notably in his famous treatise *Being and Time*, Heidegger argued that beginning with the works of the later classical Greek thinkers, Western philosophy increasingly lost sight of the most fundamental philosophical question. He called that question the "question of being," concerning the very fact of existence rather than any qualities or characteristics of phenomena. For Heidegger, humans exist only through their relations to other entities or beings, and those relations are always informed by human projects and purposes.

According to documentation provided by translator William Lovitt, the terms *enframing* (*Gestell*) and *standing reserve* (*Bestand*) first appeared in lectures Heidegger gave in 1948 and in essays published in 1954. The essays "The Question Concerning Technology" and "The Turning" make explicit use of these terms and the argument of which they are a part. Although some commentators see a sharp discontinuity between Heidegger's early writings and his later critique of science and technology, others see a basic continuity.

Arguably, enframing and standing reserve can be viewed as both outcomes and drivers of the fundamental philosophical error Heidegger critiqued in his earlier work. He proposed that "the essence of technology is nothing technological," but instead, consists of a human tendency to enframe the world as a set of resources to be appropriated and used. Enframing reduces nature to a standing reserve, a "gigantic gasoline station" and "an energy source for modern technology and industry." To illustrate this principle, Heidegger uses the example of a traditional windmill, which in his view acts together with the wind in an unobtrusive way. He contrasts the windmill with more modern techniques such as mechanized agriculture, the extraction and stockpiling of coal and uranium, and nuclear power production. In his view, these technologies "challenge" nature in intrusive ways that go far beyond traditional modes of supply and sustenance. His most poignant example is, perhaps, a vision of the Rhine River reduced from an aesthetic object and an inspiration for poetry to a mere source of water power through the installation of a hydroelectric plant. If the river remains a part of the scenic landscape, he argues, it is only as a stop for tourist visits managed by the travel industry.

Summarized here, these critiques may appear as nothing more than nostalgic reactions to change, primarily on aesthetic grounds. In the context of Heidegger's larger argument they are far more incisive, however. He argues that by enframing the world as a standing reserve, humans grow increasingly detached from their primordial relationships to the rest of the world, including each other. Not only is the Rhine River appropriated as a resource for the tourist industry, but so are the tourists themselves.

Modern activities such as extractive mining, the production and consumption of fossil fuels, factory farming, nuclear power production, and genetic engineering all fit within a general pattern of enframing and standing reserve. Accordingly, some advocates of deep ecology and biocentrism, perspectives that value nonhuman entities at least as highly as humans, have found Heidegger's concepts illuminating. However, these concepts also have the potential to help bridge the gap between deep ecology and social ecology, a perspective that emphasizes environmental protection as a means of protecting human welfare. Heidegger's concepts suggest that human welfare is indeed at stake when society engages in rampant destruction of the nonhuman world in ways that extend well beyond material safety. Viewed in Heidegger's terms, the very concept of "natural resources" is problematic: to reduce nature to a set of resources is to become thoroughly estranged from nature. When humans are reduced to "human resources" as well, serving merely as tools for each others' purposes, then that estrangement extends to an even more fundamental level. As a model and a set of practices, the human use of nature diminishes the possibilities for authentic human being.

Some critics argue that Heidegger's philosophy is irredeemably tainted by his political associations with German fascism. Those associations do, in fact, seem incongruent with his expressed concern for humanity and the state of modern society. For others, however, Heidegger's ideas have value independent of his problematic personal history, and offer unique and provocative insights regarding the connections between modern science, technology, and ecology.

See Also: Frankfurt School; Innovation; Science and Technology Studies.

Further Readings

Heidegger, M. *The Question Concerning Technology and Other Essays*. Trans. W. Lovitt. New York: Harper & Row, 1977.

Kinsella, W. J. "Heidegger and Being at the Hanford Reservation: Standing Reserve, Enframing, and Environmental Communication Theory." *Environmental Communication: A Journal of Nature and Culture*, 1/2 (2007).

Zimmerman, M. E. "Heidegger's Phenomenology and Contemporary Environmentalism." In *Eco-Phenomenology: Back to the Earth Itself*, C. S. Brown and T. Toadvine, eds. Albany: State University of New York Press, 2003.

William J. Kinsella
North Carolina State University

ENGINEERING STUDIES

Engineering studies (ES) deserves attention as it is the first conscious and viable initiative for employing academic research to critique and transform engineering practice, engineering education, and the way engineers are perceived.

ES encompasses a wide range of scholarly work that seeks to understand what it means to be an engineer and what is conveyed by "engineering work" in different historical and cultural settings through historical, philosophical, sociological, anthropological, and feminist lenses. One of ES's central findings is that the terms *engineer* and *engineering knowledge* have significantly different connotations across various national contexts. By applying the lessons drawn from the study of the genesis and development of engineering during the past two centuries, ES scholars are able to address the contemporary stakes with respect to engineering education, epistemology, and discipline formation.

Engineers are modern, qualified technical experts, and engineering—broadly defined as the preoccupation with perfecting efficient means for achieving objectives—is one of the largest occupational groups worldwide. Engineers devise tools, incentives, practices, or principles to target defined problems.

However, different communities of engineers working on similar projects define their problems differently—green engineers, for example, claim that they are able to reshape their organizational, ethical, and epistemic practices to foster different sorts of research and engineering solutions. In that regard, it is important to consider how individuals and organizations change in practice. Are engineers creating "new" sources of information about sustainability? What sort of reframing of engineering takes place? Furthermore, what sorts of resistance do practitioners encounter when they attempt to alter traditional approaches to engineering problem definition?

Contemporary contextualizations of sustainability warrant special attention by anyone concerned with ES. In point of fact, engineers may be exhorted to be more "ethical" or "sustainable," but usually this does not translate into an actual work practice. Exhortations to sustainability are typically not accompanied by the suggestion to the engineer that resources are finite and limited, so the engineer has to design and work with that constraint. Likewise, little attention is given to speaking with the potential users of technology to understand their needs and concerns, before beginning to analyze the data. This is why looking at ES work might be useful and promising. ES monitors how engineers adjust in response to societal and political pressures.

A central difficulty in ES is responding to and accounting for a disproportional appreciation, in both academic and public spheres, of the engineers' role in shaping the material contours of modernity. Moreover, the dominant professional ideology of engineering itself, which takes for granted that technology is apolitical and that engineering equals applied science, is at odds with the calls for sociopolitical leadership made by prominent spokespersons across various engineering disciplines and social critics of technology alike. ES negotiates these tensions to establish its disciplinary domain.

ES's most immediate intellectual forerunners have been active in accounting for the relationship between the technical and nontechnical aspects of engineering since the late 1980s. Subsequently, ES developed as a genre at the intersection of Science and Technology Studies (STS) and the History of Technology. The first Engineering Cultures undergraduate course was taught at Virginia Polytechnic Institute and State University in 1996. Also, the field, which was officially initiated through the establishment of the International Network for Engineering Studies (INES) followed by a workshop held at Virginia Tech in 2006, very recently launched its own academic journal.

Although it positions itself as a distinct discipline, ES maintains similarity to STS with respect to the latter's mode of argumentation and multidisciplinary audience. Whereas STS has been successful in tracing contingency in history by showing how the epistemic contents of science and technology are not independent of the functions of society and power, ES has been equally successful in revealing contingency in the coproduction of engineering knowledge and nationhood. An early challenge in ES was to rescue the engineer from invisibility—both in STS and public discourse. One of the current challenges is to account for how the transition from conceptualizing technical work as neutral to analyzing engineering design as a social and cultural process can contribute to addressing some of the most important political questions about technology.

Up until very recently, most scholars were looking at engineering communities in terms of a few features, such as technological innovation, professionalization, social responsibility, or adoption of new engineering ideas and practices. Although ES embraces a great many of those themes, it largely departs from earlier scholarship on the basis of two premises: (1) engineering is a cultural project in continuous search for cultural relevance and social legitimacy, and (2) what counts as engineering knowledge in different times and locations and what ultimately comes to constitute the relevant community of engineering practitioners are coproduced on the basis of some commonly shared stimuli, or images. The latter, with globalization being a characteristically challenging one, questions the engineers' perceptions of self as well as their understanding of engineering's role in society.

As opposed, hence, to generalizations about technology, the development of modern (post-1800) engineering appears far more complex. In fact, instead of displaying a universal, integrated narrative, the history of modern engineering consists of various subhistories marked by struggles between, on the one hand, the liberal democratic mission of engineering to determine human affairs and politics' indeterminism, as well as, on the other hand, between definitions of problems applicable to engineering and conceptions of problems appropriate to other areas of expertise. Consequently, one of the overarching challenges for engineers has been how to define a set of problems that are both socially legitimate and unique to their discipline. It is with this rationale that ES lends itself to examining the historical episodes of engineering identity (re)formation.

Engineering provides students of technology with a natural laboratory for social research. The "simple" question of "what engineering has been/is for?" acts like a prism; it is possible to see the history of technology and history proper in novel ways through that prism. When, for example, scholars explain the early-20th-century energy shift from coal to oil in terms of hegemonic sequences and conflicting labor dynamics, they largely overlook the transformation in the nature of technical work that went along with the development of technically advanced refinery techniques, thus they downplay the role of the engineer in that transformation. In fact, it may well be that the engineers' profile as docile technical experts working for the well-being of the organization favored the proliferation of corporate investment and brought operational stability in the oil system.

But, importantly, ES, like the activist vein of STS that aims to influence social scientists to make a step beyond unmasking contingency, aspires to be interventionist. Specifically, the field of engineering education becomes the main locus of intervention, where ES scholarship meets and critically converses with dominant practices in academia and the world of engineering organizations. That said, the trend toward sustainable engineering becomes a challenge to those organizational forms. Hence, another example of why ES matters lies in its potential for analyzing and affecting the implications of developing organizations that are decentralized, locally oriented, and connected to communities. This may take the form of creating "real-world" engineering design/practice labs that extend across the boundaries between universities and whatever it is that engineers identify as "in need" of sustainability engineering.

Organizations are structures designed to enforce technological projects by purposefully mediating the work of many individuals, in our case, engineers and managers. Given that organizational employment has been the norm—especially in the example of American engineers—for almost a century, the study of organizations suggests a promising arena for ES, which needs to incorporate and explore organization theory to a greater extent.

Workers, managers, engineers, and others work within the broader organizational and production chain contexts. They may, then, either reproduce dominant attitudes toward sustainability or they may emerge as critics pressing for organizational reforms. With that said, a potential question for ES would be, "How effective can the implementation of 'green engineering' practices be, given that production chains are dispersed and decentralized, so that many decisions affecting the technology's sustainability are split across the many stages of the productions chain?"

Overall, the prevalence of engineering work in our technologically mediated societies provides a practical justification for doing ES; the fact that engineering is a political activity—displaying political dimensions as the primary means of authoritatively defining the polis—underlines an ideological reason. Finally, ES contributions to how the social sciences approach technical work set up a theoretical challenge of exploring the boundaries of what is still an underdeveloped body of scholarship.

See Also: Geoengineering; Intermediate Technology; Science and Technology Studies.

Further Readings

Downey, Gary Lee. "What Is Engineering Studies For? Dominant Practices and Scalable Scholarship." *Engineering Studies*, 1/1 (2009).
Lucena, Juan. *Defending the Nation: Policymaking in Science and Engineering Education From Sputnik to the War Against Terrorism.* Lanham, MD: University Press of America, 2005.
Nieusma, Dean and Donna Riley. "Designs on Development: Engineering, Globalization and Social Justice." *Engineering Studies*, 2/1 (2010).
Wisnioski, Matthew. *Engineers for Change: America's Culture Wars and the Making of New Meaning in Technology.* Cambridge, MA: MIT Press, 2010.

Nicholas Sakellariou
University of California, Berkeley

ENVIRONMENTAL REMEDIATION

Environmental remediation is a process that aims to remove or reduce the levels of contaminants in soils, groundwater, surface water, air, or other environmental media in order to improve ecological or public health and/or facilitate the redevelopment of a contaminated site. A number of remediation technologies have been developed, some of which are routinely used and others that require further testing for effectiveness in the field. Although remediation is necessary to improve environmental conditions, there may be risks associated with environmental remediation, including the generation of secondary waste streams, the risk of future leaching of contaminants that were immobilized but not degraded, or the emissions of harmful particulate matter. The United States and Europe have implemented various regulatory and statutory approaches to facilitate the environmental remediation of contaminated sites.

Examples of situations that require remedial actions for environmental restoration include chemical and petroleum spills, abandoned landfills and industrial properties, and nuclear waste. Hazardous substances that may occur in contaminated sites include organic pollutants, such as polychlorinated biphenyls (PCBs), polycyclic aromatic hydrocarbons (PAHs) and petroleum hydrocarbons, heavy metals (e.g., mercury, arsenic, copper, lead), radioactive materials, and pesticides.

Remediation technologies can be biological, physical, chemical, or thermal and can degrade, contain, or chemically convert contaminants to less toxic or more easily extractable

Chemical spills, nuclear waste, or oil spills, like the one shown here, are cause for remedial actions for environmental restoration. They can be cleaned in situ (at the site) or ex situ (removed to another location).

Source: Damage Assessment, Remediation, and Restoration Program, National Oceanic and Atmospheric Administration

compounds. Physical and chemical mechanisms will often occur together in the remedial process. Furthermore, environmental remediation technologies can be carried out in situ or ex situ. In situ refers to the treatment of contaminants at their site of occurrence (e.g., injecting products into the subsurface to remediate groundwater), while ex situ refers to the removal of contaminated material to another location for treatment (e.g., pumping contaminated groundwater to the surface and treating in an aboveground reactor). Some remediation technologies can be applied either in situ or ex situ. Ex situ treatments are generally more expensive but faster, while impediments to in situ treatments include difficulties with the delivery of treatment to the contaminated site, uncertainties surrounding their effectiveness, and potential residual liability issues.

Soil Remediation Technologies

Soil remediation methods include excavation, bioremediation, incineration, thermal desorption, soil washing, and immobilization technologies (i.e., that limit the leaching of contaminants from media), such as the encapsulation of wastes into cements or asphalt or in situ vitrification. Limitations to the use of soil remediation technologies might result from potential solubility or stability issues, and a dependency on particular environmental conditions or soil parameters, such as pH requirements. As a result, the performance of remediation may be difficult to predict, underscoring the importance of pilot studies, comprehensive sampling and modeling, and ongoing monitoring as part of the remedial process.

Physical and/or Chemical Methods

An ex situ "dig and haul" approach excavates contaminated soil for disposal in a landfill or for subsequent treatment with remediation technology. Conversely, stabilization and solidification (S/S) are in situ techniques that, rather than removing contaminants, aim to restrict the mobility of contaminants in environmental media through both physical and chemical means. In stabilization, contaminants are converted into their least soluble, mobile, or toxic form. In solidification, there is the addition of other materials to the soil in order to encapsulate contaminated waste. For example, contaminated soil can be mixed with asphalt or concrete to make the media harden. The products must be sufficiently inert to be left in the subsurface and not deteriorate over time, leading to the rerelease of contaminants. In situ remedial techniques that employ solidification and stabilization techniques have been used broadly in the United States, although data on their long-term efficiency are limited. S/S systems are generally less costly than ex situ approaches, although barriers to using the technology include lack of technical guidance, uncertainty over the durability of the stabilizing material and rate of contaminant release, and poor past performance of the cement stabilization processes.

In situ vitrification (ISV), another immobilization technology, uses an electrical current to melt contaminated soils and reduce the toxicity, mobility, and volume of the waste material. In this process, a current is passed in the soil between electrodes to facilitate the melting of soil, and the molten soil is then cooled to form a glass. ISV can treat media contaminated with all classes of contaminants. Organic pollutants are generally destroyed or transformed into gases. They can then be collected by a gas hood and transferred to a treatment center. Inorganic contaminants, including heavy metals and radioactive materials, are encapsulated within the glass product to limit their leaching into the surrounding soil or groundwater. In situ vitrification is typically applied on a small scale because it is a time-consuming process and has high associated expenses.

Capping is another in situ remedial technique that aims to contain contaminants by creating a physical barrier between contaminated media and the surrounding environment that limits surface exposure to contaminants. Capping typically uses a cover system composed of a surface level (generally soil) that sustains vegetation, a drainage layer that supports the vegetation's growth, a low-permeability layer, and a gas-venting layer. The gas-venting layer allows for the methane gas created from decomposing organic matter to be properly vented and to avoid combustion. If the site is near water, however, contaminants could still be transferred horizontally into water bodies through contact with groundwater and soil. Research indicates that caps often fail before their designated life span as a result of ground shifts, the expansion and contraction of material, and burrowing animals. While capping may require regular monitoring and maintenance, and costs may thus be transferred to future owners, the use of caps is initially less expensive and can be completed in a short time period relative to other remedial options.

Physical-chemical remedial approaches that remove contaminants from soil, versus immobilizing them for subsequent treatment or disposal, include soil vapor extraction (SVE), supercritical fluid extraction, and soil washing. In SVE, a well is positioned in the contaminated region and a vacuum draws air down through the soil and up through the well, after which filters remove volatile and semi-volatile organic compounds (VOC and SVOC). SVE can treat large volumes of soil, although sites with complex contamination might require it be used in conjunction with other technologies. For SVE, the contaminants in the soil must have low moisture content and be permeable enough to allow for the wells to draw air through the contaminated soil at a satisfactory rate. In supercritical fluid extraction, liquefied gas (e.g., carbon dioxide) is used to extract organic contaminants, primarily PCBs and PAHs. Using high pressure and moderate temperatures, gas is compressed to a fluid state and hazardous waste can be added to a vessel that contains this critical fluid. The organics move with the critical fluid to the top of the vessel and are pumped into a second vessel. In the second vessel, the temperature and pressure are decreased and the contaminants volatilize from the critical fluid. The cost of this technology is generally high and the efficiency difficult to predict. Soil washing is a relatively new environmental remediation technique that adds a wash solution (e.g., water with or without chemical additives) to soil to extract contaminants. Contaminants are dissolved into the wash solution, and additional treatments can then be applied to the solution. Complex waste mixtures (e.g., mixes of metals and organics) make formulating the wash solution complicated; particular soil parameters may be necessary (e.g., soil with low humic content), and additional treatment steps may be necessary to remove hazardous levels of washing solution that remain in the treated area.

Biological Methods

Bioremediation refers to the process of using living organisms, such as plants (phytoremediation), fungi (mycoremediation), or bacteria to transform or degrade contaminants into their less-hazardous constituents, such as carbon dioxide, water, fatty acids, and methane. Bioremediation tends to work best for hydrocarbons, such as those found in petroleum and petrochemical products. Plants can transform inorganic compounds (e.g., cadmium) to have less mobility by accumulating them in their aboveground parts, which can then be harvested for disposal. Environmental conditions (e.g., pH, temperature, nutrients, soil moisture, and aeration) affect bioremediation, and a system can be set up for optimizing the process. This might include supplementing soils with microorganisms and an irrigation system for controlling soil moisture and soil nutrients. Bioremediation can destroy contaminants with no harmful chemical additives and little or no residual treatment required. The process, however, often requires more time relative to other treatments.

Thermal Methods

Incineration is commonly used in soil remediation and uses high temperatures to break the chemical bonds of organic compounds and reduce the toxicity and volume of nonvolatile substances. While incineration can achieve over 99 percent success rates in destroying certain contaminants, it can also create incomplete combustion products that are toxic, such as dioxins and furans. There is special concern about particulate emissions that contain chlorinated hydrocarbons, heavy metals, and large quantities of sulfur. As a result of its potential impact on air quality issues, there is often public opposition to incineration. Incineration is also energy intensive, and the U.S. Environmental Protection Agency (EPA) requires that leftover ash and material be disposed of at a hazardous waste facility.

In thermal desorption, soil is heated and organic contaminants volatilized from the soil are captured. This process happens in the absence of oxygen and at temperatures lower than in incineration. Unlike incineration, which simultaneously volatilizes and combusts contaminants, another phase of remediation is required to destroy the captured contaminants.

Water Remediation Technologies

Groundwater remediation treatments include air stripping and activated carbon adsorption, bioremediation, incineration, and specific contaminant-removal methods for certain contaminants such as mercury and radioactive elements.

Chemical and/or Physical Methods

The technique of remediating contaminated water in the ground is called "air sparging," while remediating above ground is referred to as "pump and treat." Air sparging is an in situ remediation technique where a gas, usually air, is injected into deep groundwater to create a circulation of water moving up within the aquifer that absorbs oxygen and volatilizes certain contaminants. Air sparging may be preferred by those responsible for cleanup because it is more cost effective than pump-and-treat methods. It does, however, present risks, such as contamination spreading as a result of the water table rising and the potential of off-site vapor migration.

In pump and treat, water is pumped out of the ground and treated by air stripping and/ or granulated activated carbon adsorption. Air stripping pumps the contaminated groundwater through an aeration tank where organic contaminants, such as VOCs, are volatized, and the cleaned water can then be returned to the site. Carbon adsorption filters the groundwater through "activated" carbon (processed carbon that has a large, porous surface area), which traps contaminants such as hydrocarbons and SVOCs. Spent carbon, with contaminants, must be disposed of, and its transport may require hazardous waste handling procedures.

Chemical precipitation is a method for removing heavy metals from contaminated water. Products that contain sulfur compounds can bond with heavy metals (e.g., mercury, lead) to produce a precipitate that can then be removed through settling or filtering and subsequently transported to a landfill. Adequate information is required regarding the dependence of precipitates on specific chemistry and environmental conditions, for example, the pH of water.

Biological and Thermal Methods

Bioremediation of water may occur naturally by microorganisms that are present in the medium, as well as be anthropogenically enhanced by nutrient additions. Microbial degradation of contaminants in water has been demonstrated for crude oil spills, nitrate contamination, chlorinated solvents, and gasoline contaminants. Incineration technologies, as described above, may also be applied to water, with most basic types of incinerators accepting liquid wastes.

Ecological and Health Issues Introduced by Remediation

While environmental remediation is necessary for addressing the contamination of environmental media, risks may be introduced by the remedial process. These include the release of noxious gases or air particulates, risks of contamination spreading, and the use of chemical additives during remediation that may have their own harmful effects. Furthermore, many technologies generate secondary waste streams that require proper disposal in a landfill. As a result of remedial actions, often requiring substantial energy, water, financial, and time inputs, and the risks of spreading or merely relocating pollution elsewhere, arguments have been made in favor of the precautionary principle. The precautionary principle advocates the restriction or banning of environmental pollution before its production.

Environmental remediation standards are set for the levels of dust, noise, and odor produced and the release into the surrounding environment of contaminants or remedial chemicals that are used in or are a product of remediation. Workers need to be provided with proper protective clothing, equipment, and training to limit exposure to hazardous materials, and monitoring procedures should be followed. Violations of standards on the production and release of pollutants and issues with workers' health and safety have been documented in many cases.

Regulatory Framework

Since the 1970s, state and federal governments in the United States and other countries have implemented regulations on environmental remediation. In the United States, the two most noteworthy hazardous waste statutes are the Resource Conservation and Recovery Act (RCRA) of 1976 and the Comprehensive Environmental Response, Compensation, and Liability Act of 1980 (CERCLA, commonly named Superfund). These are both enforced by the EPA and set minimum standards for national soil and groundwater cleanup policy, although most individual states have adopted their own version of the laws. While many countries in Europe have their own regulations regarding environmental remediation, the European Union is moving toward regional standards. In Canada, policy around environmental remediation occurs at the provincial level, although the federal Canadian Council of Ministers of the Environment provides general guidance through the Canadian Environmental Quality Guidelines and the Canada-Wide Standards for Petroleum Hydrocarbons in Soil.

The conservation philosophy underlying remediation regulations, such as with CERCLA and the RCRA, usually entails cleanup to the maximum extent possible and is based on the most restrictive future use of the land (e.g., residential development). Due to the large number of sites requiring remediation, technological limitations, and fiscal considerations, there has been an increasing shift in remediation policy toward risk-based decision making when establishing cleanup goals. In risk-based approaches, the objectives for remediation are to clean up sites to the extent that they no longer pose a "significant" threat to human or ecological health. This often results in partial cleanup and containment measures and more restricted uses of the land following remedial activities (e.g., groundwater may not be used for drinking-water purposes or future site use must be industrial). Contaminated sites are, however, typically concentrated in communities of color or areas where poor, immigrant, or indigenous peoples live. As these communities are faced with persistent and cumulative exposures to pollutants, there has been a critique of cleanup practices and policies that encourage cheaper remediation through lower standards or in situ approaches that do not excavate and remove hazardous materials while minimizing community and government oversight.

See Also: Environmental Science; Solid Waste Treatment; Wastewater Treatment.

Further Readings

Hamby, D. "Site Remediation Techniques Supporting Environmental Restoration Activities—A Review." *Science of Total Environment*, 191 (1996).

Whitney, H. "Cities and Superfund: Encouraging Brownfield Redevelopment." *Ecology Law Quarterly*, 30 (2003).

Wilson, S. C. and K. C. Jones. "Bioremediation of Soil Contaminated With Polynuclear Aromatic-Hydrocarbons (PAHs)—A Review." *Environmental Pollution*, 81 (1993).

Jennie Liss Ohayon
University of California, Santa Cruz

ENVIRONMENTAL SCIENCE

The interdisciplinary field of environmental science draws on ecology, geology, meteorology, biology, chemistry, engineering, and physics in its integrated study of the environment. Typically, the quantitative approach of environmental science is considered separate from environmental studies, which emphasizes the human relationship with the environment and the social and political dimensions thereof, though both fields overlap and codevelop, and their correlating university programs frequently share faculty and other resources. While environmental studies and environmental science both deal with, for instance, the problem of climate change, environmental studies would focus on the economic and political dimensions of international climate change protocols, while environmental science would quantify the effects of climate change, construct climate change models, and evaluate means of mitigation.

Though the study of the environment is as old as any human endeavor, the modern field developed from the growing public awareness and concern over environmental problems in the 1960s and 1970s, amid the publication of books like Rachel Carson's *Silent Spring* (1962) and Paul Ehrlich's *The Population Bomb* (1968), the phenomenon of nuclear proliferation even as knowledge of the long-term effects of fallout and exposure to radioactive material became widespread, and growing concerns over the anthropogenic release of toxins and chemicals into the environment, and the resulting effects on human life. While public concerns and resulting political issues furthered environmental studies, environmental science took on the task of quantifying the effects of disasters like the 1979 Three Mile Island accident or the impact of atmospheric sulfur dioxide and other emissions on acid rain, a phenomenon first noted in 18th-century London but given a name and greater public awareness in the early 1970s.

The various areas of study under the umbrella of environmental science are described below.

Ecology

The study of ecosystems—of populations of plant and animal organisms and their relationship to their environment—ecology grew out of the 19th century's natural sciences. It is an experimental science, like other biological sciences, unlike natural history, which is primarily descriptive. Ecology is concerned with the study of adaptation (in the evolutionary sense), the distribution and abundance of organisms, the development of ecosystems, biodiversity, and the movement of matter and energy through plant and animal communities. In ecology, "community" refers to two or more populations of different species in the same area, such as "the insect community of New York City" or "the animal community of the Painted Desert." One of the fundamental concepts of ecology is that of ecological succession, the orderly changes that occur to the structure of a community (called primary succession if there is no soil initially present, secondary succession if there is). These changes include the development of communities during early succession, when a new and initially unoccupied habitat is created, as well as changes to existing communities wrought by natural or man-made disturbances. Originally, succession was assumed to have an end point, a final stage of persistent equilibrium called the climax; today, though many of the concepts introduced over the course of the study of the climax stage have remained popular, it is generally considered that external disturbances and environmental changes are frequent enough to prevent a community from ever achieving equilibrium for good.

Ecology can focus on a wide range of spatial and temporal scales: the insects living in the decaying trunk of a single tree, the insects and their predators living in an entire forest, the history and experience of that insect population over many generations. Long-term ecological studies have helped delineate the complexity of ecosystems and have reinforced the importance of diversity of species, ecosystems, and genes—collectively, biodiversity—to ecological health, which in turn affirms the need to safeguard species against extinction.

Conservation biology is the study of life with the aim of reducing extinction rates and protecting species and their ecosystems. The term was coined in 1978 as the title of a conference at the University of California at San Diego to discuss tropical deforestation and species extinction; it emerged at the same time as discussions of the importance of biodiversity. Conservation biologists study the five known mass extinctions of the past—the Ordovician (440 million years ago [mya]), Devonian (370 mya), Permian-Triassic (245 mya), Triassic-Jurassic (200 mya), and Cretaceous (65 mya)—as well as the possibility of a current, ongoing mass extinction called the Holocene extinction event.

Meteorology and the Atmospheric Sciences

The atmospheric sciences include meteorology (the study of the atmosphere with an emphasis on weather processes), climatology (the study of weather conditions averaged over a period of time), and aeronomy (the study of the upper layers of the atmosphere).

Meteorology incorporates atmospheric physics and atmospheric chemistry and studies atmospheric variables like air pressure, temperature, and water vapor levels as well as their effects on weather phenomena in the troposphere. Though it is most associated with weather forecasting, this is simply meteorology's most public face. Other areas of study include the impact of weather on aviation (both flight plans and aircraft design), nuclear meteorology (the study of the distribution of radioactive materials in the atmosphere), agricultural meteorology (the study of the effects of weather on plant and animal development, crops, and the energy balance of ecosystems as well as the effect of vegetation on weather). Climatology is the *langue* to meteorology's *parole*: the study of weather over time, from years to millennia. The focus in climatology is on weather periodicity and long-term patterns, as well as on interactions between the atmosphere, land, and sea that take place on a long time scale, and the patterns of atmospheric circulation by which thermal energy is distributed across the Earth's surface.

The word *aeronomy* was formed in parallel to astronomy, and the field is a branch of atmospheric physics, which studies the atmospheric tides that transport energy inputs from the upper atmosphere to the lower atmosphere and that are thus critical in the modeling and understanding of atmospheric changes.

Environmental Chemistry

Environmental chemistry is the study of the chemical and biochemical phenomena in the environment, drawing on analytical chemistry in studying the activities of chemicals in air, soil, and water environments and the chemical effects of human activity. Those effects are usually the results of contaminants—substances introduced to a receptor that would not normally be present, or would be present in lower levels. Not all contaminants are pollutants: pollutants are specifically those contaminants that have a harmful impact. The receptor is that which has received the contaminant—either an organism, like a fish, or a medium, like the water of a lake. A receptor that retains the contaminant is a sink (such as a fish that keeps the contaminant in its body instead of passing it back into the lake through waste). Contaminants include urban runoff like gasoline, motor oil, metals, nutrients, and soil that are washed into the environment by rainstorms; nitrates and phosphorous leaching into bodies of water from farmland; and heavy metals contaminating land and water as a result of industrial activity.

Environmental chemists use quantitative chemical analysis to take measures of environmental indicators such as, in freshwater samples, dissolved oxygen, chemical oxygen demand, biochemical oxygen demand, total dissolved solids, pesticides, and heavy metals. Environmental chemistry is also used in environmental monitoring, various strategies of which are applied to monitoring the quality of the environment, especially with an eye toward preparing an environmental impact assessment. Environmental monitoring specifies the steps that need to be taken to prepare such assessments, including various measurements, the sampling methods used for such measurements, and the manner in which the data should be interpreted. Environmental chemistry helps inform these decisions because of the sheer number of chemicals present in the environment and the necessity of determining which are relevant to tracking environmental quality and change—a list that is altered over time in response to new information and especially to changing conditions, such as the chemical impact of acid rain on waterways and soil, requiring the measurement of various chemicals that previously could have been assumed to be present in irrelevant quantities if at all.

Environmental chemistry is particularly concerned with water and soil pollution. Soil pollution results from the introduction of man-made chemicals into the soil through the use of pesticides, the percolation of contaminated surface water to subsurface soil, leaks from underground storage or leaching from landfills, or direct dumping of oil, fuel, and industrial waste. Typically, these pollutants are petroleum hydrocarbons, heavy metals, solvents, or pesticides. Discovering the extent of soil pollution is a complicated task, as is the cleanup, but polluted soil poses a public health risk not only through the possibility of drinking water contamination but because some pollutants can vaporize and enter the air. Chronic exposure to some pollutants can contribute to cancer; others can lead to birth defects; others can cause chronic health problems that continue long after the pollutant is gone. Furthermore, there is damage to the ecosystem and deleterious consequences that can pass up the food chain, from the microorganisms and worms living in the soil to the small animals that feed on them to those animals' predators (which, of course, may include humans). Effects on plant life may impact crop yields or kill off local vegetation, contributing to soil erosion.

Computer models are used to inform the cleanup of soil pollution. Bioremediation and phytoremediation use living things to remove pollutants from soil: in the former, microbes digest harmful organic chemicals, while in the latter, plants pull heavy metals up out of the soil through their roots. Electromechanical systems can be used to extract groundwater and soil vapor and strip contaminants from it. Aeration can remove pollutants from soil, but may result in dispersing them in the air. In some cases, heat can raise soil temperatures high enough to volatize chemical compounds, which can then be extracted in vapor form. And the simplest form of soil pollution cleanup is simply to remove the soil entirely, or to contain it where it is, such as by paving over. The pros and cons of various methods depend on the nature of the pollution and the context of the polluted area.

Environmental chemistry also includes aquatic toxicology, the study of chemicals and other anthropogenic factors on aquatic life. Focused on more than just cleaning up water pollution, aquatic toxicology examines its effects at various scales, from the sub-cellular to large communities and aquatic ecosystems. The field has developed standardized tests that are used to evaluate the effects of discharge into surface water; in the United States these tests are used as part of the process of evaluating a wastewater permit application submitted to the National Pollutant Elimination System, managed by the U.S. Environmental Protection Agency (EPA) in accordance with the Clean Water Act.

Geology and the Earth Sciences

The earth sciences include those physical sciences devoted to the study of the Earth not already mentioned above. Geology studies the lithosphere, the rocky crust of the planet. Geophysics and geodesy study the Earth's shape, its magnetic and gravity fields, its core and mantle, and the tectonic and seismic activity of the lithosphere. Soil science studies the soil on the Earth's outermost layer of crust, while glaciology studies the cryosphere, the icy parts of the planet. Because the Earth is so physically diverse, different fields need to rely on different tools, techniques, and approaches to study different elements of it.

Environmental geology is specifically focused on applying geology to solving environmental problems and draws heavily on engineering geology, the practical side of geology that studies the geological factors affecting engineering works and human development, and vice versa. Environmental geologists work on ways to better manage geological resources such as water, land, minerals, and fossil fuels; identify the human impact on natural hazards and find mitigation strategies; and find ways to reduce pollution and better manage waste disposal. Environmental geology often overlaps with hydrogeology, which studies the distribution of groundwater in the Earth's crust and thus the transport of contaminants in that groundwater.

Environmental soil science deals with the interaction between human activity and the soil, including issues like erosion control and the loss of wetlands; septic drain fields; land treatment of wastewater; storm water; soil pollution; the effects of climate change and acid rain; the management of soil nutrients; the presence of bacteria and viruses in soil; the development of new soil microbes for use in cleaning up pollution; and the study of man-made soils.

Oceanography is often included in the earth sciences, studying as it does the majority of the planet's surface—its oceans. A broad field, oceanography draws on biology, meteorology, geology, and chemistry in its study of aquatic organisms and ecosystems, as well as physics to understand the ocean currents, waves, plate tectonics, and chemical fluxes that affect the fluctuating conditions of the marine environment. The subfield of marine chemistry is concerned with the chemistry of the ocean, the effect of anthropogenic contaminants, and the ocean's chemical interaction with the atmosphere, while marine biology studies marine life forms and their ecosystems.

Biogeochemistry is the intersection of biology, chemistry, and geology, studying the cycles of chemical elements—particularly carbon, nitrogen, sulfur, and phosphorous—that are either driven by biological activity or impact it. Carbon sequestration, the cleanup of soil pollution, and the effects and mitigation of climate change are common areas of study.

See Also: Biotechnology; Carbon Capture Technology; Clean Energy; Eco-Electronics; Engineering Studies; Geoengineering; Green Chemistry; Postindustrialism; Social Construction of Technology; Sustainable Design.

Further Readings

Carson, Rachel. *Silent Spring*. Boston, MA: Houghton Mifflin, 1962.

Cunningham, William and Mary Ann Cunningham. *Environmental Science: A Global Concern*. New York: McGraw-Hill, 2009.

Easton, Thomas. *Taking Sides: Clashing Views on Environmental Issues*. New York: McGraw-Hill, 2010.

Ehrlich, Paul. *The Population Bomb*. New York: Ballantine Books, 1983 [1968].

Friis, Robert H. *Essentials of Environmental Health*. Boston, MA: Jones & Bartlett, 2006.

Miller, G. Tyler and Scott Spoolman. *Environmental Science*. Florence, KY: Brooks/Cole, 2010.

Vig, Norman J. and Michael E. Kraft. *Environmental Policy: New Directions for the Twenty-First Century*. Washington, DC: CQ Press, 2009.

Bill Kte'pi
Independent Scholar

EUROPEAN UNION RESTRICTION OF HAZARDOUS SUBSTANCES (RoHS) DIRECTIVE

Environmental protection is integral to the development of European Union (EU) law and policies. Indeed, it provides the stimulus for a vast number of pieces of EU secondary legislation. Directive 2002/95/EC on the restriction of the use of certain hazardous substances in electrical and electronic equipment (the RoHS Directive) addresses the need to protect the environment and human health from harm posed by toxic metals and hazardous components, materials, and substances contained in electrical and electronic equipment (EEE). It does this by requiring producers of EEE to ensure that their products and components placed on the EU market comply with RoHS.

The aim of the RoHS Directive, also referred to as the "Lead Free Directive," is to restrict the use and reduce the content of six hazardous substances contained in EEE: lead, mercury, cadmium, hexavalent chromium, polybrominated biphenyls (PBB), and polybrominated diphenyl ether (PBDE). In this way, the directive supports the objectives of its sister directive 2002/96/EC on waste electrical and electronic equipment. Recital 6 states: "Restricting the use of these hazardous substances is likely to enhance the possibilities and economic profitability of recycling of WEEE." The RoHS Directive seeks to protect the environment and human health from the risks of hazardous substances contained in EEE and contributes to the environmentally sound recovery and disposal of waste electrical and electronic equipment (WEEE), which reflects the priority order of the waste hierarchy and the overarching aspirations of the Waste Framework Directive (Directive 2008/98/EC). By establishing clear maximum levels of certain substances contained in EEE, the RoHS Directive (1) facilitates greater opportunities for the recycling and recovery of WEEE, and (2) reduces the risks posed by these materials both during the manufacture of products and at the treatment and disposal stages at the end of their life. In addition, the RoHS Directive steers the focus of producers and manufacturers of such products toward prevention by encouraging reconsideration of existing product designs to incorporate greater life-cycle thinking and ecodesign improvements.

The key requirement of the RoHS Directive is established in Article 4, which bans producers from putting new EEE on the EU market if those goods contain more than certain levels of specific substances. Under the directive, *producers* are defined as any person who manufactures and sells EEE under a brand, resells under the producer's own brand equipment produced by other suppliers, or imports or exports EEE on a professional basis into a member state. All producers must comply with the requirements of the RoHS Directive and the restrictions placed upon the use of hazardous substances.

PBB and PBDE are flame retardants that persist in the environment and accumulate in living organisms. There is evidence to suggest that although they save lives and property by delaying combustion, they are highly toxic and are linked to irreparable damage of the nervous and reproductive systems as well as liver and thyroid cancer. The other four hazardous substances regulated under the RoHS Directive are metals, exposure to which through breathing and ingesting can damage organs and affect the nervous system, increase heart disease, and damage reproduction in humans, animals, and aquatic life. As a measure of the damage that can be done by these substances, the cadmium from just one mobile phone is enough to pollute 600,000 liters of water. With the amount of WEEE generated across the EU and globally growing faster than any other waste stream, it is unsurprising that the content of hazardous components in EEE has become a regulatory concern.

The scope of the RoHS Directive is currently limited to categories 1–7 and 10 of Annex I of the WEEE Directive and to electric light bulbs and luminaires in households. The EEE covered by RoHS includes small household equipment, large household equipment, IT and telecom equipment, consumer equipment, lighting equipment, electrical and electronic tools, toys, leisure and sports equipment, and automatic dispensers. While PBB and PBDEs are manufactured compounds used in flame retardants in plastics within, for example, computers, televisions, and furniture, the coverage of RoHS does not extend to the latter.

The impact of RoHS is further reduced through a number of exemptions that are also granted within the directive on the grounds of technical limitations. These exemptions include large-scale industry tools, spare parts used for repair, capacity, expansion, or upgrade of EEE, and re-use of EEE put on the market before July 1, 2006. Other exemptions are contained in the Annex to the RoHS Directive and include applications of mercury in certain florescent lamps, lead in glass for cathode ray tubes, lead as an alloying element in steel, aluminum and copper, and cadmium for plating. Exceptions appear to be allowed where either there is no alternative or substitute substance, where substitutes are excessively expensive, where there is an issue of security (national, military, or functional), or where reliability depends upon the presence of the substance.

Since July 1, 2006, all applicable new products placed on the EU market, including products shipped by companies located outside of the EU into the EU, have to pass RoHS compliance. Since directives are a form of EU secondary legislation that obliges each member state to transpose into its national laws the obligations contained in the measure, the RoHS Directive also left it to each member state to establish effective, proportionate, and dissuasive penalties for breaches of compliance.

While the RoHS Directive appears to be straightforward, its implementation proved challenging. The main stumbling block for producers was the lack of clarity in the directive, which resulted in uncertainty regarding the scope of the obligations and whether companies were operating inside or outside compliance. For example, the reference to "put on the market" contained in Article 4 of the directive was difficult to apply to all market situations. Likewise, it was unclear whether noncompliant RoHS products placed on the market prior to July 1, 2006, and used for demonstration can be sold secondhand following the deadline and whether the exclusions to military products extend to military products destined for a civilian market.

In response to the lack of clarity on legal provisions and definitions contained in the RoHS Directive and the disparities existing between member states in the approaches they have taken to implementing the requirements, the European Commission in December 2008 published a Proposal to Recast the RoHS Directive (the Proposed Recast). Along with the commitment toward better regulation through the provision of clear, simple, consistent legislation that minimizes the burdens on businesses and establishes legislative certainty, the Proposed Recast also seeks to include EEE categories 8 and 9 (medical devices and monitoring and control instruments). While the review and strengthening of the RoHS Directive should be welcomed, the proposal is still under consideration with the European Parliament and its effect, if adopted, remains uncertain.

Similar legislation restricting hazardous substances exists in a number of countries, for example, China, under the Administrative Measure on the Control of Pollution Caused by Electronic Information Products. Draft revisions to this measure were announced in August 2010 by the Ministry of Industry and Information Technology of the People's Republic of China. In Japan, RoHS compliance is established in the Law for the Promotion of Effective Utilization of Resources, and over half the states in the United States of America have or have proposed instruments for the regulation of hazardous substances. By way of an example, the Electronic Waste Recycling Act 2003 applies to California and came into force on January 1, 2007. It is modeled on the EU RoHS Directive and affects manufacturers, distributors, wholesalers, and retailers but has a more limited scope in terms of the substances it covers.

Undoubtedly, regulating for the restriction of certain hazardous substances has really only just begun. As we increase our understanding of the long-term effects on the environment and human health of substances and materials contained in products we consume, the list of substances that are regulated through outright bans or stringent restrictions is bound to increase.

See Also: European Union Waste Electrical and Electronic Equipment (WEEE) Directive; E-Waste; Science and Technology Policy.

Further Readings

European Commission. Directive 2002/95/EC on the Restriction of the Use of Certain Hazardous Substances in Electrical and Electronic Equipment. OJ [2003] L 37/19.

European Commission. "Proposal for a Directive on the Restriction of the Use of Certain Hazardous Substances in Electrical and Electronic Equipment (Recast)." COM (2008) 809 Final.

Wright, Robin and Karen Elcock. "The RoHS and WEEE Directives: Environmental Challenges for the Electrical and Electronic Products Sector." *Environmental Quality Management* (Summer 2006).

Hazel Nash
Cardiff University

European Union Waste Electrical and Electronic Equipment (WEEE) Directive

Directive 2002/96/EC amended by Directive 2003/108/EC on Waste Electrical and Electronic Equipment (the WEEE Directive) is one of a number of pieces of European Union (EU) secondary legislation that aims to attribute to producers the financial responsibility throughout the life cycle of their electrical and electronic products. The obligations contained in the WEEE Directive require producers to facilitate the dismantling and recovery for reuse and recycling of waste electrical and electronic equipment (WEEE), often referred to as "e-waste," according to their market share by product type. By doing so, the intention of such extended producer responsibility (EPR) instruments is to incentivize shifts in current patterns of production and consumption toward more sustainable, long-term, and collectively beneficial operations.

The WEEE Directive is a result of the European Parliament's review in 1996 of the EU's strategy for waste management recognition. They recognized the need to develop proposals as a matter of importance for legislation in a number of priority waste streams, including electrical and electronic equipment (EEE). The WEEE Directive seeks to address increasing volumes of WEEE by 4 percent per year, and reduce the environmental impact associated with the hazardous nature of components in EEE, for example toxic metals, hexavalent chromium, cadmium, and similar carcinogens. It also attempts to promote changes toward more sustainable patterns of consumption and production by ensuring markets for EEE send the right signals to consumers. EPR is the mechanism that has been adopted in the WEEE Directive to shape and secure these changes.

The main objective of the WEEE Directive is to prevent e-waste from arising and maximize the reuse, recycling, and recovery of wastes by improving the environmental performance of all operators in the life cycle of electrical and electronic equipment. It does so by establishing EPR obligations on producers of EEE over the course of a product's life. Extending the responsibility of producers throughout the life cycle of their products—in particular, the point at which products become waste—plays a key role in achieving the decoupling of economic growth from resource use. Internalizing the costs associated throughout the life cycle of a product—including collection, treatment, and recycling—producers are incentivized to substitute materials and alter product designs to accommodate all phases of a product's life. In this way, the aim behind EPR is to encourage producers and retailers to take responsibility for the product throughout the product's life, including shifting the focus of designers and manufacturers to include maintenance, upgrade, and after-sales service, rather than just at the moment of transaction in the store.

Environmental Issues

Design for environment has the ability to address many of the environmental issues throughout a product's life cycle—reducing resource use along the entire life cycle. This is reflected in Article 4 of the WEEE Directive, which encourages innovations in product design and manufacturing processes to facilitate greater dismantling, reuse, and recovery. However, at present, investment in design for the environment has been limited by Article 5(1), which enables producers to meet their obligations in relation to the separate collection of WEEE either through a collective take-back system or through individual take-back schemes. Similarly, Article 6 provides for the establishment of individual or collective systems for the treatment of WEEE. While the WEEE Directive includes the more burdensome individual producer responsibility option, it is not mandated. This flexibility in securing producers' compliance with their WEEE obligations has resulted in less focus on the upstream stages of product life cycles and design. This supports Ashley Deathe, Elaine McDonald, and William Amos's observation that collective programs very rarely incentivize design for the environment.

Producers and distributors of EEE are charged with responsibilities under the WEEE Directive. For the purpose of the obligations, producers are defined as (1) companies that manufacture and sell their own branded products, (2) companies that import EEE, or (3) sellers of their own branded products produced by other manufacturers. The WEEE Directive places a duty on producers to fund the collection, treatment, recovery, and recycling of household WEEE delivered to authorized treatment facilities according to their market share. This can be done by fulfilling their obligations either individually or by joining a collective scheme. Therefore, producers or third parties acting on their behalf must set up systems to provide for the separate treatment, recovery, and recycling of WEEE using best available techniques.

Distributors (any person providing EEE on a commercial basis to the party who is going to use it) have obligations to establish "bring back" systems for households whereby WEEE can be returned to the distributor free of charge. This means that consumers are able to return their WEEE on a one-to-one basis when buying new EEE of equivalent type and fulfilling the same function. However, this system is reliant on consumers being aware of the need to dispose of their WEEE separately in order that equipment is captured for separate treatment and disposal. To this end, Article 10 sets out the information that must be provided for users of EEE from private households, including information on the importance of disposing of e-waste separately, the potential effects on human health and the environment caused by the hazardous nature of components, and the return and collection systems available to them. A further tool available to help raise awareness of the environmental and human health implications associated with inappropriate disposal of WEEE is the visible fee. Article 8 of the WEEE Directive enables producers to voluntarily show consumers at the point of purchase the costs associated with the separate collection, transport, treatment, and disposal of the equipment. While this provision is not mandatory, the visible fee is one way of making consumers aware of the true environmental costs of the products that they buy and face the full consequences of their product choices.

Challenges

Considerable practical and regulatory challenges have been experienced by member states and producers of EEE in implementing their obligations contained in the WEEE Directive. Following the European Commission's review in 2006 of member states' implementation of the WEEE Directive, which identified a number of technical, legal, and administrative difficulties experienced by member states, the European Commission proposed to recast the WEEE Directive. The proposal seeks to provide a vehicle through which weaknesses within the WEEE Directive can be addressed in legislation, should the proposal be adopted by the EU following the co-decision process. As it currently stands, the proposal sets out six main actions addressing the weaknesses in the WEEE Directive. These actions include (1) harmonizing the scope of the EU WEEE and EU RoHS Directives as having their overarching objective the protection of the establishment and functioning of the internal market; (2) introducing a new Annex I to the WEEE Directive that establishes minimum monitoring requirements for the shipment of WEEE to address the current problems with exporting WEEE under the guise of reuse; (3) reducing the administrative burden on producers by establishing a common and "interoperational" registration and reporting system across member states; (4) modification to recovery and recycling targets to promote the reuse of whole appliances; (5) extending producer responsibility further by enabling member states to encourage producers to cover the entire financial burden of collection facilities for WEEE from private households; and (6) introducing new recycling and recovery targets for the separate collection of household and nonhousehold WEEE. A minimum collection rate of 65 percent for producers or third parties acting on their behalf would replace the existing weight-based minimum target of 4 kilograms (kg) per head of population.

An inherent challenge for the WEEE Directive not fully addressed by the proposed recast is the dependency of the framework (from the choice of product purchased through to disposal) on consumer behavior. While the WEEE recast proposal should be welcomed as a step toward a more effective e-waste management system, whether it goes far enough to achieve greater patterns of sustainable consumption and production is questionable. The sustained emphasis on EPR skirts the underlying problem of consumerism. The proposed WEEE Recast was scheduled for vote by the European Parliament in October 2010, but as of early 2011, had not yet been voted on. Certainly, the European regulatory framework for waste, including the sustainable management of WEEE, will continue to be a priority area for improvement at European, member state, and autonomous region levels into the future.

See Also: European Union Restriction of Hazardous Substances (RoHS) Directive; E-Waste; Science and Technology Policy; Sociology of Technology.

Further Readings

Arcadis and RPA. "Study on RoHS and WEEE Directive—Final Report." March 2008, 06/11925/AL, 213-311.

Deathe, Ashley L. B., Elaine MacDonald, and William Amos. "E-Waste Management Programmes and the Promotion of Design for the Environment: Assessing Canada's Contributions." *Review of European Community and International Environmental Law* 17 (3) (2008).

European Commission. "Communication From the Commission on the Review of the Community Strategy for Waste Management." COM (1996) 399.

European Commission. "Proposal for a Directive on Waste Electrical and Electronic Equipment (WEEE) (Recast)." COM (2008) 0810 Final.

European Commission. "The Review of Directive 2002/96/EC on Waste Electrical and Electronic Equipment (WEEE)." Brussels: European Commission, June 2006.

Institute of Environment and Human Security, United Nations University. "Review of Directive 2002/96 on Waste Electrical and Electronic Equipment (WEEE)—Final Report." Bonn: United Nations University, August 2007.

Ponting, Cerys and Hazel Ann Nash. "A Business Perspective on the Transposition of the WEEE Directive Into UK Law: Part 1." *BRASS Working Paper Series* No. 56 (2010).

Hazel Nash
Cardiff University

E-Waste

Electronic equipment that has ceased to be of value to users, such as these old computers, may contain recoverable precious materials, thereby making it a different kind of waste than traditional municipal waste.

Source: Wikimedia

E-waste is a generic term embracing various forms of electric and electronic equipment that have ceased to be of value to their users or no longer satisfy their original purpose either through redundancy, replacement, or breakage. Accordingly, e-waste includes both white goods (refrigerators, washing machines, microwaves) and brown goods (televisions, radios, and computers) that have exhausted their utility value. The information and technology revolution has exponentially increased the use of new electronic equipment. It has also produced growing volumes of obsolete products, thereby making e-waste one of the fastest-growing waste streams. The impact of global consumerism increases the usage of electronic devices as microprocessors become increasingly pervasive. E-waste contains complex combinations of highly toxic substances that pose a danger to health and environment. However, e-waste may contain recoverable precious materials, thereby making it a different kind of waste compared to traditional municipal waste.

Globally, about 40 million tons of e-waste are generated annually, comprising more than 5 percent of all municipal solid waste. Domestically, the United States accounts for 3.3 million tons whereas European Union member states generate around 8.3–9.1 million tons per year. Projections for the next 10 years indicate that the sale of electronic products in developing countries such as China, India, Mexico, Brazil, and South Africa is set to rise sharply. By 2020, in South Africa and China e-waste from old computers will have jumped by 200 percent to 400 percent from 2007 levels, and in India by 500 percent. In Senegal and Uganda, e-waste will increase four- to eightfold by 2020.

In addition, the illegal transboundary movement of e-waste in the form of donations and charity from rich industrialized nations to developing countries is turning out to be a dumping-ground exercise for e-waste. Statistics indicate that of 8.7 million tons of e-waste produced in the EU, only 25 percent is collected and treated, whereas in the United States, only 20 percent of e-waste is separated for further processing and recovery—the rest is exported to the third world. Recycling of e-waste in an informal setup for extraction and sale of metals is another reason for cross-country movement due to the availability of cheap labor in developing countries. The cost of recycling a single computer in the United States is $20, while the same could be recycled in India for $2: a gross savings of $18 if the computer is exported to India. The big recyclers of developed countries with strict environmental regimes and an increasing cost of waste disposal find exportation to small traders in developing countries more profitable than recycling in their own countries.

Unhindered by international laws, particularly the Basel Convention on the Control of Transboundary Movements of Hazardous Wastes and their Disposal, and unaffected by the plight of workers, the traders in both developed and developing countries harvest substantial profits due to lax environmental laws, corrupt officials, and poorly paid workers. As a result, there is an urgent need to develop policies and strategies to dispose of and recycle e-waste safely in order to achieve a sustainable future.

E-Waste: Classification

E-waste can be classified on the basis of its composition and components. E-waste is composed of various materials with a wide range of physiochemical properties that are toxic. Ferrous and nonferrous metals, glass, plastics, pollutants, and other are the six categories of materials reported for e-waste composition. Iron and steel constitute the major fraction in waste electrical and electronic equipment (WEEE) materials to the level of 47.9 percent as against plastics, the second largest and constituting 20.6 percent. Nonferrous material takes third place, amounting to 12.7 percent. It includes metals such as copper and aluminum, and precious metals such as silver, gold, and platinum. These have significant commercial value. Materials include lead and cadmium in circuit boards, lead oxide and cadmium in CRTs (cathode ray tubes), mercury in switches and flat-screen monitors, brominated flame retardants on printed circuit boards, plastic and insulated cables—beyond the threshold quantities, these are regarded as pollutants and, if disposed of improperly, can damage the environment.

Accordingly, the most widely accepted classification is per EU directives wherein e-waste is covered under 10 categories, including the following:

1. Large household appliances (refrigerators, freezers, washing machines, clothes dryers, dishwashers, electric cooking stoves and hot plates, microwaves, electric fans, and air conditioners)

2. Small household appliances (vacuum cleaners, toasters, grinders, coffee machines, appliances for hair cutting and drying, tooth brushing, and shaving)

3. Information technology (IT) and telecommunications equipment (mainframes; minicomputers; personal computers including CPU, mouse, screen, and keyboard; laptops; notebooks; printers; electronic typewriters; facsimile; telex; telephones and cellular telephones)

4. Consumer equipment (radios, televisions, video cameras, video recorders, stereo recorders, audio amplifiers, and musical instruments)

5. Lighting equipment (straight and compact fluorescent lamps and high-intensity discharge lamps)

6. Electrical and electronic tools (drills; saws; sewing machines; equipment for turning, milling, grinding, drilling, making holes, folding, bending, or similar processing of wood and metal; tools for screwing, nailing, welding, and soldering)

7. Toys, leisure equipment, and sporting goods (electric trains or racing car sets, video games including hand-sets, computers for biking, diving, running, and sports equipment with electric elements)

8. Medical devices (radiotherapy equipment, cardiology, dialysis, pulmonary ventilators, nuclear medicines, analyzers, and freezers)

9. Monitoring and control instruments (smoke detectors, heating regulators, and thermostats)

10. Automatic dispensers (for hot drinks, hot or cold bottles, solid products, money, and all appliances that deliver automatically various products)

About 95 percent of the total e-waste is contributed by category numbers 1 through 4 above. Individually, large household appliances contribute 42.1 percent; small household appliances contribute 4.7 percent; IT and telecommunications account for 33.9 percent, followed by consumer equipment at 13.7 percent. The remainder of the categories (numbers 5 through 10 above) vary between 0.1 and 1.4 percent.

E-Waste and Human Health

The complex composition and improper handling of e-waste adversely affect human health. A growing body of epidemiological and clinical evidence has led to increased concern about the potential threat of e-waste to human health. In developing countries such as India and China, the improper handling of e-waste affects multiple vital organs in the body due to the release of toxic chemicals. Primitive and crude methods are used by inefficient and unregulated backyard operators (informal sector) to reclaim, reprocess, and recycle materials used in electronic equipment. Processes such as dismantling components, wet chemical processing, and incineration are used. They result in direct exposure and inhalation of harmful chemicals that can affect the health of workers, particularly in the most vulnerable section of society—women and children. The informal sector is purely market driven, as it is a source of livelihood for indigent workers. It is also a business opportunity for employers and a source of secondhand, cheap goods for customers. Safety equipment such as gloves, facemasks, and ventilation fans are virtually unknown, and workers often have little idea of what they are handling.

For instance, in terms of health hazards, open burning of printed wiring boards increases the concentration of dioxins in the surrounding areas. These toxins cause an increased risk of cancer if inhaled by workers and local residents. Toxic metals and poison enter the bloodstream during the manual extraction and collection of tiny quantities of precious metals. Workers are continuously exposed to poisonous chemicals and fumes of highly concentrated acids. Recovering resalable copper by burning insulated copper wires causes neurological disorders. Evidence suggests that copper may influence the progression of Alzheimer's disease. Acute exposure to cadmium, found in semiconductors and chip resistors, can damage the kidneys and liver and cause bone loss. Long-term exposure to lead on printed circuit boards and computer and television screens can damage the central and peripheral nervous system and kidneys. Children are more susceptible to the carcinogenic effects of lead because their resistance to lead is less than that of adults.

E-Waste and the Environment

Although electronics constitute an indispensable part of our everyday lives, their hazardous effects on the environment cannot be overlooked or underestimated. The interface between electrical and electronic equipment and the environment, as discussed by A. Bandyopadhyay, takes place during manufacturing, reprocessing, and disposal of EEE. The emission of fumes/gases/particulate matter in air, discharge of liquid waste in water and drainage systems, and disposal of hazardous wastes in land contribute to the harmful impact on the environment that leads to environmental degradation.

For example, burning to recover metal from wires and cables leads to emissions of brominated and chlorinated dioxins, causing air pollution. Similarly, in the manufacture of printed circuit boards that uses lead or tin soldering, only 7 percent of the material (copper) is retained on the board; 93 percent is wasted through chemical processes that in turn can cause soil and water pollution. During the recycling process in the informal sector, toxic chemicals that have no economic value are simply dumped. The toxic industrial effluent is poured into underground aquifers and seriously affects the local groundwater quality, thereby making the water unfit for human consumption or for agricultural purposes. Atmospheric pollution is caused due to dismantling activities as dust particles loaded with heavy metals and flame retardants enter the atmosphere. These particles either redeposit (wet or dry deposition) near the emission source or, depending on their size, can be transported over long distances. The dust can also enter the soil or water systems and, with compounds found in wet and dry depositions, can leach into the ground and cause both soil and water pollution. Soils become toxic when substances such as lead, mercury, cadmium, arsenic, and polychlorinated biphenyls (PCBs) are deposited in landfills.

E-waste is a rapidly growing environmental hazard. The solution to this problem lies not in shortsighted, indiscriminate landfill use or unethical exportation but in evolving policies that include extending the responsibility of all stakeholders, particularly the producers, beyond the point of sale and up to the end of product life.

See Also: European Union Restriction of Hazardous Substances (RoHS) Directive; European Union Waste Electrical and Electronic Equipment (WEEE) Directive; Extended Producer Responsibility.

Further Readings

Bandyopadhyay, Amitava. "A Regulatory Approach for E-Waste Management: A Cross-national Review of Current Practice and Policy With an Assessment and Policy Recommendation for the Indian Perspective." *International Journal of Environment and Waste Management*, 2/1–2 (2008).

Greenpeace International. "Toxic Tech: Not in Our Backyard." http://www.greenpeace.org/international/en/publications/reports/not-in-our-backyard (Accessed September 2010).

Sarkar, Atanu. "Occupational and Environmental Health Perspectives of E-waste Recycling in India: A Review." In *E-Waste—Implications, Regulations and Management in India and Current Global Best Practices*, by Rakesh Johri. New Delhi: TERI, 2008.

Sepulveda, Alejandra, et al. "A Review of the Environmental Fate and Effects of Hazardous Substances Released From Electrical and Electronic Equipments During Recycling: Examples From China and India." *Environmental Impact Assessment Review*, 30 (2010).

United Nations Environment Programme. "Urgent Need to Prepare Developing Countries for Surges in E-Wastes." http://www.unep.org/Documents.Multilingual/Default.asp?DocumentID=612&ArticleID=6471 (Accessed September 2010).

Gitanjali Nain Gill
University of Delhi

EXTENDED PRODUCER RESPONSIBILITY

Extended producer responsibility (EPR) is a practice and a policy approach in which producers "take responsibility" for management of the disposal of products they produce once those products are designated as no longer useful by consumers. "Responsibility" for disposal may be fiscal, physical, or a combination of the two. Motivations for extended producer responsibility practices include a mixture of economic, environmental, and social factors. Extended producer responsibility shifts the economic burden of the cost of disposal from the government to the producer of the product. Within an environmental context, products must be designed for recyclability, and extended producer responsibility encourages design for recycling while discouraging the use of toxic components in the product. Finally, extended producer responsibility meets increasing consumer demand for "environmentally friendly products" that can easily be recycled and/or are manufactured using recycled content. Extended producer responsibility is a product-focused strategy that encourages environmentally friendly design and disposal of products through transfer of this responsibility to product producers.

Benefits of Extended Producer Responsibility

In extended producer responsibility, producers of products are responsible for product disposal at end of life, the point at which products are designated as no longer useful by consumers. The Organisation for Economic Co-operation and Development's (OECD) definition of extended producer responsibility identifies two specific features: the shifting of responsibility for disposal "upstream" from municipalities to producers and encouragement through incentives to make more environmentally friendly the design of products.

Rather than regulate disposal of products through traditional end-of-pipe command-and-control methods, extended producer responsibility is a preventative measure using a life-cycle or a "cradle-to-grave" perspective. Extended producer responsibility policies attempt to change how a product is produced—the "cradle"—to affect how a product may be disposed of—the "grave."

The benefits of extended producer responsibility policies may be categorized as economic, environmental, and social. The cost for the management and coordination of the disposal of solid waste is most often the responsibility of local municipalities. Rising levels of waste generation, more stringent technical requirements for the operation of landfills and incinerators, and increasing difficulty in constructing new waste disposal facilities due to public opposition all contribute to the growing cost of waste disposal. The government, and more specifically local municipalities, is generally fiscally and physically responsible for waste disposal. Extended producer responsibility is an attempt to provide an incentive to producers to design products with reduced environmental impacts while shifting the costs associated with disposal to the producer.

Environmental motivations for extended producer responsibility include increasing product recyclability, decreasing use of toxic components in products, and reducing the amount of material that is sent to a landfill or incinerator rather than reused or recycled. The recyclability of products is heavily dependent on product design. It is difficult to recycle products that are not designed for dismantling, have high levels of toxic ingredients, or have components such as composite resins that are problematic to recycle. With extended producer responsibility, designing products that can be easily recycled or reused is theoretically in the producer's best interest because they are responsible for disposal. In addition to assigning responsibility for the disposal of products, most extended producer responsibility policies also require producers to recycle a specified percentage of the collected product by weight or volume. Product recycling can reduce the amount of energy needed to manufacture a product and the creation of associated air and water pollution in comparison to producing a product from raw, virgin materials. Finally, requiring producers to take responsibility for the disposal of products reduces the volume of material disposed of through incineration or in a landfill. Decreasing landfill capacity is cited as a key motivator for extended producer responsibility policies in the European Union.

Extended producer responsibility also has social benefits. The implementation of extended producer responsibility policies may improve the public image of a company. Consumer demand for "environmentally friendly" products that can be recycled at end of life and manufactured with fewer toxic materials is increasing.

Forms of Extended Producer Responsibility Policies

Responsibility for the management of product disposal may be economic and/or physical as well as individual to companies, or collective, organized through a number of companies. With extended producer responsibility, the producer of a product finances the cost of product disposal. This may take the form of physically collecting the products at end of life or by using a producer responsibility organization (PRO). A PRO is a third-party organization that collects and processes material. With a producer responsibility organization, producers do not physically "take back" the product but instead support the process financially. In the German DSD system (described below), for example, manufacturers are required to assume responsibility for financing the disposal, through recycling, of created consumer packaging waste, but do not physically collect the products.

Extended producer responsibly practices may be voluntary or regulatory. With voluntary practices, public and private sector departments and organizations often in cooperation with nongovernmental organizations work to develop practices of extended producer responsibility in the place of formal regulation.

Examples of Extended Producer Responsibility Policies

Both the German and the Swedish governments are noted as early adopters of extended producer responsibility policies in the 1990s. Early extended producer responsibility policies in European nations and Japan targeted consumer packaging. The German Packaging Ordinance of 1991, establishing the Duales System Deutschland (DSD), is considered one of the first national-level extended producer responsibility policies. In 1994, the European Union adopted the Packaging Directive, creating union-wide targets for recycling while allowing individual countries discretion in implementation. Japanese consumer packaging policies designate responsibility for the collection and processing of packaging waste to local governments while manufacturers are responsible for the financial cost of recycling.

Extended producer responsibility policies today exist for a wide range of products. In recent years, many targeted waste electric and electronic equipment due to the growing volume and disposal and toxicity concerns. The European Union has issued extended producer responsibility directives for end-of-life vehicles, waste electrical and electronic equipment, use of certain hazardous substances in electrical and electronic equipment, and waste batteries and accumulators.

While voluntary extended producer responsibility practices in the United States exist on the national level for several products, coordinated by the U.S. Environmental Protection Agency, more than 30 states have enacted regulatory extended producer responsibility legislation for products such as electronics, batteries, mercury thermometers, and others.

Property Rights and Extended Producer Responsibility

Extending responsibility to the producer of a product for a specific externality—that of disposal—deviates from traditional notions of property rights where the responsibility of a producer ends at the point of sale of the product. The extension of property rights to the externality of waste is not unprecedented and has been used as a policy tool for hazardous wastes along with other forms of product liability laws.

Purpose and Limitations of Extended Producer Responsibility

Practices and policies of extended producer responsibility assign responsibility for disposal to producers while encouraging environmentally friendly product design. Extended producer responsibility does not reduce the volume of waste created but rather attempts to reduce the volume of material disposed of through landfilling or incineration. Unless specifically mandated, extended producer responsibility does not necessarily result in the creation of a more durable, longer-lasting product or address waste creation due to practices of planned obsolescence. Although there is a focus on reducing toxics in production, extended producer responsibility is in general directed specifically toward the disposal of a product. It is not a strategy to reduce the environmental impact of production or consumption of a product. Extended producer responsibility has been criticized as a concept that works well in theory but has yet to show strong, quantifiable results of the ability to green product design.

See Also: Cradle-to-Cradle Design; Design for Recycling; Product Stewardship; Sustainable Design.

Further Readings

Lindhqvist, T. *Extended Producer Responsibility in Cleaner Production: Policy Principle to Promote Environmental Improvements of Product Systems.* IIIEE Dissertations 2000:2. Lund: IIIEE, Lund University, 2000.
Organisation for Economic Co-operation and Development (OECD). *Extended Producer Responsibility: A Guide Manual for Governments.* Paris: OECD, 2001.
Walls, M. *EPR Policies and Product Design: Economic Theory and Selected Case Studies.* Paris: OECD, 2006.

Sarah M. Surak
Virginia Tech

FAUSTIAN BARGAIN

The term *Faustian bargain* refers to making a pact with either the devil or some other evil entity. The Faust referenced in the term is a character in German legend and literature who trades his soul for otherwise unattainable knowledge and magical powers that will give him access to all the world's pleasures—by extension, the term refers to making an agreement that delivers something highly valued today but with the expected loss in the future of something that is also highly valued. It also may carry the implication of doing business with an unsavory individual or entity (a metaphorical devil) in order to get something that you want. The phrase *Faustian bargain* was coined by Dr. Alvin Weinberg, director of the Oak Ridge National Laboratory from 1955 to 1972, to refer to the proliferation of nuclear power plants but has since been applied to other environmental compromises and proposals.

The Faustian Bargain and Nuclear Power

In 1972, Weinberg used the term *Faustian bargain* in a speech on nuclear safety, arguing that while the breeder reactor offered an almost inexhaustible source of energy that is cheap and almost nonpolluting (in contrast to energy created from fossil fuels), great risks were inherent in the technology. Furthermore, society would have to establish forms of vigilance effective far into the future, which were unprecedented in comparison to more conventional sources of power. The phrase quickly entered the vocabulary: for instance, Allen Kneese used the term in a 1973 essay in which he argued that conventional cost-benefit analysis was insufficient to answer important questions concerning the development of a society based on large-scale development of nuclear power (power based on nuclear fission technology) because it neglected the ethical dimension of those questions. It should be noted that Kneese was not arguing the nuclear fission technology definitely should not be adopted, but that if such a Faustian bargain were to be made, it should be done publicly and with complete and serious national discussion of the issues involved.

Kneese noted that nuclear fission technology offers several advantages for energy generation when compared to current technologies based on conventional power plants that burned fossil fuels. One was reduced environmental impact: except for residual heat production, the routine operation of nuclear power facilities has a lower environmental impact than those powered by fossil fuels. Another is creation of an increased, perhaps almost unlimited, energy supply not dependent on foreign oil or diminishing reserves of coal (note that he was writing in 1973, the year of an oil crisis caused by members of the Organization of Petroleum Exporting Countries [OPEC] limiting supply to the United States). Costs appeared to be similar for the two technologies while the potential of other alternatives such as geothermal energy remained largely unknown because the technologies were not sufficiently developed. Against these factors, which are fairly easy to compare in a conventional cost-benefit analysis, Kneese pointed out that as Weinberg earlier stated, adoption of nuclear fission energy production would impose a previously unknown burden: that of continuous monitoring and management of a highly dangerous material. He noted that this burden would be essentially eternal and that—even if nuclear power were used for only a few decades until other alternatives were developed—the burden of dealing with radioactive waste would remain long after the plants had ceased operation.

In addition, adoption of nuclear fission technology involves the problem of redistribution, which cannot be accommodated in conventional cost-benefit analyses. Redistribution in this case means that adoption of nuclear technology would impose severe responsibilities and possibly hazards on future generations who were not involved in the decision to adopt the technology in the first place (and might not even benefit from it, for instance, if nuclear power were no longer used but the radioactive waste still had to be stored and monitored). To cast this in terms of the Faust legend, it would be as if Faust had obtained knowledge and wealth by bargaining away the souls of generations yet to be born while incurring no such loss himself.

Two levels of hazards are involved in using nuclear fission technology for power generation: those hazards pertaining to the operation of the plants and others pertaining to the long-term storage of radioactive waste. For instance, unexpected accidents had already occurred in the limited number of test facilities in operation in the early 1970s. Because the systems involved are so complex and the possibilities for experimentation and testing are limited, it may be impossible to calculate the actual risks of critical failures or even to identify important risks. In the absence of such information, it is impossible to discuss risks and benefits meaningfully. Kneese also argued that the general public was not aware of the magnitude of disaster that could occur with a nuclear disaster (in fact, the magnitude of disaster that could occur with an accidental release of plutonium, for instance, is almost incalculable because the effects could linger in the environment for tens of thousands of years) and therefore could not be expected to make a reasoned choice about whether to permit their construction or not. It is often argued that the probability of a nuclear accident is so low as to be negligible but given the uncertainties of the new technology, Kneese argued that it would be difficult to make such a calculation, rendering the statement meaningless.

Kneese pointed to other hazards that would be difficult to evaluate numerically but should be considered in the analysis of the benefits and risks of nuclear power. One is that nuclear fission power plants require the use of plutonium, which is also used to build nuclear weapons, and that it was entirely predictable that a black market would develop diverting plutonium from power plants to weapons production (possibly by a rogue state), potentially leading to deployment of nuclear weapons. Another risk is that posed by the transportation and storage of the hazardous materials required to operate a nuclear power plant, both processes that would require continued vigilance and could expose to risk people who did not benefit from the power plant (e.g., if a rail car carrying nuclear materials derailed far from the site) and may have voted against constructing such a plan in their area. Transportation and storage issues raise additional worries if nuclear power facilities are built in less-developed countries, which may have less than adequate security and transportation networks. Finally, radioactive wastes may need to be stored in isolation for thousands of years (one estimate is at least 200,000 years), a time scale that is difficult for most humans to contemplate. The science supporting decisions regarding storage of nuclear waste is not sufficiently developed to answer basic questions such as the necessary characteristics of a secure underground storage facility over so long a length of time and political considerations (e.g., winning a contract to construct such a facility in one's legislative district) may favor short-term concerns over the safety of future generations. Analogies that have been offered to support construction of nuclear facilities, for instance, comparing their construction to that of the pyramids of Egypt or the dike system in Holland, betray a lack of understanding of time spans involved (perhaps 5,000 years for the pyramids versus 200,000 years for the safe storage of radioactive waste) and the potential for disaster caused by a nuclear accident (as opposed to the Dutch dike system, which was breached in the 1950s without major loss of life).

Other Uses

The term *Faustian bargain* has been applied to other choices that affect the environment, perhaps because imperfect compromises often result in efforts to satisfy the demands of competing parties with different priorities. For instance, David Schoenbrod of New York Law School has argued that the U.S. Environmental Protection Agency (EPA) has entered a Faustian bargain in which the role of the devil is played by the U.S. Congress. The EPA, like Faust in Schoenbrod's analysis, has traded away its soul (its responsibility to protect the environment and to produce decisions informed by the best available science) in return for being granted some powers by the U.S. Congress. Part of the price the EPA has paid for this power, Schoenbrod argues, is the willingness to provide cover for Congress by assuming the burden of implementing politically unpopular decisions (such as requiring lead to be removed from gasoline). By doing so, the EPA relinquishes some of its responsibility to protect the environment because reforms could be more quickly put into place by an act of Congress than by EPA regulations. He also notes that the EPA is more of a policy organization than one run by scientists, and that the EPA has been willing to put political considerations above scientific truth, for instance, by claiming that no danger was posed by the levels of pollutants in the air around the World Trade Center in the days following the terrorist attacks of September 11, 2001.

Anti-nuclear advocates have termed the U.S. proposal to ban underground nuclear weapons testing while maintaining nuclear weapons research and development programs a *Faustian bargain* (with the military–industrial complex involved in researching and creating nuclear weapons playing the role of the devil). The argument put forward in favor of this proposal (rather than closing down the nuclear weapons industry) is that maintaining nuclear weapons as well as the knowledge to create them is necessary as a deterrent as long as nuclear weapons exist anywhere in the world. However, anti-nuclear advocates believe that this compromise will undermine future efforts at disarmament because scientists and manufacturers will have a vested interest in continuing to work in the nuclear weapons industry and thus for the industry to continue in order to safeguard their jobs and profits.

Carbon capture and sequestration technology has been proposed as a method to combat global warming, but some scholars have termed this a *Faustian bargain* with the fossil fuel industry because, although it may mitigate pollution in the short term, it lends support to that industry and will probably extend the use of fossil fuels rather than spur the development of less polluting alternatives or of promoting energy efficiency. Ultimately, mitigating the harm caused by using fossil fuels carbon capture and sequestration, critics argue, could lead to greater environmental damage by perpetuating the use of fossil fuels.

See also: Arms Race; Carbon Capture Technology; Technology and Social Change; Unintended Consequences.

Further Readings

Kneese, Allen V. "The Faustian Bargain: Risk, Ethics and Nuclear Energy." In *The RFF Reader in Environmental and Resource Policy*, 2nd ed., Wallace E. Oates, ed. London: RFF Press, 2006.

Lichterman, Andrew and Jacqueline Cabasso. "Faustian Bargain 2000: Why 'Stockpile Stewardship' Is Fundamentally Incompatible With the Process of Nuclear Disarmament." Prepared for the Nuclear Nonproliferation Treaty Preparatory Review Conference at the United Nations, New York City, May 2000. http://www.wslfweb.org/docs/fb2000.pdf (Accessed September 2010).

Schoenbrod, David. "The EPA's Faustian Bargain." *Regulation* (Fall 2006). http://www.cato.org/pubs/regulation/regv29n3/v29n3-5.pdf (Accessed August 2010).

Thompson, Graham. "Burying Carbon Dioxide in Underground Saline Aquifers: Political Folly or Climate Change Fix?" Toronto, Ontario, Canada: Munk Centre for International Studies, 2009. http://beta.images.theglobeandmail.com/archive/00242/Munk_Centre_Paper_242701a.pdf (Accessed September 2010).

Zucker, Alexander. "Alvin M. Weinberg." *Proceedings of the American Philosophical Society*, 152/4 (December 2008). http://www.amphilsoc.org/sites/default/files/1520413.pdf (Accessed September 2010).

Sarah Boslaugh
Washington University in St. Louis

FLUE GAS TREATMENT

Flue gas is produced when fuels such as coal, oil, natural gas, or wood are burned for heat or power. It may contain an assortment of pollutants, including nitrogen oxides, particulates, sulfur dioxide, mercury, and carbon dioxide, although most of the gas consists of nitrogen. Flue gas from power plants, industrial facilities, and other sources can, if left untreated, substantially affect local and regional air quality.

Under clean air regulations, power plants and other facilities are required to use flue gas treatments to reduce the amount of pollutants that are emitted. Such approaches, which use devices such as electrostatic precipitators and scrubbers, can successfully remove 90 percent or more of certain pollutants. However, they can be very costly to install and operate, and requirements for flue gas treatment frequently provoke complex legal battles. Treatments vary

Newer technologies can remove 90 percent or more of sulfur dioxide from flue gas. However, they can be expensive to install and operate, and governmental requirements for treatment frequently provoke legal battles.

Source: iStockphoto

widely from one plant to another, and some countries have far stricter requirements than others. As a result, emissions from utilities and industries continue to be an area of concern for environmentalists.

Flue gas treatment dates back to the mid-19th century when concerns grew over the impact of sulfates on the environment. Throughout the 20th century, increasingly sophisticated devices were developed to remove pollutants by a variety of means, mostly through chemical reactions and electrostatic charges. These efforts took on new urgency as industrialized nations adapted stricter air pollution measures, including the U.S. Clean Air Act in 1970, and by subsequent regulations that imposed increasingly stringent limits on such pollutants as fine particulates.

Flue gas treatment has achieved the greatest success in reducing particulate matter, nitrogen oxides, and sulfur dioxide. Technologies to remove mercury and carbon dioxide from flue gases have lagged, but there is growing interest in these areas. Mercury is typically removed by adsorption on sorbents or by capture in inert solids as part of the flue gas desulfurization product. Carbon dioxide, a common greenhouse gas blamed for climate change, may be more difficult to remove and capture, but regulators are increasingly focusing on carbon dioxide emissions, and companies accordingly are placing a greater emphasis on efforts to remove it.

At plants that emit a number of pollutants, flue gas may go through a series of devices for cleaning. In a typical treatment process, the gas is first sent to an electrostatic precipitator. This device removes ash and other particulates by electrostatically charging them, causing them to be attracted to and deposited on plates or other collection devices. Depending on such factors as the size of the particles and the design of the electrostatic precipitator, this treatment can remove 99 percent of particulate matter. If a large amount of particulates is released into the atmosphere, they can affect the respiratory systems of people and animals, reduce visibility, and influence climate.

The gas then moves on to a denitrification unit that alters the chemical composition of nitrogen oxides through a catalytic reaction with ammonia or urea. The goal is to produce nitrogen gas, rather than nitrogen oxides. Some facilities can also reduce nitrogen oxide emissions through modifications to the combustion process. If emitted into the atmosphere, nitrogen oxides can irritate the lungs and contribute to the formation of smog.

Sulfur dioxide is removed by one of a number of processes, most of which involve scrubbers in one form or another. Most U.S. facilities rely on wet scrubbers, which use a slurry of alkaline sorbent (usually comprising limestone or lime) or seawater to clean the gases. Other technologies include spray-dry scrubbing, which also uses sorbent slurries;

a wet sulfuric acid process that recovers the sulfur in the form of sulfuric acid; dry sorbent injection systems; and a flue gas desulfurization technique known as SNOX, involving catalytic reactions, that removes nitrogen oxides and particulates as well as sulfur dioxide. Newer technologies can remove 90 percent or more of sulfur dioxide from flue gas. Sulfur dioxide in the atmosphere can aggravate respiratory illnesses and cardiovascular conditions; it also leads to acid rain, reduces visibility, and affects clouds and climate.

These flue gas technologies also have the side effect of reducing the amount of mercury, a neurotoxin that is released in both elemental and oxidized forms and that can cause brain damage and reproductive problems in women as well as wildlife. Electrostatic precipitators can remove about 30 to 60 percent of elemental mercury in the airflow, and lower amounts of oxidized mercury. Wet scrubbers can remove slightly more than half of mercury and are especially effective for oxidized mercury. Dry flue gas desulfurization scrubbers, when combined with a baghouse (a dust collector system that uses fabric filter tubes or other tools), can remove as much of 90 percent of mercury. However, since these devices are primarily engineered to remove other pollutants instead of mercury, the amount of mercury that is removed from the flue gas can vary widely from one plant to another.

Environmentalists have pressed for stricter mercury regulations, contending that coal-fired power plants, the leading source of mercury emissions, should be required to remove 90 percent of mercury emissions. It may be possible to reach such a goal by enhancing current pollution controls or by using newer methods such as power-activated carbon. The carbon absorbs vaporized mercury from the flue gas and is then collected with particulates by an electrostatic precipitator. However, mercury controls can significantly increase a power plant's operating costs. Utilities have also raised concerns about inadvertently releasing mercury into end products. Power plants recycle some of the waste from coal-fired boilers into wallboard, cement, fertilizer, and other products, while disposing of the rest of the waste into landfills. If mercury control devices diverted mercury into the wastes, it would become far more difficult to recycle or dispose of them.

Removing carbon dioxide from flue gas also presents challenges. Due to concerns about climate change, regulators are pressing to reduce emissions of greenhouse gases, including carbon dioxide, that have been blamed for causing global warming. Various technologies are being studied for removing carbon dioxide, most likely using chemical absorbents. The waste carbon dioxide could be recycled for dry ice, fire extinguishers, carbonated beverages, and a variety of other uses. However, it remains uncertain when cost-effective technologies for removing carbon dioxide from flue gas will be commercially available.

See Also: Best Available Technology; Carbon Capture Technology; Coal, Clean Technology; Electrostatic Precipitator.

Further Readings

Energy Business Review. "RWE, BASF and Linde Jointly Develop New Carbon Capture Technology." http://carbon.energy-business-review.com/news/rwe_basf_and_linde_jointly_develop_new_carbon_capture_technology_100907 (Accessed September 2010).

"Gas Recovery Cuts Energy Costs at Port Talbot Site." *The Engineer*. http://www.theengineer .co.uk/channels/process-engineering/gas-recovery-cuts-energy-costs-at-port-talbot-site/1004579.article#ixzz10YemRz1V (Accessed September 2010).

Richard, Porter C. *The Economics of Waste*. London: RFF Press, 2002.

Williams, Paul T. *Waste Treatment and Disposal*. Hoboken, NJ: Wiley, 2005.

David Hosansky
Independent Scholar

FRANKFURT SCHOOL

The Frankfurt School is a school of social theory, primarily affiliated with scholars at the Institute for Social Research at the University of Frankfurt am Main. Initially centered on the beliefs of neo-Marxists who disagreed with the narrow selection of Marx's thoughts generally made popular, the Frankfurt School broke from the dogma adhered to by traditional communist parties. Dissatisfied by both traditional capitalism and Marxism, followers of the Frankfurt School crafted an alternative path to social development. Members of the Frankfurt School were distressed by the conditions that affected social change and influenced the establishment of rational institutions. The Frankfurt School influenced greatly early proponents of the environmental movement and advocates for sustainable development, especially with their critiques of institutions, the despoliation of the planet, and the plight of workers. Although not always formally acknowledged, the Frankfurt School's theories continue to shape the beliefs of many who advocate for a greener planet.

Origins of the Frankfurt School

The Institute for Social Research was founded in 1923 in Frankfurt am Main, Germany. Affiliated with the University of Frankfurt am Main, the Institute for Social Research initially gathered together a group of orthodox Marxists. After the directorship of Max Horkheimer commenced in 1930, the Institute for Social Research soon began to focus on defining a critical theory of society, work that was published in the institute's publication *Journal for Social Research*. With the advent of the Nazi Party's power in 1933, the Institute for Social Research left Germany, eventually moving to New York City in 1934, where it became affiliated with Columbia University. With its journal renamed *Studies in Philosophy and Social Science*, the work of the Frankfurt School became noticed by a wider audience. By 1951, the Institute for Social Research returned to Frankfurt am Main under the leadership of its cofounder, Friedrich Pollock. In addition to the work of Horkheimer and Pollock, the Frankfurt School is used to refer collectively to the writings of Herbert Marcuse, Theodor Adorno, Walter Benjamin, and Jürgen Habermas.

The Frankfurt School explored the belief that neither capitalism nor Soviet-style communism provides a tenable model of societal development. Instead, the Frankfurt School suggested a third path, based on the work of Horkheimer, which provides the most equitable option for most individuals. This third way is commonly referred to as "critical theory" and is based on a rejection of such scientific concepts as positivism, pragmatism, and phenomenology. Horkheimer believed that these concepts represented a "logico-mathematical" bias that separated most theoretical explanations from actual life experience. Horkheimer, and many of the Frankfurt School, rejected the concept of "knowledge," instead seeking to be critical of any pretense of absolute truth. Indeed, the Frankfurt School was critical of both materialism and idealism, believing that both distorted reality to the benefit of a few. Instead, critical theory insisted that researchers continually look for explanations, including those that would explain what is wrong with a situation, identify actions that might result in change, and provide clear and practical goals for the future.

A second phase of the Frankfurt School shifted its emphasis to a critique of Western civilization as a whole. The traditional quest for domination of nature was examined, and its emphasis on rationality was especially found wanting. As viewed by members of the Frankfurt School, Western civilization began to be seen as a fusion of domination and technological rationality. Humans were viewed as being on a quest to bring all external and internal nature under their domination, resulting in the individual being subsumed with no possible means for emancipation. Technology, totality, and teleology were seen as hindrances to human individuality and freedom. Traditional research methods, with their emphasis on objectivity and impartiality, were eschewed and instead replaced by a desire to effect change.

Influence on Environmental Movement

The Frankfurt School's critical theory has greatly influenced the present-day green movement. Since critical theory attempts to be the ultimate social developmental theory, it asks that we look beyond major social philosophers such as Marx but also look into other areas not thought of as developmental. Because humans had traditionally attempted to control and dominate the natural world, the Frankfurt School theorists believed that this caused many individuals to lose control of their lives and thus the world. Certainly these beliefs were spurred by the horrors of World War II, but increasing awareness of the degradation of ecosystems, the social cost of pollution, and growing inequalities all created a new willingness to explore environmental issues.

Critical theory presented support for many of those critical of how the planet's resources were being used. Critical theory has among its stated goals the desire to discover and explain all problems facing human society, to determine potential agents for change, and to provide criticism and goals for change. As such, critical theory was useful to those dissatisfied with certain environmental practices of governments, corporations, and individuals. Critical theory's broadly stated goals made it seem able to conceptualize solutions to many if not all forms of social problems, including poverty, slavery, and the green movement. Critical theory and other writings of the Frankfurt School were especially popular among a generation that was skeptical of power elites and traditional forms of knowledge. To a certain extent, critical theory remains an important underpinning of environmental activism and sustainability.

Shaping Goals of Environmentalists

As environmentalism grew in popularity and breadth, certain aspects of the green movement came to believe that the movement could liberate humans from the problems that enslaved them and could assist the spread of democracy and freedom for every individual. As used by proponents of the Frankfurt School, "enslavement" refers to the many humans who are left without the power to deal with the problems facing them. Adherents to the Frankfurt School believe that those forced to work in environmentally hazardous conditions, without real choice in the matter, are slaves to larger problems beyond their control. Allowing these individuals tools to make choices regarding these issues, such as working conditions, living wages, and environmental degradation, will not only promote greater freedom but also promote greater democracy. Critical theory resulted in some powerful changes, such as a growing awareness that environmental protection was needed and that many governmental actions were being spurred by a desire to protect corporate profits at the cost of dire consequences to individuals and natural resources. Unfortunately, critical theory was not always applied to some of the solutions generated, resulting in cosmetic changes that did little to protect the environment or encourage sustainable practices.

Critical theory differs from many other approaches insofar that it does not endorse a single school of thought but instead seeks to draw upon a range of perspectives in order to reach a form of consensus. Critical theory's emphasis on considering and changing society, instead of traditional scientific ways of merely studying what occurred, made it useful to those interested in a variety of causes. The continual focus on self-reflection and emancipation, examined most deeply in Habermas's *Knowledge and Human Interests*, emphasizes that modernity has led away from enlightenment to enslavement. Much of the antidevelopment fervor encapsulated in some environmentalist groups stems from this belief.

Continuing Influence

The influence of critical theory continues, such as postmodern research, with its emphasis on situating social problems within historical and cultural contexts. This has had many benefits, such as the growing acceptance of alternative research forms, such as ethnographies, that allow holistic explorations of cultures, decisions, and situations that might be ignored using traditional inquiry methods. Critical theory also colors how research is viewed, forcing many researchers to implicate themselves in the process of collecting and analyzing data, and to relativize their findings to reflect the context in which they were reached. "Meaning" is often perceived as unstable, since the rapid transformation of social structures makes it theoretically impossible to decontextualize findings. As a result, much environmental research focuses on local occurrences rather than attempting to make broad generalizations. This rejection of objective depictions is problematic, since much of governmental policymaking depends upon data that can be used to make decisions for multiple settings. While continued reflection on the validity of perspectives is useful to a certain degree, when continued indefinitely it becomes problematic.

The writings of members of the Frankfurt School have shaped and colored the environmental movement since the mid-20th century. Rejecting both traditional political schemas and conventional methods of scientific inquiry, critical theory has allowed many passionate about the environment and sustainability to engage in actions that can bring about desired changes in our world. Limitations with the generalization of such findings, however, threaten to limit the scope and influence of such efforts.

See Also: Authoritarianism and Technology; Democratic Rationalization; Luddism; Marxism and Technology; University-Industrial Complex.

Further Readings

Habermas, Jürgen. *Knowledge and Human Interests*. Trans. Jeremey J. Shapiro. Boston, MA: Beacon Press, 1972.

Jay, M. *The Dialectical Imagination: A History of the Frankfurt School and the Institute of Social Research, 1923–1950*. Berkeley: University of California Press, 1996.

Wheatland, T. *The Frankfurt School in Exile*. Minneapolis: University of Minnesota Press, 2009.

Wiggershaus, R. *The Frankfurt School: Its History, Theories, and Political Significance*. Trans. M. Robertson, Cambridge, MA: MIT Press, 1994 [1986].

Stephen T. Schroth
Michael J. Whitt
Knox College

FUTUROLOGY

Futurologists may examine such issues as climate change and the effect of pollution (as illuminated here via a U.S. National Aeronautics and Space Administration satellite) on the environment to determine if policy intervention is required or to make long-term strategic decisions.

Source: Mark Schoeberl/U.S. National Aeronautics and Space Administration

Futurology, sometimes also termed *foresight research* or *futures studies*, is defined as "the study of the future." Practitioners of futurology may refer to themselves as futurologists, futurists, or foresight practitioners. In the context of green technology, it is the process of trying to map out the range of possible technology pathways that sustainable technology development may take and appraising which of these are likely, desirable, and probable—uncovering in the process the likely impacts of that route of development and assumptions that influence the possibility and probability of that pathway being followed.

Futurology employs a range of methodologies from across multiple disciplines and can be quantitative or qualitative in nature. Futurology research often draws together views and insights from across disciplinary boundaries to form a holistic view of possible, probable, desirable, and undesirable future outcomes.

The use of the word *foresight* in relation to futurology research can be traced back to a 1932 BBC broadcast by the science fiction author H. G. Wells; in this broadcast, he called for "Departments and Professors of Foresight." In a more contemporary account, Professor Henry David has defined futurology as "the intellectual form in which a society renders account to itself of its probable and possible futures."

Futurology in Context

There have been a number of important futures studies in the sustainability debate such as *Limits to Growth* by the Club of Rome, which aims to forecast possible future outcomes based on the notion that we inhabit a world of finite resources, with an ever-expanding population.

Futurology research involving sustainable technology often examines the technological pathways that will be necessary in order to transition to more sustainable ways of living. These studies may take into account issues of population growth and technology development to predict which sociotechnical solutions can deliver a sustainable future.

Applications of Futurology

Futurology may be used by governments and state actors to determine if policy intervention is required to help guide an industry or market toward a desirable technology outcome. It is also an approach used by some companies that aim to make long-term strategic decisions as to the direction that their company should be heading. Many nongovernmental organizations (NGOs) and charitable organizations have also used a futures studies approach to demonstrate why current political, technical, social, and economic trajectories are unsustainable and to make an appeal for more equitable alternatives.

In the field of sustainability, futurologists may examine such issues as the following:

- Climate change
- Effect of pollution on the environment
- Energy security
- Food security
- Global population growth and future demographic trends
- Innovation trajectories for sustainable technologies
- Poverty and equity
- Sociocultural issues concerned with the introduction of new technologies
- Water security

There are a range of different methodologies that can be applied to gaining insight into future trajectories and scenarios; the methods used can include (but are not exclusively) the following:

- Anticipatory thinking protocols
- Backcasting
- Causal layered analysis
- Delphi studies
- Future histories
- Futures workshops
- Morphological analysis
- Technology forecasting
- Trend analysis
- Scenario development
- Systems thinking
- Simulation
- Social network analysis

Criticism of Futurology

The challenging nature of making predictions about the future leaves much scope for imperfection; thus it can be seen that futurology is not an "exact science"; although in pursuit of knowledge, some futurologists have attempted to employ robust methods and rigor in pursuit of reliable answers. Due to the large timescales on which futurologists work, and the unpredictable nature of innovation and societal changes, it is difficult, even impossible to offer reliable foresight; so instead such predictions should be viewed as possible directions of things to come.

The scope of predictions of alternative trajectories range from the utopian vision of the world's citizens living in harmony to ecological catastrophe and devastation. What the future holds remains hard to predict; however, it is clear that futurology is a valuable tool in helping to predict—and realize—sustainable futures; as without knowledge of the future, it is difficult to make decisions in the present.

See Also: Green Markets; Innovation; Science and Technology Studies; Sustainable Design; Technological Determinism.

Further Readings

Bell, Wendell. *Foundations of Futures Studies: Human Science for a New Era.* New Brunswick, NJ: Transaction Publishers, 1997.

Bishop, Peter and Andy Hines. *Thinking About the Future: Guidelines for Strategic Foresight.* Washington, DC: Social Technologies, 2006.

Ferkiss, V. C. *Futurology: Promise, Performance, Prospects.* Beverly Hills, CA: Sage, 1977.

Malaska, Pentti. "Knowledge and Information in Futurology." *Foresight*, 2/2 (2000).

Thompson, A. E. *Understanding Futurology: An Introduction to Futures Study.* Newton Abbot, UK: David & Charles, 1979.

Wells, H. G. "Wanted: Professors of Foresight!" *Futures Research Quarterly*, 3/1 (Spring, 1932) [1987].

Gavin D. J. Harper
Cardiff University

GEOENGINEERING

Geoengineering is the idea that technology rather than restraint in consumption and production can reverse climate change. It has been defined by David Keith as "the intentional large-scale manipulation of the environment, particularly manipulation that is intended to reduce undesired anthropogenic climate change." The Council on Foreign Relations defines it as "any of a variety of strategies, such as injecting light-reflecting particles into the stratosphere that might be used to modify the Earth's atmosphere-ocean system in an attempt to slow or reverse global warming." It can also be defined as the modification of global climate by influencing the amount of heat absorbed and reradiated by the Earth without changing the amount of greenhouse gases in the atmosphere.

The active control of the Earth's climate has long been a staple of science fiction. In one of the earliest references, the 1894 story *A Journey in Other Worlds* by John Jacob Astor IV talks about straightening the axis of the Earth to combine the extreme heat of summer with the intense cold of winter and thus produce a uniform temperature for each degree of latitude year round. The idea appears in the works of Frank Herbert, Isaac Asimov, and Arthur C. Clarke, among others. The 1974 classic *The Mote in God's Eye*, by U.S. writers Larry Niven and Jerry Pournelle, visualized the use of volcanoes to mitigate climate on a distant planet, and Kim Stanley Robinson's *Mars* trilogy (1993–1996) charts the colonization of Mars by transforming its climate.

Geoengineering is different from extant attempts to combat climate change in that it is intentional and on a colossal scale. Current attempts to combat global warming are either "mitigatory" efforts, often local, to reduce human activities to change climate, or "adaptive" in that they try to adjust to climate change. Though it has been labeled a fringe science, the increasing willingness to consider it reflects mounting pessimism that global warming can be reversed through extant methods, the failure of countries to effectively cooperate to combat climate change, and the desire for a quick technological fix to the problem.

There are two kinds of geoengineering—a relatively benign, and much less expensive, effort to increase the absorption of carbon dioxide (CO_2), under the rubric *carbon dioxide removal*, such as increasing algae in the oceans, the creation of carbon sinks using trees that have been genetically modified to absorb high amounts of CO_2 from the atmosphere, or the pyrolysis or gasification of organic material (plant waste) in low-oxygen conditions to create biochar, which can be mixed with existing soil to create fertilizer and to sequester carbon. Based on European emissions of about 1.1 gigatons of carbon per year, scientists estimate that such biochar could offset around 9 percent of Europe's emissions. Substantial reductions of global warming can also be achieved by converting dark places that absorb lots of sunlight to lighter shades—for example, by replacing dark forests with more reflective grasslands.

The second, more aggressive but more effective, method is solar radiation management (SRM), which seeks to reduce the amount of climate change by reflecting some of the sun's warming rays back to space. SRM can be deployed rapidly and can cause dramatic cooling. The Earth and the atmosphere absorb about 70 percent of sunlight, while the remainder is reflected back into space. Scientists theorize that increasing the reflectivity of the planet (known as the albedo) by about one percentage point could affect the climate system powerfully enough to offset the gross increase in warming that is likely over the next century, when the amount of CO_2 in the atmosphere is expected to double. Advocates of SRM argue that it would be possible to cool the Earth back to pre–Industrial Revolution temperatures using this method. One attraction of SRM schemes is that they are relatively cheap—$100 billion could reverse anthropogenic climate change entirely, though some scientists aver that the actual cost would be much less. In comparison, it is estimated that reversing climate change through the cutting of carbon emissions would cost around $1 trillion yearly. However, all SRM methods suffer from a major defect—they try to reduce global warming without addressing the root cause of the phenomenon—the release of increasing amounts of CO_2 into the atmosphere.

Inspired Geoengineering

Geoengineering is inspired by the ability of volcanic eruptions to dramatically cool the atmosphere. The 1815 eruption of Mount Tambora, Indonesia, the largest volcanic eruption in recorded history, spewed so much sulfur dioxide into the stratosphere that farms in New England were covered in frost as late as July 1816. In the same way, the Mount Pinatubo eruption in the Philippines in 1991 cooled global temperatures by about half a degree Celsius for the 1990s. It has been estimated that just one kilogram of sulfur in the stratosphere would roughly offset the warming effect of several hundred thousand kilograms of CO_2. Other schemes include seeding bright reflective clouds by blowing seawater into the atmosphere, which will give rise to whiter and fluffier clouds that are effective at reflecting sunlight. One scheme based on ideas developed at the National Center for Atmospheric Research in the United States envisions a permanent fleet of up to 1,500 ships churning seawater and shooting the moisture into the clouds to make them whiter and fluffier, and therefore better at bouncing sunlight back harmlessly into space.

Such introduction of reflective materials into the upper stratosphere is the easiest and cheapest option. In addition to ships, such particles can be introduced by using high-flying aircraft, naval guns, or giant balloons. In addition to water, appropriate materials could include sulfate aerosols, which would be created by releasing sulfur dioxide gas. However, such ambitious efforts are more in the realm of science fiction. Actual efforts, if they are implemented, are likely to be more small scale. One method being considered is dumping iron particles in the ocean to stimulate the growth of plankton that excrete dimethylsulfide (DMS), which escapes to the air, reacts chemically, and forms an aerosol that when released into the atmosphere acts as cloud-condensation nuclei. Additional cloud cover reflects more solar energy and prevents the Earth's surface from heating as much as it would without clouds. Natural sources provide about 50 percent of the total gaseous sulfur entering the atmosphere at present (the rest comes from burning fossil fuels, although natural sources such as marine phytoplankton still dominate in the southern hemisphere). Scientists expect that the stimulation of large increases in phytoplankton productivity in the ocean might consequently increase cloud cover and reduce insolation.

A greater danger in tinkering with the albedo of the Earth is its potential impact on atmospheric circulation, rainfall, and other aspects of the hydrologic cycle. It is significant that after the eruption of Mount Pinatubo, for almost two years, rainfall and river flows dropped, particularly in the tropics. This means that SRM attempts might lead to the failure of the monsoons in South Asia and drought in large parts of Asia and Africa as well as parts of the United States. This could dangerously threaten agricultural productivity, supplies of freshwater, and the world's food supply. Indeed, some climate models already suggest that negative outcomes such as decreased precipitation over land (especially in the tropics) and increased precipitation over the oceans would certainly accompany a geoengineering scheme that sought to lower average temperatures by raising the planet's albedo.

Another danger in geoengineering is its relative cheapness that makes it tempting for a lone country, or even a lone billionaire, to try to reverse climate change unilaterally. This is all the more likely in case a tipping point is reached, when geoengineering could become an emergency shield that could be deployed if a sudden climatic shift puts vital ecosystems and billions of people at risk. However, the danger lies in the fact that it is unlikely that all countries will have similar assessments of how to balance the ills of unchecked climate change with the risk that geoengineering could do more harm than good.

This is all the more likely since the biggest stumbling block in current efforts to combat climate change is the inability of countries to come to an agreement. Thus, international negotiations to fight climate change have floundered over the precise amount of carbon emissions to be cut. While Europe has been proactive, the United States, more dependent on fossil fuels and with an economy based on enormous energy consumption, has balked at raising the bar on CO_2 emissions. Japan, while willing, has to reconcile the need for economic growth with continued dependence on an energy system powered mainly by conventional fossil fuels. The picture is complicated by the rise of new economic powers like China and India. China's emissions have surpassed those of the United States, thanks to coal-fueled industrialization and a staggering pace of economic growth. One key feature of climate change is that though it is global in nature, both in its creation and the likely impact it will have, efforts to combat it have been national. The "public good" nature of this problem means that each country has an incentive to ride on the efforts of others, discouraging effective international cooperation. Even if the United States, where total emissions have risen by approximately 14 percent from 1990 to 2008, drastically reduces its greenhouse gas emissions, it would make little difference if emissions are unchecked in the developing world. Moreover, the durability of CO_2 means that significant dissipation of the CO_2 already in the atmosphere, even if emissions can be stopped right now, will take decades. To illustrate, in 1991, Norway became one of the first nations to impose a stiff tax on emissions. However, the country has seen a net increase in its CO_2 emissions since then. In an ideal situation, merely preventing an increase of global warming levels would require a worldwide 60–80 percent cut in emissions.

Complications

Complicating the picture is that the implementation of geoengineering schemes would have to occur in the global commons such as outer space, the high seas, or the atmosphere outside sovereign territory. Public international law is applicable to activities in or affecting these commons, including climate modification schemes, but regulatory mechanisms are in their infancy. Geoengineering offers nations the tempting possibility of a technological fix to climate change, while ignoring its effect on other countries. Such concerns have led to increasing calls for global regulation of environmental law to expand to include geoengineering schemes.

The biggest danger of geoengineering is that the cure might be worse than the disease. Any attempt to tamper with an intricate and complex system such as global climate patterns could accelerate rather than reverse climate change. Even if the initial attempt to reverse global warming succeeded, it would raise the need for constant tinkering to prevent an apocalyptic relapse. For example, sulfur aerosols would cool the planet dramatically, but the moment the pumping of such aerosols stopped, they would rain down and years' worth of accumulated carbon would make temperatures surge. In other words, the world would experience the full force of postponed warming in just a couple of catastrophic years. Scientists also warn of unexpected consequences, a warning all the more poignant given the long history of innovations such as DDT and asbestos that went haywire. Sun-blocking technology could have unforeseen consequences, such as rapid ozone destruction. Finally, geoengineering schemes, once launched, are almost impossible to shut down.

Geoengineering harkens back to a time when it was believed that engineering could solve nature's "problems" or be used to interfere with nature so as to "improve" it. In fact, when President Lyndon B. Johnson received the first-ever U.S. presidential briefing on the dangers of climate change in 1965, the only remedy prescribed to counter the effects of global warming was geoengineering. Such optimistic beliefs in the ability of human engineering to solve natural problems reflected the scientific ethos of the time, which also saw Soviet scientists toy with the idea of using nuclear explosions to change the course of rivers and American scientists try to make tropical hurricanes less intense by flying aircraft into them and seeding with silver iodide, but with no clear success (Project Stormfury).

Several military projects on both sides of the Cold War also explored the use of geoengineering. However, with the adoption in 1976 of the Convention on the Prohibition of Military or Any Other Hostile Use of Environmental Modification Techniques (ENMOD) at the United Nations, such projects were abandoned. Moreover, the scientific climate had also shifted. The exposure of the damage that innovations such as DDT had done to the environment, dramatically portrayed in Rachel Carson's *Silent Spring* (1962), gave rise to an increasing skepticism about the power of science to solve problems. The belief in the benign nature of science and the Enlightenment idea of the perfectibility of man through the application of reason, through deliberate and direct action epitomized by technology, a staple of modernity and Western culture since the 17th-century Enlightenment, was replaced with a growing awareness about the dangers of careless tinkering with nature. The birth of the modern environmental movement, increasing disillusionment with the high-technology war in Vietnam, and the impact of the energy crisis created a new scientific ethos that was deeply suspicious of such megaschemes as geoengineering.

See Also: Authoritarianism and Technology; Ecological Modernization; Faustian Bargain; Technological Utopias.

Further Readings

Barrett, Scott. "The Incredible Economics of Geoengineering." *Environmental and Resource Economics*, 39/1 (2008).

Biber, Eric. "Climate Change and Backlash." *New York University Environmental Law Journal*, 17 (Summer 2009).

Caldeira, Ken. "Geoengineering: Assessing the Implications of Large Scale Climate Intervention." Testimony before the U.S. House of Representatives Committee on Science and Technology. Washington, DC: 2009.

Carlin, Alan. "Global Climate Change Control: Is There a Better Strategy Than Reducing Greenhouse Gas Emissions?" *University of Pennsylvania Law Review*, 155/6 (June 2007).

Carson, Rachel. *Silent Spring*. Boston, MA: Houghton Mifflin, 1962.

Davis, William Daniel. "What Does Green Mean: Anthropogenic Climate Change, Geoengineering, and International Environmental Law." *Georgia Law Review*, 43/3 (Spring 2009).

Eccleston, Charles H. "Can Geo-Engineering Reverse Climate Change?" *Environmental Quality Management*, 19/2 (Winter 2009).

"Every Silver Lining Has a Cloud." *The Economist* (January 31, 2009).

Hahn, Robert W. "Climate Policy: Separating Fact From Fantasy." *Harvard Environmental Law Review*, 33/2 (2009).

Keith, David W. "Geoengineering the Climate: History and Prospect." *Annual Review of Energy and the Environment,* 25 (2000).

Michaelson, Jay. "Geoengineering: A Climate Change Manhattan Project." *Stanford Environmental Law Journal, 17/1,3* (January 1998).

Niven, Larry and Jerry Pournelle. *The Mote in God's Eye*. New York: Pocket Books, 1991 [1974].

Peterson, James Edward. "Can Algae Save Civilization—A Look at Technology, Law, and Policy Regarding Iron Fertilization of the Ocean to Counteract the Greenhouse Effect." *Colorado Journal of International Environmental Law and Policy*, 6/1 (Winter 1995).

The Regulation of Geoengineering: Fifth Report of Session 2009–10 Report, Together With Formal Minutes, Oral and Written Evidence. London: The Stationery Office, 2010.

Robinson, Kim Stanley. *Red Mars (Mars Trilogy)*. New York: Bantam Spectra, 1993.

Victor, David G., et.al. "The Geoengineering Option: A Last Resort Against Global Warming?" *Foreign Affairs*, 88/2 (March/April 2009).

Wood, Graeme. "Re-Engineering the Earth." *The Atlantic* (July/August 2009).

Sahil Francis
University of Leipzig, Germany,
École Normale Supérieure, Paris

GEOGRAPHIC INFORMATION SYSTEMS (GIS)

The field of geographic information systems (GIS) is a rapidly emerging field that seeks to collect, store, manipulate, analyze, and display information about spatially distributed phenomena. It combines hardware, software, data, people, procedures, and institutional arrangements, and helps institutions with inventory, decision making, and problem solving.

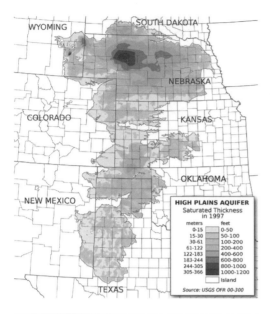

This digital map, showing the saturated thickness of the High Plains aquifer, is an example of how GIS uses geospatial information. The points, lines, and polygons showing boundaries and landmarks are part of the vector system; the shading showing the aquifer thickness is a raster type on a tessellation of cells.

Source: Wikimedia

A closely related term—less frequently used—to GIS is *geomatics*, originally introduced in Canada. An important component of GIS relates to handling geospatial data. The geospatial data are collected by various means, processed, and stored, or disseminated for later use by GIS. In essence, GIS is used to manipulate, summarize, query, edit, and visualize the geospatial information. It is increasingly being used as a decision-making tool for natural resource management and several other applications.

Data Representation

In order to represent diverse geographic phenomena in GIS, a uniform framework is needed. The geographic phenomenon can be categorized into two different groups: objects and fields. Objects refer to the phenomena that have definable crisp boundaries (discrete existence), whereas fields refer to the phenomena that do not have sharp boundaries, but are rather fuzzy in their presence and occur at all places (continuous existence). Fields can be again continuous or discrete according to their fuzziness in representation. Examples of object-like phenomena are rivers, buildings, volcanoes, and islands. Examples of discrete fields include land use maps, soil maps, and geology maps, and continuous fields include factors such as temperature, elevation, and humidity. Both the object as well as field type information can be represented in GIS using points, lines, and polygons (vector format), or raster data layers. In addition, GIS stores two kinds of information: the spatial data and the attribute data. The spatial data represent features that have a known location on the Earth. The spatial data of maps in GIS are structured as points, lines, or polygons, which together constitute geographic objects as they represent geographic features. Attribute data are the information linked to the geographic features (spatial data) that describe the features. Attribute data are thus the nongeographic information associated with the point, line, or polygon elements in a GIS. An attribute table can store a variety of data, including ratio scale, interval, and ordinal or nominal scale.

Data Types

The geographic phenomena in GIS are represented by two different data types—vector and raster. The vector method stores point locations of geographic features using an X, Y coordinate system and is represented by points, lines, and polygons. Thus, it is possible to represent cities, roads, administrative units, or other features using vector concepts. However, there are several features, such as temperature, elevation, and humidity, that cannot be represented in vector format. The tessellation concept in GIS helps in representing such nonvector data. By dividing the space into uniform grid intervals (tessellation), a variety of data can be stored. The tessellated space resembles a two-dimensional (2D) array of grids/cells, which is called a raster. Each array position is defined by a line and column number and is referred to as a "cell." Cell values can be of many types ranging from 1-bit (binary) to 128-bit (complex numbers). Each data storage method, either vector or raster, has advantages and disadvantages. An obvious drawback with raster storage is the necessity to store the entire matrix of rows and columns. Storing the entire matrix may include unwanted data. In contrast, vector data usually require less storage space in the computer, but may require more sophisticated programming and processing time. The raster structure, by virtue of its matrix, has a built-in ability to perform neighborhood-type analyses more easily than the vector format because of its row and column structure. Further, remote sensing data are raster data, and hence can be directly adopted into GIS models for further analysis. In a vector system, coordinates may be stored at any desired scale, but in a raster format, the resolution is fixed by the pixel size. Vector storage allows for easy variable resolution of the data and is good for printing accurate representations. Considering these advantages and disadvantages, modern GIS packages handle both the raster and vector models effectively.

Data Management in GIS

One of the important characteristics of GIS is the ability to manipulate data by theme, usually called "data layers." A data layer consists of a set of logically related geographic features and attributes. For example, map data can be separated into several data layers, that is, based on the data type and geographic themes (points representing cities or villages, roads or streams as lines, and areas as polygons). The concept of layers applies to both vector and raster models. The layers can be combined with each other in various ways to create new layers that are a function of the individual ones. An important characteristic of each layer within a layer-based GIS is that all locations within each layer may belong to a single aerial region or cell, whether it is a polygon bounded by lines in a vector system or a grid cell in a raster system. Such a database bundle is called a geodatabase. However, it is possible for each region to have multiple attributes. By storing the information in thematic data layers or objects registered to a common geographic coordinate system, any combination of these layers or objects may be overlaid to form a new dataset to be used to answer a question or analyze a problem.

Projections in GIS

Since Earth is three-dimensional (3D) in nature, the representation of Earth featured in GIS as layers in 2D format needs a mechanism that can help in transition between 3D to 2D. This is achieved through projections. The entire geospatial database generated under GIS must be in a planimetric coordinate system, in other words, it must be represented in a 2D reference frame so that area and length estimates are accurate. There are many ways by which a 3D globe can be represented as a 2D cylinder, cone, or plane. Conversion is done by wrapping the Earth with paper in these shapes, transferring the global features onto the paper, and then unwrapping it. The process can be mathematically represented using functions. Converting from 3D to 2D introduces some loss in either area, shape, or direction measurement. Based on the property it preserves, the projection receives its name (like "equal area projection" or "conformal projection"). Further, as the Earth is not a smooth surface and has gravitational undulations, replicating the exact position and height of a location accurately on a map is difficult. So assumptions about the shape of our Earth as either ellipsoid or spheroid have to be made so that mathematically it can lead to projection. Every country has adopted its local mean sea level surface reference for height measurement, which is called "vertical datum," and adopted a mathematical surface (ellipsoid) that fits better for its portion of the globe, which is called "horizontal datum." Thus, projection parameters differ from one region to another and from place to place, so care should be taken to project the spatial data accurately in GIS, prior to spatial analysis and area calculations.

GIS Software and Trends

One of the most popular and widely used commercial GIS software systems is ArcGIS from Environmental Systems Research Institute (ESRI). ArcGIS organizes vector data into shape files and raster data in a variety of formats, including commonly used image formats such as "jpeg" and "tiff." Other popular commercial GIS software includes Autodesk, IDRISI, Geomedia Professional, MapInfo, RegioGraph, Manifold System, and TNTmips. In comparison to these commercial GIS packages, over the past 10 years free and open-source software development has gained popularity in the GIS community. As a result, highly advanced, user-friendly free software is being developed that is useful for several purposes ranging from Internet map server applications (like the MapServer project), to spatial database management systems to store geographic data (like PostGIS), to desktop GIS for data editing and analysis (like QGIS and SAGA). Further, with the rise of Internet technologies and the World Wide Web, new Internet protocols such as the hypertext transfer protocol (HTTP), new markup languages (like HTML, DHTML, XML, and GML), as well as easy to use interfaces, tools (like Flash), and languages (like .NET, Java, and scripts), Web-based GIS technology is rapidly developing and gaining popularity. An example application is Google Maps (http://maps.google.com), which is useful for accessing a variety of spatial information around the globe.

GIS Applications

The potential of GIS lies in integrating a variety of information useful for cartography (creating maps) and retrieving attribute information when needed. The information can then be used in the decision-making process useful in building green technologies or applying those technologies at wide spatial scales. GIS can be used to perform a variety of spatial analysis, including overlaying combinations of features and recording resultant conditions, as well as analyzing network routes, proximity analysis, and buffer analysis. The GIS operations can be performed on either single map layers or multiple layers. GIS can generate a new set of maps by query and analysis. Increasingly, GIS-based decision support systems (DSS) are designed and implemented by various organizations to serve as an advanced modeling system for natural resource management problems. In addition, GIS is being widely used for environmental monitoring and assessment, population census analysis, service provision, hazard mapping, health sector mapping, agricultural development, land evaluation, change detection, forest fire assessment, soil resource mapping, crop acreage assessments, soil resource mapping, and wasteland mapping and assessment. Further, with the advancement in mobile communication technology, GIS technology has already penetrated into mobile phones, by which users can find nearest bank locations, theaters, hospitals, or restaurants. Such applications are called location-based services (LBS) and can address several user needs and aid rapid retrieval of information useful for effective decision making. Recently, GIS is being used for designing intelligent transportation systems (ITS).

The main objective of ITS is to make the transportation systems efficient and safer. ITS requires significant geographic information in the form of a nationwide map database organized into GIS for effectively managing, analyzing, and visualizing transportation data through ITS. In addition, GIS has immense potential in various other fields and for addressing natural resource problems, including the rapidly emerging green technologies. Green technologies involve all those approaches and technologies that can lead to environmentally friendly solutions. Green technologies are gaining popularity as they are supposed to reduce the amount of waste and pollution that is created during production and consumption. Effective natural resource management through green technologies can lead to conservation of resources. To this end, resource inventory along with decision making is a must. GIS can help in identifying the source, location, and extent of adverse environmental impacts, and can help devise practical plans through green technologies for monitoring, managing, and mitigating environmental damage.

See Also: Sustainable Design; Systems Theory; Technology and Social Change.

Further Readings

Burrough, P. A. and R. A. McDonnell. *Principles of Geographic Information Systems.* Oxford, UK: Oxford University Press, 1998.

Clarke, K. *Getting Started With Geographic Information Systems.* Englewood Cliffs, NJ: Prentice Hall, 1997.

Cox, Allan B. and Fred Giffort. "An Overview to Geographic Information Systems." *Journal of Academic Librarianship* (1997).

Davis, D. E. *GIS for Everyone: Exploring Your Neighborhood and Your World With a GIS.* Redlands, CA: ESRI Press, 2000.

DeMers, M. N. *Fundamentals of Geographic Information Systems,* 2nd ed. Hoboken, NJ: Wiley, 2000.

Goodchild, M. F. and S. Gopal, eds. *Accuracy of Spatial Databases.* Oxfordshire, UK: Taylor & Francis, 1989.

Harder, C. *Serving Maps on the Internet: Geographic Information on the WWW.* Redlands, CA: ESRI Press, 1998.

Nyerges, T. "Understanding the Scope of GIS: Its Relationship to Environmental Modeling." In *Environmental Modeling With Geographic Information Systems,* M. Goodchild, L. Steyaert, and B. Parks, eds. Oxford, UK: Oxford University Press, 1993.

Krishna Prasad Vadrevu
University of Maryland, College Park

GREEN BUILDING MATERIALS

Green building, which emphasizes an environmentally friendly approach to development and construction, relies in large measure on building materials that are associated with low levels of pollution and sustainability. Such materials may incorporate recyclable components, emit low levels of harmful gases, or be harvested or manufactured in a way that has comparatively few environmental side effects.

Green buildings are generally designed, sited, built, maintained, and even demolished with an eye toward minimizing environmental impacts. The main goals are efficiently using energy, water, and other resources; reducing waste and pollution; safeguarding the health of occupants; and, in many cases, improving worker productivity. Green building may use a range of techniques to meet these goals. These may include building designs that minimize the use of lighting and temperature control; use highly energy-efficient and water-efficient fixtures; employ alternatives to conventional construction materials; and use landscaping that requires relatively little water and chemicals.

The use of green building materials is a priority for environmentalists and many policymakers because structures have a significant impact on resources, human health, and the economy. Studies by the U.S. Environmental Protection Agency (EPA) and other agencies, for example, have found that construction activities use 60 percent of raw materials in the U.S. economy, excluding food and fuel. The EPA estimated that in 2002, buildings accounted for about 40 percent of the total U.S. energy consumption, 68 percent of electricity, and 12 percent of water. Buildings, along with adjoining parking lots and other infrastructure, are also associated with the paving over of natural surfaces. This creates runoff that washes pollutants and sediments into streams and lakes, often generates a heating influence on local temperatures, and results in the loss of animal habitat.

Construction materials represent a major proportion of building impacts on the environment. This is partly because the construction, renovation, and demolition of buildings generates about 170 million tons of waste annually, amounting to almost 60 percent of the nonindustrial and nonhazardous solid waste, excluding food and fuel, in the U.S. economy. Salvaging building materials and reusing them cuts down on waste and the need to create new landfill space. It also reduces greenhouse gas emissions that may be generated from waste decomposition or by the need to need to extract and process raw materials.

Some of the concepts that green building draws upon date back centuries. Many ancient structures were designed to maximize heat in the winter or cooling in the summer; the Anasazi designed villages in the U.S. Southwest with homes that received as much solar heat as possible during the winter. More recently, the environmental movement of the 1960s and 1970s, along with energy shortages, spurred a renewed focus on more energy efficient and environmentally sustainable construction practices. The American Institute of Architects, with funding from the EPA, published the *Environmental Resource Guide* in 1992 to help builders evaluate the environmental impacts of building materials, products, and systems. The U.S. Green Building Council, founded the following year, launched its Leadership in Energy and Environmental Design (LEED) program later in the decade. The program certifies buildings or communities that are designed to certain standards that emphasize energy efficiency and minimize environmental impacts.

Elected officials have lent support for green building as well. The U.S. city of Austin, Texas, introduced the first green building program in 1992, and other local and state governments have passed laws that require public buildings to meet certain environmental standards—sometimes providing incentives for private buildings as well. With energy legislation in 2005 and 2007, U.S. Congress imposed energy efficiency and sustainable design requirements for federal offices as well as certain other buildings. In addition, several executive orders and agency rules have promoted green building since the 1990s.

However, green building is not without some challenges. Buying sustainable materials and designing a building to specifications that minimize energy and water use often drives up the initial cost of the project. Over time, however, a carefully designed green building can deliver cost savings in terms of lower utility costs and more durable materials. The buildings can also exceed the value of conventionally designed buildings, according to various matrices. For example, studies of commercial real estate have found that LEED- and Energy Star–certified buildings tend to command higher rental or sale prices and achieve higher occupancy rates than noncertified buildings. Green buildings may also deliver difficult-to-measure benefits, including higher worker productivity and improved occupant health. Studies have indicated that buildings designed with an emphasis on environmental quality can reduce rates of respiratory disease, allergy, asthma, and sick building symptoms.

Green buildings generally deliver their greatest benefits in terms of lower operating costs when initially designed by an integrated group of architects and other specialists. This is because the building can be conceived as a single, unified system rather than a group of disconnected units. When buildings are designed for maximum efficiency, for example, the heating, ventilation, and air conditioning (HVAC) equipment can sometimes be downsized, which may produce a significant savings.

Costs aside, designers of green buildings sometimes need to make difficult decisions over what constitutes environmentally friendly materials and construction standards. Determining the comparative environmental impacts of numerous products, ranging from lumber to asphalt, can be enormously complex. Trying to make that task easier, chemical, steel, and cement companies have ramped up spending into research and development of more environmentally friendly building materials.

A comprehensive green building project often begins with careful consideration to the construction site. Because the structure should have minimal impact on the environment, green buildings may be located within densely developed corridors, sometimes in industrial or commercial developments known as brownfields. When possible, planners try to design green buildings in areas that offer ready access to mass transit systems. Green building also emphasizes the importance of protecting existing landscaping and natural features.

Contractors generally look for materials that are produced in ways that have minimal environmental impacts. Natural and manufactured materials often associated with green building include trass, linoleum, panels made from paper flakes, adobe, baked or rammed earth, clay, vermiculite, flax linen, sisal, seagrass, cork, expanded clay grains, coconut, sheep wool, wood fiber plates, calcium sandstone, and Roman self-healing concrete. Such materials generally include one or more of the following characteristics:

- *Sustainable sources*: Builders turn to plant materials with a small environmental footprint. These can include rapidly growing plants such as bamboo and straw, as well as lumber from forests certified as sustainably managed.
- *Recyclable parts*: Such materials may include recycled stone, recycled metal, and other products that, ideally, are both constructed largely from recycled parts and can also be recycled themselves.
- *Efficiently manufactured products*: The materials, when manufactured, should come from processes that reduce energy consumption, minimize waste, and generate relatively little pollution.

- *Locally available*: Projects emphasize materials that are found locally, thereby minimizing the resources needed for transportation to the building site.
- *Salvaged, refurbished, or remanufactured*: Some materials, instead of being shipped to landfills, can be used in part or completely to make a product that is often highly functional and even desirable. Such products, much in demand by the green building industry, may include recycled industrial goods, such as coal combustion products, foundry sand, and demolition debris.
- *Reusable or recyclable*: Builders prefer materials that can be easily dismantled and reused or recycled at the end of their useful life.
- *Recycled or recyclable product packaging*: The materials, in addition to having a small environmental impact, should ideally be wrapped in recycled content or recyclable packaging.
- *Durable*: By using materials that last as long or longer than conventional products, builders can reduce the future waste stream.
- *Minimal toxins*: Materials in green buildings should emit minimal carcinogens, reproductive toxicants, or irritants.
- *Minimal chemical emissions*: The materials should produce minimal emissions of harmful chemicals, such as formaldehyde and volatile organic compounds (VOCs). These chemicals, often emitted by paints, coatings, and cleaning and maintenance products, can affect the health and productivity of occupants.
- *Low-VOC assembly*: In addition to not emitting VOCs, the materials should be installed with compounds that produce minimal VOCs, or though mechanical attachment methods that do not involve any emissions.
- *Moisture resistant*: Products and systems in a green building should be resistant to moisture, thereby inhibiting the growth of biological contaminants. If successful, this can create a healthier environment for the occupants.
- *Easily maintained*: Simple cleaning methods should be sufficient to maintain the materials, as opposed to the need for toxic methods of cleaning.

The materials are generally manufactured offsite. This minimizes waste and provides more opportunities for recycling, as well as emitting less noise and dust at the building site.

Once materials are selected, a green building project also focuses on certain design techniques to maximize efficiency. For example, builders often implement passive solar building designs, orienting windows and walls to maximize solar heat in the winter. They may use awnings, porches, and trees to shade windows and roofs from the hot summer sun. White roofs can minimize the need for air conditioning in the summer. The effective placement of windows can also allow more natural light into the building. This technique, known as daylighting, minimizes the need for electric lighting. Studies have shown that daylighting is beneficial to occupants, helping with productivity and a feeling of well-being. Similarly, the building may incorporate an approach known as solar water heating that takes advantage of sunlight to heat up water, typically in a storage tank that is mounted above solar collectors on the roof.

To further reduce energy demand, green buildings often contain high-efficiency lighting systems with advanced lighting controls. Motion sensors may be tied to dimmable lighting controls. Instead of general overhead lights, individual task lighting may be utilized. The use of light colors for roofing and wall finish materials can also reduce demand for air conditioning; and high-efficiency insulation helps to retain heat.

Such energy-saving techniques can be augmented by on-site generation of renewable energy. Solar panels, wind turbines, hydropower, or biomass can minimize a building's reliance on the power grid. In some cases, a building may not require any energy at all from the local utility. However, such power-generating features can be highly expensive and sometimes impractical.

Green building also focuses on reducing water consumption and protecting water quality. Critical water consumption problems can occur when the demands on local aquifers exceed its ability to replenish itself. For that reason, sustainable builders try to increase the use of water that is collected, purified, and recycled on-site. Strategies to minimize water use include the following:

- Dual plumbing, which recycles water in toilet flushing
- Water conserving fixtures, such as ultra-low flush toilets and low-flow showerheads
- Rainwater collectors to gather water for various on-site needs
- Paperless toilets that use bidets to reduce or eliminate the use of toilet paper, thereby reducing sewage and making it more possible to recycle water on-site
- Point-of-use water treatment and heating, which improves water quality and energy efficiency while reducing the demand for additional water
- Irrigation with greywater, which comes from such sources as dishwashing or washing machines. The greywater can also be used for additional nonpotable purposes, such as flushing toilets and washing cars
- Landscaping with native or drought-tolerant plants, which can lessen the need for watering lawns and gardens

Another major goal of green building is to improve indoor air quality. This is a substantial concern, given the amount of time that most people spend indoors and the high levels of certain pollutants within many buildings. Indoor air quality can be affected by microbes, such as mold and bacteria; gases, such as carbon monoxide, radon, and VOCs; and dust and other particles in the air. Buildings rely on carefully designed heating and air conditioning systems to ensure proper ventilation and air filtration, as well as isolating areas that may affect air quality through certain operations, such as printing, dry cleaning, or even cooking facilities. Indoor air quality can also be improved by using construction materials and interior finishing products that emit little or no VOCs or formaldehyde. Some green buildings also emphasize the circulation of fresh air, sometimes with the most basic of approaches: allowing occupants to open office windows.

However carefully designed, a green building must also be carefully operated and maintained in order to have a minimal environmental impact. Attention is given to maintaining the building's heating and air conditioning system to ensure that it is running efficiently, and cleaning facilities with a minimum of water and chemicals. In addition, maintenance staff and the building occupants are encouraged to reduce waste and to recycle and compost whenever possible.

Well-known examples of green buildings include the following:

- *The William and Flora Hewlett Foundation Headquarters Building, Menlo Park, California*: The first LEED Gold Building in California, 64 percent of its construction materials (by cost) contain at least 20 percent postconsumer and/or 40 percent postindustrial recycled content in aggregate, and it features such environmentally friendly materials as wheatboard and soyboard for countertops, denim insulation, and rubber tires in fitness center flooring. A third of its materials (by cost) were manufactured within 500 miles of the site; most of its wood came from forests certified by the Forest Stewardship Council, and over half of its paved surfaces use pine-pitch and rosin-based paving material that is cooler than asphalt, and does not leach hydrocarbons into the Earth. The building also relies on a variety of energy conservation strategies, including daylighting and motion sensors, while deriving some energy from rooftop photovoltaic panels; and it uses waterless urinals and low-flow fixtures to reduce water consumption by 15 percent. It is served by three bus lines and has bike racks and shower facilities for bicycle commuters.

- *30 St. Mary Axe, London, UK*: Also known as the Swiss Re Tower and the Gherkin, this prominent London structure uses energy-saving methods to cut its power consumption by about half. A natural ventilation and lighting system was incorporated by constructing six shafts in gaps in the floors that create a double glazing effect, sandwiching air between two layers of glazing and insulating the office space within. The shafts act to allow warm air to escape from the building in summer while allowing in sunlight during the winter that warms the building. The natural lighting also keeps electric costs down. The tower, made largely from recycled and recyclable materials, also features a system to capture rainwater and reuse it. In addition to filtering incoming air, the building cleans outgoing exhaust.
- *60L Green Building, Melbourne, Australia*: This commercial office building was constructed within the shell of a 1870s brick warehouse, allowing for the reuse of brick walls and other materials. It was designed to emit near-zero greenhouse gas emissions while using only about a third as much energy as a comparable traditional commercial development. Features include natural lighting and reflective areas that draw light into the building core, as well as energy-efficient lamps and ballasts. A large central atrium allows air to flow into offices and is then vented to the atmosphere through four thermal chimneys. Rooftop photovoltaic arrays collect solar energy that provides about 10–30 percent of the building's common area power needs, while providing surplus energy into the city's power grid on weekends and holidays when the building is largely empty. Its water consumption is less than half that of a comparable conventional building, largely because of the use of rainwater and recycled water as well as water-efficient fixtures.

See Also: Green Roofing; Greywater; Passive Solar; Recycling; Zero-Energy Building.

Further Readings

Froeschle, Lynn M. "Environmental Assessment and Specifications of Green Building Materials." *The Construction Specifier*. http://www.calrecycle.ca.gov/greenbuilding/materials/CSIArticle.pdf (Accessed January 2011).

Gumbel, Peter. "Building Materials: Cementing the Future." *Time*. http://www.time.com/time/magazine/article/0,9171,1864315,00.html (Accessed January 2011).

Kim-Carberry, Susie. "Green Building 101: How Space Defines Us." Reuters. http://www.reuters.com/article/idUS40381426620110106 (Accessed January 2011).

U.S. Environmental Protection Agency. "Green Building." http://www.epa.gov/greenbuilding/pubs/about.htm (Accessed January 2011).

David Hosansky
Independent Scholar

GREEN CHEMISTRY

Green chemistry is a new scientific field that focuses on the design, manufacture, and use of chemical processes that could prevent pollution and at the same time improve yield efficiency. Its creation dates from 1991, when the U.S. Environmental Protection Agency (EPA) launched the Alternative Synthetic Pathways for Pollution Prevention research program under the auspices of the Pollution Prevention Act of 1990. This program marked a radical departure from previous EPA initiatives in emphasizing the prevention of pollution, as opposed to managing it after it is produced. The new research program was expanded shortly thereafter to include the development of greener solvents and safer chemicals, and was finally given the name *green chemistry* in 1993.

The goal of the Pollution Prevention Act of 1990 was not simply to regulate the quantity and type of emissions, but to manage the entire structure of industry to reduce the amount of pollution it generated. The main mission of green chemistry has been to develop the scientific knowledge necessary to meet this goal. It was Paul Anastas, one of the principal founders of green chemistry, who claimed early on that by improving the synthetic process of chemicals, it might be possible to prevent the generation of pollutants at the source level. So, synthetic chemistry, aimed at improving the synthetic procedure of chemicals, has been the main approach of green chemists to address environmental problems, as opposed to the purification of produced pollution and pollution management of traditional environmental chemistry.

To help define a more specific research agenda, the 12 principles of green chemistry were formulated by Paul Anastas and John Warner in 1998. They are (1) prevention, (2) atom economy, (3) less hazardous chemical syntheses, (4) designing safer chemicals, (5) safer solvents and auxiliaries, (6) design for energy efficiency, (7) use of renewable feed stocks, (8) reduction of derivatives, (9) catalysis, (10) design for degradation, (11) real-time analysis for pollution prevention, and (12) inherently safer chemistry for accident prevention.

Of these principles, "atom economy" emerged early to become a central concept among researchers in green chemistry. Originally suggested by Barry Trost in 1973, atom economy was designed to overcome the limitations of the traditional concept of "yield" used for calculating the efficiency of chemical reactions. To calculate the yield, chemists traditionally considered only the amount of the main chemical product they intended to produce ("target molecules") and not the by-products, which might include environmentally hazardous materials. In contrast, atom economy takes into account all reactants and all products and hence provides a more reliable indicator of whether or not the reaction produces undesirable by-products—that is, a pollutant. It seemed almost a utopian idea when the concept was first proposed by Trost, but green chemistry has since demonstrated that the elegant efficiency of high atom economy is indeed achievable.

To achieve high atom economy, Trost suggested several types of chemical reaction, such as hydrogenation, metathesis, and cyclo-addition, as they are technically called. These reactions have been important research topics for green chemists, and some have been very successful in making substantial progress.

Ryoji Noyori, for example, won the 2001 Nobel Prize in Chemistry for his research in hydrogenation reactions. Hydrogenation reaction adds the hydrogen atom into unsaturated chemical bonds and generally produces molecules known as optical isomers. Optical isomers are twin molecules that have the same chemical formula but have different geometrical shapes, like the relationship between the right hand and the left hand. Since these optical isomers have different chemical properties, chemists who need to synthesize a molecule with hydrogenation reaction should carefully separate the target molecule from its optical isomer. Noyori developed catalysts and solvent systems of hydrogenation reaction that can selectively produce the target molecule and thereby help increase the efficiency and efficacy of hydrogenation reaction.

Similarly, the 2005 winners of the Nobel Prize in Chemistry—Yves Chauvin, Richard Schrock, and Robert Grubbs—researched metathesis reaction, which has been acclaimed as a great step forward for green chemistry. Metathesis reaction is the chemical process that accompanies the exchange of chemical bonds between reactant molecules. As the metathesis reaction only changes the configuration of the atoms in two different molecules, it does not produce by-products theoretically. To produce this high atom economy reaction, the above named green chemists have also developed various catalysts and solvent systems.

Green chemists believe that pollution prevention is not only better for the environment but also can have substantial economic benefits. They insist that new synthetic pathways and the use of alternative raw materials to prevent pollution are economically preferable for industry because they reduce the cost of pollution management and increase the efficiency of manufacturing processes.

Thus far, they have been proved right. In fact, green chemistry has been able to recruit many industrial partners early on, who have become active participants in its development. Many of the recent Presidential Green Chemistry Challenge Awards, for example, have gone to international chemical and pharmaceutical firms. Merck Corporation won a Greener Synthesis Pathways Award in 2005 and again in 2006 because of the new synthetic procedures it developed for some of its drugs. Thanks to a new hydrogenation catalyst, Merck increased by nearly 50 percent the yield of the synthetic process of sitagliptin, which is the active ingredient in the type 2 diabetes medicine Januvia. Merck's new process for the synthesis of Emend reduced 41,000 gallons of waste per 1,000 pounds of the drug that it produces. Other international chemical firms that won the competition have been BASF, Dow Chemical Company, DuPont, Pfizer, Bayer, and Monsanto. Bayer won the Greener Reaction Condition Award in 2000 and the Greener Synthetic Pathways Award in 2001 for developing a water-based nontoxic agent that is an alternative to organic feedstocks. Pfizer improved its Zoloft production line, doubled product yield, and reduced the use of raw material by 20–60 percent and won the Greener Synthesis Pathways Award in 2002. Monsanto received the first Greener Synthetic Pathways Award in 1996 for increasing the yield of the disodium iminodiacelate (DSIDA) synthesis process.

In contrast to its partnership with industry, government support for green chemistry has not kept up with the expansion of the field and its growing needs. There was a legislative proposal in 2004 to elevate the support of green chemistry to a new level. The Green Chemistry Research and Development Act proposed by Representative Phil Gingrey (R-GA) was to make an interagency working group in charge of coordinating federal green chemistry research and development activity funded by the National Science Foundation, Environmental Protection Agency, National Institute of Standards and Technology, and the U.S. Department of Energy. The act passed in the House of Representatives by an overwhelming majority, 402 to 12, on April 21, 2004, with a few amendments, but the Senate refused to vote on it. Gingrey reintroduced the bill in 2005 and 2007, and Senator Olympia Snowe (R-ME) introduced the same act in the Senate in 2008, but with the same outcome.

Despite this legislative setback, green chemistry continues to thrive as an academic field with its own journals and institutions. Since 1999, the Royal Society of Chemistry has published the journal *Green Chemistry*. The American Chemical Society (ACS) founded the Green Chemistry Institute in 1997. Many universities are establishing their own free-standing programs in green chemistry. At Yale University, the Center for Green Chemistry and Green Engineering was founded in 2007 as a part of the School of Forestry and Environmental Studies. Other academic programs are the program in Green Chemistry and Chemicals Policy at the University of California (United States), the Green Chemistry Centre of Excellence at the University of York (United Kingdom), the Center for Sustainable and Green Chemistry at the Technical University of Denmark, and the Green Chemistry program at Tel Aviv University (Israel).

Green chemistry also has been acknowledged by numerous awards created in the field. The American Chemical Society (ACS) and the National Institute of Standards and Technology have sponsored the Kenneth G. Hancock Memorial Award since 2006, and Dow Chemical Company has presented the ACS Award for Affordable Green Chemistry since 2007. The most prestigious award in green chemistry, however, remains the Presidential Green Chemistry Challenge Award, created in 1995 as a result of President Bill Clinton's Reinventing Environmental Regulation Initiative. There are five awards in the Challenge: the Academic Award, Small Business Award, Greener Synthesis Pathways Award, Greener Reaction Condition Award, and Designing Greener Chemicals Award.

See Also: Engineering Studies; Environmental Science; Science and Technology Studies; Technology Transfer.

Further Readings

Anastas, Paul and John Warner. *Green Chemistry: Theory and Practice*. New York: Oxford University Press, 1998.

Lancaster, M. *Green Chemistry: An Introductory Text*. London: Royal Society of Chemistry, 2010.

Woodhouse, Edward and Steve Breyman. "Green Chemistry as Social Movement?" *Science, Technology and Human Value*, 30/2 (2005).

June Jeon
Independent Scholar

Green Markets

Green markets are markets for sustainability. As such, they are less related to buying and selling green products and more related to businesses fundamentally shifting the way they operate. Green markets incorporate both technology and psychology. Goods and services in green markets are not only produced in sustainable ways, they are also evaluated according to whether they should be produced at all. As such, green market production decisions are based upon whether the product contributes to the well-being of society and the ecosystem, rather than being based solely upon whether the product is profitable. Green markets have various characteristics, drivers, and types, along with a number of green incentives. A dichotomy of green markets also exists and is characterized by two ends of the green spectrum.

People, Planet, Profits

Many companies are realizing that "less bad" is not good enough; they must strive to leave a positive footprint on the Earth, which may include responding to the biggest global challenges of our time, such as poverty, climate change, species loss, and water crises. Similarly, sustainability is no longer viewed as simply a smart idea, trend, or accepted practice; it is becoming the expected practice, and seen as a critical component for corporate operations.

It can not only enable business to anticipate social and environmental trends, but also catalyze innovation and be more responsive to all stakeholders—communities where they operate, employees, suppliers, and the Earth.

Green markets have emerged from two primary movements: sustainable development and corporate social responsibility. In the business world, the three pillars of sustainable development (social, ecological, and economical) often translate into a "triple bottom line." While "green" may characterize the ecological component of sustainability, and corporate *social* responsibility places the emphasis on people, there is growing recognition that what is good for the planet is often best for people, too. The significance of profits in the social, ecological, and economic sphere is hotly debated. Some people believe they should be equally weighted, while others question the fundamental idea of profits being the primary imperative of a corporation. These kinds of questions inspire eco- and social entrepreneurs—people who address environmental and social issues by delivering products or services that generate social change.

Green Market Types

Green markets are often associated with the green economy, green business, eco- and social entrepreneurship, corporate social responsibility, and design for environment approaches. Companies of all sizes, as well as cooperatives and social enterprises, offer products and services across a range of sectors. These sectors include sustainably harvested forest and fish products, energy (renewable and efficient), green buildings, retrofitting, agriculture, tourism, waste management, low-carbon transportation, and information technology components. The "market" for green cities, for example, encompasses: green buildings in which to live and work, including affordable housing; low-carbon transportation (public transit, bicycle infrastructure); cogeneration of heat and energy; "blue infrastructure" for rainwater management; and space for urban agriculture.

Carbon markets represent one niche of green markets, with research, development, and investment focusing on carbon credits and offsets, as well as technology such as carbon capture and storage. However, as pointed out by a recent report from Oxfam, adapting to climate change is also shaping green markets. While mitigation efforts are carbon focused, water is the primary driver behind adaptation measures, as many of the most urgent impacts are—and will be—felt through changes in global and local water regimes. As a growing sector of the green economy, investments in the Blue Economy focus on research and innovation in the following areas: flood and drought defenses; water conservation, storage, and treatment; source water protection; efficient irrigation; and data collection, analysis, and reporting tools that enable adaptive water management to help reduce the future impacts of extreme events and natural disasters.

Green Incentives

Government can provide the policy framework of incentives to help indicate a general direction for the market. The policy toolbox includes taxing the "bads" (e.g., British Columbia's carbon tax); supporting innovation; encouraging collaboration between industry and academic institutions in such areas as chemical engineering, waste-to-energy production, ecodesign, and biomimicry; and promoting innovative clean technologies such as renewable energy, green chemistry, recycling of e-waste, and achieving zero waste across sectors.

Government and business combined can also play a key role in making "green" the easy, affordable choice. For example, safety and quality standards for cars, pesticides, and baby bottles demonstrate a handful of precedents where governments and businesses have conscientiously crafted market choice in favor of public health and safety. Customers in the global north no longer have the option of buying cars without seatbelts, and farmers around the world can no longer buy the pesticide DDT. Canadian parents cannot purchase a baby bottle made with the toxin bisphenol A (BPA), nor can Washington parents find BPA-containing sippy cups, food and beverage containers for children under 3, or sports bottles on store shelves. Pre-selection has occurred even before consumers set foot in the store. As choice editors, government and business are now being asked to make it easier for the consumer to choose sustainable products and services. By ensuring that only sustainable goods and services reach consumers in the first place, the choice at the store becomes a range of sustainable options.

Where sustainability is a requirement for market entry, companies are motivated to invest in research and development to keep their products accessible. Both competition and cooperation can help reduce costs of innovation, helping to make "green" affordable. For example, Walmart has set ambitious targets to green its products and supply chain while ensuring affordability. While its scope of purchasing power enables enormous economies of scale, the ripple effect throughout its global supply chains may offer learning opportunities for other businesses.

In today's market with abundant options, claims, and counterclaims, labeling initiatives are one way to verify—quantify and qualify—"greenness" across sectors. For example, WaterSense-certified toilets are becoming commonplace in homes across Canada and the United States. Organic and fair-trade bananas, chocolate, and coffee are widely found, with fair-trade coffee now featured in numerous mainstream coffee shops and campuses. However, in its "State of the World 2010 Report," the Worldwatch Institute suggests providing product information, such as ecolabeling schemes, has little effect.

Getting Green Into the Market

Companies operating in green markets are redesigning their products and manufacturing processes to encompass people, planet, and profit objectives. One of the most influential opportunities presented to industry has been the cradle-to-cradle (C2C) approach advanced by William McDonough and Michael Braungart in their 2002 book *Cradle to Cradle: Remaking the Way We Make Things*. C2C emphasizes "optimizing positive impacts" of industry rather than minimizing its negative impact. From the outset, materials and products are designed with life cycles that are safe for both human health and the environment. Using nature's "biological metabolism" as a model for developing a "technical metabolism" flow of industrial materials, products and materials can be reused continuously through biological and technical metabolisms. C2C-inspired operations and stakeholder relationships are based on the principles of "waste equals food," maximize use of renewable energy, and "celebrate diversity" by respecting human and natural systems.

Zero Emissions Research & Initiatives (ZERI) is a global network of creative thinkers who are inspired by science and nature's design principles. Viewing waste as a resource, the ZERI Foundation addresses world challenges. It has identified 100 innovations that it claims could generate 100 million jobs within 10 years and is collaborating with the United Nations Environment Programme to help disseminate them across the globe.

Along a similar vein, the Biomimicry Institute catalyzes the development of radical new technologies and products inspired by nature. The institute has partnered with the ZERI Foundation to create the Innovation for Conservation Fund. Modeled on the conventional system of brokers receiving commissions for successfully raising capital for corporations, the fund is intended to protect the biodiversity that is inspiring a new industrial age. A commission will be paid to the fund by companies benefiting from 3.8 billion years of evolution. The two organizations will identify market-ready, biologically inspired technologies and encourage all company fundraising to commercialize such technologies to contribute to the preservation of species and their habitats.

"Celebrating diversity" also defines collaboration among diverse, often historically opposed, actors. Companies, investors, and nongovernmental organizations (NGOs) are increasingly commonplace. For example, such partnerships often characterize the development of codes of practice and standards for companies and their products.

Ceres is an American network of investors, environmental organizations, and other public interest groups working to integrate sustainability into capital markets. By adopting the 10-point (or comparable) code of corporate environmental conduct offered by the Ceres Principles, companies formalize their commitment to environmental accountability and dedicate themselves to continuous improvement, discussion, and public reporting.

The International Organization for Standardization (ISO) offers another set of standards and practical tools for green-market companies. ISO 14000 standards address environmental management systems, as well as performance evaluation, life-cycle analysis, labeling, auditing, and communication. Organizations assess where their activities have an environmental impact, and benefit by increased savings in consumption of materials and energy, lowered costs of waste management and distribution, and improved corporate image in the eyes of the public, customers, and regulators.

Green Spectrum

There are multiple approaches and debates within green markets. Presented as "Green Market 1" (GM1) and "Green Market 2" (GM2), respectively, this dichotomy characterizes two ends of the spectrum along which actors operate. The term *global south* describes people living in countries outside of North America, Europe, Australia, and Japan, while those countries are referred to as the *global north*.

Green Market 1: Mint Green

In GM1, neoliberal economics are the dominant paradigm, tweaked to incorporate what were traditionally considered externalities, such as pollution and waste. Businesses operating in this type of market tend to view sustainability as a technical challenge that can be solved by increasing research, innovation, and competitiveness. They tend to focus on discrete green projects as opposed to integrating sustainability into their corporate charters. Key drivers for change in GM1 include lack of public trust in business; high-profile shareholder activism, as well as leading companies demonstrating operational and competitive benefits; urgency to cut costs (both fallout from the recession commencing in 2008); and growing expectations among consumers that their lifestyle should not be "costing the world."

Some companies are drawn to GM1 because they realize that corporate social responsibility can have a positive effect on their bottom line through improved efficiency and asset utilization—reducing energy and waste bills—as well as risk reduction. When companies focus on the triple bottom line, they also realize many intangible benefits—not only are they better able to maintain social license to operate, but they reap human capital benefits in the form of greater satisfaction and loyalty among employees, attracting and retaining talent.

Others companies are reluctant participants. They have been the focus of shareholder activism or other external forces, and "greening" is perceived as necessary to preserve or rescue a tarnished corporate image. Shareholder interests are the dominant concern, but there is increasing discussion around the interests of other stakeholders such as employees, customers, communities where companies operate, and the Earth.

Marketing in GM1 primarily focuses on selling a "green lifestyle" to middle- and upper-class shoppers across the global north that are increasingly willing to pay more for "green" goods (like organic food, or Fair Trade coffee) and services (like ecovacations, eco-event planning for conferences, concerts, and weddings). However, Walmart's entry into the green market is challenging green exclusivity, galvanizing discussion around affordability and accessibility of green markets. The future of GM1 will be shaped by metrics that quantify key "green" (sustainability) attributes of companies as a whole, including the life cycles of their entire product line.

Green Market 2: Dark Green

GM1 asks the question, "How can this X be produced in a sustainable way?" while GM2 asks an entirely different question: "Is it desirable to produce X at all and if so, under what type of conditions?" GM2 places a greater emphasis on values, re-examining the psychology of "the good life," and challenges two key assumptions: (1) greater consumption (GDP) increases well-being, and (2) markets must constantly grow. GM2 takes measurement of well-being into consideration, sometimes referred to as genuine progress indicators (GPI).

GM2 believes prevailing worldviews, institutions, and technologies fail, and perhaps are incapable of meeting humanity's most urgent needs. Climate change, deteriorating soil quality, rapid loss of biodiversity, pandemics, rising food prices, pollution, and dwindling oil supplies—these are the issues markets must respond to. These "environmental" problems are viewed as human problems that require a different mentality and cultural shift to address. Legal, political, and economic structures established years ago must now adapt their worldview from a world that is empty to a world that is full. This means challenging collective behaviors of extreme consumption, unsustainable environmental practices, and cultures that prioritize individual private benefit above public or community gain. GM2 emphasizes public goods and services (like healthy air) over private ones. It sees a stable global climate and ecological resilience as global public goods, requiring global solutions arrived at through cooperation.

GM2 explicitly acknowledges that traditional markets have failed to value a critical service sector that must now be valued: ecosystem services. The ecological system is the framework within which an economic system must operate, and GM2 not only "internalizes externalities," such as pollution, but it properly accounts for the depletion of the Earth's natural capital, on which all other capital depends. Local governments and communities play a significant role planning regions, while national and regional governments collaborate in their choice editing to establish large markets with a range of sustainable choices. Marketing focuses not on shifting consumption, but reducing it in the global north where excessive consumption has been the cultural norm. Social marketing is directed instead toward promoting sustainable behaviors. It follows that efficiency increases are not necessarily encouraged, as these can often lead to more consumption, not less. GM2 also includes alternative currencies, such as local currencies, bartering, lending circles, and forms of exchange outside of the traditional money economy.

GM2 places a greater emphasis on sustainability as a whole, and the social aspects of a sustainable market. This market values traditionally marginalized groups, such as women, indigenous peoples, or the poor, and prioritizes green and decent jobs—livelihoods over lifestyles. GM2 is seen as a key mechanism to meet key objectives such as the Millennium Development Goals.

Climate change is a strong driver of GM2, though the carbon footprint receives as much attention as the ecological footprint. Adaptation is a significant part of the equation, specifically as it relates to the livelihoods of many people in the global south. Blue Market opportunities for climate change adaptation focus on increasing access to water for the world's poorest people in the most water-scarce regions. GM2 builds a resilient economy that is capable of surviving and adapting in response to unexpected changes and catastrophes.

GM2 markets of the future may look very different than the ones that currently exist. With an emphasis on global equity, climate justice, and resilience, these markets might be characterized by fewer goods and services that employ more people. People in the global south may be working more, for higher wages, while employees in the global north may be working fewer hours, for less pay. Because GM2 views healthy people and ecosystems as the goal of an economy, metrics of the future GM2 may measure components of well-being and ecological integrity, rather than production or recycling processes, as is the focus for GM1. Institutions, in turn, might shift to serve the goal of achieving sustainable prosperity.

See Also: Carbon Market; Cradle-to-Cradle Design; E-Waste; Extended Producer Responsibility; Green Technology Investing; Sustainable Design.

Further Readings

Daily, Gretchen and Kathy Ellison. *The New Economy of Nature: The Quest to Make Conservation Profitable*. Washington, DC: Island Press, 2002.

McDonough, W. and M. Braungart. *Cradle to Cradle: Remaking the Way We Make Things*. New York: Northpoint Press, 2002.

Oxfam. "The New Marketplace for Adaptation: Climate Change and Opportunities for Green Economic Growth" (2009). http://www.oxfamamerica.org/files/the-new-adaptation-marketplace.pdf-1 (Accessed September 2010).

United Nations Environment Programme (UNEP). "Green Jobs: Towards Decent Work in a Sustainable, Low-Carbon World" (2008). http://www.unep.org/PDF/UNEPGreenjobs_report08.pdf (Accessed September 2010).

Worldwatch Institute. "State of the World 2010: Transforming Cultures." http://www.worldwatch.org/node/6368 (Accessed April 2011).

Stephanie Lepsoe
Independent Scholar

GREEN METRICS

Green metrics are quantified measures of an organization's environmental impact and performance, and—alongside social and financial information—are important pillars in the broader field of sustainability accounting. Developing the capacity to measure environmental impacts offers benefits in benchmarking performance, offering data against which management can take action, documenting progress over time, evaluating the effectiveness of management initiatives, and optimizing the design of products, processes, and portfolios. Additional commercial benefits have been identified as decreased organizational risk and enhanced stakeholder attractiveness.

While an organization might develop and apply its own green metrics in a private fashion, a number of overarching frameworks have evolved, such as the Global Reporting Initiative (GRI), the performance guidance offered by the International Organization for Standardization (ISO) in the ISO 14000 series of environmental management standards, and—more specifically—the ISO 14040 Life Cycle Analysis approach. Such frameworks offer programs of established metrics, which cover areas such as consumption of input resource (like raw materials, water, or energy), process or product outputs (like solid waste, emissions, water discharge, or spills), management and compliance (like fines or sanctions for noncompliance, or cost of pollution prevention), and environmental conditions (like concentration of contaminants in surface soil or ambient air, or area of land rehabilitated annually). Green metrics are commonly featured as an addendum to financial performance reports, and are usually accompanied by qualitative descriptions of activities and aims. A full quantitative reconciliation of environmental impact into a financial equivalent, alongside that of social impact and financial data, forms the basis of the embryonic Triple Bottom Line (TPL or 3BL) report.

Business Metrics

Companies have long relied on financial indicators such as turnover, profit, and earnings ratios to communicate their health and achievements to a range of stakeholders, including employees, customers, regulators, and shareholders. However, financial measures alone are somewhat retrospective—reflecting on events that have already occurred—offering the manager little useable data to proactively improve performance. The past 20 years have seen the development of management tools—notably the Balanced Scorecard—for measuring business performance more comprehensively. Rather than simply reflecting results, the introduction of customer, process, and employee perspectives aims to focus decision making, improve management response time, and motivate employees to achieve the strategic goals supported. Such performance measurement structures have a strong disposition toward overarching financial goals, and environmental concerns have been difficult to express unless they can be translated into financial equivalents. Where operating costs are reduced by lowering energy consumption, the relationship is straightforward, although this would have still found difficulty making it onto the scorecard in competition with metrics that held greater leverage. Considerations such as the carbon dioxide emitted from transporting materials to site did not readily feature in the metrics of the late 20th century.

The landscape has changed significantly in the 20th and 21st centuries. A growing awareness of the often negative impacts of globalization has engendered ethically aware consumers and investors, and the emergent field of corporate social responsibility (CSR) has risen to address some of these concerns. Environmentally, the international benchmarking of atmospheric carbon dioxide levels to 1990 levels through the United Nations Framework Convention on Climate Change at Rio in 1992 and subsequent procession toward national targets via the Kyoto Protocol in 1997 have brought the quantification of environmental impact into sharp focus, and organizations play a crucial role in achieving these targets. In his 1997 article in the *Harvard Business Review*, Stuart Hart reflected on the critical role of businesses, highlighting that although these benchmarks are set politically, and consumption occurs at the individual level, companies—particularly large ones—are in the unique position of having the motivation, resources, and agency to direct and drive a sustainability agenda.

Advantages and Difficulties

Managing and reporting an organization's environmental impact offers benefits above and beyond cost reduction and compliance with environmental legislation. Proactive approaches engender a position of influence in government policy, achieve an inherently lowered business risk via readiness for future legislative scenarios, and preferentially attract talented employees. Green metrics also substantiate an agenda of corporate responsibility that becomes a marketing commodity for employees, customers, and investors. However, the skeptical proclamation of environmental priorities used purely as a marketing tool, disjointed from the executed priorities of the company, is considered a cynical policy commonly called "greenwashing." There is also a lively debate as to whether companies would deliver a greater overall societal benefit by maintaining their "raison d'être" and maximizing monetary surplus. Suggesting that attention to environmental management is a distraction from the core function, skeptics highlight that a maximized profit sustains economies through taxation and in essence provides the financial resources for governments to go on to ameliorate the environmental impact. However, the demand for support in developing green metrics through the frameworks in the following section is suggestive of the variety and scale of companies that are exploring potential benefits at hand.

Example Frameworks

Although an emerging and evolving field, there are many examples of companies using and reporting green metrics. Examples include Credit Suisse (financial services), Fujitsu (information technology), General Motors (automotive), Rio Tinto (mining), and Shell (energy). The nature of such measurement varies widely and can usefully be considered as occupying a spectrum that reflects how comprehensive the metrics are and the resources (including information management systems) required to support them. There exist sector-specific structures for measuring environmental performance (such as LEED in the building industry) and a number of overarching projects for wide adoption, which include life-cycle assessment, ISO 14000, and the GRI G3 Guidelines.

The ISO 14000 series of international environmental management standards offers specific guidance on the measurement aspects of environmental activity and performance. This guidance is considered critical to having the ability to review the stage of the Plan-Do-Review-Act management cycle inherent in the standards. In particular, ISO 14031 focuses on methods of evaluating environmental impact and the rigorous development of performance metrics that are grouped as follows:

- Environmental condition indicators (ECI), such as concentration of contaminants in ambient air µg/m^3)
- Environmental performance indicators (EPI), which can be either management focused (MPI, such as number of audit findings identified and resolved), or operationally focused (OPI, such as wastewater discharged per unit of production, e.g., l/1000 units)

Furthermore, ISO 14040 offers specific guidance on the product life-cycle assessment (LCA) approach (also known as cradle-to-grave analysis). Popular in the United States, LCA involves identifying and quantifying the holistic environmental impact caused by the inception, carrying out, and conclusion of a business activity, whether manufacturing a product or delivering a service. Thus, an environmental responsibility is assumed that vertically extends beyond the organization's own boundaries—upstream into the embedded energy of raw materials and downstream to the actions necessary for responsible product disposal. This more comprehensive approach to sustainability allows organizations to environmentally optimize products and processes during their design.

However, critics of the LCA approach point to the difficulties involved in the normative judgments required at the assessment stage, where some degree of ranking must take place in order that action can be prioritized; for example, is reducing energy consumption more or less critical than reducing water consumption? Other difficulties involve the ongoing process innovations within companies that necessitate the use of very current data for LCA comparisons and the assumptions that will have to be made in comparing alternative designs. The newest models of vehicles or white goods can differ significantly in their material content from their predecessors, and energy utilizations in improved manufacturing processes will be based on estimates. Nevertheless, LCA represents a first step toward the premise of full life-cycle costs; for example, where the disposal costs of products might be built into purchase decisions rather than being carried separately by society (or indeed future generations).

Example Metrics

Both the GRI G3 Guidelines and ISO 14031 contain specific guidance on indicators of performance, although ISO takes the more specific view on the relative role of different data types. First, direct measures are described that have much in common with the emphasis in the G3 Guidelines. "Direct metrics" are understood to be those directly based on International System of Units (SI) of measurement, such as kilograms (kg) of raw material consumed, megajoules (MJ) energy consumed, or quantification of spills in liters. These base measures are subject to modification to produce further types of data that offer greater utility for management and benchmarking.

"Relative metrics" flex a base measure against another variable in a normalizing manner, for example, in liters of water consumed per 1,000 units of production, total kilograms of hazardous waste per 1,000 units of production, or contaminant concentrations expressed as µg/m^3. This flexibility offers a good deal of utility to operational managers who are able to make comparisons across product lines or by tracking metrics when output is variable. Indexed metrics involve converting a base measure to a predetermined benchmark, which might be a historic measure or an agreed target. The most frequent format of the indexed measure is a percentage, such as the percentage of recycled feedstock, or the percentage of increase/decrease in algae blooms. The final data type described in ISO 14031 is that of "aggregated data" where data describing an activity is collected from different locations and articulated as a single value. Examples include total emissions by type, total water withdrawal, or total person-hours spent reactively responding to environmental occurrences.

Reporting

Companies using green metrics might report them on an informal basis, as information that accompanies the financial report, or in a manner consistent with the metrics framework adopted. It is usual for all formats of green metric reporting to be complemented by qualitative descriptive data. Internationally, the most widely used formal sustainability reporting framework is that of the Global Reporting Initiative (GRI), an independent not-for-profit collaborating center of the United Nations Environment Programme. The project's most recent framework—the G3 Guidelines, published in 2006—includes guidelines for reporting economic, social, and environmental performance and defines a foundation of 30 green metrics across nine themes. The themes cover use of resources, including materials, energy, transport, and water. Impact themes include biodiversity, level of emissions, waste, and spills and remediation effort. Each of the measures is accompanied by a rationale for its inclusion, guidance on how the metric should be compiled, a definition of terms, and a guide to associated documentation.

These foundational metrics are enhanced by "sector supplements" for the financial services, metals and mining, automotive, logistics and transportation, public agency, tour operating, telecommunications, and electric utility sectors. Sector supplements map metrics that are more specific to that industry against the foundational framework with the aim of increasing their adoption by making them more applicable. Per the main corpus of measures, the sector supplements are developed and annually reviewed by a network of industry representatives rather than being designed and imposed top-down; this ownership is an important tenet of the GRI approach. An example of the sector supplements include six tailored green metrics in the metals and mining sector, covering such pertinent issues as measures for land rehabilitation, management of large volumes of process waste, and agreed criteria for establishing whether biodiversity management plans are necessary.

Perhaps the most ambitious reporting method is that of full triple bottom line (TBL, or 3BL) accounting. TBL collates environmental, financial, and social measures to produce a wider descriptive of an organization's success. Although TBL reporting carries with it a suggestion that green metrics are objectively translated into a single value, this is difficult to apply; the appropriateness of netting detrimental economic impact against ameliorative measures is conceptually problematic. A TBL methodology that offers something innovative yet rigorous remains elusive and is likely to feature in the landscape of green metrics in coming years.

See Also: Carbon Footprint Calculator; Cradle-to-Cradle Design; International Organization for Standardization (ISO); LEED Standards.

Further Readings

Global Reporting Initiative. http://www.globalreporting.org (Accessed September 2010).
Group of 100 Incorporated. "Sustainability: A Guide to Triple Bottom Line Reporting." (2003). http://www.group100.com.au/publications/G100_guide-TBL-reporting2003.pdf (Accessed September 2010).

Hart, Stuart. "Beyond Greening: Strategies for a Sustainable World." *Harvard Business Review*, 75/1 (1997).

International Organization for Standardization. "A Guide to the ISO14000 Series of Standards." http://www.iso.org/iso/theiso14000family_2009.pdf (Accessed September 2010).

Beverley J. Gibbs
University of Nottingham

GREEN NANOTECHNOLOGY

One of the frontier areas of contemporary technology is the emerging ability to manage and control matter at a molecular level, popularly known as nanotechnology. Widely seen as one of the most exciting technologies of the 21st century, nanotechnology has been hailed as the next industrial revolution and the basis of an industry worth billions of dollars by the year 2020. Nanotechnology refers to the use of materials at an atomic or molecular level—*nano* literally means "extremely small" in Greek. Nanotechnology is the ability to measure, see, manipulate, and manufacture things, usually between 1 and 100 nanometers. A nanometer is one billionth of a meter; a human hair is roughly 100,000 nanometers wide. At the nano level, material takes on very different properties. Thus, for example, when nanoparticles of antimony tin are incorporated into coatings, they offer transparent protection from ultraviolet radiation, unlike larger-size particles.

Green nanotechnology can be defined as a set of principles that call for a conscious effort to eliminate or minimize pollution from the production of nanomaterial. Thus, it is a classic example of anticipatory governance that can be defined as a distributed form of emerging political order with an emphasis on long-term political thinking. There are two key norms that govern green nanotechnology—the production of nanomaterial without harming the environment, and the creation of new nanoproducts that can address extant environmental problems. The key principles behind the application of green technology to nanotechnology are the prevention of waste, energy efficiency, and the prevention of air pollution. It is hoped that these aims can be achieved by the use of nanotechnology for efficient and controlled manufacture that dramatically cuts down on the waste produced in manufacturing, and second, through the use of nanomaterials as catalysts to decrease and eliminate toxic by-products. It is also done by the incorporation of "green" principles right at the inception stage.

In many ways, nature has already perfected green nanotechnology—for example, the ribosome of a cell. Moreover, since some studies have questioned the safety of carbon-based nanotechnology (see below), some have argued in favor of "wet nanotechnology" based on organic materials in contrast to the "dry nanotechnology" that is based on carbon or its derivatives. Scientists are now exploring the use of biological organisms to grow nanomaterials. In nature, numerous inorganic materials are synthesized by living organisms, the best example of which is photosynthesis. Nanoscale manipulation by nature is also what allows color variation in animals to attract female attention or for purposes of camouflage and what gives multilayered seashells their superior strength.

One of the central worries regarding nanotechnology is that it will lead to unanticipated consequences, and so in the creation of nanotechnology, scientists are trying to anticipate problems and to be proactive in addressing them. This is in direct contrast to earlier technologies, many of which were developed without an anticipation of their negative impact (oil), in secret (nuclear), or in a spirit of progress (chemical). The postwar Green Revolution, for example, increased the production of food manifold, but it was only later that the environmental consequences of the heavy use of pesticides and fertilizer to boost agricultural output became evident. The electronics and information technology industry, at its inception, was deemed to be relatively "clean." However, the disposal of computers that become obsolete at a rapid pace has led to the seepage of heavy metals such as lead, mercury, and cadmium, as well as chlorofluorocarbons and brominated flame retardants, from landfills into water supplies. If incinerated, these pollute the atmosphere. The manufacture of semiconductor industry components is also resource heavy with a single dynamic random access memory (DRAM) that weighs a few grams being the end product of 1.7 kilograms of raw material.

Moreover, the fear of nanotechnology has been heightened by anticipations of disaster in fiction, such as Michael Crichton's famous novel *Prey* or Bill Joy's warning that it would be better to abandon nanotechnology to prevent the end of the world in "grey goo." Therefore, one of the key aspects of green technology is the aspirational goal of anticipatory governance—to avoid the mistakes of earlier technological transformations, such as coal, oil, or even microelectronics and information technology, and develop clean, nontoxic, and environmentally friendly procedures for the synthesis and assembly of nanoparticles.

The use of technology as a value-neutral concept that could have only positive results is a reflection of the scientific attitude that prevailed from the triumph of reason over faith that marked the 17th-century Enlightenment until the mid-1960s, when a series of works exposed the dark side of technology. Conceptually, the subject is framed by the principles of green chemistry and green engineering. This means that nanomaterials must be produced with minimum impact on the environment, including a life-cycle approach to nanotechnology, identifying what is not "green" in current approaches, and ensuring that nanoproducts incorporate green methods of manufacture right from the design stage. These ideals include the designing of new nanoproducts so that they do not generate waste, the creation of safer chemicals and chemical products, more efficient chemical synthesis processes, the use of renewable raw materials such as green algae, the avoidance of chemical derivatives, and the use of catalysts to provoke reactions, all of which conserve energy and generate minimal waste. One indication of the interest in green chemistry was the awarding of the 2005 Nobel Prize in chemistry to a group of scientists who worked on metathesis, or the changing of the order in which reactions take place.

The Next Growth Industry

Nanotechnology is seen as the next growth industry and federal policy in the United States and many developed countries, in reaction to the perils of climate change, is slowly moving toward a milieu that is favorable to the development of clean and green technology. Moreover, it is widely held that 21st-century technology, unlike 20th-century technology, will have to be environmentally friendly. In 2009, global investment into clean energy rose by 4.4 percent, and in 2008, investment exceeded $150 billion for the first time. In September 2006, the U.S. House of Representatives passed the Green Chemistry Research and Development Act of 2005, which called for federal coordination of green chemistry research. The bill would ensure that life-cycle assessments (LCAs) are carried out on new products, and that "trade-offs" between the environment and efficiency be considered. However, the bill, which supporters hoped would lead to a coordinated interagency effort to develop green technology, was not passed in the Senate. Given the litigious climate in many advanced countries, especially in the United States, corporations are eager to prevent future lawsuits that could potentially lead to millions of dollars in losses. All this has led to intense efforts to develop "green nanotechnology" or nanoproducts that are as clean and green as possible right from the start.

Significantly, nanomaterials have many properties that lend themselves to green technology. One of their key properties is their high surface-area-to-volume ratio that makes it possible to reduce the amount of material needed to manufacture things in comparison to current methods. For example, tiny amounts of platinum and other precious metals can be used in automobile catalytic converters or fuel cells, making new technology much cheaper and much greener. Other nanotechnology applications relevant to green technology are the use of nanomaterial as catalysts for greater efficiency in current manufacturing processes by minimizing or eliminating the use of toxic materials (green chemistry principles); the use of nanomaterial and nanodevices to reduce pollution (e.g., water and air filters); and the use of nanomaterial for more efficient alternative energy production (e.g., solar and fuel cells). New sources of energy may even be discovered using the principles of nanotechnology. Scientists in South Korea, for example, have demonstrated that it is possible to use sound as a power source to drive nanogenerators based on piezoelectric nanowires. This could mean that sound, whether that of music or of traffic, may be able to be converted into electricity.

Nano-enhanced green industry technologies are one growth area. The use of mesoporous silica nanoparticles (MSN) in combination with catalytic systems to synthesize biodiesel from free fatty acids in vegetable oils, the use of nanogold and nanopalladium catalysts to remove carbon monoxide from the air, especially in mine shafts, and the use of nanoporous absorbents to remove mercury and other toxic heavy metals from wastewater, particularly in the oil and gas industry, are all promising examples of green nanotechnology. In addition, there is the field of green nanomanufacturing, which combines tools from microelectronics manufacturing and organic chemistry, such as the rapid, error-free reproduction of nanoparticles, and the creation of nanomaterials for biomedicine. Biometric nanofibers that have antimicrobial properties could be used in the packaging industry to create wrappings that can "breathe" and are biodegradable. Both these innovations would address a problem that has bedeviled the packaging industry, namely, the huge amounts of biopolymer waste that the chemical, food, and pharmaceutical industries create.

In the sense of applications, green nanotechnology can also be defined as nanotechnologies that provide energy, clean water, and a good environment in a sustainable way by addressing environmental challenges in areas such as energy, water, food, and increasing population. Among the environmentally significant innovations of nanotechnology are the atomic-level synthesis of new and improved catalysts for industrial processes, the self-assembly of molecules to create new chemicals and materials, the construction of microscale/nanoscale reactors, and advanced solar and fuel cells. These ideas, some of them still at the theoretical stage, translate into the various branches of green nanotechnology.

Green technology principles are also evident in the way nanotechnology is used and exploits the differing properties of matter at the nanoscale. Some applications of this emerging science try to reengineer and reduce the environmental impact of current technologies such as computers, automobiles, and batteries. For example, a newly developed nanomaterial could replace the tin-lead solders that are used as interconnects in electronic products. Others create new energy-efficient methods of power such as new solar cells, and yet others try to make current manufacturing methods more environmentally friendly. For example, in the auto industry this would mean the use of nanomaterials to build car bodies that would cut down on the amount of fuel required to power them, the use of nanosensors to monitor fuel use, the use of self-cleaning nanocoatings on car exteriors that would mean less cleaning time and solvents, and the use of embedded nanomaterials in tires to allow them to last longer. Many of the advantages in this area come from the lighter properties of nanomaterials. Using nanotechnologies to fabricate materials that are lighter and stronger than conventional materials, as in the case of carbon nanotubes, translates to clear fuel efficiency gains in cars or planes. It has been estimated that using nano-sized catalysts in car engines (e.g., substances that speed chemical reactions) results in using 70–90 percent less of the same catalyst in bulk form. Other applications of nanotechnology that have implications for the environment are the greater storage capacity, lifetime, and safety of batteries using nanotechnology, which will dramatically increase energy efficiency. One promising area for green nanotechnology is the discovery of processes by which energy use can be made incredibly efficient, thus contributing to cutting down carbon dioxide emissions and reducing global warming. For example, a futuristic use of nanotechnology would be to use thin layers of polycrystalline silicon on nanostructured solar cells so that some layers can be used to power the building, while others can be "switched off" to keep the building cool. Advanced nanotechnology is likely to make it easier to switch to renewable sources of energy such as wind and solar power, which currently are insufficient to address energy needs or too expensive. Engineers at Stanford University in the United States have demonstrated that even light behaves differently at scales of around a nanometer. By creating solar cells thinner than the wavelengths of light, it is theoretically possible to trap the photons inside the solar cell for a longer time, increasing the chance they can be absorbed, thereby increasing the efficiency of the solar cell. Scientists suggest that by properly configuring the thicknesses of several thin layers of films, an organic polymer thin film could absorb as much as 10 times more energy from sunlight than predicted by conventional theory. The ultimate ideal for green technology would be molecular nanotechnology, or the ability to manufacture anything by assembling molecules rather than reducing raw material to the finished product, which generates zero waste.

Nanotechnology and Green Engineering

"Green engineering" is another emerging branch of nanotechnology that focuses on the design of products incorporating energy efficiency and biodegradable materials. At the center of this process is LCA, which studies the environmental effect of any product throughout its life cycle and take steps to ensure that the environmental impact is minimal. For example, nanofibers created from starches and proteins could be manufactured using electrospinning technology. The ultimate in green engineering is a cell where abundant raw material is used in a benign manner, waste is recycled, and energy is used in an efficient manner. This branch of engineering focuses on the design of products incorporating energy efficiency and biodegradable materials. Scientists are looking at the possibility of using nanoengineering to create ultrafine membranes that can extract carbon dioxide from the emissions of fossil-fuel power plants and even recycle it into other nanotechnology-enabled processes that can generate biodiesel fuel from algae. Another approach of green nanotechnology is the improvement of existing nanotech processes. Scientists at the University of Waterloo in Canada are studying silicon nanowires and silicon nanocrystals that can absorb sunlight in more energy-efficient ways, allowing it to be converted into electric current.

Not only would such methods be cleaner, they could also prove to be cheaper. For example, traditional methods to manufacture gold nanoparticles that are essential in electronics and medical imaging use toxic solvents in a resource-intensive method that costs around $300,000 per gram of gold nanoparticle, while innovative synthesis methods based on the principles of green nanotechnology that use nontoxic solvents, catalysts, and purification by nanoporous filtration, reduce the cost to $500 per gram of gold nanoparticle. Studies have shown that many bacteria and plants can create metal ions from soils and solutions during detoxification processes, leading to the formation of insoluble complexes with the metal ion in the form of nanoparticles. One example of this type of bioreduction and nanoparticle production is magnetostatic bacteria that can synthesize magnetic nanoparticles. In the same way, gold nanoparticles (2 to 20 nm) could be created inside alfalfa seedlings.

One of the key applications of green nanotechnology would be in water purification that would enable the use of water that is currently wasted because it cannot be treated properly. Scientists theorize that nanotechnology-enabled physical filters with nanometer-scale pores can remove 100 percent of bacteria, viruses, and even prions from water. Another advantage of green nanotechnology in water purification would be the elimination of downstream pollution, and advanced nanotechnology could even purify the chemical-laden water that is the by-product of industrial processes. One of the most attractive aspects of green nanotechnology is that, given its scale, it can be designed cheaply. This would address a problem that has affected green technology—renewable sources of energy such as wind and solar power are expensive. With nanotechnology, elegantly designed filter materials and smaller actuators will allow even the smallest filter elements to be self-monitoring and self-cleaning. Self-contained, small, completely automated filter units can be integrated into systems scalable over a wide range. Some companies have looked at the possibilities of nanotech in catalytic filters that can be used to clean exhaust from anything from small engines such as those on leaf blowers to coal- and natural gas–fired electricity generating stations. The products are based on extremely thin nanofibers coated with a catalyst that reacts with nitrogen oxides and other harmful substances. Finally, in the area of printing technology, a new type of toner used in printers and copiers significantly reduces the amount of energy used in toner manufacture and also roughly halves the amount of toner needed to print a page. This new process is estimated to use 25 percent less energy and, because the particles are smaller, it takes only half as much toner to print a page.

Success or Failure?

Environmentalists, however, have pointed out that nanotechnology (like other technologies that were hailed as a panacea to global environmental problems) is now seen as a savior but that such expectations may not materialize. They point out that the extant environmental costs of nanomaterial production (such as increased energy and water demands) and unknown dangers may outweigh any perceived benefit of green nanotechnology. At the center of any evaluation process for new technology is LCA, which studies the environmental effect of any product throughout its life cycle and takes steps to ensure that the environmental impact is minimal. The most worrisome aspect is that although green nanotechnology incorporates the principle of "safety by design," proper life-cycle analysis and validated nano-specific risk assessment methodologies may take as long as

15 years to be valid. Moreover, earlier technologies such as asbestos revealed their environmental impact long after they had been used extensively. With nanotechnology, the danger is compounded because nanoparticles are so small that they can be widely dispersed, even within human beings.

Critics argue that rather than reverse environmental degradation, nanomaterials, in their potential to contribute to global warming, ozone layer depletion, and environmental or human toxicity may, per unit of weight, have an impact that is 100 times greater than those of conventional materials like aluminum, steel, and polypropylene. The most disquieting feature of nanotechnology is that even if used in smaller quantities than conventional chemicals, nanomaterials may have a greater toxicological burden. In 2006, the Woodrow Wilson International Center for Scholars' Project on Emerging Nanotechnologies (PEN) estimated that 58,000 metric tons of nanomaterials will be produced worldwide from 2011 to 2020. Given the greater potency of nanomaterials, this could have an ecological impact equivalent to 5 million metric tons—or possibly even 50 billion metric tons—of conventional materials.

Though the evidence is far from conclusive, there are indications that nanomaterials could create a new generation of toxic chemicals that are far more potent given their minuscule size. Some studies have seriously challenged the promises of nanotechnology, and there are fears that in spite of the laudable intentions of green nanotechnology, the technology itself is too dangerous. For example, some experiments have shown that carbon nanotubes, when introduced into rats, cause inflammation, granuloma development, fibrosis, artery "plaque" that causes heart attacks and DNA damage, and even cancer. This has led to carbon nanotubes being derided as the "new asbestos," referring to the lung disease that asbestos caused in the 20th century.

There are also indications that carbon nanotubes diminish rice yields and make wheat more vulnerable to other pollutants. Specifically, two types of carbon nanomaterials—C70 fullerenes and multi-walled nanotubes (MWNT)—delayed rice flowering by at least one month. Some scientific studies warn that nanomaterials now in commercial use can damage human DNA, negatively affect cellular function, and even cause cell death.

Alarmingly, some nanomaterials are toxic to commonly used environmental indicators such as algae and invertebrate and fish species, and some nanomaterials impair the function or reproductive cycles of earthworms, which play a key role in nutrient cycling that underpins ecosystem function. Most recently, disturbing new evidence has shown that nanomaterials can be transferred across generations in both animals and plants. Seeds exposed for only two weeks to C70 fullerenes passed these on to the next generation of seeds. Exposure to carbon nanotubes also made wheat plants more vulnerable to uptake of pollutants. Carbon nanotubes pierced the cell wall of wheat plants' roots, providing a "pipe" through which pollutants were transported into living cells. This is particularly worrying, since rice and wheat are staple crops that are essential to the food supply of a growing population, and any negative impact on this, especially one that can be carried across generations, would have major consequences.

One key argument for green nanotechnology, as seen above, is that minimal quantities of more potent nanomaterials can theoretically achieve the tasks of much larger amounts of conventional materials, and materials such as carbon nanotubes allow lighter industrial components to be less energy and resource intensive. However, some early LCAs indicate that the manufacture of nanomaterials in fact has an unexpectedly large ecological footprint. Highly specialized production environments, high energy and water demands of processing, low product yields, high waste generation, the production and use of greenhouse gases such as methane, and the use of toxic chemicals and solvents such as benzene all contribute to this conclusion. Finally, these studies argue that manufacturing nanomaterials and nanodevices (including nanomaterials to be used in energy generation, storage, and conservation applications) is extremely energy intensive. These studies have challenged the notion that nanoapplications will save energy.

Even those areas in which green nanotechnology seems particularly promising have been challenged. Critics have questioned the actual extent to which nanosolar panels are effective, pointing out that thin film cells in nanosolar panels have a 14 percent efficiency when compared to conventional wafer-based crystalline silicon panels. Extant silicon panels remain much more efficient at around 25 percent and are much cheaper. Moreover, many of the dramatic changes brought about by nanosolar applications, such as energy-generating plastic-based paint that can harvest infrared (nonvisible) light, are theoretical, and research remains at an exploratory stage. Further, there is disturbing evidence that nanoparticles such as cadmium, quantum dots, and silver and titanium dioxide nanoparticles that are used in solar nanotechnology can be very toxic. The same is true of carbon nanotubes that are at the heart of new supercapacitors, batteries, and lightweight super parts for planes and cars. A National Institute of Standards (NIST) study in the United States found that DNA-wrapped single-walled carbon nanotubes (SWCNTs) shorter than about 200 nanometers could easily penetrate human lung cells, posing an increased risk to health. In agricultural applications, even if smaller quantities of nanochemicals are used in agriculture, because of their far greater potency, they could still pose a greater toxicological burden. Others argue that nanopesticides will only prolong dependence on chemical methods of fertilization and insect control when the whole concept of chemical control of agriculture is increasingly called into question. Thus some environmentalists warn that given the power of nanomaterials, extant problems could become worse. Ironically, the unknown environmental costs may one day outweigh the potential environmental gains for the environment, and despite the good intentions of green nanotechnology, the cure might be worse than the disease.

See Also: Biotechnology; Eco-Electronics; Green Chemistry; Science and Technology Studies; Technological Utopias; Unintended Consequences.

Further Readings

Adams, Wade and Linda Williams. *Nanotechnology Demystified*. New York: McGraw-Hill, 2006.

Berger, Michael. "Truly Green Nanotechnology—Growing Nanomaterials in Plants." *Nanowerk* (April 2, 2010).

Committee to Review the National Nanotechnology, Initiative, and Council National Research. *Matter of Size: Triennial Review of the National Nanotechnology Initiative*. Washington, DC: National Academies Press, 2006.

Fildes, Jonanthan. "'Asbestos Warning' on Nanotubes." BBC. http://news.bbc.co.uk/2/hi/7408705.stm (Accessed June 2010).

Friends of the Earth (FOE). http://www.foe.org/sites/default/files/090713-OECD-environmental-Brief.pdf (Accessed June 2010).

Goldman, Lynn and Christine Coussens. *Implications of Nanotechnology for Environmental Health Research*. Washington, DC: National Academies Press, 2005.

"Green Nanotechnology." http://crnano.typepad.com/crnblog/2004/02/green_nanotechn.html (Accessed June 2010).

Poland, Craig A., et al. "Carbon Nanotubes Introduced Into the Abdominal Cavity of Mice Show Asbestos-Like Pathogenicity in a Pilot Study." *Nature Nanotechnology*, 3/7 (2008).

Quick, Darren. "Nanoscale Solar Cells Absorb 10 Times More Energy Than Previously Thought Possible." http://www.gizmag.com/ultra-efficient-nanoscale-solar-cells/16517 (Accessed June 2010).

Schmidt, Karen. *Green Nanotechnology: It's Easier Than You Think*. Washington, DC: Woodrow Wilson International Center for Scholars, Project on Emerging Technologies, April 2007.

Seung, Nam Cha, et al. "Sound-Driven Piezoelectric Nanowire-Based Nanogenerators." *Advanced Materials* (August 30, 2010).

Sabil Francis
University of Leipzig, Germany,
École Normale Supérieure, Paris

GREEN ROOFING

Green roofing represents a range of technological solutions to conventional roofing that provides improved energy efficiency and uses more environmentally friendly or sustainable materials. Commonly available roofing materials such as slate, metal, and clay have higher environmental benefit when selected over asphalt and even wood due in part to their reusable or recycled content and their longevity. The addition of composite roofing products made exclusively from reclaimed and recycled materials such as tires, plastics, and sawdust provides a variety of green roofing options. The system most commonly referred to as a "green roof" replaces roofing tiles with planted vegetation as a way to provide a variety of environmental benefits, including energy efficiency, insulation, storm water management, pollution reduction, and increased wildlife habitat and green space.

Comparison of Roofing Materials

When evaluating a roofing material, it is important to consider its source of raw materials, life cycle, maintenance, and disposal. The majority of American roofs use asphalt shingles, which, while cost effective, have no insulating qualities and a life span averaging 15–20 years. Asphalt shingles are made from petroleum-based products, a nonrenewable resource, and are rarely recyclable.

Other conventional roofing materials include wood (cedar shake), metal, tile, and slate. The environmental benefits of choosing wood shingles come from their higher insulative values and from the availability of sustainably managed forest materials. Yet, the fact that wood is a renewable resource is offset by the need for treatment with toxic chemicals in order to meet building fire codes. Wood also falls short on durability, having issues over time with cracking, mold, and other moisture problems that lead to decay.

Metal roofing can be a green alternative if it is made from recycled materials. Metal roofs are typically made from copper, aluminum, or stainless steel. They are more insulating and durable, lasting up to 50 years. Light-colored metal roofs have a high solar reflectance that

"Green roofing" most commonly describes planted rooftops, such as this on the Environmental Protection Agency's Region 8 Office in Denver. Benefits of green roofs are heightened in urban environments, where environmental problems like storm water runoff, excessive radiant heat, and limited green space are exacerbated.

Source: U.S. Environmental Protection Agency

reduces the amount of heat passed into the building. Metal roofs are being combined with other technologies to create hybrid green roofs integrating, for example, solar panels. Metal roofing can likewise be an integral part of a building's rainwater harvesting system by efficiently channeling rainwater across its surface without the toxins rain can pick up across an asphalt roof.

Slate is the most durable roofing choice. Using salvaged slate roofing materials reduces the energy expenditure of removing this nonrenewable resource, making it a greener roofing solution. Clay tiles offer a durable roof that can last more than 50 years. Clay, like metal, can create a highly reflective cool (or white) roof if chosen in a light color.

Composite recycled roofing materials mimic the look of a natural roof using products made of recycled plastics, rubber tires, and cellulose fibers from recycled wood products. These manufactured tiles have a lighter weight load than slate or clay, require little to no maintenance, and are durable to 50 years or longer. More recycled roofing materials are being introduced as technologies develop and hold great promise for the green building industry.

The Living "Green Roof"

Although green roofing can refer to a variety of environmentally beneficial roofing solutions, it most commonly describes planted, or vegetated, rooftops. Green roofs are being used in residential and industrial architecture and can provide credits in green building initiatives (such as LEED) under storm water management, water efficiency, energy and atmosphere, heat island mitigation, materials and resources, innovation, and design categories. The benefits of green roofs are heightened in urban environments where the majority of land area is covered by impervious surface, exacerbating environmental problems caused by storm water runoff, excessive radiant heat, and limited green space.

The basic composition of a green roof is a series of layers placed over a typical or reinforced roof structure. The layers are, in order: vapor barrier, thermal insulation (installed above or below the membrane), cover board, waterproof membrane, root barrier, drainage layer, filter membrane (allows water to flow out while retaining fine particles from the growing medium), growing medium, and vegetation. Simpler versions consist of only four layers: insulation, waterproofing membrane, growing medium, and vegetation.

Green roof systems are labeled extensive or intensive based on variables including depth of growing medium, desired activity, and types of plants grown. An extensive roof is a low-profile roof with growing medium ranging in depth for an average two to six inches.

This creates a roof weight of 10–50 pounds per square foot when fully saturated. The medium is engineered with approximately 75 percent inorganic materials to 25 percent organics, which supports low-growing plants up to two feet tall. Ideal plants for an extensive green roof are sedums and other succulents, alpine plants, and herbs or grasses with a high tolerance for drought, wind, frost, and heat. Extensive roofs can be installed on slopes up to 30 degrees, with some companies using a grid structure successfully on higher-sloped roofs and even on vertical walls. Extensive roofs only require regular irrigation during the first year or two, converting to a low-maintenance system as the plants establish and mature. Green roof retrofits are typically extensive.

Intensive systems employ a deeper growing medium of eight inches or more to support the variety of plants used and are more typical of a green roof that meets recreational requirements such as vegetable gardening, social spaces, and even ponds. Intensive roofs look like traditional rooftop gardens without plants in planters, but rather installed in a full-scale garden. The engineered growing medium is composed of approximately 45 percent organic materials to 55 percent mineral. Due to their deep growing medium and larger plants, these roofs can weigh 80–120 pounds per square foot fully saturated and require more structural support than the extensive systems. The weight factor alone can limit the feasibility of an intensive green roof solution, depending on the underlying structural integrity and support of an existing roof or architecture. Intensive roofs also require a low to zero slope and higher maintenance, typical of a garden.

As seen in many green technologies, a green roof is more expensive to install than a conventional roof, estimated at $10 per square foot and higher, but the longer life and reduced energy requirements for heating and cooling the home make the cost comparatively lower than traditional roofs when annualized over the life of the roof. As part of a sustainable building design, green roofs can be enhanced by the use of a rainwater harvesting system and solar-powered irrigation methods. They can also add sustainability by becoming productive food sources, like traditional gardens.

Environmental Benefits

A green roof absorbs more than 60 percent of all rainfall across its surface. This figure may increase over time as plants mature and the system increases in efficiency. Rainfall absorption is influenced by the depth of the growing medium, types of plants, and intervals between rain events. In other words, less initial moisture content and higher volume of available medium lend increased capacity for rainfall capture and storage per event. By intercepting and delaying runoff, a green roof is beneficial in reducing overall hydraulic loads on septic and sewage systems and can help minimize flood events by replacing the impervious surface of a typical roof. Over an expansive roof surface, a green roof can substantially mitigate storm water runoff.

Water retained by a green roof is taken up by the plant roots and released through its leaves in the process of evapotranspiration. This release of moisture has a cooling effect on the air at the roof surface and has been shown to reduce ambient air temperatures by up to 70 degrees Fahrenheit. Because rooftops account for up to 25 percent of total surface area in dense urban settings, green roof technologies hold promise as one solution to the heat island effects observed in cities where dark, impervious surfaces of rooftops and pavement release enough absorbed radiant heat to create significant spikes in ambient air temperatures.

Green roofs act as vegetated filters by trapping both airborne and nonpoint source pollutants during evapotranspiration, improving runoff water quality. Soil in the growing medium traps and binds heavy metals found in rainwater, capturing over 95 percent of cadmium, copper, and lead and 16 percent of zinc, according to a 1993 study by the London Ecology Unit. It is estimated that 1,000 square feet of green roof can remove 40 pounds of particulate matter from the air in a year, and models predict that broad-scale use of green roofing in urban areas could amplify pollutant reduction currently accomplished by existing vegetation by up to 45 percent.

These same beneficial qualities have positive effects on the buildings underneath as well as on the environment around them. Green roofs provide shade and insulation that protect roof membranes, at least doubling the life of a traditional roof, significantly reducing heat fluctuation across the system, and even adding acoustic insulation.

A normal roof creates positive heat flow during the day by absorbing solar radiation and increasing the flow of heat through the house. At night, the roof releases the same radiant heat as it pulls heat away from the house, creating negative heat flow. A green roof moderates this heat flow cycle, requiring less manufactured heating and cooling internally. This increases the energy efficiency of the building and in turn reduces greenhouse gas emissions. A normal roof shows heat fluctuations in summertime of up to 80 degrees Fahrenheit. A green roof reduces that fluctuation down to about 11 degrees Fahrenheit. This efficiency is most significant in summer, as a green roof is more effective at reducing heat gain. In winter, a frozen green roof loses performance and insulating capacity unless covered with snow, resulting in minimal reduction of heat loss.

As modified green spaces, green roofs offer aesthetic benefits as well as sources of habitat for wildlife. Considered to be extensions of edge habitat or patch habitats, green roofs are being studied for their positive effects on wildlife. Benefits are increased when vegetation selection includes species native to the building's environment. Birds and butterfly species have been shown to use habitats up to 20 stories above ground level, and green roofs may even provide habitat free from ground-level predators.

The aesthetic aspects of a green roof have the same effect as natural landscapes in reducing stress and enhancing quality of life. Whether as rooftop green spaces in an urban setting or the natural extension of a wild landscape, a green roof offers environmental, aesthetic, and psychological benefits that hold promise for more widespread future use.

See Also: Green Building Materials; LEED Standards; Rainwater Harvesting Systems; White Rooftops.

Further Readings

Oberndorfer, Erica, et al. "Green Roofs as Urban Ecosystems: Ecological Structures, Functions, and Services." *BioScience*, 57/10 (2007).
"The Resource Portal for Green Roofs." http://www.greenroofs.com (Accessed August 2010).
U.S. Environmental Protection Agency. "Reducing Urban Heat Islands: Compendium of Strategies: Green Roofs." http://www.epa.gov/heatisld/resources/compendium.htm (Accessed August 2010).

Dyanna Innes Smith
Antioch University New England

GREEN TECHNOLOGY INVESTING

Green technology is usually regarded as an enormous and amorphous category, but it is clearly related to "climate change themes." The 2009 summit on climate change in Copenhagen tried to set a target of ensuring that global warming does not exceed 2 degrees Celsius by cutting greenhouse gas emissions and proliferating the development and deployment of green technology. The goal was to reduce carbon dioxide (CO_2) levels as well as to increase energy efficiency. For this, the World Economic Forum (WEF) 2010 Report indicates that under a business-as-usual "Reference Scenario," global investment on clean energy must reach $500 billion per annum by 2020. On the other hand, the International Energy Agency World Energy Outlook 2009 contains a "450 Scenario" that requires an investment of $38 trillion between now and 2030. The WEF 2010 Report has envisioned a map of clean energy technologies by stage of maturity (Research & Development and Proof of Concept, Demonstration and Scale-Up, Commercial Roll-Out and Diffusion, and Maturity) along with private funding sources for each stage. It envisions the role of venture capital in the first three stages. Private equity is expected to start covering from second stage to the last stage. Public equity and credit/debt markets are expected to cover the last two stages. Besides investing, the multiple green technologies at different stages of maturity also require different policy instruments. In the following paragraphs, we explore the feasibility of such a vision.

We begin by outlining the green technology fields that require funding. According to WEF 2010, the key renewable energy sectors are onshore wind, offshore wind, solar photovoltaic power, solar thermal electricity generation, biomass, municipal solid waste-to-energy, geothermal power, small-scale hydro, sugar-based first-generation biofuel, and cellulosic, algal, and other second-generation biofuels. Nuclear power is another possible field, but not something that the report explores. All such green technology fields could contribute to creating associated fields that use green technology, such as green buildings with Leadership in Energy and Environmental Design (LEED) certification provided by the U.S. Green Building Council.

While investing in all of the above-mentioned fields is critical from the supply side, the demand side also remains problematic in terms of persuading energy users to avail themselves of opportunities to save money. On top of that, there are critical enablers needed where investment is essential. These enablers include "smart grid" systems with smart meters that can track data about energy prices and demand patterns and thereby create an interactive flow between suppliers and consumers in order to increase efficiency in use. The use of lithium-ion batteries for utility-power storage can further increase this enablement. Advanced transportation through use of electric or hybrid vehicles and associated development of requisite infrastructure can also be a good enabler. Finally, a critical enabler is carbon capture and storage through advanced technologies that employ precombustion capture, postcombustion capture, and oxycombustion. These enablers would help the energy system to absorb and use all of the renewable energy it can produce and still meet the world's energy needs.

According to WEF 2010, the Copenhagen Summit failed to reach a global mechanism for price emissions, but it did create new pressure for effective, nationally focused bottom-up initiatives. Though significantly insufficient, the $100 billion pledge as part of Quick Start funds for investing in green technology in the developing world to reduce energy poverty is a good start. All national policy initiatives based on these halfhearted global attempts are critically intertwined with initiatives in "market-dependent" investing initiatives. But the global economic crisis has created many barriers to reaching Copenhagen's goals and depressed investment activity. The issue of policy and investment is further complicated by the existence of different types of capitalism—Anglo-Saxon liberal market economies, European coordinated-market economies, Chinese-style socialist–capitalist economies, hybrid versions of state-led and market-led economies of emerging markets like India, Brazil, and others.

The above-outlined issues create enormous complexities in structural barriers and agents' decision-making environment. Multiple conflicting signals from states and markets further increase the uncertainty and risk factors for investors in a globalized market. Given the fact that green technology is a vast field that has a history of volatility and mixed results of sustainability in terms of growth, lack of consistent profit returns, and long-term success of marketability of technological innovations, the ebb and flow of investor enthusiasm is not surprising. But the massive global economic downturn of 2008 has definitely made matters worse.

However, the good news, according to WEF 2010, is that the highly capital-intensive green technology sectors have shown a drop in cost of renewable energy and the debt spreads for projects' return to long-term levels show an increasing trend toward leveling. Hence, slowly the cost reductions can be passed on to clients, thereby incentivizing demand and providing further momentum to the supply side. Over time, one can hope for the development of increasing vertical and horizontal integration in the various related sectors and creation of an ecosystem with supportive networks and integrated supply-chain logistics. However, in the meantime, there are conflicting interests in investing patterns. Whether the investment is government directed through fiscal stimulus or from credit markets or private or public equity markets or from market units like venture capital, institutional investors, corporate investors, angel investors, among others, there are difficult structural barriers and high risks that create agent dilemmas in decision making.

Defining the Market

Given that climate change is a global issue, two lead institutions are constantly defining the boundaries of green technology or clean energy technology and investing in it. These are states and markets. Hence, all types of investing are intertwined with local, national, and global policy initiatives. Global policy initiatives include financial regulations to stabilize and increase transparency in financial markets. For example, Basel 3 global capital rules require a critical source of investment lending—banks and their securities arms—to increase their stock of capital holdings in comparison to their lending. This is designed to reduce dubious risky lending and excessive leveraging. The expectation is that such regulations will reduce global market volatility, panic, and a massive worldwide crash.

National policies to incentivize growth of the highly capital-intensive and high-risk green technology sector, ironically, also include directives for investing by public institutional funds, such as pension funds, which are designed to get a higher return but without taking excessive risk. In Europe, institutional investors bear the expectation-weight to carry out investments that are considered to be beneficial not only to the companies but also to the society. But without clear policy backing in the form of legislation, this can become difficult. In the United States, the state of California has decided that a part of the pension fund capital from the California Public Employees Retirement System (CalPERS) should be invested in clean tech.

There are several current issues in the United States and California that have made this situation complex. According to *The Economist* (2010), the futures market believes that American interest rates (real and nominal) will continue to be below 1 percent and will likely continue until 2012 in the hope of stimulating borrowing. This has a paradoxical consequence for saving instruments such as pension schemes. Schemes are expected to build up capital, which is then used to buy an income in retirement, say in the form of an annuity. Low rates increase the liability of pension plans as they require a larger amount of capital to buy a given level of income. In addition, deflation (an issue constantly debated in the context of the U.S. economy) is a hidden risk for pension schemes. If it occurs, then it will cut the nominal income of those companies or public-sector bodies that have to fund future pensions. This creates a potential gap between assets and liabilities, thereby creating the necessity of putting aside more money as savings instead of investing.

The above-outlined irony is further manifested when one considers the situation of CalPERS. It has a history of leading a crusade to improve the quality of corporate governance in America. But the economic downturn revealed that it lost billions for failing to run a "tight ship." Since its assets and reputation took a hit, it is working hard to fix the issue. However, if it fails to achieve investment returns on its $200-billion portfolio needed to help pay for workers' health and retirement benefits, then the state's taxpayers will have to pick up the cost. Given that California is already facing a $19 billion budget deficit, this is an alarming prospect. This has put the pension fund in a vicious bind. It is undertaking reforms and being cautious by investing in conventional stocks and bonds. However, this fails to bridge CalPERS' long-term assets–liabilities gap. Hence, it is being pushed to invest in riskier assets such as emerging-market stocks, private-equity funds, and big infrastructure projects. This kind of complexity shows how investment flow from institutional investors, despite the Obama presidency's directives for more focus on green technology investing, could be difficult.

The Barack Obama presidency, through the passing of the American Recovery and Reinvestment Act, has made green investing a priority in fiscal stimulus and has provided subsidy incentives in the form of grants and tax credits. The critical goal is to generate employment, enable the development of new skills, and become an export leader of products in this market. Green technology is seen as the future of development trends in both the industrial and service sectors. However, according to WEF 2010, a key weakness of the stimulus approach is that only around $25 billion—14 percent of the total allocated—actually reached clean energy technology providers or project developers during 2009. A sizable portion of it was earmarked for smart grid technology. This funding will increase to $60 billion through 2010–2011 before receding. This opens up the question of how to close the stimulus funding for clean energy in due course without causing the industry to collapse. Since the U.S. government is looking for significant support from venture capitalists, the function of this source of funding is critical for assessing the sustainability of the clean energy sector.

According to WEF 2010, 2009 saw a significant drop in volume of venture capital and private equity. Between 2004 and 2007, investments had soared with about 1,573 funds targeting technology investments. With credit crisis at its height, the initial public offering (IPO) route was cut off for some late-stage technology companies. These companies had to turn to private equity funds or trade sales. Overall, the major hits were taken by the solar and biofuel sectors. Hence, asset financing that typically is required to build wind farms, solar projects, and biofuel plants was hit hard too. However, the good news continues with venture capital (VC) investing in green building. The leading investors are

Regenerative Ventures and Khosla Ventures. Calera Corp., for example, is developing a processor that captures the carbon released during the cement-making process, and the by-product of this process can be used as a recycled material in concrete. Proponents of green building claim that companies that engage in green building benefit significantly from such activities. They improve companies' reputations and attract and retain employees and customers. By proactively accepting higher environmental standards, companies can avoid the high costs of adjustments resulting from future pro-green legislations. Ironically, they also point out that a substantial number of firms in the oil and financial services industries are the largest occupiers of green office buildings. Another favorable outlook for green investing can be seen in the fact that a number of mutual funds devoted to socially responsible investments are emerging.

Critiques: Mismatch?

Critiques of venture capital in the context of American deployment are concerned about the mismatch between the character of VC and the demands of green technology. Most green technology consists of turnkey projects with high upfront capital requirements. This requires more of an industrial policy with the government picking "winners." VCs cannot compete with the lobbying of corporate giants with deep pockets of capital and strong connections to lawmakers. Industrial policies or government dependence are typically anathema to proponents of the Anglo-Saxon liberal market economy since it can distort markets by locking out better alternatives.

The liberal market proponents point out that the hype of alternative energy always goes up during periods of oil crisis in the United States. VC involvement during these times has typically failed to deliver disruptive innovation that leads to Schumpeter's much-vaunted "creative destruction" of incumbents. The prominently successful U.S. green technology VC First Solar that produced solar photovoltaic was founded in 1984, sold to the Walton family in 1999, and its stock's initial public offering was in 2006. This is too long a time period from incubation to IPO. Hence, the gloomy prediction is that the current green technology investment boom may be an unsustainable bubble. However, one needs to remember that a liberalized market economy has its own problems that can affect investing patterns. For example, the CalPERS caper opened up the market-distorting actions of "placement agents" who help financial firms win investing assignments from big pension funds in exchange for a fee. This has led to the creation of a culture where outside firms win contracts based not necessarily on merit but on whom they know. This then becomes picking "winners" of a different kind.

Proponents of the Silicon Valley model of VC investing point out that given the huge diversity in the green technology field, VC's strength of developing deep knowledge of the evolutionary trajectories of technologies and markets cannot be leveraged. Moreover, green technology fails to deliver huge returns, as has been typical of VC activities in the information technology (IT) field. Such activities led to creation of giants like Google and Intel that delivered unimagined profits to the venture capitalists. Even a political-problem-solver entrepreneur like George Soros is willing to offer $1 billion from his hedge fund for clean energy projects, but with the stringent criterion of profitability.

As a 2009 study of hedge funds shows, they performed best when, during the economic crisis, the U.S. Securities and Exchange Commission forced transparent registration of investors. This implies that—more than risky undertakings of hedge funds—the quality of educated investors and managers is critical. Educated investors and managers who do not have "control aversion" are willing to rely on the quality of information being provided by the institution. They know that the institutions have a reliable asset base provided by knowledgeable investors. Hence, they are confident about the risky but well-calculated strategies of the hedge funds. Liberal market VCs should consequently look toward hedge funds for investment in green technology. Their involvement with hedge funds would increase their reputation for competence and credibility for taking risks in the clean tech sector.

Without huge returns, VCs cannot raise more money for future investments from their public institutional investors like pension funds. They become a risky liability for public institutional investors who have to look out for the interests of their public investors. The WEF 2010 report acknowledges that, historically, private investors have failed to benefit from investing in energy efficiency. However, all is not doomed. Over the past few years, VC and private equity have been slowly trickling into specialist technology, such as lighting (VC vantage point for light-emitting diodes [LED]), light-responsive windows, energy-efficient building products, industrial-technology related software, and the like.

But the U.S. push for green technology faces many political obstacles. The domestic public's appetite for out-of-control budget deficits is dwindling as unemployment in the United States keeps rising despite the stimulus. On top of that, outsourcing of jobs and immigration have become hot-button issues in an economy that is facing a long, uphill battle for recovery. The global consulting firm Accenture also reports that emerging economies like China and India are far more likely than American consumers to purchase environmentally friendly devices as they are more likely to be first-time consumers of advanced gadgets.

From an American perspective, the inability to swiftly rebuild its economy is to be seen against the backdrop of the titanic rise of the BRICS (Brazil, Russia, India, China, and South Africa), especially Asian giants China and India. These two giants have an insatiable appetite for fossil fuel usage and are unwilling to bear the burden of the costs of consequent severe environmental degradation. They blame the developed world, particularly the United States, the biggest polluter, and expect it to carry the largest share of the environmental burden. Consequently, an accord on the Kyoto Protocol is still a mirage. Without clear commitment to a carbon-reduction target, policies and incentives become muddled. There is no clear political will to discourage investments in fossil technology. This scenario is compounded by the fact that instead of abating fuel usage and looking for alternative sources, Russia—another major emerging economy—is currently in the process of staking out its claim in uncharted areas of the Arctic that have huge reserves of oil and natural gas. Latin America is also not far behind with its huge oil and mineral reserves. Here, there are contradictory tendencies. While the price of oil rises and falls, the price of minerals is rising at meteoric paces due to China's and India's demand. Mineral reserves like lithium are critical for green technology. Chile, China, and Bolivia have the largest lithium reserves. In addition, Brazil is the market leader in first-generation biofuels. America's ambition to be a global leader in green technology is thus faced with severe competition.

The policy environment becomes complicated as China keeps following a "beggar-thy-neighbor" policy through currency manipulations and keeping its market mostly closed to foreign companies even as it touts that it welcomes foreign direct investment. In 2008, China's own company China Energy Conservation Investment Corporation entered into a joint venture with the United Kingdom's Carbon Trust to explore clean energy technology opportunities. In mid-2009, APG Asset Management, which manages the Dutch civil servants' pension plan, created a $100 million fund. This fund targets energy-efficiency investment in China's manufacturing sector by backing Chinese firms directly. China, according to WEF 2010, is now a leader in LED manufacturing, representing a third of global production. Furthermore, like India, China also provides loan guarantees designed to reduce the cost of private lending and improve project economics in order to spur private financial institutions to lend on terms not available otherwise. This has enabled the rise of local green start-ups. India and China have also developed networks of microfinance to provide low-cost, clean-energy equipment such as solar lanterns, biodigestors, irrigation pumps, and so on. Like India, which has thousands of LEED projects, China now also has a three-star rating system and has 300 registered LEED buildings.

According to China's Greentech Initiative, China potentially could be a $500 billion to $1 trillion a year market for green technologies. However, American banks have consistently failed to convert their initial foothold into China—by buying into chunks of Chinese state banks—into major domestic market shares. Their VC limited partnerships—the dominant form of formal venture investing globally   and the firms they backed in joint ventures enabled China to get access to capital, technology, management skills, and the prospect of better corporate governance. China now has a megabase of wind farms as well as grid extensions to some of the areas where excess renewable energy is currently being produced. The enabling firms made a lot of money in the rising Chinese stock market. However, they failed to become solo leaders or huge monopolies and to capture access to China's huge domestic market. This would have ensured creation of a diversified portfolio and continuity of diversified profit base for the American banks.

Domino Effect

This limited performance has a domino effect. Institutional investors such as the pension funds became leery of continuing to advance of money to VCs. According to Martin Kenney (2009), "The basis of VC industry is to invest in firms early in their life cycle and then sell these investments to others later in the life cycle—hopefully with capital gains." But in order to do so, VCs first have to specialize in and deeply understand the business space. This would enable a community of increasingly experienced investors to come together. This would further enable the VC community to find economic spaces within a nest of new firms. They would be able to look into deeper possibilities of technological evolution and create a firm to occupy the technological space before other adjacent firms can react. In short, they can have the advantage of a market leader. However, in foreign markets such as China, where even getting a toehold in the market independently is a problem, institutional investment sources for VC are likely to dry up due to long-term losses.

The story gets even more complex when one considers the fact that China is emerging as a major leader in green technology through its own stimulus of loans and grants. According to WEF 2010, in 2009, China unveiled a $46.9 billion low-carbon stimulus package. The majority of the fund was targeted for ecology projects like waste and water treatment and reforestation; $7.3 billion was destined for energy savings. The government allocated $1.5 billion for developing clean vehicles; $19 billion was allotted for grid infrastructure and advanced technology. At the same time, the government removed polysilicon and 2-MW turbines from the Encouraged Import List. This provided a boost to China's domestic clean technology industry.

In comparison to the U.S. Silicon Valley model, European VCs followed a different model. In any specific industry, collaboration among multiple dedicated small firms that could tweak their existing knowledge base to reorient toward new requirements, slow growth, lesser return, and incremental jumps in technology were acceptable to investors that were mostly banks. The green technology industry is mostly manufacturing oriented and, hence, countries like Denmark and Germany with significant manufacturing expertise became leaders in green technology. Hence, the German company Siemens has been the market leader in the engineering field of green technology. The focus of the European Union (EU) is more internal, and indigenous growth in green technology has been quite strong.

Moreover, the EU has a long history of involvement in green technology. This is evident from its policy initiatives like feed-in tariffs (FIT) for renewable electricity generation. FITs are either market dependent or market independent. Market-dependent modes are based on premiums over spot market prices and benefit large companies that can afford to take the risk of transaction costs. Market-independent modes follow a fixed-pricing system of contracts enforced by the government. This enables risk-averse small investors and communities to access clean technology at cheap prices. Hence, of all the European green stimuli, those targeting communities are likely to be the fastest in reaching the sector. Currently, market-independent policies are proving to be stronger and more cost efficient. Spain has tried to come up with a good market-dependent policy mix through its introduction of a cap and floor policy for premiums. This is intended to avoid the problem of over- or under-compensation in cases of fixed premiums.

Japan and South Korea with their *kiratsu* and *chaebol* have similar models of green technology development. Hence, the German car industry, as well as the Japanese and South Korean car industries, is already making great strides in the hybrid and electric car sector. According to WEF Report 2010, South Korea has taken the lead among the green fiscal stimuli released in the post–economic crisis era. The government plans to increase the Korean share of overseas clean energy markets by 8 percent, mainly through export of low-carbon technologies.

In contrast to all the above-mentioned continents and countries, in most developing countries, especially those in Africa, making energy available to far-flung rural areas is of immediate urgency. Even though significant entry of Western and Chinese banks into African states can be tracked in recent times, they need to focus on funding firms that follow the Indian multinationals' policy of marketing to the bottom of the pyramid. This essentially entails meeting needs rather than creating wants through development of simple cost-effective and efficient small-scale innovations. This ensures a healthy profit through volume of sales. It also helps nonprofit nongovernmental organizations (NGOs) to get funding and slowly generate enough profit to spread the scale of their work. For-profit firms, through linkages with NGOs, can increase their profit levels significantly. Hence, investing institutions need to come up with new models of investing.

Green technology investing in a world of rising concerns over global warming should represent a desperately needed change from business-as-usual attitudes. While politics, markets, and societies all have their problems, good investors are those who can negotiate around the messy empirical problems and come up with winning solutions. Such solutions should take into account contingent global and local sociopolitical scenarios.

See Also: Clean Energy; Green Chemistry; Green Markets; Green Nanotechnology; Hybrid/Electric Automobiles; Information Technology; Innovation; Intellectual Property Rights; Sustainable Design.

Further Readings

"Africa's Banking Boom: Scrambled in Africa." *The Economist* (September 18, 2010).

Andreasson, H. Oscar and Daniel Karlsson. "Pension Capital in Cleantech? A Study of Factors Obstructing Swedish National Pension Funds From Investing in Clean Technology." Master's thesis in Environment Management, University of Gothenburg School of Business, Economics, and Law. Spring 2010.

"Another Paradox of Thrift: Why Low Interest Rates Could Also Encourage Savings." *The Economist* (September 18, 2010).

Couture, Toby and Yves Gagnon. "An Analysis of Feed-in-Tariff Remuneration Models: Implications for Renewable Energy Investment." *Energy Policy*, 38 (2010).

"Finance After the Crisis: Investor, Heal Thyself." *The Economist* (September 18, 2010).

"Financial Regulation: Basel's Buttress." *The Economist* (September 18, 2010).

"Foreign Investment in China: Even Harder Than It Looks." *The Economist* (September 18, 2010).

Frumkin, Dvir and Donald Vandegrift. "The Effect of Size, Age, Beta and Disclosure Requirements on Hedge Fund Performance." *Journal of Derivatives and Hedge Funds*, 15 (2009).

Gottfried, David and Malik Hriday. "Perspectives on Green Building." *Renewable Energy Focus*, 10/6 (November/December 2009).

Kenney, Martin. "Venture Capital Investment in the Greentech Industries: A Provocative Essay." BRIE Working Paper 185, July 18, 2009. http://brie.berkeley.edu/publications/wp185.pdf (Accessed October 2010).

UPFRONT, Peoples, Projects, and Programs. "News From the Field." *Sustainability: The Journal of Record*, 2/6 (December 2009).

UPFRONT, Peoples, Projects, and Programs. "News From the Field." *Sustainability: The Journal of Record*, 3/4 (August 16, 2010).

World Economic Forum. "Green Investing 2010: Policy Mechanisms to Bridge the Financing Gap." Geneva: World Economic Forum, January 2010.

Pallab Paul
University of Denver

Greywater

On a global scale, there is growing apprehension over the declining reserves of groundwater. Mounting demands for water worldwide, with a growing world population and over-burdened sewage treatment plants, have kindled interest in the use or recycling of greywater. Greywater derives its name from its cloudy appearance. It is neither whitewater (groundwater or potable water) nor polluted (blackwater) sewage. Any wash water that has been used in the home, except from toilets, is called greywater. Greywater comprises 50–80 percent of residential wastewater generated from all of a home's sanitation equipment excluding the toilets. Greywater is wastewater generated from domestic activities such as laundry, dishwashing, and bathing. The usual sources of greywater are water from bathroom and kitchen sinks, showers, tubs, and washing machines.

Water generated from domestic activities such as laundry, dishwashing, and bathing comprises 50–80 percent of residential wastewater.

Source: iStockphoto

Approximately 26 billion gallons of water are used every day in the United States. The U.S. Geological Survey states that the average American uses between 80 and 100 gallons of water each day. As water sources are depleted, the usual development is to overuse existing supplies of water. Historically, those supplies have been obtained through the building of reservoirs, dams, municipal deep wells, and, for the homeowner, private wells. Municipalities and water companies are often required to enforce water use restrictions on their users, especially in times of drought. These restrictions oftentimes are in direct conflict with population growth and can create problems and price increases for local industries. Homeowners may object to water restrictions as this can affect their lifestyles as well as the care and maintenance of homes and landscapes.

Finding new supplies of water can be expensive and is never guaranteed. With an increasing awareness of the environment and the impacts of global climate change, a potential solution is to lower domestic demand of water on a year-round basis. This can avoid depletion of water supplies, ensuring a more abundant supply during times of drought. As a consequence, this policy can possibly eliminate the need for water use restrictions and ensure adequate supplies for additional population growth. This can be accomplished in a variety of ways and centers around a philosophy based on water conservation. Water conservation is the responsibility of water users and suppliers. In homes, wise water use includes the installation of low-flow water fixtures and appliances, taking shorter showers, running only full loads of laundry and dishes, and keeping plumbing fixtures in good repair. Outside the home, methods to conserve water include greywater reuse, rainwater collection, and water-conserving landscaping and irrigation practices.

Greywater holds great promise for achieving the goals of ecologically sustainable development. Greywater recycling is a method of water conservation that can help conserve precious water resources. If greywater contributes 50–80 percent of the total wastewater flow to domestic sewers, then this great volume of water can be recycled and reused for other purposes both inside and outside most conventional homes. Inside the home, greywater from showers and bathtubs can be used for flushing toilets. Outside, greywater can be used to irrigate lawns, trees, shrubs, and garden plants. Why is this important and why should it matter? It's a waste to irrigate plants with drinking water when plants can thrive on used water that contains bits of organic matter. Greywater reuse is a part of the fundamental solution to many ecological problems. The benefits of this type of approach include lower freshwater use, less strain on septic systems, potential lower energy and chemical use, groundwater recharge, increased plant growth, reclamation of otherwise wasted nutrients, and increased awareness of and sensitivity to natural cycles.

Government regulation governing domestic greywater use for landscape irrigation is still an evolving concept. It continues to gain support as the actual risks and benefits are considered and put into clearer perspective. By definition, greywater is considered in some jurisdictions to be sewage, which includes all wastewater, including greywater and toilet waste. In the United States, states that adopt the International Plumbing Code state that greywater can be used for subsurface irrigation and for toilet flushing. States that adopt the Uniform Plumbing Code state that greywater can be used in underground disposal fields that are like shallow sewage disposal fields. Policy varies from state to state. Where greywater is considered sewage, it is governed by the same regulatory procedures enacted to ensure properly engineered septic tank and effluent disposal systems. These systems are designed for long life and to control the spread of disease and pollution. As a consequence, the use of greywater has suffered or was not permitted. With water conservation becoming a necessity in a growing number of states, especially the arid ones, there is pressure on regulators to seriously reconsider the risks against the benefits of using greywater.

Systems to distribute greywater vary. Washing machines are typically the easiest source of greywater for use because greywater can be diverted without cutting into the existing plumbing. These machines have an internal pump that automatically pumps out the water. This can be used to advantage to pump the greywater directly to the landscape. Simple homeowner systems may involve a small storage tank to collect greywater from a washing machine. More involved systems or most recycling systems involve a diverter valve that separates greywater from blackwater. The blackwater is sent to a conventional wastewater treatment system. The greywater is run through a filter, usually a sand filter, to remove any organic matter, and then relocated on site either to be treated by a separate process or used as is. Common treatment includes chlorination or iodine for disinfection. Biological treatment systems are also available, including constructed wetlands involving water hyacinths or other plants for disinfection. These aquatic plants remove pollutants by directly absorbing them into their tissues. They also provide a suitable environment for bacteria to transform pollutants and reduce their concentrations.

Whatever system is used, the same basic greywater guidelines apply. Greywater is different from freshwater and requires different guidelines for it to be reused. Do not store greywater for more than 24 hours. If greywater is stored, the nutrients in it will start to break down, creating bad odors. Minimize contact with greywater. A greywater recycling system should be designed for the water to soak into the ground and not be available for people or animals to drink. Infiltrate greywater into the ground; do not let it pool or run off. Pooling greywater can provide mosquito breeding grounds as well as a place for human contact. Keep the system as simple as possible; avoid pumps and filters that need upkeep. Simple systems last longer, require less maintenance and energy, and cost less money. Install a three-way valve for easy switching between the greywater system and the septic/sewer system. Match the amount of greywater your plants will receive with their growing requirements.

The benefits of greywater recycling meet the goals of ecologically sustainable development of both homes and grounds. These embrace lower freshwater use, less strain on septic or treatment plants, less energy and chemical use, groundwater recharge, enhanced plant growth, reclamation of otherwise wasted nutrients, and an increased awareness of and sensitivity to natural cycles on the part of the consumer.

See Also: Carbon Footprint Calculator; Desalination Plants; Rainwater Harvesting Systems; Sustainable Design; Wastewater Treatment; Water Purification.

Further Readings

Kinkade-Levario, Heather. *Design for Water: Rainwater Harvesting, Stormwater Catchment, and Alternate Water Reuse.* Gabriola Island, British Columbia, Canada: New Society Publishers, 2007.

McLamb, Curtis. *Graywater: The Next Wave.* Suwanee, GA: GRAYWATER Resource Inc., 2004.

U.S. Geological Survey. http://ga.water.usgs.gov (Accessed September 2010).

Carl A. Salsedo
University of Connecticut

High-Efficiency Particulate Air (HEPA) Systems

High-efficiency particulate air (HEPA) filters were developed in the early 1940s and used first by the Manhattan Project to contain the spread of airborne radioactive contaminants. Introduced commercially in the following decade, "HEPA filter" refers not to a specific filter design but to a specific level of efficiency, as defined by the Department of Energy (DOE) in the United States and the European Committee for Standardization in the European Union. The DOE standard defines a HEPA filter as one that removes at least 99.97 percent of airborne particles 0.3 micrometers in diameter (the Most Penetrating Particle Size, or MPPS). The European standard is similar, but defines five HEPA classes—H10 through H14—of increasing efficiency. In contrast, the typical pleated filter used in a home furnace in the 21st century—itself as much as five times more effective than conventional fiberglass filters—traps less than 10 percent of particles smaller than 1 micrometer.

HEPA filters grew in popularity and necessity as the Cold War technology boom saw the growth of industries in need of highly efficient air filters: computers, electronics, aerospace, and nuclear power, for example. Existing industries such as hospitals and pharmaceutical manufacturers also kept the filters in high demand. U.S. Department of Transportation studies have shown that, thanks to the use of HEPA filters on aircraft, the air in modern passenger planes contains about the same level of airborne fungi and bacteria as the average home—far less than would be expected given the passenger population. The filters help to slow down the spread of colds, flu, and other diseases from city to city. Even more critical is the use of HEPA filters in hospital settings and other medical uses, where they are typically rated higher than the DOE standard—often 99.995 percent efficiency, the HEPA H14 class in the European system—and equipped with high-energy ultraviolet lights that kill any bacteria and viruses that the filters trap.

Unlike membrane filters, HEPA filters do not work like sieves or strainers, catching anything larger than a certain size while the rest passes through holes. Instead, they typically involve a pleated sheet of randomly arranged fiberglass fibers with diameters usually between 0.5 and 2 micrometers, with corrugated aluminum separators between the pleats. The sheet is then bonded to a sturdy base. A motorized fan passes air through the filter, where particles are trapped when they adhere to the fibers or become embedded in them. The smallest particles collide with gas molecules; this slows their passage through the filter and increases their likelihood of becoming trapped. While a sieve would be most effective with particles over a certain size, with the typical HEPA filter design, the MPPS does not actually refer to the smallest-sized particle the filter traps but to a middle ground: both smaller and larger particles are more easily trapped by the filter, while the MPPS particles are neither small enough to be slowed by gas molecules nor large enough to be easily trapped. This is the reason why a HEPA filter's efficiency is measured according to the retention of MPPS particles and thus represents its minimum efficiency. It is important to remember, though, that this is a rating of the filter's efficiency—not the filtration system, which must be scaled appropriately to the size of the space where air is being filtered and equipped with a fan to ensure that all air passes through the filter.

HEPA filters are tested by passing an aerosolized product of a known size through them in order to test the filter's efficiency. Dioctyl phthalate and polyalphaolefin are common test substances. Particle counters can easily detect leaks in the filter, while a photometer or aerosol detection device measures the mass of particles in the airflow upstream and downstream of the HEPA filter, and the results are compared to determine the filter's efficiency.

For a sense of scale, consider the sizes of many of the particles owners of HEPA filters are seeking to eliminate from the air:

- *Viruses and bacteria*: 0.0011 micrometer. However, viruses usually travel through the air as part of larger particles, 0.3 micrometers or larger, composed for instance of mucous or other material
- *Smoke (cigarette or exhaust)*: 0.011 micrometer
- *Household dust particles*: 0.011 micrometer
- *Pet dander*: 0.11 micrometer
- *Airborne fungi and mold spores*: 150 micrometers
- *Pollen*: 10,000 micrometers
- *Human hair*: 50,000 micrometers

The human respiratory system filters out particles as small as 35 micrometers, but anything smaller can contribute to respiratory problems, in addition to the risk to human health, depending on the nature of the particle.

It is important to note that filters only retain particles, not neutralize them, which is why medical-use filters use the aforementioned ultraviolet lights. Furthermore, filters have to be replaced periodically, at a cost most homeowners consider prohibitive for home use. Asthmatics and other individuals with respiratory problems may find HEPA filters worthwhile for home use because of their medical benefit (or may have an insurance plan that compensates them for the expense). An alternative to an always-on HEPA filter large enough for a residential home (or multiple smaller filters) is the use of a HEPA vacuum cleaner, but their efficacy varies. While many vacuums are equipped with HEPA filters and may help somewhat with allergenic conditions by trapping fine pollen and dust mite feces particles, a really effective vacuum needs to be built so that all the air passing through the machine is forced through the filter; this requires a more powerful motor because of the density of the HEPA filter compared to standard vacuum dust filters.

Similar to the HEPA rating is the ULPA: ultra low particulate air. The ULPA standard is defined by the Institute of Environmental Sciences and Technology (IEST), a nonprofit technical organization relied upon for technical standards in clean-room and controlled-environment practices. ULPA filters remove at least 99.999 percent of airborne particles of 120 nanometers (0.12 micrometer) or larger. ULPA filters are usually used by the semiconductor industry, where the particles that most need to be filtered out are especially small; in medical and home-use contexts, there is no benefit to using an ULPA filter instead of HEPA, which remains the industry standard in medicine and biology.

Air-purifying systems equipped with a HEPA filter usually employ other processes as well to complete their work. For instance, there is the high-energy ultraviolet light mentioned above for degrading bacteria and viruses, and a fan or other forced-air system to ensure the air passes through the filter. Activated carbon is often used to adsorb small volatile chemical molecules, converting them to a solid state from a gaseous one—this has the additional effect of odor control. Ionizer purifiers, frequently sold for home use, generate electrically charged gas ions that attach to airborne particles and cause them to stick to a collector plate, usually as an alternative to HEPA filtration.

See Also: Environmental Remediation; Green Chemistry; Green Nanotechnology; Membrane Technology.

Further Readings

Friis, Robert H. *Essentials of Environmental Health.* Sudbury, MA: Jones & Bartlett, 2006.
Miller, G. Tyler and Scott Spoolman. *Environmental Science.* Florence, KY: Brooks/Cole, 2010.
Vig, Norman J. and Michael E. Kraft. *Environmental Policy: New Directions for the Twenty-First Century.* Washington, DC: CQ Press, 2009.

Bill Kte'pi
Independent Scholar

HYBRID/ELECTRIC AUTOMOBILES

Although hybrid/electric vehicles (HEV) make up only a small portion of the market (approximately 5 percent), sales have been steadily increasing. J. D. Power predicts U.S. hybrid sales will triple between 2008 and 2015. In Canada, sales of hybrids tripled from 2004 to 2006. The U.S. Department of Energy (DOE) points out that while HEVs have a long history in the automobile and motorcycle niche market, their recent popularity is the result of advances in electric-drive technologies that have enabled the major automakers to mass produce these vehicles by integrating internal combustion engines (ICE) with standardized batteries and electric motors. The draw for the consumer is that HEVs typically get better city and combined fuel economy than their equivalent ICE counterparts and roughly the same or somewhat less highway economy. The Toyota Prius presently has the top fuel mileage according to the U.S. Environmental Protection Agency (EPA), garnering 51-miles-per-gallon (mpg) city and 48-mpg highway ratings. At the same time, hybrid motorcycles achieve more than double the best HEV automobiles. As a result, hybrids have developed a market niche as "green" or "environmentally friendly" vehicles, a reputation that this entry will demonstrate is not presently supported by empirical life-cycle analysis (LCA).

The Toyota Prius, the most well-known hybrid, is a parallel hybrid. Parallel hybrids use two engines: an internal combustion engine (ICE) and an electric motor. They are also the only hybrids mass-produced by major automakers.

Source: Wikimedia

Notwithstanding, hybrids are just one in a range of automotive technologies aimed at reducing the car's ecological and carbon footprints. Dennis Simanaitis (2009) identified 10 different types of automotive technologies aimed at the environmentally conscience buyer, including the following:

- *Advanced gasoline*: Technological improvements to the ubiquitous gasoline ICE (e.g., direct fuel injection, variable valve timing, turbo/super charging) have improved fuel mileage and reduced carbon and emissions associated with photochemical smog (relative to an equivalent-sized vehicle), even though automobiles in general have become significantly heavier. For example, a 1960s Austin Mini weighed in around 1,400 pounds while a 2010 BMW new Mini weighs about 2,500 pounds. Both Minis achieve approximately the same fuel mileage, despite the weight differential. However, the new Mini achieves significantly lower emissions, has much better performance, and is more comfortable, reliable, and safer due to technological advancements.
- *Diesels*: Diesels typically get better fuel mileage than equivalent-sized gasoline ICEs, and their emissions have dropped significantly with new technologies such as "urea injection." Diesel sales are much larger in Europe than North America.
- *Flex fuel*: These vehicles use E85 (85 percent ethanol, 15 percent gasoline) but can switch to standard gasoline. In the United States and Canada, ethanol is typically distilled from corn, and much controversy exists about the use of "food for fuel" and the life-cycle impacts of corn ethanol. The DOE claims 0.87 million Btu of fossil energy is required to make 1 million Btu of ethanol at refueling stations, while standard gasoline needs 1.23 million Btu for 1 million Btu of gasoline at the station. On the contrary, Harro von Blottnitz and Mary Ann Curran (2007) reviewed 47 published assessments that compare bio-ethanol production to gasoline on a life-cycle basis. They found contradicting results, with some studies favoring ethanol in terms of emissions and others finding greater impacts over various environmental parameters. Further, many critics, activists, and environmentalists have blamed the production of ethanol from food sources for the shortages and soaring costs of basic foods and the resultant riots that occurred in various locations around the world.
- *Micro-hybrids*: These vehicles use a larger 12-volt alternator/motor that works in tandem with the ICE for start-and-stop driving. Typically, they use regenerative braking to help recharge the batteries.
- *Parallel-hybrids*: These are the most popular and are the only hybrids mass produced by the major automakers; they are readily available in dealerships around the world. Honda's Insight (now Civic) was the first to be introduced to the market in 1999, and the best-selling Toyota Prius is the most well-known example. Parallel-hybrids are the more sophisticated versions of HEVs and use two engines—an ICE and an electric motor. "Full" or "strong" variants can run only on the electric motor for a limited range (e.g., Toyota) and recharge their batteries through regenerative braking and onboard ICE recharging systems. Some systems use the ICE and electric motor running simultaneously (e.g., Honda), never running solely on their electric motor.

- *Parallel plug-in hybrids*: This HEV can run 5 to 10 miles on the electric motor only. They are plugged into the grid to recharge their battery packs. Emissions associated with the electric range are determined by the local hydrogrid. Because 45 percent of the grid in the United States is generated by coal, carbon and photochemical smog emissions are directly correlated to pollutants generated by electrical generating plants on the grid. None of the major manufacturers widely available has plug-ins on the market to date. However, the Nissan LEAF, with a claimed 100-km-plus range, is available in small numbers to a few select markets and if an online reservation is made.

- *Series plug-in hybrids, or extended range electric vehicles*: The ICE engine in these hybrids serves only to act as a generator to charge the batteries—the car drives solely on the electric motor. Trade-offs have to be made in terms of battery size, weight, range, and cost. Chevrolet released its plug-in hybrid electric Volt model in December 2010—the most fuel-efficient car sold in the United States as of early 2011. However, the vehicle has limited availability. Widespread sales across all of Canada and the United States was expected in mid-2012.

- *Battery electric vehicles (BEVs)*: the major auto manufacturers are planning to release BEVs to the market in the near future. Presently, only small niche manufacturers are in this market. BEVs have no internal combustion engines and operate on a plug-to-plug basis. In 2010, the range was short and performance was modest with the then-current technology. Emissions associated with BEVs are determined by the local hydrogrid. If all vehicles in the United States were converted to BEVs, essentially almost half of all the cars would be running on coal, which is not a "clean" fuel. Furthermore, the significantly larger battery packs over a standard ICE vehicle are associated with increased LCA impacts from raw material extraction and processing. Notwithstanding, electric motors are more efficient mechanically than internal combustion engines, which will offset some of the increased emissions, especially when charged by renewable energy sources. The niche market Tesla Motors (2010) claim their BEV Roadster is 88 percent efficient while most conventional ICE-equipped automobiles are 20 to 25 percent efficient (BEV = 1.14 km/MJ, Hybrids = 0.56 km/MJ, conventional ICE car = 0.48km/MJ). Further, recharging BEVs using a "smart grid" electricity network and using vehicle-to-grid when they are parked will also do much to reduce impacts.

- *Fuel cells*: unlike batteries that store electricity, fuel cell vehicles typically convert hydrogen to create electrical energy. There are some fuel cell demonstration automobiles and commuter buses in use but no commercially available automobiles. In the operation phase of a vehicle life cycle, fuel cells generate electricity inside the cell through reactions between a fuel and an oxidant, triggered in the presence of an electrolyte. Hydrogen fuel cells use atmospheric oxygen as their oxidant. Fuel cells produce zero or very low emissions, depending on the source of the hydrogen.

Unfortunately, most hydrogen is produced from applying steam to methane, with the steam sourced from burning fossil fuels. Joseph Romm (2010) points out the following:

> For this reason, we do not escape the production of carbon dioxide and other greenhouse gases. We simply transfer the generation of this pollution to the hydrogen production plants. This procedure of hydrogen production also results in a severe energy loss. First we have the production of the feedstock methanol from natural gas or coal at a 32 percent to 44 percent net energy loss. Then the steam treatment process to procure the hydrogen will result in a further 35 percent energy loss.

Hybrids: The Fashionable Environmental Statement

Notwithstanding all of the above, parallel-hybrids, henceforth in this article referred to as hybrids or HEVs, have become de rigueur or fashionable among celebrities and politicians who are buying the car for the environmental statement it makes more than any other reason. What is clear from the scholarly research is that the increased popularity of hybrids is largely based on an image of environmental friendliness, rather than empirical evidence of environmental superiority. Heffner et al. (2007) points out the following:

> Many households acknowledged purchasing their HEVs as a response to environmental concerns. However, most had only a basic understanding of environmental issues or the ecological benefits of HEVs. Rather than buying their HEVs with measurable environmental goals in mind, most of the individuals in this study bought a symbol of preserving the environment that they could incorporate into a narrative of who they are, or who they wish to be.

Unfortunately for those who bought hybrids in order to make an environmental statement, a recent study by CNW Marketing Research (2007) indicates that the environmental benefits of HEVs are overstated, by a significant margin, when an LCA is used to benchmark their environmental performance. The emphasis to date has been on relying solely on fuel mileage as the only indicator of net environmental impacts. MacLean and Lave (2003) pointed out in their work that only one previous LCA had been conducted on automobiles and that 73 percent of carbon emissions associated with a Ford Taurus resulted from the use phase; the remainder was associated with the manufacturing, recycling, and disposal phases of the product life cycle.

CNW's LCA, claimed to have no outside organization, company, or enterprise fund in whole or in part any portion of their study, compared over 100 makes and models of cars and trucks. CNW used 4,000 "data points" for each car in a "cradle-to-grave" approach, including the following:

- Energy consumed in research and development
- Energy consumed in junkyard disposal
- Energy needed to produce parts
- Greenhouse gas emissions
- Fuel mileage
- Vehicle life span

These data were compiled and normalized into currency to make the result more understandable. The top five vehicles in terms of life-cycle impacts include the following:

1. *Smart*: $0.583

2. *Ion*: $0.621

3. *Focus*: $0.621

4. *Cavalier*: $0.655

5. *Wrangler*: $0.656

All but one of these vehicles are lightweight subcompacts with small, fuel-efficient three- or four-cylinder engines. Oddly, the much heavier four-wheel-drive Jeep Wrangler, equipped with a 3.8 liter V6 engine, made the list—CNW reports that the Jeep uses less energy to manufacture, has a longer life span, and is easier to recycle. The worst offenders included the Rolls-Royce ($11.83), the VW Phaeton ($14.26), and at rank 283, the Maybach ($15.97).

CNW found that hybrids fared poorly because of a lower than average vehicle life span and because of the increased complexity of their battery technology and extra electric motor. Much controversy exists from critics of the "Dust to Dust" study in terms of assigning the Prius a life span of 109,000 miles. CNW also points out that in order to use nickel metal hydride (NiMH) batteries, significant environmental impacts from raw materials are accrued in the LCA. Toyota buys nickel mined and smelted in Sudbury, Ontario, Canada. This nickel is shipped to Wales for refining, then to China for further processing, and then to Toyota's battery plant in Tokyo—a 10,000-mile trip, mostly by petrol-powered container ships and diesel-powered locomotives. These LCA impacts are expected to be mitigated somewhat by switching to lithium-ion batteries in the near future.

The top HEV on the 2008 list was the Toyota Prius, ranked 140th at $2.91 per mile. Hence, it is difficult from these data to make an argument that the Prius is "green" as the Toyota entails 3.8 times more life-cycle impact than the Mercedes Smart. These types of results are incongruous with the advertising campaigns associated with HEV. As part of a larger research project for presentation at a peer-reviewed conference, the author conducted an analysis of English-language hybrid vehicle television commercials found online in March 2009 by Charles Hostovsky. He conducted a Google and YouTube search and downloaded all available commercials. The study was qualitative, relying on thematic analysis, using grounded theory. A total of 19 commercials from four manufacturers were found and reviewed in detail, including Toyota (n = 9), Honda (n = 5), Ford (n = 4), and GM (n = 1). Examined were dominant images, themes, and slogans (catchphrases) used in the advertisements.

The largest category at 58 percent (n = 11) was dominated by images that involved the hybrid set in nature. Themes including images of the deep woods were common and also included scenes of mountains, glaciers, and lush river valleys as well as images of biodiversity (i.e., species at risk, rich and varied bird and plant life). A subset of 26 percent (n = 5) had a strong ecological restoration thematic, including images of the following:

- A seed planted grows into a tree bearing HEVs as fruit that fall to the ground and start to drive away
- Flowers follow the car and the city greens
- North America becomes green from space from HEV use
- Everything the hybrid passes on the road transforms to green
- Seedlings/flowers emerge through the asphalt freeway pavement

Some of the slogans associated with the hybrid in nature include Kermit the Frog saying, "It's easy being green." The most outrageous statement: the HEV represents "harmony between man (sic), nature and machine."

The second largest theme, 26 percent (n = 5), involved "transformative technology." Images of da Vinci, hybrid engines, and other transportation modes dominated these commercials. Typical slogans included: "one small step on the accelerator, one giant leap for mankind (sic)"; and "creating new alternatives … ones that won't change the way we live, but rather the impact our lives have on the environment around us."

Only 16 percent (n = 3) of the HEV commercials contained no environmental theme. Here the automakers' marketers appealed to the vehicles' utility as a hatchback and their excellent fuel mileage. The ads contained images of young adults loading their cars, parking in the city, and everyday driving. Slogans included "the hybrid for everyone is here" or "most fuel-efficient car in its class."

The author's analysis suggests the marketing of HEVs involves greenwashing. "*Greenwashing* is a term derived from whitewashing by environmentalists who claim that some corporations want to present an environmentally responsible public image by misleading consumers regarding their environmental practices or the benefits of their products or services," according to Hostovsky. Connie Davidson (2009) points out that an important characteristic of greenwashing is the suggestive and manipulative use of information. Although poor environmental performance itself is not necessarily greenwashing, a false representation of a poor performance as environmentally friendly can be considered to be so. Clearly, in the case of television ads, HEVs' environmental friendliness appears to exploit the common perception of hybrids but does not reflect the empirical LCA evidence—evidence that suggests HEV environmental performance is "middle of the pack" relative to other automobiles.

Hybrids and Transportation Demand Management

One of the major issues concerning HEVs for transportation planners revolves around transportation demand management (TDM). TDM, at its core, tries to get people out of their cars (i.e., personal occupancy vehicles [POV]) in order to walk, bicycle, use transit, carpool, create urban car-free zones and car-free days, combine trips, and so forth, thereby reducing vehicle miles traveled (VMT). However, TDM strategies can be thwarted by the "conservation rebound effect." This phenomenon occurs when the consumer chooses to use more of the resource (increase demand) instead of realizing the energy and environmental savings. For example, a person with a more efficient home furnace may choose to raise the setting on the thermostat or a person driving a more efficient car may drive more often and farther. Since the commuter perceives an HEV as an "environmentally friendly" car, he or she may also choose to drive and switch from taking transit, a phenomenon known as "induced/latent travel demand." Induced travel demand reduces freeway levels of service (LOS) and increases traffic congestion, resulting in greater carbon and photochemical smog emissions. Lorna Greening et al. (2000) reviewed the literature in terms of the conservation rebound effect. He reviewed 23 studies involving fuel-efficient automobiles and found that the conservation rebound effect ranged between 10 and 30 percent. Kenneth Small and Kurt Van Dender (2005) found short- and long-run rebound effects 5 percent and 22 percent with fuel-efficient vehicles. Very little research has been conducted to examine if the CRE exists for HEVs in particular. Peter De Haan et al. (2006) surveyed 367 buyers of the Toyota Prius 2 in Switzerland. They found that vehicle size did not increase when replaced with hybrids and that average household vehicle ownership remained stable. However, they were unable to determine if hybrid drivers will, in the future, drive more. In a follow-up study, J. Dickinson and de Haan (2009) found no increase in VMT with Swiss Lexus hybrid owners. Since this study involved a large SUV model equipped with HEV technology, and considering that European driving behavior differs substantially from the low-density sprawl-dominant North American context, it is difficult to extrapolate from Dickinson and de Haan for North American transportation planners.

Another problem for TDM strategies are misguided policies based on the perception that HEVs are "environmentally friendly." Some governments are offering economic incentives by providing rebates or allowing single-occupancy vehicles in freeway high-occupancy vehicle (HOV) lanes. Some business and government offices have created HEV parking spaces in prime locations in parking lots, unfortunately creating initiatives to promote hybrids that may be counterproductive to transportation demand management (TDM) and transit-oriented development (TOD), that is, getting people out of their cars, which is fundamental to smart growth planning.

In summary, despite HEVs' achieving among the highest fuel mileage of commercially available automobiles, miles per gallon is only one factor in a vehicle's life-cycle ecological footprint. There is no empirical LCA–based evidence that hybrids are more environmentally efficient than standard cars with gasoline-powered internal combustion engines. The only comprehensive LCA conducted to date places HEV impacts as middle-of-the-pack in terms of the life-cycle phases of manufacturing, life span, operation (i.e., fuel economy), recycling, and disposal.

Hybrid marketing through television commercials exhibits a strong tendency toward greenwashing. Evidence of eco-smugness, likely as a result of deep corporate greenwashing, may dissuade commuters from practicing transportation demand management strategies and contribute to increased traffic congestion through induced travel demand. Unfortunately, governments and business appear to be reacting to the popularity of these vehicles with misguided programs, subsidies, and policies aimed at promoting HEVs, which work against TDM and smart growth. Much more research is required, especially in terms of the relationship between hybrids and the conservation rebound effect and induced travel demand.

Some of the important strategies of smart growth planning—transportation demand management and transit-oriented development, in particular—fundamentally require commuters to "get out of their cars" more often. When we can reasonably reduce our North American auto dependency, HEVs can play an important role by reducing carbon emissions, since driving is indeed a necessary evil. In order to fulfill that role, hybrid/electric vehicles need to improve their longevity as well as reduce the ecological footprint associated with their manufacturing processes.

See Also: Carbon Footprint Calculator; Sustainable Design; Technology and Social Change; Unintended Consequences.

Further Readings

CNW Marketing Research, Inc. "Dust to Dust: The Energy Cost of New Vehicles From Concept to Disposal—The Non-Technical Report" (2007). http://www.cnwmr.com/nss-folder/automotiveenergy (Accessed August 2010).

Davidson, Connie. "Green-Washing." In *Green Politics: An A-to-Z Guide,* Paul Robbins and Juliana R. Mansvelt, eds. Thousand Oaks, CA: Sage, 2011.

de Haan, Peter, Michel G. Mueller, and Anja Peters. "Does the Hybrid Toyota Prius Lead to Rebound Effects? Analysis of Size and Number of Cars Previously Owned by Swiss Prius Buyers." *Ecological Economics,* 58/3 (2006).

Dickinson, J. and P. de Haan. "Analysis of Potential Direct Rebound Effects Associated With Hybrid Lexus RX400h." Rebound Research Report No. 5, ETH Zurich, IED-NSSI, report EMDM 1472 (2009). http://www.uns.ethz.ch/res/irl/emdm/ETH_RRR05_LexusRX400h_EMDM1472.pdf (Accessed September 2010).

Even-Har, Meirav and C. Hostovsky. "The Montreal Car Free Day: A Catalyst for Multimodal Transportation Planning." *Plan Canada,* 46/2 (2006).

Greening, Lorna A., David L. Greene, and Carmen Difiglio. "Energy Efficiency and Consumption: The Rebound Effect—A Survey." *Energy Policy,* 28 (2000).

Haddock, Vicki. "Oh, So Pious, Prius Drivers/Smugness Drifts Over the Warming Earth—Is That a Bad Thing?" *San Francisco Chronicle* (July 15, 2007).

Heffner, Reid R., Kenneth S. Kurani, and Thomas S. Turrentine. "Symbolism in California's Early Market for Hybrid Electric Vehicles." *Transportation Research Part D,* 12 (2007).

Hostovsky, Charles. "Greenwashing." In *Green Consumerism: An A-to-Z Guide,* Paul Robbins and Juliana R. Mansvelt, eds. Thousand Oaks, CA: Sage, 2011.

Hostovsky, Charles. "Popularity of Hybrid Vehicles: Implications to Transit-Oriented Development and Travel Demand Management." Canadian Association of Geographers Conference, CFHSS/FCSH Conference System, (2009). http://ocs.sfu.ca/fedcan/index.php/cag2009/cag2009/paper/view/1647 (accessed September 2010).

MacLean, Heather L. and Lester B. Lave. "Life Cycle Assessment of Automobile/Fuel Options." *Environmental Science & Technology,* 37/23 (2003).

Romm, Joseph J. "The Hype About Hydrogen: Fact and Fiction in the Race to Save the Climate." In *Culture Change,* published by Sustainable Energy Institute. http://www.culturechange.org/hydrogen.htm (Accessed August 2010).

Simanaitis, Dennis. "The Tech Choice Is Yours: Diesel? Plug-In Hybrid? Battery Electric? Which Makes Sense for You?" *Road and Track*, 60/9 (May 2009).

Small, Kenneth and Kurt Van Dender. "The Effect of Improved Fuel Economy on Vehicle Miles Traveled: Estimating the Rebound Effect Using U.S. State Data, 1966-2001." *Policy & Economics,* UC Energy Institute (2005).

U.S. Department of Energy. "Ethanol: The Completer Energy Life Cycle Picture" (March 2007). http://www.transportation.anl.gov/pdfs/TA/345.pdf (Accessed August 2010).

U.S. Department of Energy. "Hybrid Electric Systems: Goals, Strategies, and Top Accomplishments" (June 2010). http://www1.eere.energy.gov/vehiclesandfuels/pdfs/hybrid_elec_sys_goals.pdf (Accessed August 2010).

U.S. Environmental Protection Agency. "2011 Most and Least Fuel Efficient Vehicles (ranked by city, then highway)" http://www.fueleconomy.gov/feg/bestworst.shtml (Accessed September 2010).

von Blottnitz, Harro and Mary Ann Curran. "A Review of Assessments Conducted on Bio-Ethanol as a Transportation Fuel From a Net Energy, Greenhouse Gas, and Environmental Life Cycle Perspective." *Journal of Cleaner Production*, 15 (2007).

Charles Hostovsky
University of Toronto

Information Technology

Green information technology (green IT) can be thought of fundamentally as a new round of technological innovation and a set of organizational practices that green the IT infrastructure and/or that use IT to green other domains such as building and supply chains. Given the comprehensive scope, dimensions, and meanings of green IT, it should also be treated as a condensed concept that encompasses many pieces, including technological innovation, business strategy and philosophy, organizational conduct, relevant policy, and a movement. Green IT can benefit not only our economy by increasing energy efficiency while saving costs, but also our natural environment by reducing greenhouse gas (GHG) emissions.

Green IT was born in a historical time when the negative environmental impact of significant IT growth, significant energy demands from IT and other domains, and the increased urgency of dealing with climate change were recognized, and the economic crisis and oil price increases were looming large. A couple of milestones, including the U.S. Environmental Protection Agency's (EPA) data center report to Congress and the formation of the Green Grid, also contributed to the emergence of green IT. Development of green IT has begun to reverse the negative impact of IT on the environment. Green IT is currently associated with different meanings and different priority goals by different parties. There is also a lack of a common standard and metrics to capture the comprehensive nature of green IT and to measure its success. These pose challenges to the development of green IT. At this stage, green IT should be treated as an emerging technological innovation and social phenomenon.

A Sketch of Green IT

Although often unknown to many, green IT has emerged particularly since 2007 and experienced impressive growth since then as a new business and technological and social phenomenon. *Green IT* is still an evolving term that is subject to various interpretations by different parties, and at this writing there is no commonly agreed upon definition. It can, however, generally be understood as how information technology is planned, designed, deployed, developed, used, and disposed of in a environmentally and socially responsible and economically sustainable manner. Green IT includes not only a set of new technological innovations such as data center greening, server virtualization, energy-efficient computers, and cloud computing, but also a set of new organizational practices and strategies such as e-waste management, telecommuting, IT recycle and remarket, and power management.

Green IT is a comprehensive term encompassing multiple dimensions as each of the aspects mentioned above (and oftentimes multiples of them) can be associated with different technological innovations and/or organizational strategies. For instance, on the "dispose" aspect of green IT implementation, organizations can develop a set of responsible e-waste management programs to properly recycle their IT equipment and/or redeploy it by giving it to other departments or organizations for reuse. In another example, on the "plan," "deploy," and "use" aspects, the green data center has emerged as a key green IT innovation to store, process, and manage data and servers for higher energy efficiency and lower environmental impact. To reach these goals, the green data center often implements a set of new technologies and strategies such as using virtualization and storage tiering technologies to enhance the server capacity, low-emission building materials to decrease the carbon footprint of data centers, and alternative energy to increase cooling of servers. Server virtualization is a major technology used to green the data centers. It involves defining and combining several virtual machines in one powerful physical box, thus giving access to different people and applications. This could increase the capacity utilization of the physical server and eliminate the need for and energy cost of using several physical servers.

In terms of areas and scope of application, green IT has recently expanded from green IT 1.0 to green IT 2.0. The commonly referred to green IT 1.0 focuses on the traditional application of green IT in the IT domain itself—how to reduce energy and carbon footprints of IT equipments that include computers, printers, servers, data centers, and so on. More recently, the potential of IT's role in climate change has been experimented with and is starting to be implemented from the IT domain to virtually all other non-IT areas and industries such as water, transportation, and city. In the new green IT 2.0 innovation, IT's potential role has been explored to make the rest of the world efficient and green. Examples include the design and use of smart meters using IT in smart grids that help homes and offices to be more energy efficient, and the design and use of intelligent IT systems to help cities better manage their traffic systems, thus reducing carbon dioxide (CO_2) emissions from unnecessary vehicle use and transportation. Another area that differs from both green IT 1.0 and 2.0 is electronic waste management, which is more concerned with the human health issues related to IT equipment. This area is concerned with how to recycle IT equipment; how not to use toxic materials in the manufacturing of computers, mobile devices, and other IT equipment; and how to design IT products that are recyclable.

Birth and Drivers of Green IT

Why did green IT arise recently, and what are the possible drivers for it? The emergence of green IT partly reflects the historical burden of IT on the environment. IT not only helped expedite the expansion of world trade and industrialization, but it has also increased the chance of circulation of waste and contamination. Two-thirds of computers end up in landfills, generating health risks to the public. While the information age often leaves us an image of streamlined businesses with neat and clean, high-speed, intelligent computers and information systems, what we do not always see is the environmental impact of unsustainable development of IT. In *High-Tech Trash*, Elizabeth Grossman revealed the not-so-clean side of high tech. For instance, although the microchips in our computers are as tiny as a pinky fingernail (Pentium 4 chips), their production may cost as much as 1,600 grams (i.e., 3.5 pounds) of fossil fuel and chemicals each.

The fast growth of unsustainable IT not only produces e-wastes that are detrimental to health, it also causes increasing energy demands. IT has largely facilitated the process of globalization and advance of world economy. Moreover, wide application of IT in virtually every business has also pushed business to reach maximum capacity of energy sources and to induce soaring costs that need urgent solutions. It is estimated that the number of PCs in use will double between 2008 and 2014, and mobile voice and data traffic will increase four times from 2007 to 2012. Even a relatively recent user application, the Second Life virtual world, now requires more than 9,000 servers, and its data storage needs will grow by over 69 times in the future, based on prediction of IBM consultants. Gartner Inc. predicted that continued power demands will top 80 percent of data centers running at the maximum available power and cooling levels by 2010.

While increasing energy demands of IT and e-waste, along with an increased concern about climate change, could be the main drivers of green IT, the recent economic crisis and increases in the price of oil also helped trigger the first round of green IT innovation. With increasing needs from all types of businesses, communication, academic, and governmental organizations, the data center is the central nerve system of IT used to store and process data. A milestone event related to data centers played a key role in initiating green IT. Around 2006, the U.S. Congress took an interest in IT energy use and passed a law requesting EPA to study data center energy use and efficiency. After months of research, EPA released a milestone report that evaluated the energy use and the consumption of data centers and servers in the United States and suggested potential opportunities for energy-efficiency improvement. The study found that due to increasing demands for computer resources, data centers in the United States have had significant growth: energy consumption of these data centers and the power and cooling infrastructures that support them more than doubled from 2000 to 2006, and would double again by 2011. It was estimated that the electricity use of these data centers would reach 61 billion kilowatt-hours in 2006 (costing $4.5 billion annually), and this number would go up to more than 100 billion kilowatt-hours, representing a $7.4 billion annual electricity cost by 2011.

This EPA study, along with other events, caused increasing attention to the soaring energy and electricity costs of growing data centers. As newer generations of servers with higher performance come out, their power consumption grows even more. A server uses so much power for maintenance that for every kilowatt that it uses, it also needs 1 kilowatt of generator capacity, more air conditioning, and more UPS capacity for uninterruptable power supply. Such fast increase of power-related costs has pushed the industry to become seriously interested in energy efficiency of IT use. Manufacturers and users began to think about new technologies or new ways to increase the capacity utilization of servers, data centers, and other IT equipments—green IT. Since the release of the EPA report, more companies—especially server designers such as IBM and Dell—and manufacturers of data center equipment have paid attention to the energy efficiency and cost issues of their products. One of the results was the formation of the Green Grid in 2007, an industry consortium that works together to advance energy efficiency in data centers and business computing ecosystems. Having joined forces with almost 200 member organizations as of this writing, it seeks to standardize a set of metrics and methods to promote green IT practices. For instance, it popularized the most widely used metric to measure energy use efficiency of data centers, the Power Usage Efficiency (PUE). PUE is calculated by dividing the amount of power coming into the entire data center by the amount of power consumed by the IT infrastructure in it. The closer it is to 1, the more efficient it is.

Green IT's Positive Impact on the Environment

The above drivers and critical events triggered the green IT movement, which started to recognize and emphasize the role of IT as a solution to rather than a destroyer of environmental sustainability. New inventions such as eco-software and virtualization were made, efficiency data center technologies were reprioritized, many existing IT technologies started to be added to new "green" dimensions to increase IT efficiency and thus cut costs.

Global e-Sustainability Initiative reports that the information and communications technology (ICT) carbon emission in the United States is expected to increase from 2.5 percent to 2.8 percent between 2007 and 2020. The implementation of green IT can help reduce at least 2.5 percent of CO_2 in the United States. But this number may be underestimated given the fact that IT is embedded in almost all kinds of business and nonprofit transactions now, and its indirect impact on CO_2 has not been accurately captured. Many expect that, beyond its own domain, IT can play an even larger role in reducing GHG emissions and enabling energy efficiency in almost all other industries such as urban planning, transportation, and water. As estimated in the recent SMART 2020 report, ICT-based solutions can help reduce CO_2 emission in most business sectors by 13–22 percent annually by 2020, which is about five times the footprint of its own sector. This could save gross energy and fuel costs of $140–$240 billion.

Players in Green IT and Some Initial
Signs and Results of Implementation

Green IT has become an emerging trend and reality, especially for the IT industry, which is the first mover to produce green IT before popularizing it to the other vendor industries. Due to the fact that its development is still in an infant stage, there is a lack of systematic study and statistics regarding actual green IT implementation and its initial results on environmental impact. However, some anecdotal evidence is adding up and showing part of the picture. In a recent survey of IT professions in multiple industries published by Gregor Harter and colleagues, more than 50 percent of companies indicated that they already had some green IT solutions or would be implementing them within a year. News about individual IT companies' green IT innovations is mushrooming. For instance, IBM recently announced its design of a new Power 575 supercomputer, equipped with IBM's latest POWER6 microprocessor. Using water-chilled copper plates on the microprocessor to remove heat from the electronics, it can reduce typical energy consumption for cooling the data centers by 40 percent. As one of the early and popular applications of green IT, the green data center has been implemented in leading firms such as Google, Cisco, and IBM. So far, server virtualization solutions in various industries have helped save about 8.4 billion kilowatt-hours of electricity each.

Some nonprofit organizations have also played an active role in green IT innovation by challenging businesses or working collaboratively with them to establish industry and technology standards for the twin goals of saving energy costs and protecting the environment. Basal Action Network and Silicon Valley Toxics Coalition specialized in fighting e-waste production. EPA launched Energy Star as a voluntary labeling program in 1992 to promote energy efficiency in computer hardware and other electronic equipments. The

Green Electronics Council developed the Electronic Product Environmental Assessment Tool (EPEAT) in 2007 to evaluate, compare, and recommend computer products based on a series of 51 environmental attributes that categorize products into bronze, silver, or gold standards. By promoting computer products with a low environmental footprint, these organizations helped push the emergence of a green IT industry that works not only on energy-efficient computing but also on data centers, power management, responsible disposal, and so on.

A number of voluntary consortiums (usually nonprofit organization [NPO]–business alliances) similarly played a key role in promoting green IT. Promoting more than just green IT standards and metrics, these consortiums also aim for real-world change starting with their own member organizations. The Green Grid is a classic example. Another example is Climate Savers Computing (CSC). Initiated by Google and Intel in 2007, CSC is a consortium made up of eco-conscious businesses, conservation organizations, and consumers, which pledged to cut down GHG emissions collectively through increased use of power management features in computer systems. A CSC 2010 report stated that the practice of green IT has reduced GHG by about 36.8 million metric tons since 2007, which is about 60–70 percent of its original estimate of saving 54 million by 2010. In another voluntary environmental advocacy community, Sustainable Silicon Valley (SSV), by practicing environmentally sustainable technological practices, high-tech firms such as Cisco, Hewlett Packard (HP), and Sun all pledged to reduce their carbon dioxide emissions toward a goal of 20 percent below their 1990 levels by 2010, and some members have achieved this goal. SSV's 2009 annual progress report summarized that SSV partners together have cut CO_2 emissions by 758,000 tons from 1990 to 2008, which is equal to removing 125,942 cars from the roads for one year.

Challenges of Green IT

The "green" component in green IT is especially subject to different interpretations, depending on the parties involved. This and the multiple dimensions of green IT make it challenging to become a standard term or to obtain a standard definition. This has also created many alternative names for green IT such as green computing, green ICT, greener IT, efficient IT, and sustainable IT. Yet these different names have the common goals of energy efficiency, cost savings, and carbon footprint reduction. Ideally, green IT should address all of the triple bottom lines—economic profit, ecological impact, and social responsibility. When implemented by different parties, certain priorities are emphasized, and not all three are always clearly addressed. For business organizations, the primary goals for engaging in green IT are often energy efficiency and cost savings, while environmental and social responsibility are often treated as the secondary goals. This poses some doubt as to how truly green the green IT can be. For nonprofit organizations that emphasize e-waste issues, the primary goal is often social responsibility—how human health should be protected from toxic materials in nongreen IT equipments. Such splitting of priorities poses challenges to the development of green IT toward its ideal achievement—taking care of profit, planet, and people—even though all three goals of green IT could be addressed regardless of priority and serve as a common ground for different types of organizations to work together.

The lack of both a common understanding and definition of green IT also indicates that there is a lack of common understanding of how green IT should be measured. It is yet unclear what the baselines for measuring the success of green IT are, and there is a lack of strong successful green IT best practice cases to support such. While there are some voluntary standards such as Energy Star to measure the energy efficiency of computer products and PUE to measure the power usage efficiency of data centers, we do not yet have one strong common standard to measure all aspects of green IT and to evaluate how green the comprehensive nature of green IT is. For instance, no metrics exist to measure how energy efficient and green the process of making IT products is, the efficiency of renewable energy use in a data center, and the efficiency of useful IT work done per kilowatt of power. The missing standard on one hand speaks of the novelty of green IT, and on the other hand it speaks of many barriers. One of the barriers is the missing price tag on carbon in the United States. As long as this remains unchanged, energy always appears cheaper than it actually is, and pollution is often deemed free by some. Another barrier that tends to be ignored is that green IT should also be considered a set of environmentally responsible practices in addition to a technological innovation, simply because no matter how efficient the green IT technology can be made, if we keep producing more and do not control our behavior of wasting resources, green IT cannot be truly green.

See Also: Design for Recycling; Electronic Product Environmental Assessment Tool (EPEAT); E-Waste; Innovation; Technology and Social Change.

Further Readings

Global e-Sustainability Initiative. "SMART 2020: Enabling the Low Carbon Economy in the Information Age: United States Report Addendum." http://www.smart2020.org/_assets/files/Smart2020UnitedStatesReportAddendum.pdf (Accessed July 2010).

Greenpeace. "Switching on to Green Electronics." http://www.greenpeace.org/international/en/publications/reports/Switching-on-Green-Electronics (Accessed July 2010).

Grossman, Elizabeth. *High Tech Trash: Digital Devices, Hidden Toxics, and Human Health*. Washington, DC: Island Press, 2006.

Hejmanowski, Ken and Gill Friend. "Climate Savers Computing Initiative (CSC 2010 Progress Report)." White paper produced by Natural Logic, Inc. http://www.climatesaverscomputing.org/docs/2010-Progress-Report.pdf (Accessed July 2010).

Lamb, John. *The Greening of IT: How Companies Can Make a Difference for the Environment*. Indianapolis, IN: IBM Press, 2009.

U.S. Environmental Protection Agency. "Report to Congress on Server and Data Center Energy Efficiency: Public Law 109-431." http://www.energystar.gov/ia/partners/prod_development/downloads/EPA_Datacenter_Report_Congress_Final1.pdf (Accessed September 2010).

Jingfang Liu
University of Southern California

INNOVATION

Innovation is generally defined as the creation of a new way of doing something, whether it is concrete (e.g., the development of a new product) or abstract (development of a new philosophy or theoretical approach to a problem). Innovation plays a key role in the development of sustainable methods of production and of sustainable methods of living because in both cases it may be necessary to create alternatives to conventional ways of doing things that were developed before environmental consideration was central to most people's framework for making decisions.

Because innovation plays a central role in business success as well as in scientific progress, considerable research has focused on specifying the working conditions that are likely to produce useful innovations. In general, scholars have noted that the best model for producing useful knowledge about the empirical world (i.e., knowledge based on observation and experimentation rather than theory or belief) is to foster the work of many relatively autonomous specialists whose work is judged by its merits rather than its conformity to pre-existing beliefs or traditional ways of doing things. This reflects the attitude that enables the creation of modern scientific practice, an attitude that may be traced back to 17th-century Europe.

Several attitudes and practices from that period also apply to fostering modern scientific and technical innovation. First of all, scientific or innovative contributions should be evaluated based on impersonal criteria (i.e., do they accurately describe the world or does the new process work more efficiently than the old) rather than according to who produced them or the personal characteristics (e.g., race, gender, nationality) of the person who produced them. Second, knowledge should be shared rather than kept secret so others can apply it to their work and the general level of knowledge can increase. Third, scientists should act in a disinterested manner, seeking to increase knowledge rather than focus purely on personal gain. Fourth, scientific claims cannot be made on the basis of authority but are open to challenge and should hold up under scrutiny. Of course, some of these rules are somewhat modified in the modern world—for instance, people do profit from their own discoveries, both directly in terms of holding patents and indirectly in terms of career success—but the basic principles hold true.

Scientific Innovation

In *The Structure of Scientific Revolutions*, Thomas Kuhn made a distinction between what he termed *normal science* and *revolutionary science*. He defined normal science as the process of solving puzzles within the paradigms currently established for one's particular science. For instance, in astronomy, it was believed for centuries that the planets orbited around the Earth (the geocentric model) and complex models and calculations were developed to try to explain the observed movements of the planets within this model. Revolutionary science in contrast involves challenging or changing the paradigm, as Copernicus did when he proposed a heliocentric universe in which the Earth as well as the other planets orbited around the sun. Most science in any time period is normal science, with people working within an existing framework that includes methods, assumptions about nature, symbolic generations, and paradigmatic experiments. Even observations that do not seem to fit the existing paradigm will be explained within it (as planetary motion was for centuries in the geocentric model) or ignored as anomalies. However, at some point, the contradictions and anomalies may become too obvious and can trigger a scientific revolution, as happened in the 16th century in Europe (notably not recognized by a powerful social institution, the Catholic Church, until centuries later).

Most scientists and technical employees today are analogous to normal scientists, working to discover practical applications or to illuminate small areas of knowledge within a given scientific model. For instance, many scientists in the United States are employees of corporations, government agencies, and so on, and are expected to work within accepted models rather than challenge them. This leads to conflict between the scientist's desire for autonomy and the organization's desire for practical results, and can stifle innovation that could lead ultimately to greater breakthroughs. One way this problem is dealt with is to have people specialize in either basic or applied science, with different evaluative criteria for each, and to have part of an organization's budget reserved for basic research that may challenge the existing paradigm rather than work within it.

Another conflict for scientists and technical employees, particularly those working in for-profit companies, is their desire to communicate their discoveries to others versus their employer's desire to keep such discoveries confidential in order to protect their profitability. Patent law is intended to allow both desires to be met. The purpose of the patent system is to stimulate scientific and technical invention by reserving the right to profit from a discovery for a period of years to the patent holder (which may be an individual or organization such as a company or university), while also making the information from the discovery public so others may learn from it. The patent holder may sell or license the right for others to use his or her discoveries and collect fees from them.

Facilitating Innovation Within Organizations

Changes in organization may be less dramatic than scientific discoveries, but are equally important in terms of promoting efficiency and productivity. For instance, an organization may innovate in the way it operates or delivers services, resulting in greater efficiency, fewer errors, faster speed of production, and so on. Sandford Borins identifies several characteristics typical of organizations that are successful at innovation. First, top management supports innovation and provides leadership in this area. Second, individuals who push for innovation are rewarded. Third, the organization dedicates resources specifically to innovation rather than expecting it to happen as a matter of course. Fourth, the organization has a diverse workforce and welcomes ideas from outside the mainstream. Fifth, the organization's bureaucratic layers are closely connected so that innovations can be easily communicated and implemented. Finally, the organization is willing to experiment with different ways of doing things with the understanding that not all will be successful.

Borins notes that some of these characteristics are the opposite of what is seen in many government organizations and companies. For instance, in many organizations, people who suggest or enact innovation may be subject to sanction or dismissal, and the organization may display no interest in testing different ideas to see which are useful and practical. Some organizations have a superficial commitment to innovation in the sense that they eagerly embrace whatever the current trendy solution is, but do not display the commitment to evaluate the usefulness of the new ideas or conduct any kind of measurement to see if they produce the desired results. Both approaches stifle effective innovation (as they would stifle effective scientific progress) because they are based on received beliefs and authority rather than on empirical observation and testing.

Industrial and Technological Innovation

Joseph Schumpeter used the term *creative disruption* to describe change of the economy from within. He viewed entrepreneurs, who invent new goods and new ways of doing things, as essential to keeping an economic system constantly evolving. New products or ways of doing things necessarily disrupt existing markets: For instance, the department and catalog store Montgomery Ward was once a major retailer but went out of business in 2001, due in part to loss of market share to low-price competitors such as Kmart and Walmart. To take another example, the instant-film camera developed by Polaroid was a popular consumer product for several decades but ceased production when it was surpassed by digital cameras. Schumpeter saw the process of creative disruption as positive in the long run because it promoted economic growth and rewarded innovation and improvement, while also noting that individuals and corporations could also suffer when their particular skills or products were no longer demanded by the market.

Clayton M. Christensen coined the term *disruptive technology* (later *disruptive innovation*) to describe innovations that improve a product or service in ways that disrupt an existing market (as opposed to a "sustaining innovation," which improves an existing product and reinforces the position of leading manufacturers in the field). The disruptive innovation often has characteristics that the traditional customer base does not care about, and may even be inferior compared to existing products, but will appeal to a different set of customers with different priorities. The innovation is "disruptive" not to the consumer (who, at least at first, has the choice to buy either the existing or innovative product) but to businesses that may be doing a good job supplying an existing product and yet see their market disappear as the new technology becomes widespread. An example is downloadable music files that offer the convenience of buying music online and playing it from one's computer, as well as the ability to purchase individual songs. This appealed first to young people who were quite comfortable with computers and MP3 players (versus older consumers more used to fixed stereo systems and the concept of songs collected into albums), and severely cut into the market for compact discs.

Cooperation between manufacturers and other institutions such as universities can facilitate innovation. Martin Kenney coined the term *university-industrial complex* to describe, in the biotechnology industry, the flow of resources among universities (which provide knowledge and skilled labor), multinational corporations (that produce products), and venture capital firms (that provide financing to both research and production). He notes that university-employed scientists have provided most of the research that formed the basis of the biotech industry, that scientists often move between employment in academia and the corporate sector, and that many university graduate programs have been created or enlarged specifically to train students for the biotech industry. Development of the biotech industry was facilitated in large part by increased federal funding for science, with grants awarded on a competitive basis, which rewarded innovation while also facilitating the creation of well-equipped research labs at universities as well as within corporations. Other sciences have also followed the biotechnology model, with close relationships between the university and corporations becoming the norm, such that many universities now have "technology transfer" offices to facilitate the process.

Regional methods of organization can also influence innovation. AnnaLee Saxenian looked at the differing fortunes of two areas once noted for their high-technology industry: Silicon Valley (south of San Francisco, California) and the Route 128 area (near Boston, Massachusetts). In the 1970s, both were noted as centers of innovation in the electronics industry, fueled in part by university research and military spending, and both faced downturns in the early 1980s. However, Silicon Valley recovered, with the help of new start-ups such as Sun Microsystems as well as the continued prosperity of established companies such as Intel and Hewlett-Packard (HP), while Route 128 companies such as Wang and Digital Equipment Corporation went out of business, and other area companies declined. Business investment in Silicon Valley increased by $25 billion between 1986 and 1990 while only increasing by $1 billion in the Route 128 area, and by 1990, Texas and southern California had both surpassed Route 128 as centers of electronics production.

Saxenian attributes these differences to differing regional industrial organization. Silicon Valley has a network-based industrial system with dense social networks and open labor markets, which promote experimentation and collective learning so that competitors can learn from each other. The boundaries between individual companies and other institutions such as universities remain fluid. In contrast, Route 128 was characterized by a small number of large, hierarchical firms with barriers to information sharing between different firms as well as between firms and other institutions. The network system was better able to adapt to change (e.g., recovering from the loss of silicon chip manufacturing to Japan) while the Route 128 manufacturers were not able to respond when the industry shifted from microcomputers to workstations and personal computers. In general, Saxenian argues that industrial organization based on independent firms (the Route 128 model) can flourish when markets are stable and technology changes slowly because they can capitalize on economies of scale, but in a rapidly changing industry, firms may find themselves saddled with obsolete technology and a workforce with outdated skills, and are less able to access external sources of information. In contrast, the regional network type of organization is more flexible in responding to change and better able to promote collective technological advance.

Facilitating Green Innovation Across Organizations

In contrast to the usual business practice of reserving the right to use discoveries, some companies have chosen to make green business innovations publicly available. One example is the Eco-Patent Commons, which places patents for some clean technologies in the public domain. The Eco-Patent Commons was created in 2008 by IBM, Sony, Nokia, and Pitney-Bowes in conjunction with the World Business Council on Sustainable Development. As of 2010, 100 "IP-free" technologies were available on the Eco-Patent Commons, meaning that the patent holders waived their intellectual properties rights to the technologies. The patents may be accessed through a searchable database on the World Business Council on Sustainable Development website. Examples of the type of patents in the Eco-Patent Commons include a method for using a surfactant for separating chemicals from wastewater; a more efficient method of creating olefins (used in packaging, electronics, adhesives, and other products) by reducing waste and extending the life of catalysts; and a technology that uses a magnetized solution in refrigeration in place of refrigerants, which can damage the Earth's ozone layer, and also reduces the energy required to run the compressors.

A second technology sharing group, the GreenXchange, was created by Nike, Best Buy, and other companies in partnership with the Creative Commons for the purpose of sharing intellectual property related to green product design, packaging, manufacturing, and similar processes. The concept is the same as with the Eco-Patent Commons: companies make some of their innovations publicly available, forgoing patent fees, in the interest of making manufacturing and business practices more environmentally sustainable. Nike began the project because it had an active program in environmental sustainability and noticed that many other companies were conducting research along parallel lines, while at the same time large gaps in research remained: their idea was that by sharing ideas and resources, each company would be able to produce more innovative research rather than duplicating discoveries already made elsewhere. GreenXchange has three levels of commitment. At the most basic level, the contributor allows its patents to be used in basic academic research. At the second level, certain patents (including those for commercial applications) may be made available for sustainability uses under a standard license. At the third level, unpatented discoveries may be made available by registering them in the knowledge database.

A more recent venture is the Innovation Exchange, established by the Environmental Defense Fund for the purpose of sharing information about best practices with regard to environmental issues including energy, water, and climate. Although the information in the Innovation Exchange may have been previously available, it was often not easy to find, as it was buried in reports or on company websites. Information on the Innovation Exchange is organized by industry sector (e.g., food and agriculture, manufacturing, financial services) and topic (e.g., clean water, energy efficiency, paper and packaging) and currently includes several types of information, including case studies, research reports, recommendations, and research tools.

An Example of Innovation: The Green Revolution

The Green Revolution refers to the initiative to apply modern farming methods, including the use of high-yield hybrid seeds, synthetic pesticides and chemical fertilizers, and irrigation, to developing countries in Latin America, Asia, and Africa. This effort is strongly identified with the agronomist Norman Borlaug, who received the Nobel Peace Prize in 1970 for his efforts in combating world hunger. Although the technologies used in the Green Revolution were not new, they were applied in a new context, and also led to the establishment of local programs of agricultural research to continue improving the work begun by Borlaug.

In addition, the Green Revolution required a new way of thinking about the world's food supply. For instance, the regular occurrence of food shortages in much of the world was accepted before the Green Revolution, and India was on the brink of a famine in 1961 when Borlaug began his work there. Improved varieties of crops plus the use of irrigation, fertilizer, and pesticides increased crop yields by several hundred percent (reportedly as much as 10 times higher in some cases) and not only removed the immediate threat of famine but also created a surplus that allowed India to become a food exporter.

The Green Revolution has not been without its detractors, although few would dispute its success in forestalling widespread hunger. The problem is that major innovations in an ecosystem can have consequences beyond those that were intended, and some of those consequences may be negative. For instance, heavy use of chemical fertilizers contributes to the death of fish stocks in some countries, and may have adverse effects on human health as well. The exclusive use of a few high-yielding plant types creates monoculture farming, which is highly susceptible to disease, in contrast to traditional agriculture, which incorporates more diversity. The high-yielding seed varieties require intense irrigation, in some cases depleting local water supplies, and the use of patented seed varieties, fertilizer, and pesticides requires farmers to buy their seed from multinational companies annually.

Current Examples of Environmentally Friendly Innovations

Every year, the Environmental Defense Fund (EDF), an organization that partners with businesses to achieve environmental goals, publishes a review of noteworthy environmental innovations of the previous year. The organization's stated goal is to "Make green business the new business as usual," and its emphasis is on fostering innovations that are environmentally friendly but also economically profitable for companies. In 2009, the companies honored included major organizations such as Cisco Systems, Coca-Cola, Google, and Walmart, as well as smaller organizations such as the Iowa Soybean Association and the Zocalo Community Development organization in Denver, Colorado. Innovations are assessed on the criteria of producing environmental benefits (e.g., reducing pollution, conserving water, or restoring natural resources), providing business benefits (e.g., increasing revenues or cutting costs, reducing liability, or enhancing investment opportunities), replicability (i.e., they should be able to be adopted by other companies), and innovativeness (i.e., they should be original and not yet commonly employed in the business world).

Companies like Enterprise Rent-A-Car and Verizon have achieved energy savings by adopting thin client computers rather than personal computers (PCs) in their computing networks. Thin client computers are stripped-down machines that store data and run programs on central servers (and hence are sometimes describes as "dumb machines") as opposed to personal computers, which have their own processors and memory. Thin clients are commercially available products that have been used for years in basic applications such as terminals in bank branches and in call centers, but can also be used in more traditional office settings. Each thin client terminal uses considerably less electricity than a PC (generally 6 to 20 watts for a thin client versus 70 to 150 watts for a PC), and even though they require a more elaborate centralized computing facility, the use of thin clients can results in energy savings of 25 to 50 percent for a company. Thin clients also require fewer materials to manufacture and last longer than PCs, reducing the amount of solid waste created.

Several firms were commended for adopting labeling or other informational practices about their product. Patagonia, a clothing manufacturer based in California, has created a website that allows consumers to track products and their use of resources from design through delivery. For instance, a Patagonia T-shirt is designed in California, produced in Los Angeles with cotton grown in Turkey, then shipped from a warehouse in Reno, Nevada, traveling about 7,840 miles, requiring about 4.7 kilowatt-hours to manufacture and generating 3.5 pounds of carbon dioxide (CO_2). Walmart has introduced a limited form of supply chain tracing for a new jewelry line launched in 2008 called "Love, Earth." The gold and silver used in the products in this line are traced back to a specific mine, an attempt to distinguish it from gold and silver that comes from unknown or more environmentally damaging sources. The British retailer Tesco in 2008 began to label some of its products with their carbon content as a means to encourage process improvements within the company as well as to inform consumers of the environmental impact of different products. Twenty products were included in the first round of this process, including light bulbs, detergent, orange juice, and potatoes, and Tesco reports that it has already reduced the energy use per kilogram of crisps (potato chips) by one-third.

The pharmaceutical company Novo Nordisk was recognized for adopting a driver training program for its sales fleet that was aimed at increasing energy-efficient driving. In six months, the program increased fuel efficiency from 19.0 to 19.5 miles per gallon, savings that, if applied to all corporate fleets in use, could prevent the creation of 1 million tons of CO_2. Improving driver training has the advantage of being immediately applicable to fleet operations without requiring major changes, such as switching to hybrid vehicles, and of increasing safety because many of the energy saving techniques are also safe driving habits (e.g., plan routes in advance, anticipate stops, and avoid speeding). Several different systems of driver training were also noted, including online training courses and products that monitor driving habits.

The Iowa Soybean Association, an organization representing 6,000 growers, has developed a network to share information among farmers working with similar soils in similar weather conditions. A major goal is to reduce the use of nitrogen fertilizer, which is expensive and pollutes waterways, to the least amount possible, which produces the best economic and environmental results. Although broad guidelines are available from agronomy experts regarding nitrogen use, more specific and useful information was gained by farmers conducting experiments in their own fields and by sharing information with farmers working in similar conditions. Many farmers involved in this project have reported that they have been able to reduce nitrogen use (sometimes as much as 30 percent) while maintaining the same profitability per acre.

Another agricultural innovation recognized by the EDF was the use of smart irrigation systems to reduce water use and runoff. Automatic irrigation systems are used on farms, real estate developments, and office parks (not a minor use; landscaping consumes about 58 percent of urban water): nationally, about 60 million such systems are in use. Automatic systems generally operate on timers so they turn on at regular intervals for a set length of time without regard for current weather conditions or recent amounts of rainfall. Commercial smart irrigations, such as those manufactured by PureSense and HydroPoint, gather information about weather and soil conditions and adjust irrigation accordingly. Advantages of a smart system include reduced operating costs (primarily because of lower water consumption) as well as deceased landscape runoff, and in some cases increased crop yields due to appropriate levels of irrigation.

The Bon Appetit Management Company developed a program called the Low Carbon Diet for its on-site restaurant services to over 400 corporations and colleges. The goal of this program is to reduce the amount of greenhouse gas emissions associated with its food service (globally, food production and transportation is believed to produce about a third of the world's greenhouse gases) while keeping customers happy. This was achieved through menu changes, use of different supply chains, and changing some operational practices. For instance, Bon Appetit set goals that include reducing the use of beef by 25 percent, obtaining all meat from North America, reducing the use of cheese by 10 percent, reducing food waste by 25 percent, composting whenever possible, and reducing packaging by 10 percent. In addition, Bon Appetit has a requirement that at least 20 percent of all foodstuffs are purchased within 150 miles of the kitchen where they are prepared.

Intel was recognized for providing economic incentives for reducing environmental impact. Part of every employee's compensation in 2008 included bonuses based partly on company-wide success in meeting three environmental goals: increasing energy efficiency of products, improving the company's reputation for environmental leadership, and the purchase of green power and completion of several renewable energy projects. (Intel was the largest purchaser of green power in the United States in 2008, according to the U.S. Environmental Protection Agency [EPA].) Currently, Intel is the only major corporation that includes environmental goals in computing compensation for all employees, and although the environmental components play only a minor role in computing the bonuses, Intel management indicates that including the environmental goals helps shift awareness toward long-term goals such as increased use of renewable energy.

Power management software produced by companies such as 1E and Verdiem, as well as free software from the EPA, allows companies, universities, and other organizations with larger numbers of computers to achieve significant energy savings from computer usage by allowing central IT departments to shut down computers remotely at night. Many users leave their computers on 24 hours a day so they do not have to wait for them to power up in the morning, and IT departments often instruct users to leave them on so maintenance can be performed and software installed after office hours. However, eliminating usage of the electricity required to run computers when they are not being used can result in considerable saving: for instance, the University of Wisconsin, Oshkosh, reports that it saves about $9,000 annually in electricity from shutting down computers at night. Verizon Wireless reports that it achieved a 24 percent reduction in power consumption and CO_2 emission by using shutdown software, saving about $1.3 million annually.

Key card systems, commonly used in European and Asian hotels, have been installed in many U.S. hotels, from large chains like Westin to single family–owned hotels. Key card technology is available from companies like DBS Lodging Technologies, and is based on a simple concept: lighting, heating, and cooling a hotel room is controlled by a key card that the occupant puts in a slot when in the room and removes when leaving the room. When the key card is not in the slot, electricity is automatically turned off, resulting in energy savings of between 25 and 45 percent, paying for the costs of installing the system within one to two years.

Zocalo Community Development, a real estate developer in Denver, Colorado, built Colorado's first LEED-certified condominium (condo) development, RiverClay. This development has many environmental features, from bike racks to Energy Star appliances, and includes rooftop solar photovoltaic panels, which condo owners have the option to buy. The panels are designed to supply 20 to 30 percent of a condo's electricity and cost about $11,000, but owners are allowed to pay for them over 30 years as part of their mortgage, and they are also eligible for utility rebates and federal tax credits. Energy generated by the panels that is not used by the owners is sold to the local utility company, a process known as net metering.

See Also: Design for Recycling; Green Nanotechnology; Information Technology; Intellectual Property Rights; Participatory Technology Development; Science and Technology Policy.

Further Readings

Bleischwitz, Raimund, Stefan Giljum, Michael Kuhndt, and Friedrich Schmidt-Bleek, et al. *Eco-Innovation: Putting the EU on the Path to a Resource and Energy Efficient Economy.* Wuppertal, Germany: Wuppertal Institute for Climate, Environment and Energy, 2009. http://www.wupperinst.org/uploads/tx_wibeitrag/ws38.pdf (Accessed September 2010).

Borins, Sandford. *Innovating With Integrity: How Local Heroes Are Transforming American Government.* Washington, DC: Georgetown University Press, 1998.

Christensen, Clayton M. *The Innovator's Dilemma: When New Technologies Cause Great Firms to Fall.* Boston, MA: Harvard Business School Press, 1997.

Eco Innovation. http://www.eco-innovation.net (Accessed September 2010).

Environmental Defense Fund. "Innovation Exchange." http://innovation.edf.org/home.cfm (Accessed September 2010).

Kenney, Martin. *Biotechnology: The University Industrial Complex.* New Haven, CT: Yale University Press, 1986.

Makower, Joel. "Green Innovation Becomes a Great Idea: The State of Green Business 2010." *GreenBiz* (February 5, 2010). http://www.greenbiz.com/news/2010/02/05/state-green-business-2010-green-innovation-becomes-great-idea (Accessed September 2010).

Organisation for Economic Co-operation and Development (OECD). "Eco-Innovation in Industry: Enabling Green Growth" (2009). http://browse.oecdbookshop.org/oecd/pdfs/browseit/9209061E.PDF (Accessed September 2010).

Saxenian, AnnaLee. *Regional Advantage: Culture and Competition in Silicon Valley and Route 128*. Cambridge, MA: Harvard University Press, 1994.

Schumpeter, Joseph A. *Capitalism, Socialism and Democracy*. London: Routledge, 2010 [1943].

World Business Council for Sustainable Development. "Eco-Patents Database." http://www.wbcsd.org/templates/TemplateWBCSD5/layout.asp?type=p&MenuId=MTU2MQ&doOpen=1&ClickMenu=LeftMenu (Accessed April 2011).

Sarah Boslaugh
Washington University in St. Louis

INTELLECTUAL PROPERTY RIGHTS

Intellectual property rights (IPR) can be defined, in essence, as an attempt to extend tangible concepts such as property ownership to intangibles such as movies, music, knowledge, and even genes. Intellectual property is traditionally divided into two branches: industrial property and copyright. The Convention Establishing the World Intellectual Property Organization (WIPO), concluded in Stockholm on July 14, 1967 (Article 2(viii)) provides that "intellectual property shall include rights relating to: literary, artistic and scientific works, performances of performing artists, phonograms and broadcasts, inventions in all fields of human endeavor, scientific discoveries, industrial designs, trademarks, service marks and commercial names and designations, protection against unfair competition, and all other rights resulting from intellectual activity in the industrial, scientific, literary or artistic fields." The areas mentioned as performances of performing artists, phonograms, and broadcasts are usually called "related rights," that is, rights related to copyright. The areas mentioned as inventions, industrial designs, trademarks, service marks, and commercial names and designations constitute the industrial property branch of intellectual property.

Several key conflicts over intellectual property rights have taken place in India, such as the challenge against the patenting of the healing properties of turmeric. The black turmeric pictured here is famous for its unique medicinal properties and is on the verge of extinction due to bio-piracy.

Source: Ramesh Raju/Wikimedia

With reference to green technology, there are three areas where IPR is important: (1) the patent rights of new technology that is environmentally friendly, (2) the ownership of the cultural products and traditional knowledge of indigenous people that is in the intellectual property commons, which means such knowledge is not exclusively owned, and (3) contests over traditional farmer crop varieties or know-how developed over generations.

The key organization for the enforcement of IPR is the World Trade Organization (WTO) and the key instrument for the enforcement of IPR is the Trade Related Aspects of Intellectual Property (TRIPS), an international agreement that allows countries to take up disputes at the WTO. The other major agreement is the Convention on Biological Diversity (CBD) signed by 192 nations with three main objectives: (1) the conservation of biological diversity, (2) the sustainable use of the components of biological diversity, and (3) the fair and equitable sharing of the benefits arising out of the utilization of genetic resources.

Most of the conflict over IPR revolves around concerns about cost. While developed countries insist that innovations must be protected by IPR, many in the developing world argue that this drives up the costs of products, whether software or drugs, to unaffordable levels. Opposition to IPR is either official (e.g., lobbying at the WTO by nongovernmental organizations [NGOs] that in 2001 were able to incorporate a declaration that reaffirmed the right of member countries to produce generic versions of patented drugs) or unofficial (e.g., the rampant levels of software piracy in the developing world).

With regard to green technology, the pattern is likely to be similar. The development of green technology is imperative because existing technologies contribute heavily to global warming. However, since innovation in green technology is centered in the developed West—and IPR for economic, legal, and cultural reasons is central to the Western world—patent law and IPR over green technology is a major stumbling block. At the Copenhagen summit on climate change in 2009, India and China demanded that they be given exemption from licensing fees for green technology. The two nations argued that new green technology be subject to "compulsory licensing," an exemption of the licensing fee granted by the WTO for the manufacture of generic drugs in emergency situations where patent-protected pharmaceuticals were seen as prohibitively expensive. The Thai government used the mechanism to allow local medicines factories to produce HIV drugs at a fraction of their cost. China and India contend that climate change is a similar emergency.

Moreover, since contemporary modes of patenting are piecemeal, meaning that different innovations are in the hands of separate companies, the licensing fee compounds. For example, several patents for fuel cells, wind energy, and carbon sequestration are scattered among several U.S. companies. Such complex patterns of ownership make it increasingly hard to synthesize different technologies to create new products. For instance, Chinese car manufacturers are unable to produce low-cost hybrid cars, crucial for the environment, since they would have to pay American companies licensing fees. On the other hand, U.S. corporations have aggressively demanded that patent law be protected, and in June 2009 were able to get the U.S. Congress to pass legislation that ensured that U.S. policy, "with respect to the United Nations Framework Convention on Climate Change, shall be to prevent any weakening of, and ensure robust compliance with and enforcement of, existing international legal requirements for the protection of intellectual property rights related to energy or environmental technologies."

However, the sheer size of the market in the developing world, especially in countries like China and India where old polluting technology is the norm, is likely to be tempting. The market for green technology in China is estimated to be $500 billion to $1 trillion annually. If the technology, because of intellectual property rights, is too expensive, much like software, it is likely to be reverse-engineered or copied. Therefore, just like in the case of software, it is likely that companies will be faced with the choice of offering their technology at expensive rates and not getting any share of the market or offering the technology at much lower rates in a bid to capture the market, as Microsoft did with its Windows operating system in China.

Innovations are also likely to come from the developing world. China, in 2008, spent more than the United States on research into green technology. The country also has technology transfer pacts with U.S. companies. In 2004, China paid about $5 billion for four nuclear reactors incorporating the latest technology from Pennsylvania-based Westinghouse Corp. In exchange, Westinghouse agreed to transfer the technology and know-how to Chinese state-owned firms, and has an agreement with Arizona-based First Solar to build the world's largest solar-panel farm—25 square miles in Mongolia that will provide power to 3 million households.

At its core, intellectual property is a system of permission-based restrictions. Those who "own" the property set the default limits for those who wish to use it, subject to certain public policy constraints such as fair use. Those in favor of intellectual property rights argue that unless creators and innovators are materially rewarded, they have no incentive to create. Those who oppose intellectual property rights, such as advocates of open-source software, argue that the desire to create is not dependent on material reward, and that knowledge, culture, and their products are the heritage of mankind.

This idealist argument has its own pitfalls. For example, given the fact that indigenous people are not aware of the intricacies of Western property law and that their traditional knowledge could be patented, thereby excluding them from benefiting from their knowledge, such as the famous case when the pharmaceutical giant Pfizer used the traditional knowledge of the Hoodia tribe in the Kalahari Desert, who knew that chewing the leaves of certain plants would suppress hunger, to create a weight-loss drug. Later, under pressure from a coalition of NGOs, Pfizer agreed to share part of the proceeds with the tribe. Incidents such as this have led to even proponents of the knowledge-is-free paradigm to advocate restricted access to traditional knowledge, seeking at least a limited protection for cultural knowledge. One possible way around the dilemma has been the adoption of a "sui generis" mode tailoring IPR paradigms to suit the unique needs of the nation, with the country emphasizing issues like biodiversity protection, community rights, and sustainable use. Several key conflicts over IPR have occurred in India; for example, the successful challenge of the Indian government to the patenting of the wound-healing properties of turmeric, a popular Indian spice.

New developments have changed the landscape of IPR. Globalization has meant that traditional knowledge, which had been of limited relevance, has suddenly become extremely important. This includes the key role of biotechnology in an emerging knowledge economy: the powerful position that the private sector has in the emerging "nano-cogno-bio-info" knowledge society. The consolidation and integration of pharmaceutical, chemical, industrial, and other sectors, symbolized by a wave of mergers and acquisitions driven by competitive pressures, prohibitively expensive biotechnology research and development (R&D), the potential for knowledge conglomerates and the super-research university—all underline this process. IPR practices also privilege broad patents that have the effect of driving competitors out of the market and, by deterring entry, increase consolidation. What makes traditional knowledge significant is that pharmaceutical research is increasingly based on indigenous knowledge, especially in identifying beneficial plants. All plant-based drugs in use are derived from fewer than 90 plant species. With more than 250,000 species of plants on Earth, the commercial potential is enormous.

With regard to green technology and the environment, one of the key points of contention is ownership of the traditional knowledge and culture of indigenous peoples. Some of the bitterest fights over IPR are over the right to "privatize" knowledge. Exclusive paradigms of knowledge protection, a heritage of Western intellectual property rights traditions, lay claim to the whole of knowledge and enclose it within a legal framework. On the other hand, environmental knowledge such as the beneficial properties of several plants, especially among indigenous peoples, has no clear ownership title. Instead, the knowledge is in the commons.

Key to any definition of intellectual property rights is the nature of property. Traditional property law applied to physical objects, if taken, would deprive the owner of his or her property; IPR tries to apply the same concept to intangibles that can be infinitely reproduced without depriving the original owner of his or her property, or to things in the commons. Thus, a music file can be digitally reproduced an infinite number of times without depriving the original creator of his or her property. Opponents of IPR argue that there is nothing that is completely new—each new innovation is based on inventions that came before.

See Also: Green Technology Investing; Innovation; Science and Technology Policy; University-Industrial Complex.

Further Readings

Areddy, James T. "Optimism Over China's Green Technology Market." *Wall Street Journal* (September 10, 2009).

Convention on Biological Diversity. http://www.cbd.int/convention (Accessed July 2010).

Drexler, Eric K. *Engines of Creation: Challenges and Choices of the Last Technological Revolution.* New York: Anchor Press/Doubleday, 1986.

Euractive Network. "China, India Push for 'Patent Free' Green Tech." http://www.euractiv.com/en/innovation/china-india-push-patent-free-green-tech/article-187567 (Accessed May 2009).

Farnsworth, N. R. "Screening Plants for New Medicines." In *Biodiversity*, E. O. Wilson and F. M. Peters, eds. New York: Academic Press, 1988.

Hamilton, Chris. "Biodiversity, Biopiracy and Benefits: What Allegations of Biopiracy Tell Us About Intellectual Property." *Developing World Bioethics*, 3 (2006).

Haunss, Sebastian and Kenneth C. Shadlen, eds. *Politics of Intellectual Property: Contestation Over the Ownership, Use, and Control of Knowledge and Information.* Cheltenham, UK: Edward Elgar, 2009.

Kansa, Eric C., Jason Schultz, and Ahrash N. Bissell. "Protecting Traditional Knowledge and Expanding Access to Scientific Data: Juxtaposing Intellectual Property Agendas via a 'Some Rights Reserved' Model." *International Journal of Cultural Property*, 12 (2005).

Maskus, Keith E. *Intellectual Property Rights in the Global Economy.* Washington, DC: Institute for International Economics, 2000.

Shellenberger, Michael and Ted Nordhaus. "The Revolution Will Not Be Patented." *Slate* (April 21, 2010).

Sabil Francis
University of Leipzig, Germany,
École Normale Supérieure, Paris

INTERMEDIATE TECHNOLOGY

Intermediate technology refers to tools, basic machines, and engineering systems that poor farmers and other rural people can make or readily buy to better support themselves and improve their well-being. It is a step between traditional tools and systems that are inefficient and advanced technology that is both highly expensive and impractical for the needs of impoverished areas.

Proponents of intermediate technology say that items easily adaptable for rural needs can lead to greater productivity while minimizing the dislocation often associated with technological changes. Much intermediate technology can be built and serviced using locally available materials and knowledge. It generally is seen as more harmonious with the environment and with traditional ways of life, designed to focus on people rather than making them servants of machines.

Intermediate technology tends to favor labor-intensive solutions over capital-intensive ones, although labor-saving devices are also used where this does not mean high capital or maintenance cost. In practice, appropriate technology is often described as using the simplest level of technology that can effectively achieve the intended purpose in a particular location. Its proponents generally demur from the belief that technological development is inherently synonymous with progress.

Although generally associated with relatively basic devices, often made out of old machine parts or cloth or wood, it can also involve advanced technology, such as energy-efficient light bulbs or small adsorption refrigerators, that can provide a benefit to poor, rural populations. It sometimes combines cutting-edge research with simple materials. For example, medical research into the spread of cholera has led to the use of cloth filters, sometimes made out of old saris or other articles of clothing, to collect water in a way that substantially reduces pathogens in poor villages where disinfectants and fuel for boiling are not readily available. At the other end of the spectrum, high-efficiency, white light-emitting diode (LED) lights are used in remote areas of Nepal, replacing kerosene lamps or wood fires that emitted pollutants and posed a fire risk.

The term *intermediate technology* is sometimes used interchangeably with "appropriate technology." However, appropriate technology more often refers to the use of environmentally friendly and energy-saving approaches in industrialized countries, as opposed to the application of useful devices in rural, less-developed areas.

Ernst Friedrich Schumacher, a British economist and statistician who critiqued Western economies, is credited with coining the term *intermediate technology*. His ideas became widely distributed through his influential 1973 book *Small Is Beautiful.* Schumacher later created the Intermediate Technology Development Group, which assisted rural communities to develop simple and practical technologies. Schumacher's emphasis on sustainability came at a time of increasing ecological concerns and energy shortages, and he became highly regarded within the environmental movement.

The emphasis on intermediate technology can also be traced to Mohandas Gandhi, who advocated small, local, mostly village-based technology to help India's villages become self-reliant as part of their struggle against the British and wealthy Indians. Gandhi favored simple devices such as the bicycle and the sewing machine. He believed the powers of technology should be produced and used artfully, benefiting local economies. He became known for championing the spinning wheel, or *charka*, which his followers used to produce cloth locally to cause the British monopoly on textiles to collapse. Concerned about systems that favored production over workers, he said, "It is better for a machine to be idle than a man to be idle."

When Schumacher and his colleagues created the Intermediate Technology Development Group, they focused on gathering information about tools and methods suitable for poorer rural areas and small towns in developing nations. An example of their work is a cassava grinder in Nigeria. Cassava, an important food source in Nigeria, has to be ground and dried before it is cooked. Residents who lacked access to good mechanical grinders often hurt themselves or spent many hours at grinding with such tools as tin cans with holes punched in them. The intermediate technology solution in this case was a cassava grinder built out of old bicycle parts and hacksaw blades. Highly efficient, it enabled people to grind about 16 pounds of cassava root in five minutes. The group also helped develop low-tech agricultural tools, building methods, bicycle-drawn ambulances, and even egg trays.

Although such simple technologies are generally preferred by intermediate technology proponents, they also see a role for carefully designed modern industrial plants. An example would be the construction of a costly factory to process sugarcane and maize stalks into fiberboard. Even though it may employ relatively few people, it could benefit many farmers whose stalks would otherwise have gone to waste as well as villagers who could use the fiberboard for housing and clothing.

Although intermediate technology tools developed in conjunction with local residents are generally popular, there have been cases in which they failed to win over villagers. For example, engineers with a nongovernmental organization (NGO), Compatible Technology International, attempted to improve the lives of Guatemalan women who labored many hours to hand-shell corn. The engineers produced a corn sheller that consisted of a piece of wood with a hole in the middle. By pushing a cob of corn through the hole, the women could shave off the kernels far more quickly. But the women preferred to continue hand-shelling corn, explaining to the engineers that they enjoyed the time together. The lesson for intermediate technology developers: sometimes time-saving devices are not welcome.

In many cases, however, intermediate technology innovations have made a great difference in developing countries. Examples of intermediate technologies follow.

Building Materials

Locally available materials, such as adobe, compressed earth blocks, and straw bales can be used to build durable structures. Those who live near volcanoes can mix volcanic ash with lime to make a type of cement that does not need heating. In some cases, products from a nearby city may provide building materials. New technologies can provide simple and inexpensive options as well. For example, a product manufactured in the United Kingdom known as Concrete Canvas provides a sturdy alternative to tent cities. Builders pump up the air bladder to provide temporary scaffolding, lay out the canvas on top of the air bladder, and pour water on the canvas. After the canvas dries, the air bladder can be removed, leaving a basic structure as large as about 600 square feet that can last for several years. The structure is waterproof and fireproof.

Energy Savings

Biomass, sometimes made from bio oil, waste organic matter, vegetable oil, or even feces, can be used for power. Human energy can also be tapped in innovative ways. A nongovernmental organization in Guatemala, Maya Pedal, has developed a series of machines that are powered by bicycles. For example, engineers fitted modified bikes to hand-powered grinding mills and corn threshers, resulting in devices that can mill three pounds of any type of grain per minute.

The sun is also a convenient energy source, especially in remote areas that receive plentiful sunlight. Portable outdoor stoves known as solar cookers are increasingly popular in developing countries. They often use a combination of devices to harness solar heat, such as a mirror or reflective metal to concentrate heat into a small cooking area and a plastic bag or glass cover to trap the heat inside. The sun can also be used to power specially designed LED bulbs. Such devices can provide several hours of light when the battery is fully charged, reducing the need to burn scarce supplies of wood.

Increasing energy efficiency can be as important as finding sources of energy. Rocket stoves and similar wood stoves, for example, burn wood efficiently through a combination of controlled use of fuel, complete combustion, and carefully designed ventilation. Such stoves have been used for cooking in a number of sites in developing countries, including Rwandan refugee camps; they are also used for space heating and providing hot water.

Another innovative device for remote areas that lack electricity is the Pot-in-Pot refrigerator, used to preserve food. It consists of a smaller clay pot inside a larger one, with regularly moistened sand separating the two. The inner pot is cooled by evaporation. Vegetables and fruits can stay fresh as long as several weeks.

Water Quality

With more than 1 billion people lacking access to proper drinking water, the elimination of waterborne diseases is a major focus of intermediate technology. Engineers have pursued various methods to purify water, focusing on the use of locally available materials. For example, clay, diatomaceous earth, or sand can be used to filter water. Chemical treatments may rely on aluminum salts or crushed seeds from certain plants. Water can also be irradiated with ultraviolet light, including light from the sun. One of the most successful, and simplest, water purification devices, known as the LifeStraw, can be worn around the neck. Water is sucked up a straw and through a filter that catches almost all pathogens.

Gathering and transporting water can be as challenging as purifying it. Devices for collecting water in arid areas include large pieces of vertically arranged canvas for capturing water droplets from condensed fog, structures known as air wells that promote condensation from air, and plastic sheets that gather water from dew or frost.

One of the most successful intermediate-technology devices for transporting water is the Hippo Roller. Designed to replace the five-gallon jugs that women would carry on their heads to carry water to their homes from distant rivers and lakes, the Hippo Roller consists of a 24-gallon barrel attached to a large handle. The barrel can be pushed along the ground in the style of a rotating steamroller.

Agricultural Tools

Engineers have developed a plethora of simple devices for agricultural purposes. Examples range from knapsack crop sprayers to pedal-powered maize shellers. The Universal Nut Sheller, a simple hand-powered tool that relies on a crank, is capable of shelling more than 100 pounds of certain types of peanuts per hour.

Public Health

Rather than try to supply outlying regions with professionally trained doctors, some NGOs favor training villagers to treat many ailments. The trained villagers treat a majority of health problems, with more severe cases going to small, local hospitals or, less commonly, to larger and more expensive urban media centers. Herbal medicines are sometimes used in lieu of synthetic drugs in cases where their efficacy has been demonstrated. Medical products associated with intermediate technology include the phase-change incubator, which can test cultures at a lower cost than a laboratory or portable incubator, and the Jaipur leg, a rubber-based prosthetic that was developed for victims of land mines. Engineers have also produced a number of wheelchair designs, often made out of recycled parts, that are far less expensive than newer wheelchairs in industrialized countries.

Technological Devices

Companies have designed a number of low-cost computers, typically dust-resistant and durable, for use in developing countries. In some countries, children have been given laptop computers, some as small as textbooks, with screens easily readable in bright sunlight for those who go to school outdoors. Those who lack Internet access may be able to use educational CDs and DVDs.

Other devices include wind-up radios for areas without electricity and village phones—cell phones that are sold to residents in remote areas who are trained in using them as a payphone for their villages. Cooperative computer networks provide free or inexpensive Web and e-mail services, while services such as Loband strip out high-resolution images from webpages and render them as simple text, thereby making them accessible in areas with slow download speeds.

The field of intermediate technology is constantly evolving. Numerous NGOs and other groups continue to create new devices and strategies to help rural residents in developing countries, providing an alternative to both traditional tools and larger-scale Western technological approaches.

See Also: Appropriate Technology; Passive Solar; Rainwater Harvesting Systems; Science and Technology Policy; Technology and Social Change.

Further Readings

Listverse. "Ten Cases of Appropriate Technology." http://listverse.com/2010/06/12/10-cases-of-appropriate-technology (Accessed September 2010).

Manas Journal. "What Is Intermediate Technology?" http://www.manasjournal.org/pdf_library/VolumeXXVIII_1975/XXVIII-13.pdf (Accessed September 2010).

Ward, Olivia. "$20 Billion Promised at UN for Maternal, Child Health." *The Toronto Star*. http://www.thestar.com/news/world/article/865172--40-billion-promised-at-un-for-maternal-child-health?bn=1 (Accessed September 2010).

David Hosansky
Independent Scholar

INTERNATIONAL ORGANIZATION FOR STANDARDIZATION (ISO)

The International Organization for Standardization (ISO) is the leader in developing voluntary standards for for-profit, not-for-profit, and nongovernmental organizations (NGOs). Based in Geneva, Switzerland, at the Central Secretariat, this international organization boasts a network of 163 countries, both developed and developing nations. Two well-known certification standards—ISO 9000:2000 and ISO 14000:2004—have been successfully integrated as global standards for management, in terms of quality and environmental sustainability, respectively.

According to the 2009 report "Selection and Use of the ISO 9000 Family of Standards," the core process links management, resources, and stakeholders in a continuous feedback loop at every step of the production process. ISO 9001:2008 is perhaps the most well-known and validated "approach to managing the organization's processes so that they consistently turn out product that satisfies customers' expectations."

ISO 14000 is a comprehensive and systematic resource for organizations to manage, assess, and evaluate their environmental impact. "Environmental Management: The ISO 14000 Family of International Standards" (2009) highlights existing and emerging standards for the following:

- Life cycles of services and products
- Product labels and environmental claims
- Evaluation of performance
- Cost production and control of resources and materials
- Communicating with external constituencies and research findings
- Greenhouse gas emissions and carbon footprinting

Within the past decade, however, a new conversation emerged regarding guidelines for social responsibility (SR): "one that looks beyond product safety and reliability to consider the social impact of production" (www.consumersinternational.org). The end result of this dialogue was the creation of a working group (WG) responsible for developing a new standard for corporate social responsibility—ISO 26000.

The WG, chaired by the Associação Brasileira de Normas Técnicas (ABNT) and the Swedish Standards Institute (SIS), was charged with listening to the concerns of diverse categories of stakeholders in order to draft the language and parameters of the new standard. According to a report issued by the Central Secretariat, task groups (TG) were established to complete different parts of the drafting process. TG1 coordinated participation of all stakeholders (consumers, government, industry, labor, etc.). TG2 was concerned with all communication activities: transparency, dissemination, and flow. TG3 provided internal guidance for the ISO and Technical Management Board (TMB). The last three task groups (4, 5, and 6) reviewed and edited drafts and provided feedback on the core principles and implementation of SR.

First Plenary Meeting: Salvador de Bahia, Brazil, March 2005

According to a news release on ISO's website, the first meeting held in Brazil was to focus on determining the following:

- Structure and terms of the reference of the working group and SR guideline standard
- A project plan
- Appointing conveners and secretariats for its subgroups

In this meeting, Task Group 1 began its process of connecting with various stakeholders, and the next meeting was scheduled.

Second Plenary Meeting: Bangkok, Thailand, September 2005

The focus of this meeting was to lay the foundation or design specifications for ISO 26000 on Corporate Social Responsibility. The following sections were formalized verbatim into the working draft: introduction, scope, normative references, terms and definitions, the SR context in which all organizations operate, SR principles relevant to organizations, guidance on core SR subjects/issues, guidance for organizations on implementing SR, guidance annexes, and a bibliography.

Other areas of concern surfaced in terms of how to communicate this information online and to broaden participation of all stakeholders.

Third Plenary Meeting: Lisbon, Portugal, May 2006

In this meeting, participating members reviewed feedback on the first draft of the standards and continued assessing the methods and means of communicating with stakeholders. Prior to this meeting, a brochure titled "Participating in the Future of International Standard ISO 26000 on Social Responsibility" was published in April 2006, offering a summary of the process so far.

Fourth Plenary Meeting: Sydney, Australia, February 2007

This meeting was perhaps the most important as participation by experts in developing countries increased and several key decisions and agreements were reached via consensus. At this meeting, Task Groups reported progress toward the following goals and objectives:

- TG1 (stakeholder engagement) proposed a trust fund—approved by ISO Council—to increase participation of stakeholders via sponsorships, allocation of funds, and donations.
- TG2 (communication) secured two outcomes: continued improvement of communication activities and a draft of framework for the ISO Communication Action Plan Model.

- TG3 operationalized internal procedures and delineated the roles of researchers and experts.
- TG4 achieved its task of revising and editing the design specifications, specifically introduction, scope, principles, definition of SR, and contexts.
- TG5 reviewed the second draft, and identified seven core thematic issues of social responsibility, verbatim: (1) environment, (2) human rights, (3) labor practices, (4) organizational governance, (5) fair operating practices, (6) consumer concerns, and (7) community involvement and society development.

Fifth Plenary Meeting: Vienna, Austria, November 2007

A record number of experts, stakeholder groups, and members from developing countries actively engaged in the process during this meeting. Based on comments from the third draft of the standard, a new task force—Integrated Drafting Task Force (IDTF)—was created to take over the writing of the fourth draft. More awareness was raised about ISO 26000, and more participants were able to attend via the ISO Trust Fund approved in the last meeting.

Sixth Plenary Meeting: Santiago, Chile, September 2008

The process of creating a standard for ISO 26000 reached a milestone in this meeting. The draft of the standard was approved for circulation as a Committee Draft. The Integrated Drafting Task Force (IDTF), in collaboration with International Labour Organization and the United Nations Global Compact, identified the following key topics that needed to be addressed at this meeting, verbatim:

- International norms of behavior
- Nature of reference to social responsibility initiatives, and government
- Sphere of influence
- Relevance, significance, and prioritization

Seventh Plenary Meeting: Quebec, Canada, May 2009

The Committee Draft of the circulated document was approved by multiple stakeholders (300 experts, 60 countries, 20 liaison organizations) via consensus and was upgraded in status to Draft International Standard. The thematic issues identified by TG5 were still being discussed, and members were continuing to work toward resolving them.

Eighth Plenary Meeting: Copenhagen, Denmark, May 2010

ISO released a statement on its website announcing that the Final Draft International Standard (FDIS) will be made available after it has been voted on by member countries.

Implications for Green Technology

The new standard for ISO 26000 does not specifically address technology. However, technology is represented as a sub-concern within the core issues of SR; in particular, the protection of consumers' data and privacy was discussed in a draft of a chapter on consumer issues. For more specific information, consult the ISO's website about its ISO/IEC 27001:2005 standard on Information Technology.

Other aspects of technology include transfer across borders and nanotechnologies (Working Draft Text of the core issue of environment); technology as an issue under Community Involvement and Societal Development; and technology as a barrier to participation in the creation of ISO 26000 and other business arenas. More dialogue about the impact and implications of technology will continue to unfold as the new standard is adopted and implemented, specifically in reference to "supply chain management standards to reduce the risk of terrorism, piracy and fraud." This series or family of standards is applicable during any stage of the supply chain process, and offers specifications, best practices, audit and certification requirements, and implementation guidelines for keeping people and cargo safe across borders.

See Also: Green Metrics; Science and Technology Policy; Technology and Social Change.

Further Readings

Consumers International. "Corporate Social Responsibility and Standards" (August 2009). http://www.consumersinternational.org (Accessed July 2010).

"Governance as a Form of Social Responsibility." *Journal for Quality & Participation* (October 2009). http://www.asq.org/pub/jqp/past/2009/october (Accessed April 2011).

International Organization for Standardization (ISO). "Chapter 6 7 Consumer Issues Draft 2" (April 2007). http://isotc.iso.org/livelink/livelink/fetch/8929321/8929339/8929348/393583 7/4591396/4591399/6308820/1st_draft_on_Chapter_6_7_consumer_issues.pdf?nodeid= 6309358&vernum=-2 (Accessed July 2010).

International Organization for Standardization (ISO). "ISO/IEC 27001:2005—Information Technology." http://www.iso.org/iso/iso_catalogue/catalogue_tc/catalogue_detail.htm? csnumber=42103 (Accessed July 2010).

International Organization for Standardization (ISO). "Participating in the Future of International Standard ISO 26000 on Social Responsibility" (April 2006). http://www.iso .org/iso/iso26000_2006-en.pdf (Accessed July 2010).

Perera, Oshani. "How Material Is ISO 26000 Social Responsibility to Small and Medium-Sized Enterprises (SMEs)?" International Institute for Sustainable Development. http:// www.iisd.org/pdf/2008/how_material_iso_26000.pdf (Accessed April 2011).

Schwartz, Birgitta and Karina Tilling. "'ISO-Lating' Corporate Social Responsibility in the Organizational Context: A Dissenting Interpretation of ISO 26000." *Corporate Social Responsibility and Environment Management*, 16 (2009).

Cory Lynn Young
Ithaca College

LABOR PROCESS

Labor process theory is a Marxist theory about the organization of work in capitalist societies and specifically critiques the scientific management of Fredcrick Taylor, which revolutionized factory and corporate organization in the early 20th century. Labor process theory draws principally on Marx and on American economist Harry Braverman, who wrote primarily in the 1960s and 1970s. Also important to the theory is the notion of two kinds of subsumption of labor. In formal subsumption, existing forms of labor production are assimilated, such as when a company purchases the produce grown by a traditional farmer. In real subsumption, methods and processes of production are transformed by capital, such as in the modern agriculture industry's giant greenhouses or factory-like chicken farms.

Marx had, of course, been interested in the socially organized process by which labor is put to work in order to meet human needs and the way it varied according to different modes of production. Under capitalism, the relationship between the workers and the owners of the means of production (the capitalists) impacts the management of the labor process. Braverman reexamined Marx's ideas in light of the developments in capitalism in the 20th century, particularly "Taylorism," the scientific management style first popularized by mechanical engineer Frederick Taylor in his books *Shop Management* (1903) and *The Principles of Scientific Management* (1911).

Just as modern factories had been enabled by the development of standardized interchangeable parts and the assembly line, Taylor wanted to increase the productivity and efficiency of factories by standardizing processes. This meant not only standardizing which two pieces connected together in building a widget, but the exact physical steps a worker would take to connect those pieces, and the steps for bringing the pieces to the worker and moving them along down the line. Time and motion studies helped to analyze workers' motions and to streamline them. Of course, being able to control a worker's actual motions—as well as his or her breaks and time management—required a high level of control at the management level and a higher ratio of middle management to workers than had ever been seen before. Though the modern workplace is not considered Taylorist as such, many of the important ideas of scientific management and its emphases on efficiency, time management, and standardization of practices have become absorbed by modern-day work culture to such a degree that it can be difficult to remember that these ideas ever needed introduction. Taylorism was responsible for increasing social tensions between the blue-collar and white-collar classes, especially the resentment toward middle management workers who neither "did the actual work" as laborers did nor provided ideas, design, or starting capital as upper management did, but rather acted as babysitters.

It was this that Braverman saw as one of the most important changes to work culture since Marx's time, and much of his writing was devoted to the "degradation of work" caused by strict management control over the labor process. Braverman was writing in the late 1960s and early 1970s, at a time when U.S. dominance in the global market and at the forefront of technological achievement was no longer a secure thing, when countries like West Germany and Japan were making serious strides in their own industrial and technological development, and when inflation and oil crises raised the question of just how the 20th century was going to work out for the U.S. economy. In Braverman's view, "skilled" work had declined sharply since the introduction of scientific management. As management increases control over the workforce, increasing productivity, profits, and political power over the working class, a separation occurs between conception and execution. Managers appropriate innovation, planning, and design while workers are relegated to the task of simply performing routinized, preprogrammed tasks.

The process of transforming a work task into something routinized and preprogrammed like Braverman described and Taylor prescribed is called "de-skilling" and has been the focus of conversation among many social scientists since the 1974 publication of Braverman's *Labor and Monopoly Capital*. Complex labor processes are broken down into constituent parts of small, simple, unskilled tasks that can be performed without innovation or understanding, without qualitative variation, and without the worker needing to know why the task is being performed. Skilled craftspersons are displaced by unskilled task-performers, with the advantage to management that the job requires no training, and the disadvantage to laborers that they acquire no skill in the performance of their job—after years of experience, they are neither better nor worse than they were when they started, and their value in the labor market is no different. This keeps wages down, makes employment less secure, and reduces the options available to workers who become unemployed (which also makes them more dependent on their current employer). Furthermore, people whose work requires no skill—requires nothing significant of them—become alienated from their work and disinterested in the fruits of their labor.

Of course, this is not true of the entire labor market, simply because Braverman is principally talking about the manufacturing sector, which has shrunk in the highly industrialized world as the service sector has grown. But Taylorization of a sort is certainly visible in the service sector world, as customer service representatives and technical support operators work from scripts handed down to them by middle management describing the various scenarios they are likely to encounter, for instance. Technology has long since transformed the job of cashiering, turning the cash register from a tool to assist in the work into a machine that does most of the work itself, in conjunction with bar-code scanners, coupon printers, and receipts that say "Have a nice day." Even occupations we think of as highly skilled such as psychological counseling and legal representation have been broken down into constituent tasks in order to reduce costs and delegate more of the labor to lesser-skilled workers. The tasks of one-on-one therapy and psychiatric prescription writing, for instance, are often divided among two or more workers, so that the most expensive worker (the psychiatrist) can divide his or her labors among the greatest number of customers. And more and more companies are outsourcing legal work to other countries, where low-paid legal researchers and paralegals, with little to no training outside the workplace, can prepare legal documentation that is then reviewed and approved by a practicing attorney who is thus able to take on a caseload 10, 20, or 30 times greater than he or she could handle in a traditional law firm environment.

Braverman is primarily discussing one approach to management and has been criticized for applying it universally—though as his supporters and successors point out, that does not make him wrong in the cases where that management style is in place. More significantly, though, he glosses over the possibility and effects of worker opposition to deskilling and strict management control, especially through labor unions (the primary force ensuring that the laborer with 20 years of experience is valued more highly than someone without). But the fact that his work continues to be debated and addressed decades later demonstrates its influence, particularly as the world has shifted away from the manufacturing sector that initially concerned him.

See Also: Democratic Rationalization; Marxism and Technology; Technological Determinism.

Further Readings

Braverman, Harry. *Labor and Monopoly Capital: The Degradation of Work in the Twentieth Century*. New York: Free Press, 1974.

Kaplan, David M., ed. *Readings in the Philosophy of Technology*. Lanham, MD: Rowman & Littlefield, 2004.

Postone, Moishe. *Time, Labor, and Social Domination: A Reinterpretation of Marx's Critical Theory*. Cambridge, UK: Cambridge University Press, 1993.

Scharff, Robert C. and Val Dusek, eds. *Philosophy of Technology*. Hoboken, NJ: Wiley-Blackwell, 2003.

Veak, Tyler J., ed. *Democratizing Technology: Andrew Feenberg's Critical Theory of Technology*. Albany: State University of New York Press, 2006.

Bill Kte'pi
Independent Scholar

LEED Standards

Buildings are large contributors to greenhouse gas (GHG) emissions and global warming, in addition to being major consumers of nonrenewable resources. There is concern that rapidly expanding urbanization worldwide will accelerate the pace of global warming and result in destructive climate change. In response, one school of thought promotes the way to passive methods of building and living by dematerializing the systems and services that are energy and resource intensive, and by adopting a simpler approach to the way people live. The other choice, mostly preferred by the building industry, is to change the way we build our towns and cities by making the built environment "green." In order to measure the effectiveness of this attitude shift, we need to define "green buildings" and measure their output in terms of carbon savings and prevention of resource depletion.

U.S. National Aeronautics and Space Administration's (NASA's) Propellants North Administrative and Maintenance Facility building team is hoping for platinum, LEED's highest rating.

Source: Jim Grossmann/U.S. National Aeronautics and Space Administration

In response to this requirement, the U.S. Green Building Council (USGBC) has developed Leadership in Energy and Environmental Design (LEED) Standards. The LEED Green Building rating standards are a market-driven framework for evaluating the environmental performance of buildings. These ratings are beginning to influence the asset and rental values of buildings, and in addition, confer green credentials that are valuable for the professional reputation of the owners (landlords, developers, and owner-occupiers), professionals (architects, engineers, building services engineers, contractors), and operators (facility managers, property managers, vendors) of the buildings. This means that market opportunities are already in and will henceforth become more and more conducive for technologies and policies that promote efficient use of energy and resources in buildings.

LEED Rating Schemes

The LEED rating schemes address specific building typologies and project scopes. LEED schemes for New Construction (NC), Core & Shell (CS), Schools, Neighborhood Development (ND), Retail, Healthcare, Homes, and Commercial Interiors (CI) generally address new project developments or major renovations. LEED for Existing Buildings (EB)—Operations and Maintenance (O+M) specifically looks at the ongoing environmental performance of existing buildings and requires recertification up to every five years. The assessment process is generally carried out across five categories: sustainable sites, water efficiency, energy and atmosphere, materials and resources, and indoor environmental quality, thus interrogating the environmental credentials of the building or project across a range of impact areas, and confers Certified, Silver, Gold, or Platinum rating to it.

Practicalities

It is important to remember that the LEED series of sustainability rating standards are frameworks that enable the professionals to design and operators to deliver postoccupancy performance of their green buildings, but LEED by itself cannot deliver the desired results. LEED is a tool that can be used at every step of the design, construction, and operations processes to deliver green objectives. For example, to ensure that a green building will perform as intended for many years after construction, the designers can use LEED-NC to specify the standards to which the building is conceptualized, but also specify the standards in LEED-EB to which it is expected to perform in its lifetime. The continuity from designed to actual performance by the successful transition of "design intelligence" from the professionals to the operators and occupiers of the building through commissioning is a prerequisite in LEED-NC, which can be enhanced by the creation of an ongoing program for continual maintenance, monitoring, and improvement in the green-use strategies through the LEED-EB scheme. Generally, the new LEED 2009 schemes now require an undertaking from the building owners/occupiers to share whole building energy and water usage data with the USGBC for at least five years after getting certified. This is a minimum project requirement for any project's eligibility for LEED certification from 2009. LEED frameworks are thus being evolved to spearhead operations and maintenance thought processes to begin at concept design by allowing the design process to interrogate maintenance regimes and vice versa, in order to eliminate ineffective but well-intended green design features or inadequate operations and maintenance procedures.

Some studies have been carried out by the USGBC to evaluate whether new buildings certified under LEED-NC are performing as intended in their operations phase. The results show that the best-performing buildings are those in which LEED principles were integrated from the early project design stage. The study also found that there is a shift between the modeled and actual energy performance in some buildings that can indicate inaccurate design energy modeling, lack of performance monitoring, lack of maintenance and improvement to the building, or ineffective transition of information about design features to the maintenance and operations team. This underlines the fact that LEED is a tool that will perform according to the expertise of the professionals using it, and focuses on the importance of the commitment of the whole team from design to operations to make it work as intended.

Worldwide Applicability

LEED ratings have been formulated primarily for construction projects in the United States. In order to cater to projects in diverse cultures with different environmental priorities, the LEED schemes allow their implementation to existing local standards or the referenced U.S. standards, whichever are more stringent. Some countries like India and Canada have adopted the LEED standards and modified them to suit the conditions and regulations prevalent in their own regions. There has also been a move to adopt the U.S.-based American Society of Heating, Refrigeration, and Air-Conditioning Engineers (ASHRAE) methodology for simulating energy performance of building projects by some country-specific schemes (other than LEED) instead of their whole-scale replacement by LEED. This move toward having more of a common ground between different country-specific schemes will ultimately make it easier to compare some aspects of building performance like GHG emissions or energy consumption per unit area. Companies that operate globally and that would like to demonstrate their commitment to green operations will find these comparisons useful when selecting office premises to rent or calculating their operations costs in situations different to their home operations.

Best Practice

Usually, design and construction schedules are tightly aligned with little or no scope for accommodating deviations to any of these timetables. It is obvious that the green design concept cannot be an afterthought, but must be one of the defining requirements of the project. The targets that may be set at the concept design stage may be influenced by several factors like occupancy patterns, functional requirements, procurement strategies, construction phase practicalities, costs, and aesthetics, among others. In addition, a decision for the building to be designed to zero carbon emission standards throughout its life will imply that the design must invest heavily in renewable technologies, energy management, intelligent building fabric and services, shading devices, green roofs, and other such strategies that will help achieve this target. This decision will also bear directly upon factors like up-front and continual costs, procurement strategies, aesthetics in some cases, and construction and operational practicalities, to name a few. The earlier the decision to build to zero carbon emission standards is made, the less disruptive it will be for the design decisions already in place. Again, suitability of green design strategies may differ from project to project, and others may find it beneficial to focus on sustainable procurement, ventilation, and indoor environment or water-saving technologies and still achieve the desired LEED certification level with less investment in zero-carbon technologies.

The flexibility afforded by LEED to the design team to choose the best sustainability strategies may draw comments about the preference to choose the most easily achievable LEED points for a building design, whereas the more difficult or expensive to achieve categories may be ignored. The pointed criticism is that indoor air quality (IAQ) credit points are generally not prioritized, resulting in lesser improvement in living and working environments in LEED buildings as compared to non-LEED buildings. This may be because the technologies associated with the points in the IAQ credits may not be as accessible or widely known as those applications in the energy and atmosphere category, which has been the main focus of sustainability efforts in recent years due to the emphasis on achieving low- or zero-carbon buildings. Therefore, it is advisable to obtain advice from suitably qualified and experienced LEED accredited professionals as early as possible, and for such a professional to become part of the design team from the earliest stage will ensure that waste-eliminating decisions are made and optimum green design strategies that help achieve a balanced outlook across all LEED categories are selected.

The economic and time costs of incorporating green design to LEED standards cannot be underestimated. A realistic assessment about what is achievable, given project constraints, must be made at an early stage. The mandatory prerequisites for some credits may be time bound, for example, an environmental site assessment (ESA) before construction commences is a mandatory prerequisite for a school project, and if such a project is to begin construction without conducting an ESA, this prerequisite may be unachievable and hence the project ineligible for LEED certification. Then the trade-off between costs of delaying the project timetable so as to accommodate the requirements of this prerequisite and the value of the LEED certification to the project will need to be evaluated. Similarly, the Energy and Atmosphere section can deliver almost a quarter of the required points for a Platinum rating in most schemes. However, the costs of procuring and installing a highly efficient heating and ventilation system and its design and maintenance services will need to be incorporated into the project budgets. For this, value engineering and life-cycle assessments (LCAs) should become standard processes in a project's design phase. LCAs help to evaluate the environmental and economic impact of buildings or their components from cradle to grave, bringing not only direct financial benefits, but also indirect benefits from preventing pollution, thus reducing the need for costly remedial measures. This is seen as the new carbon currency—designs are set to be value engineered not only for direct economic reasons but also for the costs of remedying the effects of carbon emissions to air and other environmental pollution.

Triple Bottom Line

The triple bottom line is the financial, environmental, and social effects of policies and actions that determine the viability of sustainable buildings. The triple bottom line in the construction sector will need to show environmental, economic, and social benefits of the attitude shift toward green buildings. On a larger environmental scale, resource depletion will slow down, forests will be managed and virgin rainforests preserved for future generations, impact of hazardous materials in landfill will be understood, and steps will be taken to stop their use in buildings. It is also important to show that user comfort and health are at the top of the social criteria driving the green building movement and that more attention is being paid to the quality of the indoor environment, resulting in healthier spaces to live and work. For the market, the adoption of LEED standards will see a rise in businesses supplying green technologies to the construction industry. Expertise will be required from building physicists, simulation experts, architects, services engineers, and other professionals in the green building industry, and technology advancements in renewable energy products and materials will start new manufacturing activity of these products. With green design capability gaining more and more value in terms of professional credentials, all participants in the design and construction of a green building will aspire to get known for being experts in this field. The payback from designing and building green will thus not only be gained directly from markets but also from inculcating the practice of sustainable building and living in future generations to come.

See Also: Green Building Materials; Green Markets; Microgeneration; Sustainable Design; Zero-Energy Building.

Further Readings

Allen, Edward. *How Buildings Work: The Natural Order of Architecture*. New York: Oxford University Press, 1995.

Behling, Sophia and Stefan Behling. *Sol Power: The Evolution of Solar Architecture*. Munich: Prestel, 1996.

Fowler, Kim M. and Emily M. Rauch. "Assessing Green Building Performance: A Post Occupancy Evaluation of 12 GSA Buildings—Case Study." Pacific Northwest National Laboratory and U.S. GSA (2008). http://www.gsa.gov/graphics/pbs/GSA Assessing_Green_ Full_Report.pdf (Accessed April 2011).

Trusty, Wayne. "Integrating LCA Into LEED—Working Group A (Goal and Scope) Interim Report." Athena Institute and USGBC (2006). http://www.usgbc.org/ShowFile.aspx? DocumentID=2241 (Accessed July 2010).

Turner, Cathy and Mark Frankel. "Energy Performance of LEED for New Construction Buildings." New Buildings Institute and USGBC (2008). http://www.usgbc.org/ShowFile .aspx?DocumentID=3930 (Accessed July 2010).

Swati Ogale
Independent Scholar

LIGHT-EMITTING DIODES (LEDs)

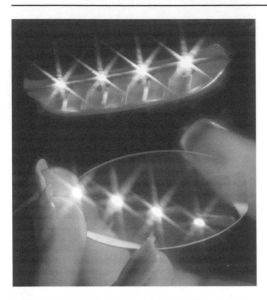

Early light-emitting diode (LED) designs emitted a bluish-white light; white LEDs, based on blue LED technology, arrived on the market in the 1990s. LED bulbs generally have long life spans and contain a tiny flake of semiconducting material with a surface area of less than 1 square millimeter.

Source: Randy Montoya/Sandia National Laboratory, U.S. Department of Energy

Light-emitting diodes, also called solid-state lighting or, simply, "LEDs," are highly efficient, durable, and long-lasting lighting devices. The technology has improved enormously since the 1960s when the first LEDs came to market. LEDs are now the industry standard in a variety of specialty lighting markets, and the popular bulbs are rapidly entering the general illumination market. LED bulbs are more energy efficient and last longer than incandescent, halogen, and fluorescent bulbs, but their up-front costs are higher. Since a fifth of electrical generation goes toward lighting, LEDs hold the potential to greatly reduce energy use. However, energy efficiency rebound effects could partly or entirely offset these savings.

Designers and product engineers have long appreciated LEDs for their extended life spans, which makes them ideal for indication lighting in electronic devices. Early bulbs were dim and expensive. Over subsequent decades, the industry successfully improved both light output and energy efficiency. During the 1970s, the cost per lumen of light output from a standard LED bulb was about $10. This quickly dropped to about $1 per lumen in the 1980s, and to just 10 cents per lumen in the 1990s.

LEDs from the 1960s and 1970s could only emit yellow-green, orange, red, or infrared light. White LEDs, based on blue LED technology, arrived on the market in the 1990s. Early LED designs for general illumination emitted a bluish-white light. Developers worked to improve white LED light quality for general illumination and specialty lighting applications. Subsequently, a more complete spectrum of LED illumination became available to lighting engineers, who then tinkered with the subtleties of "warm" or "cool" white LED output. In the 2000s, LEDs became standard for aircraft, ship, and automotive lighting as well as emergency lighting, signage, flat screen display backlighting, operating room lighting, supermarket freezer lighting, and a variety of other specialty lighting applications. In 2010, numerous LED manufactures released bulbs designed for general illumination priced at around $60. Prices have since dropped but up-front cost for the bulbs still poses a barrier to broader LED adoption.

LEDs contain a tiny flake of semiconducting material with surface area that is often less than 1 square millimeter. These semiconducting materials, such as aluminum-gallium-arsenide (AlGaAs) and gallium nitride (GaN), are not conductive in their pure form. However, they become conductive when doped with impurities to create n-type material (containing extra negatively charged particles) or p-type material (containing extra positively charged particles). When n- and p-type materials are joined, the charged particles migrate and stabilize to create a nonconductive zone. Like a typical diode, the nonconductive zone blocks electrical current in one direction. However, applying current in the other direction pushes the charged particles out of the nonconductive zone. The gate then regains conductivity and current can freely flow. In a light-emitting diode, this current induces electroluminescence. Semiconductor properties determine light output color, which corresponds to the energy of the released photons. Integrated optical and reflective components shape the light output.

Lighting engineers determine the efficiency of lighting devices by measuring the lumens of light output per watt of electrical input. The most efficient incandescent lights can convert one watt of electricity into 17 lumens of light output. Compact fluorescent bulbs yield about 60 lumens per watt, and linear fluorescent bulbs produce about 80 lumens of light output per watt. In 2006, commercially available white LEDs surpassed the efficiency of linear fluorescent technology. Since then, researchers have developed prototype LEDs with a luminous efficacy of over 200 lumens per watt.

Beyond energy efficiency, LEDs offer users several other distinct benefits, as well as a few drawbacks and limitations. Since they do not contain filaments or glass, they are generally more durable than incandescent and fluorescent bulbs. LED bulbs generally have long life spans, ranging from 25,000 to 100,000 hours—compared to 15,000 hours for fluorescents, and 2,000 hours for incandescent filament bulbs. The long life span of LEDs makes them especially ideal for applications where bulb failures are difficult to replace or create dangers.

For instance, LEDs are now standard for municipal traffic signals. LED signal lights cost much less to power than incandescent signal lights. Furthermore, since the bulbs fail less frequently, traffic intersections updated with LED signals are safer for motorists and pedestrians. However, LED signal lights do have several drawbacks. Blowing snow can completely cover the signal lamps, since the bulbs do not emit enough heat to resist icy buildup, and when first introduced, green LED signal lamps frequently failed under real-world conditions. Most LED lamps contain multiple LEDs, so when failure does occur, it typically emerges incrementally over time rather than all at once.

LEDs provide adjustable and high-quality lighting options, making them ideal for backlighting various types of flat screens, from mobile phones to televisions. LEDs are not prone to the flickering or humming associated with fluorescent technologies. Likewise, their ability to quickly and reliably switch on and off makes them well suited for fiber optic communication devices. LEDs produce a highly directional light that is ideal for task lighting, streetlights, and flashlights, but is more challenging to integrate into bulbs for general illumination. Visible spectrum LED bulbs produce comparatively little ultraviolet radiation or heat compared with other types of bulbs. However, even small amounts of heat trapped inside the LED's electronics can cause the devices to fail if not adequately dissipated. LED products offer lighting designers flexibility since the bulbs are small, dimmable without color shifts, and shock resistant. Furthermore, environmentalists value LEDs because they consume less energy than other lighting options and they do not use regulated toxic substances, such as the mercury found in fluorescent lighting.

The total economic benefits of switching to LEDs are only realized over their long life span. Even though LEDs are less expensive to operate over time, the up-front costs are higher than other available alternatives. Despite this restriction, LED bulbs are displacing conventional bulbs in numerous lighting markets. Over the past decade, the market for high-brightness LEDs has doubled every two to three years. LED demand for traffic signals, vehicle lighting, signage, and backlighting drove this exponential growth. Demand for LEDs is set to remain strong over coming years.

According to the U.S. Energy and Information Administration, lighting applications in commercial buildings consume more energy than air conditioners, refrigerators, computers, and office equipment combined. Presumably, broad LED adoption could cut electrical demand for lighting by half or more, but only in certain contexts. Even though LEDs use less energy, they do not necessarily hold the potential to reduce energy consumption on a large scale unless countries or regions first institute stops to prevent the energy efficiency rebound effect, or Jevons Paradox, from negating efficiency gains. For instance, if LEDs make illumination less expensive, then homeowners, organizations, and municipalities may increase their use of lighting services as a result, thereby offsetting the gains achieved through higher device efficiency. Or, as energy users save money on lighting, they may choose to spend those savings on energy-intensive purchases that would otherwise have not been made. Therefore, LEDs are only valuable as energy reduction mechanisms in regions where backstops, such as increased energy taxes or regulations, are in place to prevent the rebound effect from taking hold. Otherwise, deploying LEDs may simply shift energy consumption from the lighting sector to other sectors.

See Also: Appliances, Energy Efficient; Eco-Electronics; Green Building Materials; Unintended Consequences.

Further Readings

Hadhazy, Adam. "17 Projects Shaping the Future of LED Lights." *Popular Mechanics* (January 2010).

Herring, Horace and Steve Sorrell, eds. *Energy Efficiency and Sustainable Consumption: The Rebound Effect.* New York: Palgrave Macmillan, 2009.

Pimputkar, Siddha, et al. "Prospects for LED Lighting." *Nature Photonics*, 3 (2009).

U.S. Department of Energy. "Solid-State Lighting." http://www1.eere.energy.gov/buildings/ssl/projects.html#2008portfolio (Accessed September 2010).

Ozzie Zehner
University of California, Berkeley

LUDDISM

Luddism, or in terms of contemporary usage, *neo-Luddism*, is a term that describes a wide range of attitudes, dispositions, and actions toward an equally diverse range of technologies. All forms of Luddism share a critique of some aspect of technology, of the relationship of technology to society, or of specific technologies. Luddism is applied both as derogation and as compliment. That such variety of thought and action is explicitly linked, by both advocates and critics alike, to the historically specific Luddism of the early 19th century gives the topic much of its interest.

Between roughly 1811 and 1816 in the Midlands and North of England, groups of disaffected weavers, cloth shearers, and other textile-industry employees banded together in secret groups for the purpose of pursuing acts of intimidation, violence, and sabotage. They targeted mill owners, some of the machinery, particularly knitting frames, and the mills themselves, identifying them as both the agents and the symbols of the oppression they felt as the textile industries eroded their livelihoods and working practices. They referred to themselves as Luddites and, alongside the practical and often-violent prosecution of their protest, they invoked a series of their own symbolic and mythical antecedents: Ned Ludd, the idiot youth who had broken frames some years before; King Ludd, and other self-consciously pompous titles that aped and provoked the social niceties of the period; and most potently for the Luddites of Nottinghamshire and Yorkshire, the figure of Robin Hood. It is this symbolic aspect of Luddism that finds fulsome development in contemporary neo-Luddism.

It is, nonetheless, for the destruction of the hated knitting frames themselves that the Luddites are best remembered. However, there was little consistency of action in this regard—sometimes frames and other machinery were destroyed, sometimes they were not. The character and attitude of the targeted mill owner was important, and some were killed in attacks and planned assaults. Luddites themselves were killed both in the actions and as the result of juridical proceedings. Luddism was secretive precisely because it was illegal and the activities associated with membership of such groups, such as secret oath swearing, were punishable by death as was frame breaking itself from 1812 onward. The British state certainly devoted a huge effort and deployment of forces to try to undermine Luddism.

Significantly, the secretive, mass, and pointedly specific Luddism of the early 19th century did not articulate a general or critical philosophy of technology in general, a point that much of the public, individualistic, and diffuse neo-Luddism of recent decades often overlooks in its articulation of just such a general critique of technology. Indeed, the term *technology* was not in widespread use during the period of the original Luddites, certainly not in the sense that the term is understood today. The Luddite destruction of technology could then be contextualized by locating it in a history or tradition of industrial sabotage; frame breaking and other destructive acts preceded it while the Captain Swing rural English riots, complete with the breaking of agricultural threshing machines, of the early 1830s provide an additional stepping-stone to the consolidation of widespread industrial sabotage throughout the 19th and 20th centuries. It is this that partly legitimizes a "conservative" designation of Luddism as mindless vandalism and willful rejection of technology, a perverse anti-progressivism, a characterization that dominates mainstream media use of the term.

Despite this, Luddism or neo-Luddism has survived, developed, and become a self-conscious expression of disdain, critique, or rejection of both specific and general contemporary technology. Given that such neo-Luddism thrives most strongly and vocally in postindustrial, wealthy, advanced welfare societies, this self-declared lineage to the historical Luddites is intriguing. For some commentators, including Kirkpatrick Sale and Nicols Fox, the congruities between historical Luddism and contemporary neo-Luddism are startling, resonant, and notable. Nicols Fox connects the two variants via a "hidden tradition" that covers arts and literature, individual protest, and industrial action. Sale's historical Luddites were "rebels against the future" from whom contemporary campaigners and protestors can learn both practical and ideological lessons. Other commentators are more cautious in assigning historical antecedence; Steven E. Jones argues that historical

Luddism is specific enough in terms of its scope, outlook, and aspirations to be regarded as a labor subculture whose historical specificity must always be noted and respected. Such differences of opinion resonate with the contrasting views of previous generations of historians who have attempted to claim a particular motivation, rationale, or philosophy for the Luddites, a task complicated by the paucity of written documents left by the Luddites themselves.

While disputes concerning the validity of viewing the historical Luddites as direct or indirect antecedents of contemporary neo-Luddism have been developed, others have concentrated on the content of Luddism in both its historical and contemporary forms. Of interest here is the way in which many modern-day neo-Luddites see the historical Luddites as articulating a nascent environmentalism, defending an organic pre-industrial nature against the ravages of industrialism, and promoting a general anti-technology worldview. Contemporary neo-Luddites have developed these ideas such that for many of them there is a direct line of development from the historical Luddites to their own campaigns and actions, for example, anti–genetically modified (GM) crop demonstrations, actions against nuclear power, and critiques of the "surveillance" society.

Neo-Luddites tend, in contrast to the historical Luddites, to emphasize technology as a ubiquitous and autonomous force largely out of control of the citizenry, democratic institutions, and unmanageable even beyond the intentions of those who designed and developed it. This view of technology as an abstract social force tends to license among neo-Luddites a series of rather overarching critiques, for example, condemning the genetic modification of species tout court or seeing only negative and apocalyptic consequences if nanotechnology continues to be developed. For some, it has been a short step to link the comprehensive rejection of a particular technology to wider social critique, for example, anti-globalization, and to ally neo-Luddism to political ideologies such as anarchism or deep ecology. Luddism has, then, been advocated by many as a unifying banner for a wide range of disparate and critical stances toward technology. Neo-Luddism also articulates, often less explicitly, a view of a benign alternative to that which it attacks and often includes an emphasis on the communal, the rural, the organic, the holistic, and appropriate technology. However, detailed definitions of the content of these terms, as with technology, are often left wanting.

One key feature of historical Luddism that is modified in much contemporary neo-Luddism is that of violence and violent intimidation. One key difference is that neo-Luddite violence is largely conducted on behalf of those deemed to be affected by it rather than those whose livelihoods are under direct threat from it. The "monkey-wrenching" of the technologies of the society-wide "big machine," the ecosabotage of genetically modified crops, and the destruction of land-clearing vehicles is usually done in this somewhat remote fashion. Indeed, reflecting some of the content of neo-Luddism as a knowing and ironic disposition toward autonomous technology, many acts of neo-Luddite violence are more potent symbolically than literally. The public events at which computer monitors or television sets are destroyed are a powerful example of this. However, some neo-Luddites have also come very close to endorsing more murderously violent reactions, for example, the campaign of terror conducted by the "Unabomber" Theodore Kaczynski on the basis of a critique of industrial society that he penned and that contains many ideas that resonate with much neo-Luddite thinking.

One recent implicit assault on ecological neo-Luddism comes from Stewart Brand, founder of the *Whole Earth Catalog* and a long-standing advocate of alternative, appropriate, and "slow" technology. While sharing and indeed pioneering many of the foundations of the ecological critique upheld by many neo-Luddites, Brand now advocates a radically different approach to technology. Global climate change and the need to manage the planet as a whole make necessary for Brand the adoption of technologies including nuclear power and genetic engineering. Such a call is anathema to many of a neo-Luddite disposition that insist that autonomous technology is the biggest threat to ecological sustainability and who maintain that the original, historical Luddites provide a model of thought and action appropriate to countering its perceived menace.

See Also: Anarchoprimitivism; Appropriate Technology; Earthships; Marxism and Technology; Technological Autonomy.

Further Readings

Binfield, Kevin, ed. *Writings of the Luddites*. Baltimore, MD: Johns Hopkins University Press, 2004.

Brand, Stewart. *Whole Earth Discipline*. New York: Penguin, 2009.

Fox, Nicols. *Against the Machine: The Hidden Luddite Tradition in Literature, Art, and Individual Lives*. Washington, DC: Island Press, 2002.

Jones, Steven. E. *Against Technology: From the Luddites to Neo-Luddism*. London: Routledge, 2006.

Sale, Kirkpatrick. *Rebels Against the Future: The Luddites and Their War on the Industrial Revolution: Lessons for the Computer Age*. Reading, MA: Addison-Wesley, 1995.

Neil Maycroft
University of Lincoln

M

Maglev

Maglev, or maglev, is an abbreviation for magnetic levitation, a method of suspending an object in the air supported by magnetic fields. One well-known use of maglev is in transportation: maglev trains have been in use since the 1980s, with several operating today and extensive networks proposed for the future. Maglev trains incorporate a basic fact about magnets: like poles repel each other. Maglev trains use rare earth magnets, which produce a stronger magnetic field than ferrite (iron compound) or alnico (alloys of iron, aluminum, nickel, cobalt, and copper) magnets to lift and guide the train cars over a guideway: the charge of the magnetized coils of the guideway repel the charge of magnets on the undercarriage of the train so that it levitates slightly

This magnetic levitation train in Shanghai, China, is the fastest passenger train currently in service (431 kilometers per hour). While lack of friction from wheels allows maglev trains to travel faster, air resistance means that they are only slightly more efficient than conventional trains.

Source: iStockphoto

(typically in the range of 1–10 centimeters) above the guideway. Once levitated, the train is moved forward by propulsion provided by the guideway coils, which are constantly changing polarity due to alternating electrical current that powers the system.

Maglev trains eliminate a key source of friction, that of train wheels on the rails, although they must still overcome air resistance. This lack of friction means that maglev trains can reach higher speeds than conventional trains; current maglev technology has produced trains that can travel at speeds over 500 kilometers per hour (310 mph), which is twice as fast as a conventional commuter train and slightly faster than the TGV (*Train a Grande Vitesse*) in use in France. However, air resistance means that maglev trains are only slightly more energy efficient than conventional trains.

Maglev trains have several other advantages compared to conventional trains. They are less expensive to operate and maintain because the absence of rolling friction means that parts do not wear out quickly (as do, for instance, the wheels on a conventional rail car). This also means that fewer materials are consumed by the train's operation because parts do not constantly have to be replaced. The design of the maglev cars and railway makes derailment highly unlikely, and maglev rail cars can be built wider than conventional rail cars, offering more options for using the interior space and making them more comfortable to ride. Maglev trains produce little to no air pollution during operation because no fuel is being burned, and the absence of friction makes the trains very quiet (both within and outside the cars) and provides a very smooth ride for passengers. Finally, maglev systems can operate on higher ascending grades (up to 10 percent) than traditional railroads (limited to about 4 percent or less), reducing the need to excavate tunnels or level off the landscape to accommodate the tracks.

Besides the costs of construction, one factor to be considered in developing maglev rail systems is that they require the use of rare earth elements (scandium, yttrium, and 15 lanthanides) and while these elements are not truly rare, concentrated deposits that make their use economically feasible are rare. Rare earth elements are in high demand for various technological applications, and currently almost all the world's supply comes from China, a fact that could be an issue in global development of maglev systems should China choose to reduce mining of rare earths, although there are deposits in other countries that could be exploited.

Several train systems using maglev have been developed over the years, most operating over relatively short distances. The first commercial maglev train was developed in Great Britain as a shuttle between the Birmingham airport and a nearby rail station. It operated for 11 years but has since been replaced with a cable liner system. Germany constructed a maglev train in Berlin (the M-Bahn) that began operation in 1991 to overcome a gap in the city's public transportation system caused by the Berlin Wall; however, when the wall was destroyed, the M-Bahn was dismantled. The 1986 World's Fair (Expo 86) in Vancouver, British Columbia, Canada, included a short section of maglev train within the fairgrounds. Today, experimental maglev facilities exist or are under construction in California, Georgia, and Virginia in the United States; in Chengdu, China; in Emsland, Germany; and in Yamanashi prefecture, Japan (where the current world record speed was achieved).

Several commercial (i.e., not experimental) maglev systems are currently in operation around the world. In Aichi, Japan, near Nagoya, a maglev train built for the 2005 World's Fair is still in operation. It is about 9 kilometers long with nine station stops over that distance and reaches speeds of about 100 km/h. This maglev train was constructed as a prototype system to replace conventional urban transportation train travel rather than long-distance rail travel. The Korean Rotem Maglev was also built with urban transit rather than long-distance travel in mind; it runs between the Daejeon Expo Park and National Science Museum, a distance of 1 km. The longest current commercial maglev system is in Shanghai, China; it covers about 30 km and runs from downtown Shanghai to the Pudong International Airport.

The United States currently has no commercial maglev trains but several prototype systems exist or are under construction, and the United States invested over $70 million over the years 1999–2006 for research and development into maglev systems. The U.S. Department of Transportation calculates that maglev trains would be most useful over distances of 150 to 500 miles; at shorter distances, the faster speed of maglev trains over conventional trains would not make a substantial difference in travel time, and for longer trips, the higher speed of air travel compensates for the greater travel times to the airport (versus rail stations, which are typically located in city centers).

The Department of Transportation also calculates that the capital investment necessary to create a maglev system is higher than for its two major competitors in ground transportation—incremental high-speed rail (which upgrades existing railways for high-speed travel) and new high-speed rail construction. However, these higher costs would be compensated in some circumstances by higher potential speeds: a trip from Boston to New York would take about an hour on a maglev train operating at the highest speeds currently achievable, which is similar to the time required on regularly scheduled airline transportation. The Department of Transportation also calculated that the increased costs would be outweighed by the benefits of maglev transportation in only two long, heavily populated areas of the country: the northeast corridor and California.

See Also: Best Available Technology; Clean Energy; Hybrid/Electric Automobiles; Technological Momentum.

Further Readings

Bonsor, Kevin, "How Maglev Trains Work." http://science.howstuffworks.com/transport/engines-equipment/maglev-train.htm (Accessed September 2010).

Federal Railroad Administration, U.S. Department of Transportation. "Magnetic Levitation (MAGLEV)." http://www.fra.dot.gov/Pages/200.shtml (Accessed August 2010).

Federal Railroad Administration, U.S. Department of Transportation. "Report to Congress: Costs and Benefits of Magnetic Levitation" (September 2005). http://www.fra.dot.gov/downloads/RRdev/maglev-sep05.pdf (Accessed September 2010).

International Maglev Board. "Maglev—What Is It? And Is It Relevant?" http://magnetbahnforum.de/index.php?en_what-is-maglev (Accessed September 2010).

Sarah Boslaugh
Washington University in St. Louis

Marxism and Technology

According to the philosophy of Karl Marx (1818–1883), a German writer and theorist who cofounded communism along with Friedrich Engels, technology is one of the major forces that drive history as it assists in the formation of economic, political, and legal institutions. The level of technological knowledge, in Marx's view, is instrumental in shaping the economic structures of any given society because the dominant technology controls the mode of production in that society. Marx saw history as constantly in flux. Thus, he believed that the bourgeoisie (capitalists) of his own time were revolutionaries who had overthrown the preindustrial class that had formerly controlled all modes of production. Once the bourgeoisie

Marxist theory was critical of the fact that capitalism produces inherent inequalities. In developed nations, schoolchildren have access to computers, the Internet, and phones, yet more than 3 billion people in the world lack access to a telephone.

Source: iStockphoto

were in charge, however, Marx believed that they began to exploit the proletariat (the working class) in order to further advance the technologies needed to fuel capitalism's rise to power. This exploitation resulted in the alienation of the proletariat, setting the stage for the ultimate overthrow of capitalism.

While Marx acknowledged that technology is necessary to improve productivity and efficiency in a capitalist society, he was critical of the fact that capitalism produces inherent inequities that lead to class conflict between the bourgeoisie and the proletariat. Differences in the two groups are heightened by the fact that the bourgeoisie controls the development of technology and all access to it. Over time, Marxian philosophy has been used as the basis for mandating redistribution of income and the establishment of an all-powerful state in a number of communist/socialist countries. Despite the fall of Soviet communism in the 1990s and the democratization of postcommunist societies, Marx continues to drive many debates about the benefits and detriments of technology and their propensity for creating unequal access to those benefits.

Marxism and Environmentalism

Karl Marx was not an environmentalist, even according to a 19th-century concept of the word. He believed that nature played an essential role in human production. Neither did Marx buy into the classical liberal concept of "scarce resources." He believed that socialism would be capable of preventing natural resources from ever being exhausted. Technology was seen as a means of assisting humans in their quest to control nature. Ultimately, as technologies progressed, Marx accepted the notion that nature would be viewed as an entity in its own right. Modern Marxists tend to break up into schools of thought on how Marxism can be used to deal with environmental issues. Eco-Marxists who embrace humanism argue that technologies that are proved harmful to the environment should be abandoned. They see it as the responsibility of communist governments to take on the role of environmental protector. Orthodox Marxists, on the other hand, tend to blame capitalism for environmental crises, and they maintain that science and technology should be used to solve the problems that they helped to create.

Amid the contemporary debate over the extent to which global capitalism is responsible for impoverishing large numbers of people in developing countries and for creating an international environmental crisis, debates over whether Marxism can be used to solve existing problems continue to be heated. Pro-Marxists claim that it can, but detractors call Marxist views "ecologically unsustainable."

Technology and the Masses

In the 21st century, inequities in access to essential communication technologies have continued to present problems in many countries around the world. In the United States and other developed nations, most schoolchildren have access to computers and the World Wide Web. However, many of them lose that access once they return home since many parents and grandparents have no idea how to even access the wealth of information available online. Even if parents and grandparents had the necessary knowledge in developing countries, scores of families have no telephone access to connect them with a service provider. Although the lack of Internet access is generally limited to low-income individuals and families in developed countries, it is more widespread in poor nations. Peter Lurie, an attorney with Virgin Mobile USA, and Chris Sprigman, a fellow at the Stanford Law School Center for Internet and Society, suggest that the solution is "broadband Marxism," which involves the privatization of telecommunication networks in many of those countries where bridging the digital divide is currently almost impossible. Increasing access to technology would ultimately benefit governments and capitalists as well as individuals because many developing countries, including some with high rates of unemployment, cannot fill jobs because the pool of applicants lacks the technical skills necessary to perform required tasks.

While Internet access may be deemed a luxury, the telephone is almost universally considered a necessity. Yet, it is estimated that millions of people around the world lack access to basic telephone services. With private companies controlling access to telephones, the cost of land lines may be unaffordable for many families. Those who can afford it often turn to expensive mobile phones, which are more readily available to fulfill communication needs. Those without the financial means to afford mobile phones are forced to do without this essential instrument of communication. As a result, approximately 1.5 million residents of low-income nations have no reliable access to telephone services. Unlike the United States, where there are 667 phones for every 1,000 people, nations such as Burma (4/1,000), Nigeria (5/1,000), Pakistan (23/1,000), and Guatemala (65/1,000) have extremely limited telephone access. The borough of Manhattan in New York City has more available telephone lines than the whole of Africa.

While many developing countries began preparing to breech the so-called digital divide by laying fiber-optic lines and transoceanic cables and erecting satellite relay stations in the 1990s during an economic boom, they currently lack the resources to connect existing lines to remote customers. Some experts believe the answer to the problem is to forgo old technologies for new ones, establishing the wireless-fidelity networks known as Wi-Fi and granting Internet-telephone capabilities via VoIP. These measures would negate the use of land wires that sometimes cost up to $300 a foot. Constructing a Wi-Fi station at a cost of approximately $10,000 could provide Internet access to an unlimited number of families within a two-mile area simply by installing a relatively inexpensive home-based antenna.

Some observers believe that modern technology is beginning, at least in part, to take the mode of production away from the capitalist and place it in the hands of consumers. As proof, they cite the rise of home studios and the popularity of YouTube as proof that musicians no longer have to depend on record companies to give them a shot at fame and fortune. Cadence Weapon (Rollie Pemberton) got his start by posting his early efforts on the "open stage" of the Internet, as did Great Britain's Lady Sovereign and Canada's Arcade Fire and Broken Social Scene. Established recording stars such as Moby and

Leonard Cohen are also producing some of their efforts in home recording studios. Would-be filmmakers, such as Shane Felix, who produced the space opera *Revelations*, are also bypassing the big studios to create their own films at home with video cameras and computers. Even big-name producers, including Don McKellar, the director of *Childstar*, are using cell phones with cameras to complete projects. Some producers of low-cost movies such as *The Blair Witch Project*, which were created by more traditional methods, are now using the Internet to become international phenomena. Would-be reality stars are also finding that the Internet provides them with a perfect venue for turning their personal lives into opportunities for winning their proverbial 15 minutes of fame.

Through the so-called iRevolution, consumers are now able to create their own playlists on their iPods and MP3 players, bypassing radio stations and record company output. In this way, consumers who choose to do so can refuse to pay exorbitant prices for a CD or tape just to get a single song. Instead, they download it from iTunes, Amazon, or one of hundreds of other sources for less than $2, a portion of which goes to pay royalties to recording artists. In an article in *Maclean's* magazine, Brian Johnson suggests that such innovations are leading to a major "democratic renaissance in the arts."

Technological Aftermath in Postcommunist Societies

Karl Marx saw technology as a major influence on war-making, and the validity of that view has been proven as the paraphernalia associated with waging war has become ever more sophisticated and ubiquitous. Despite Marx's claims that communist societies are better suited than capitalist societies to protecting natural resources, the devastation left in many Eastern European countries after the withdrawal of the Soviet Union offers ample proof that communism in practice has often exhibited little concern for the environment of member countries. Many postcommunist countries have been forced to struggle to deal with the removal of abandoned war equipment that was left behind by the Soviet war machine. Since they are also struggling economically, officials are often overwhelmed by the sheer magnitude of environmental problems.

In addition to dealing with debris from the Soviet war machine, many postcommunist countries are also dealing with the aftermath of irresponsible agricultural practices and unregulated use of pesticides that threatened the environment. In Bulgaria, for instance, officials were forced to deal with large amounts of organochlorine pesticides that were stockpiled after they were banned by the government. From the 1970s to the 1990s, HCB (hydrochlorobenzene) was used extensively in Bulgaria, and many rivers are still polluted. In Estonia, the major problem following Soviet withdrawal was the aftermath of massive amounts of jet fuel that had contaminated the soil, particularly in the area near Tapa, where six square kilometers of land had been saturated with the fuel. Toxic chemicals, explosives, and weapons had also been recklessly disposed of in the inland waters of Estonia, and a uranium plant had discharged 1,200 tons of uranium and 750 tons of thorium into the Gulf of Finland. Restoration costs in Estonia were estimated at 3.5 billion EKR.

During Soviet occupation, most countries within the bloc were apathetic about environmental protection. After the breakup, officials began passing new legislation and creating new governmental entities designed to deal with existing problems and to prevent further ecological damage. This is particularly true in those countries that applied for membership in the European Union, which has stringent requirements for membership, calling for all member nations to commit to dealing with the realities of climate change, engage in emissions trading, encourage biodiversity, improve overall environmental health, and take steps to correct any issues that have been identified, and promote sustainable development.

As the 21st century progresses, Karl Marx's views on technology will certainly continue to be hotly debated, and existing technologies will continue on their course of reshaping the characteristics of capitalist production.

See Also: Frankfurt School; Labor Process; Postindustrialism; Technological Utopias.

Further Readings

Allenby, Braden R. *Reconstructing Earth: Technology and Environment in the Age of Humans*. Washington, DC: Island Press, 2005.

Bloom, Bernard S. "Stretching Ideology to the Utmost: Marxism and Medical Technology." *American Journal of Public Health*, 79/12 (1979).

Burkett, Paul. "Marx's Vision of Sustainable Human Development." *Monthly Review* (October 2005). http://monthlyreview.org/1005burkett.htm (Accessed September 2010).

Calhoun, Craig, ed. "Karl Marx." *Dictionary of the Social Sciences*. New York: Oxford University Press, 2002.

DeGregori, Thomas R. *The Environment, Our Natural Resources, and Modern Technology*. Ames: Iowa State Press, 2002.

European Union. "Environment." http://europa.eu/pol/env/index_en.htm (Accessed September 2010).

Foster, John Bellamy. *The Ecological Revolution: Making Peace With the Planet*. New York: Monthly Review Press, 2009.

Freedman, Robert. *Marx on Economics*. New York: Harcourt Brace, 1961.

Johnson, Brian D. "Someone Call Karl Marx: The Means of Production Is in the Hands of the Masses and a Revolution Is Under Way." *Maclean's*, 118/51 (December 19, 2005).

"Karl Marx." *Stanford Encyclopedia of Philosophy*. http://plato.stanford.edu/entries/marx (Accessed September 2010).

Kirsch, Scott and Don Mitchell. "The Nature of Things: Dead Labor, Nonhuman Actors, and the Persistence of Marxism." *Antipode*, 36/4 (2004).

Lurie, Peter and Chris Sprigman. "Broadband Marxism." *Foreign Policy*, 141 (March/April 2004).

Marx, Karl and Friedrich Engels. *The Communist Manifesto*. New York: Penguin, 1967.

McLellan, David, ed. *Karl Marx: Selected Writings*. New York: Oxford University Press, 1977.

Schor, Juliet B. and Betsy Taylor. *Sustainable Planet: Solutions for the Twenty-First Century*. Boston, MA: Beacon Press, 2002.

Torgerson, Douglas. *The Promise of Green Politics: Environmentalism and the Public Sphere*. Durham, NC: Duke University Press, 1999.

Wetherly, Paul. "Marxism and 'Manufactured Uncertainty' and Progressivism: A Response to Giddens." *Historical Materialism*, 71/1 (2000).

Wood, Allen W. *Karl Marx*. London: Routledge, 1981.

Elizabeth Rholetter Purdy
Independent Scholar

MATERIALS RECOVERY FACILITIES

A materials recovery facility processes recyclable materials. The materials are then sold to manufacturers as raw materials for new products. Materials recovery facilities, also known as materials reclamation facilities or materials recycling facilities, are generally classified as either "clean" or "dirty."

A clean materials recovery facility takes in commingled, recyclable materials that have been separated from municipal solid waste. The most common clean facilities are single stream, where all recyclable materials are mixed together, or dual stream, where recyclables are delivered in mixed streams, with one typically consisting of mixed paper and the other of such materials as glass, nonferrous metals, and plastics. The recyclables are sorted and then prepared for market through baling, shredding, crushing, or other techniques. Nonrecyclable materials are separated out, but they generally amount to less than 10 percent of the total stream of waste taken in by a clean facility.

A dirty materials recovery facility takes in a broad stream of solid waste and separates out recyclable materials through manual and mechanical sorting. The recyclables are then processed for market while the nonrecyclable materials are sent to a landfill or other disposal facility. The amount of recyclable materials can vary widely but typically is at least 5 percent and sometimes almost 50 percent. A dirty facility offers the advantage of recycling more materials since it processes the entire waste stream and does not rely on residents or businesses to correctly identify recyclable materials that might otherwise be disposed in the trash. But dirty facilities are also more expensive to build and operate.

The construction of modern materials recovery facilities began in the 1970s. Although recycling is largely overseen by state and local governments, the U.S. Congress in 1970 passed the Resource Recovery Act, which authorized grants for recovering useful materials from solid waste. The 1976 Resource Conservation and Recovery Act authorized funding for grants and demonstration projects on converting waste to energy and recovering materials such as glass and plastic. Recycling has become increasingly popular ever since in the United States, a trend that is mirrored overseas.

Materials recovery facilities use a series of conveyers that carry recyclable materials over sorting screens or other mechanisms that divide the materials. As single-stream recycling becomes more popular, more facilities are designed to accept and separate various types of recyclable materials. Automated systems can sort a number of materials simultaneously, such as paper, cardboard, aluminum, plastic, and glass, using such tools as magnets and ultraviolet optical scanners. The mechanized process is augmented by workers who sort items by hand. Small materials recovery facilities process less than 10 tons of waste per day, while large ones can handle 500 tons or more.

Melting, shredding, and pulping are all used to prepare objects for recycling. Glass is often chipped and melted for use into new glass objects, although some facilities offer bottle reclamation, in which bottles are sterilized for reuse. Salvaging is also a form of recycling in which a product is stripped of valuable components, such as the removal of lead from car batteries. Shredding is used to package plastic, metal, and paper for processing, while pulping is used to convert paper products into a slurry that can be made again into paper.

The materials, once broken down, are shipped to facilities that specialize in using recycled goods for manufacturing. Demand for recycled materials fluctuates, and sometimes a materials recovery facility ends up with a backlog of materials. The facility, in some cases, may have to landfill excess material if it runs out of storage space.

The recycling process has become increasingly sophisticated. Some systems integrate water into a materials recovery facility, washing recyclables while crushing and dissolving biodegradable organics. New mechanical biological treatment facilities combine a sorting facility with a form of biological treatment such as composting or anaerobic digestion. By processing the biodegradable waste, the technology reduces emissions of carbon dioxide. Such a facility can also process the waste to produce refuse-derived fuel, a high-calorific fuel, generally derived from plastics and biodegradable organic waste that can be used in cement kilns or power plants.

As landfill space becomes scarcer, materials recovery facilities play an important role in reducing the waste stream. Recycling also cuts down on the demand for raw materials and lessens pollution associated with the manufacturing of new products. One disadvantage, however, is that recycling can sometimes be more costly than manufacturing new products.

Recyclable materials can be used in a wide variety of products, from building materials and furniture to jewelry and paperboard. Common household items that contain recycled materials include paper products such as newspapers and napkins; soft drink containers made from aluminum, plastic, or glass; steel cans; and plastic laundry detergent bottles. Recycled glass can be turned into "glassphalt" and used as an alternative to bituminous asphalt pavement; recycled plastic can be used in carpeting, benches, and pedestrian bridges.

See Also: Design for Recycling; Recycling; Solid Waste Treatment.

Further Readings

City of San Jose: Environmental Services. "Materials Recovery Facility (MRF)." http://www
    .sjrecycles.org/residents/mrf.asp (Accessed September 2010).
Dubanowitz, Alexander J. "Design of a Materials Recovery Facility for Processing the
    Recyclable Materials of New York City's Municipal Solid Waste." Master's thesis,
    Columbia University, 2000. "http://www.seas.columbia.edu/earth/wtert/sofos/dubanowitz_
    thesis.pdf (Accessed September 2010).
Swartzbaugh, Joseph T., et al. *Recycling Equipment and Technology for Municipal Solid
    Waste: Material Recovery Facilities* (Pollution Technology Review) (No. 210). Park Ridge,
    NJ: Noyes Data Corporation, 1993.

David Hosansky
Independent Scholar

MEMBRANE TECHNOLOGY

Membrane technology is an advanced technology used to purify substances by separating out unwanted materials using semipermeable membranes rather than by using chemicals during a filtration process. When they were first introduced in the 1980s, all microfiltration systems were too expensive for general use; but over the following decade, these systems became more readily affordable. The most commonly used membrane is a hollow fiber wall made up of thousands of bundled fibers containing billions of microscopic

Two maintenance workers replace filters in one of the filtration units at a reverse osmosis water treatment plant. Reverse osmosis is one of the four types of membrane filtration technologies used in municipal water treatment plants.

Source: iStockphoto

holes (pores). Pressure or vacuum is used to force the material to be filtered through the membrane, trapping unwanted substances among the fibers. The chief advantage to membrane technology is that it uses much less energy than is expended in traditional processes such as flocculation, sediment purification techniques, sand filters, and ion exchangers. By the early 21st century, membrane technologies were widely used in both domestic and commercial settings, particularly in the area of water filtration and reclamation. As a result of ongoing research in the field, improvements to existing technologies were constantly being made.

The four types of membrane filtration technologies are classified according to the pore size of the filter: microfiltration (MF), 0.1 to 1 micron; ultrafiltration (UF), 0.003 to 0.1 micron; nanofiltration (NF), 0.001 to 0.003 micron; and reverse osmosis (RO), approximately 0.0005 micron. The latter is also known as hyperfiltration. The first two types are commonly used to remove large particles, but the removal of salt from seawater generally calls for the use of either nanofiltration or reverse osmosis. In the first case, productivity is high, but pressure differences are low. In the second, the need for increased pressure results in lower productivity. Membrane system modules are commonly either tubular or plate and frame, and the factors of cost, suitability for a particular project, and reliability determine which system is used. The configuration of the filter may be either hollow fiber, spiral wound, or pleated. The hollow fiber configuration looks like what it is: a bundle of several thousand fibers ranging in size from 0.5 to 1.0 microns. The spiral wound configuration, on the other hand, has been described as looking like a paper towel with the fiber serving as the "paper" and the permeate tube acting as the cardboard tubing. The pleated configuration looks like a traditional filter tube, such as the one used in older vacuum cleaners. Because membrane fouling is common during the process, cleaning solids away from filters is often necessary, using forward flushing, backward flushing, chemical cleaning, or various combinations of methods.

In Practice

Because membrane technologies have proved so effective in purifying water, they are widely used in municipal water treatment plants to provide drinking water and for reclamation purposes. Membrane technology has been particularly useful in countries with inadequate water sources. Such is the case in Singapore, which has no natural aquifers or groundwater.

After a strenuous study conducted by the Public Utilities Board (PUB) and the Ministry of Environment, it was determined that reclaimed or NEWater could safely be used to furnish residents with a supply of safe drinking water while promoting sustainable development. In December 2002, the city began using a combination of microfiltration, reverse osmosis, and ultraviolet disinfection to reclaim water. Fully confident in the membrane technology used in NEWater, the city subsequently erected four other facilities for industrial and commercial use. In 2008, the PUB announced that Black and Veatch, a firm based in Kansas City, Missouri, had been hired to design two additional plants. Completed in 2010, the Changi Plant became the largest such facility in Singapore. The second facility consisted of a membrane bioreactor built at the Jurong Water Reclamation Plant. Local firms were hired to construct both facilities. The PUB continues to pursue technological innovation to solve its problems with inadequate water supplies by using membrane technology to reclaim and desalinate water in addition to collecting rainwater and importing water from Malaysia. Cities that do not lack water resources are also using membrane technologies to filter and/or reclaim water. For instance, in New South Wales, Australia, Eraring Energy uses a combination of microfiltration and reverse osmosis to provide one fourth of all water to the city.

In the city of Arlington in Washington state in the United States, city officials were forced to implement a more effective wastewater system to protect its expanding population and the eight species of salmon living in the Stillaguamish River and its tributaries. They chose the flat-plate membrane system and opted for the largest membrane pore size available because it cut down on maintenance, saved energy, and was more efficient for their purposes.

Rising concerns about global warming have resulted in the need for industries, such as chemical manufacturing, which emit large amounts of carbon dioxide, to reduce such emissions while improving overall energy efficiency. It has been estimated that from 40 to 70 percent of capital and operating costs are devoted to separation operations. One way of solving this problem is through the use of membrane technology, which is relatively low-cost, saves energy, is compact and lightweight, requires a limited amount of maintenance, and is capable of multistage operations.

Innovations

The membrane bioreactor (MBR) is considered one of the most successful innovations of membrane technology. These reactors have been used to solve acknowledged limitations with the conventional activated sludge process. MBR performs solid/liquid separations through either microfiltration or ultrafiltration, eliminating the need for the gravity separation commonly used in the past. Activated sludge, clarification, filtration, and MF/UF are performed by a single unit. Advocates argue that the chief advantages of MBRs are their compact footprints and the perfect solid barriers that prevent sludge settleability. The external membrane bioreactor (EMBR) is used most often, but the use of submerged membrane bioreactors (SMBRs) is on the rise. In 2005, two pilot-scale submerged membrane reactors were tested in San Diego, California, over the course of a year. The results demonstrated that advantages of SMBRs over sludge processes included minimal fouling, high-quality effluent production, and energy savings. The study also revealed that SMBRs allowed for improved removal of organics that had long been considered difficult to degrade and expanded the range of possible operations.

The U.S. government has become extremely interested in promoting membrane technology, awarding a number of grants through its Small Business Innovation Research Program. Charles Russomanno, who oversees such grants as part of his position with the Office of

Energy Efficiency and Renewable Energy, a branch of the U.S. Department of Energy, has publicly stated that he sees membrane technology as a "viable alternative" to conventional separation methods. Fully aware of continuing problems with fouling, instability, low flux, low separation factors, and poor durability, the Department of Energy is constantly on the lookout for projects devoted to improving or surpassing existing membrane technologies.

A number of individuals in the academic and industrial sectors have been working on improving membrane technology. William N. Gill, a Russell Sage Distinguished Professor of Chemical Engineering at Rensselaer Polytechnic Institute in Troy, New York, has developed a composite polyamide membrane that has the ability to remove 99.5 percent of oxides and chemicals during MF and UF applications. In Albany, Oregon, Osmotek, Inc., invented the High Solid Ultrafiltration System, which won "Best in Show" at the American Institute of Chemical Engineers' Process Industries Expo in 2001. This system is reported to repel 97 percent of oils and solids as they flow through the membrane system, greatly reducing chances of the solid buildup known as fouling. Dr. Andrew Benedek, a Canadian who founded ZENON Environmental, Inc., has invented a hollow-fiber membrane that looks like a strand of spaghetti. This membrane can be placed directly into the liquid mix (liquor), doing away with additional separation steps. It can also be retrofitted to existing tanks. In a world ever more conscious of the impact of global warming, the need for membrane technology has become abundantly clear.

See Also: Innovation; Intermediate Technology; Technology and Social Change; Water Purification.

Further Readings

Allenby, Braden R. *Reconstructing Earth: Technology and Environment in the Age of Humans*. Washington, DC: Island Press, 2005.

DeGregori, Thomas R. *The Environment, Our Natural Resources, and Modern Technology*. Ames: Iowa State Press, 2002.

Ebner, Armin D. and James A. Ritter. "State-of-the-Art Adsorption and Membrane Separation Processes for Carbon Dioxide Emitting Industries." *Separation Science and Technology*, 44/6 (2009).

"Emerging Membrane Technologies." *Pollution Engineering*, 33/6 (2001).

Foster, John Bellamy. *The Ecological Revolution: Making Peace With the Planet*. New York: Monthly Review Press, 2009.

Kelsey, Chris, et al. "Energy Efficient Membrane Plant Protects Sensitive River. *WaterWorld*, 24/12 (2008).

Krukowski, John. "Opening the 'Black Box': Regulations and Recycling Drive Use of Membrane Technologies." *Pollution Engineering*, 33/6 (July 2001).

Landers, Jay. "Singapore Plans Two More Reuse Plants, Increasing Its Reliance on Reclaimed Water." *Civil Engineering*, 78/8 (2008).

Lenntech. "Membrane Technology." http://www.lenntech.com/membrane-technology.htm (Accessed September 2010).

Muilenberg, Tom. "Water Management With Membrane Technologies: The Benefits of Microfiltration." *Pollution Engineering*, 35/5 (2003).

Puming, Pang. "Membrane Technology Saves Money." *Pollution Engineering*, 38/2 (2006).

Russell, R. Shane, et al. "Process Limits of Municipal Wastewater Treatment With Bioreactor Submerged Membrane." *Journal of Environmental Engineering*, 131/3 (2005).

Schor, Juliet B. and Betsy Taylor. *Sustainable Planet: Solutions for the Twenty-First Century.* Boston, MA: Beacon Press, 2002.

Spear, Mike. "Separations in Flux." *Chemical Business*, 24/1 (2010).

Torgerson, Douglas. *The Promise of Green Politics: Environmentalism and the Public Sphere.* Durham, NC: Duke University Press, 1999.

Trussell, R. Shane, Samer Adham, and R. Rhodes Trussell. "Process Limits of Municipal Wastewater Treatment With the Submerged Membrane Bioreactor." *Journal of Environmental Engineering*, 131/3 (2005).

Wong, Joseph M. "The Low-Down on Low-Pressure Membranes." *Pollution Engineering*, 38/7 (2006).

Elizabeth Rholetter Purdy
Independent Scholar

MICROGENERATION

Microgeneration is the generation of heat and power on a small scale to suit the needs of small communities, businesses, or residences. Typically, this approach contributes to a lower carbon footprint and less environmental impact by relying more heavily on alternate energy sources like solar cells, wind turbines, and hydroelectric power. The specifics of microgeneration vary greatly from locale to locale. For instance, in the urbanized developed world, a residence or business may keep its connection to the traditional power grid but operate some alternate means of power generation, so that it is drawing from the grid only when something goes wrong or when its alternate means are insufficient.

In some areas, people may stay connected to the power grid but only draw from the grid when their alternate means are insufficient. Public utilities charge the microgeneration buildings for the energy used.

Source: iStockphoto

Under the 2005 Energy Policy Act, all U.S. public electric utilities are required to make net metering available to customers on request. Net metering records energy inflows and outflows, billing only for the difference. Whether or not credits in the customer's favor—when more energy has been generated than consumed—roll over to the next billing cycle varies from state to state. In most states, credits roll over month to month but do not carry over from one billing year to the next. State law also varies in whether electric utilities can put a limit on the percentage of subscribers who are signed up for net metering, whether there is a power limit on energy inflows, and how customers whose accounts end the billing year in a credit are compensated. (There have been several attempts to introduce federal law on net metering.) In order to protect workers repairing power lines, generators that are hooked into the grid must be built so that they will disconnect their feed in the event of grid failure—otherwise, there would be a risk of electrocuting those workers. And of course, because power lines carry alternating current, generators of direct current must use an electrical inverter to convert their output, a step usually already taken by people generating direct current at home, since conventional house wiring and appliances use alternating current.

The legal environment relevant to microgeneration also varies. There are federal and, in many cases, state income tax credits available for the use of renewable energy; and in some states where electric utilities are required to use renewable sources for a certain portion of their energy generation, they may fund private microgeneration efforts to meet this obligation, virtually paying a customer to go off the grid. The zoning laws applicable to building wind turbines and other sources of power also vary. In the United Kingdom, the Microgeneration Certification Scheme has been adopted as a certification scheme intended to cover all microgeneration technologies, the first of its kind and the foundation of the country's Low Carbon Buildings Program, which rewards green building with government grants to offset the initial costs.

A variety of technologies are usable in microgeneration. In addition to a connection to the grid, if applicable, there must be a power plant and infrastructure for the storage and conversion of energy. The system as a whole is usually called a "balance of system." The energy storage apparatus is necessary for efficiency and to make surplus energy available when the demand is greater than the supply currently being generated. Batteries are the most common solution, but hydrogen fuel cells, flywheel energy storage, and pumped-storage hydroelectric are possibilities, depending on the nature of the balance of system. Power conditioning equipment is used to invert energy for the grid and to convert energy from stored DC to usable AC. Surge protectors, switches, and groundings constitute the necessary safety equipment, while meters are useful for power consumption, power fed into the grid, and energy storage. There are additional equipment needs specific to the type of power generation, such as various types of mounts for solar panels, a sturdy foundation and grounding system for wind turbines, and so forth.

The technologies used in microgeneration are generally chosen with sustainability and appropriate technology in mind. Typically, one of the goals is to reduce one's environmental impact and carbon footprint, though some have suggested that it is more cost effective to reduce one's carbon footprint by purchasing carbon credits. If waste heat is generated by the power plant, it can be diverted for household or hot-water heating. Microgeneration may also be one element in a larger self-sufficiency scheme, including a greenhouse for growing vegetables, rainwater harvesting to provide a water supply, composting toilets, and so on. Microgeneration is necessary for autonomous buildings, for instance, which operate independently from the local infrastructure not only in terms of power generation but separate from the natural gas grid, communication systems, water systems, and the sewage treatment system. In some parts of the world, the principal benefit to autonomy is not environmental responsibility but the ability to continue functioning when the grid is unreliable. In the developed world, an autonomous residence is sometimes called "a house with no bills." Because start-up costs are high, microgeneration must be planned carefully and thoughtfully to be economically feasible. Certain technologies like wind turbines and solar panels benefit from economies of scale: they are popular enough that the relevant equipment is cheaper than newer technologies or those that have not been widely adopted.

Microgeneration Power Sources

Wind turbines need to be located in areas of high wind, away from obstructions like buildings. They are essentially tall windmills, freestanding, with heavy foundations to keep them rigid. Steady, heavy winds can produce significant amounts of electricity with no carbon impact and little waste heat, but it can be difficult to know the right spot on which to locate a turbine on a property, and in some cases subsequent developments—such as a building later constructed on an adjacent property—can ruin the long-term investment. Freestanding wind turbines are widely considered one of the best renewable energy sources, and the equipment is available from a wide variety of manufacturers and specifications, though in the long term it may be more effective when used at wind farms than at private residences. Building-mounted "micro wind turbines" are available but have little to recommend them in most cases: their usefulness is highly proportionate to the windiness of their location, and in most cases where they are useful, a freestanding turbine would be a better choice. The urban homes for which micro turbines are most convenient are unlikely to benefit from them.

Solar power has long been a popular alternative energy source. It has the benefit of a number of installation options, including directional mounts so that the position of the equipment can be adjusted seasonally to maximize solar energy collection, and rooftop solar panels to put roof space to its most efficient use. Solar energy lends itself well to hot water heating as well. Biodiesel (made from various forms of biomass, including recycled cooking oil from restaurants) and biogas (made from animal waste) are generally not considered practical for homes, but farms, college campuses, and many businesses can provide some or all of their electricity by these means. Heat can also be produced by biomass boilers, which are more practical for many homes; this is essentially a more efficient woodstove, frequently using a hot-water tank to store heat (in the form of heated water) when demand is low. However, while they are cost effective over their life span, the installation cost in an existing building is quite high.

Micro hydro refers to hydroelectric power plants producing up to 100 kilowatts of power, enough for a home or a small community. Because of the nature of the power source—often a dammed pool at the top of a waterfall, with hundreds of feet of pipe, a generator, and a Pelton wheel (when a waterfall of more than 50 meters is available) or Francis or propeller turbine (in low-head scenarios)—the construction and installation need to be tailored specifically to the site, and so the project is a more difficult and daunting one than installing a wind turbine or solar power. In many cases, there are zoning issues pertaining to one's rights or lack thereof to dam a body of water passing through one's property, but political initiatives calling for more green homes and green communities are helping to make this easier. Homes and other buildings can also be heated through use of heat pumps, which extract low-grade heat from the ground or the air and convert it into high-grade heat for heating interior spaces and hot water.

See Also: Appliances, Energy Efficient; Appropriate Technology; Clean Energy; Cogeneration; Composting Toilet; Smart Grid; Sustainable Design.

Further Readings

Farr, Douglas. *Sustainable Urbanism*. Hoboken, NJ: Wiley, 2007.

Page, Tom. *Prospects for Microgeneration: Energy Use, the Environment, and the Wider Community*. Saarbrucken, Germany: Lambert Academic Publishing, 2010.

Parker, Dave. *Microgeneration: Low-Energy Strategies for Larger Buildings*. New York: Architectural Press, 2008.

Bill Kte'pi
Independent Scholar

Occupational Safety and Health Administration (OSHA)

On December 29, 1970, U.S. President Richard M. Nixon signed the Occupational Safety and Health Act, which led to the introduction of the Occupational Safety and Health Administration (OSHA), an agency within the U.S. Department of Labor whose strategic and operational activities are the responsibility of the Deputy Assistant Secretary of Labor. OSHA's aim is to "develop and enforce workplace safety and health regulations."

The signing of the act occurred among growing alarm at then-current occupational health and safety statistics; 14,000 health and safety fatalities, 2.5 million work-related disabilities, and 300,000 emerging reports of work-related illnesses. Both domestically and internationally, these figures were considered morally unacceptable and reflected badly on the administration at that time.

OSHA is recognized throughout the United States as "the resource" for occupational safety and health issues. Its mission is to ensure the health, safety, and welfare of working men and women by issuing enforcement standards as specified within the act. In addition, the agency provides occupational safety and health information, education, training, and research in the pursuit of proactive occupational safety and health practices and standards.

OSHA is the outcome of a 19th-century U.S. economy, which historically focused on economic profit over its moral obligation to safety. Initially, the United States, like many other countries at that time, failed to either appreciate or embed occupational safety and health as a key priority to sustained economic development.

In 1877, the nation's first safety and health legislation was passed in the state of Massachusetts. This key legislation established legislative control measures associated with guarding factory fire exits and safe systems for elevators. The introduction of factory inspectors and restrictive controls for health hazards followed in 1890. However, these initiatives were not consistent throughout all states.

In 1903, increased concern about the occupational health and safety failing within the economy led the United States to start to publish key statistics on occupational fatalities and additional health and safety data, often industry specific. In 1913, the Department of Labor was established. In addition, key health and safety pioneers such as Dr. Alice Hamilton in the 1930s instigated statewide job-related safety measures and legislative changes. In 1933, Frances Perkins was appointed Secretary of Labor—the first female member of the cabinet—whose remit was to assure the workplaces were as "safe as science and law can make them." This led to the introduction in 1934 of the Bureau of Labor Standards to police the promotion of health and safety and administrative efficiency of legislation. Such responsibilities would become the sole responsibility of OSHA.

In 1971, OSHA was under significant pressure to both administer and enforce the Occupational Safety and Health Act's "standards package" in federal standards, country-wide standards in construction, industry, maritime, and other fields. The task was significant and required a focused approach. Employers were given a 90-day period of grace to become compliant. To monitor this compliance, 10 regional offices were established and 49 area offices. Enforcement officers were recruited and operated within this structure. Their remit was to educate, set standards, and enforce legislative guidance. The "special assistance position" was created to support small businesses; it also provided significant funding and advice to ease the burden and to incentivize small business compliance. This was initially known as the New Directions training and education grants and later known as the Susan Harwood Grants. Over half a million small businesses have participated in the consultation program to date.

Similar to the United Kingdom's (UK's) Lord Roben's report, which instigated the Health and Safety at Work Act of 1974, the U.S. Standards Deletion Project set about eliminating standards that were deemed irrelevant. In doing so, it provided safety rules that were considered more user-friendly to stakeholders. In addition, greater powers, rights of entry, and penalties were imposed against serious breaches as negligence and falsifying safety statistics.

Unlike the United Kingdom, the United States was able to ease the bottleneck of pending court hearings by granting OSHA officers the power to negotiate out-of-court settlements. However, the passage of the 1990 Omnibus Budget Reconciliation Act witnessed the largest financial impact on willful violations.

OSHA inspectors also underwent extensive training and had to operate within quality assurance systems; they strived to nurture a culture that fostered collaboration and cooperation with employers, such as the Voluntary Protection Programs, the Nationally Recognized Testing Laboratory Program, the "Safety Health Program Management Guidelines," the "OSHA Technical Manual," the OSHA Training Institute Education Centers, and the OSHA Outreach Training Program.

OSHA also worked very closely with other U.S. government agencies, such as the U.S. Bureau of Alcohol, Tobacco, and Firearms and the U.S. Department of Energy (DOE), sharing expertise and experience to enhance its service. The relationship with the DOE resulted in the signing of a memorandum of understanding in 1992 to strive for continuous health and safety improvement. OSHA also revalidated its "Field Operations Manual," which is now referred to as the "Field Inspection Reference Manual." This led to the creation of the Strategic and Rapid Response Team, which was to be located in each OSHA office.

Historically, critics of OSHA stated that its authority was too bureaucratic, expensive to comply with, confrontational with employers, and expected savings did not compensate for the cost of legislative compliance. This argument was further fueled by rising inflation and the state of the economy. Other Western organizations experienced similar criticisms, including the United Kingdom's Health and Safety Executive (HSE) and Canada's Canadian Centre for Occupational Health & Safety (CCOHS).

The U.S. administration's reaction was gradual though radical, with the introduction of the President's Regulatory Relief Plan, which resulted in the downsizing of OSHA and greater emphasis on compliance assistance and employer education. The U.S. House of Representatives, U.S Supreme Court, and the Circuit Court of Appeals also ruled against several OSHA standards, for example, via the Congressional Review Act.

OSHA proactively invested heavily in IT systems to enhance its auditing and reporting systems. The National Reinvention Steering Team and the Negotiated Rulemaking Act added further momentum in direct response to the Government Performance Results Act. The recording of accurate dates was imperative in monitoring industry-specific sectors and those organizations that repeatedly breached legislative and enforcement controls.

The millennium witnessed a new era of challenges that tested the foundations of OSHA. Most notable were the terror attacks of September 11, 2001, the BP Texas City refinery, and more recently in 2010, the BP Deepwater Horizon oil spill, which emphasized the importance and mutual benefits of proactive collaborative and open communication between government departments and sector-specific organizations to learn from disasters and to continually question the validity of previous data and statistical evidence.

Prior to this, the U.S. administration responded with the introduction of key legislation such as the Needlestick Safety and Prevention Act and further exercised the Congressional Review Act, which repealed ineffective regulations. In 2003, an Enhanced Enforcement Program (EEP) was introduced and later revised in 2008. Its aim was to target violators with previous violations. However, since its introduction, OSHA has faced continued criticism about how it is implementing its strategic goal operationally. The current system is considered by many as still overly bureaucratic and failing to target and prosecute a significant number of repeat offenders. Thus, OSHA is once again facing pressure to revise specific standards. The EEP in 2010 was replaced with the Severe Violator Enforcement Program.

OSHA in 2010 was not without its critics. However, OSHA 2010 is a far cry from its original inception. It fully appreciates that legislative guidance is instrumental, though not the complete remedy in forging a willing adherence to health and safety best practice. It now more fully appreciates that organizations need continued support and guidance if real commitment and conformance are to become normal practice. The past 10 years have witnessed new challenges and an ever-increasing need for OSHA to be more responsive to major incidents.

OSHA now has the benefit of cutting-edge information technology and fast-track communications and has developed a highly interactive website. Its effectiveness was clearly demonstrated during Hurricane Katrina in 2005, when OSHA demonstrated its leadership by playing a critical role in the coordination and monitoring of safe systems of work. This has not created a sense of complacency, and OSHA appreciates that its task is still significant and requires an even more focused approach.

There has been significant progress in offering crucial support systems in educating U.S. organizations on the legal, economic, and moral benefits in adhering to health and safety standards. Since its inception, OSHA has established over 28 Occupational Training Institute Centers that serve the nation—and where over 150,000 students and 520,000 employees have been competently trained. OSHA's role within the U.S. economy has been beneficial in fostering a health and safety culture.

See Also: International Organization for Standardization (ISO); Labor Process; Science and Technology Policy.

Further Readings

Occupational Safety and Health Act 1970. http://www.legalarchiver.org/osh.htm (Accessed August 2010).

Occupational Safety and Health Administration (OSHA). "All About OSHA." OSHA 3302-06N. Washington DC: U.S. Department of Labor, 2006.

MacLaury, Judson. *The Occupational Safety and Heath Administration: A History of Its First Thirteen Years, 1971–1984.* Washington, DC: Department of Labor, 1984.

Stender, John H. "Enforcing the Occupational Safety and Health Act of 1970: The Federal Government as a Catalyst." *Law and Contemporary Problems,* 38 (Summer/Autumn 1974).

Derek Watson
Independent Scholar

OFFSHORE OIL DRILLING (GULF OIL SPILL)

The worst accidental marine oil spill in history began on April 20, 2010, with an explosion on the Deepwater Horizon, a semi-submersible offshore drilling rig owned by Transocean and leased by British Petroleum (BP). The rig was drilling for oil in the Macondo Prospect in the Gulf of Mexico, about 50 miles off the coast of Louisiana, and at the time of the accident a production casing was being installed by the Halliburton Energy Company. The explosion and subsequent fire resulted in 11 deaths and numerous injuries to men working on the platform while also setting off a gusher that poured an estimated 4.9 billion barrels of oil into the ocean before being successfully capped on July 15.

The oil from this spill first made landfall in Louisiana, and by June had also reached Mississippi, Alabama, and Florida. The oil slick (oil on the surface of the water) dissolved more rapidly than expected, but because oil also spreads underwater the extent of the damage caused by the spill has not yet been determined. A study in August 2010 suggested that there was a huge plume of dispersed oil in the Gulf of Mexico, which could pose a threat to wildlife for years to come. The oil spill also imposed a severe economic burden on people living on the Gulf Coast, with the hospitality, fisheries, and energy industries being particularly affected.

Offshore Drilling

The practice of drilling for oil under the ocean floor dates back to the 1890s in the United States with the early wells connected to the shore by wooden piers. Beginning in the 1930s, oil companies began drilling from movable barges and in 1938 the first freestanding ocean well was constructed in the Gulf of Mexico, one and a half miles off the Louisiana coast. Offshore drilling increased after World War II as demand for petroleum increased and technologies developed during wartime such as sonar and radio positioning aided in offshore exploration. Technological advances allowed wells to be drilled under deeper and deeper water: in the 1950s and 1960s, the deepest wells in the Gulf of Mexico were a few hundred feet under water, by 1980 some were drilled under more than 2,000 feet of water and by 2000, this had increased to over 10,000 feet of water.

Over time, oil transforms from a liquid to a "mousse" and then to tarballs, like this one on Okaloosa Island in Fort Walton Beach, Florida, from the 2010 Deepwater Horizon oil spill.

Source: Wikimedia Commons

The Outer Continental Shelf Act in 1953 clarified jurisdiction over offshore drilling in the United States. Jurisdiction belongs to individual states for three international nautical miles off the coast, with the exception (for historical reasons) of Texas and the Gulf Coast of Florida, where the state jurisdiction extends three marine leagues seaward, and Louisiana, where it extends three nautical miles seaward. Federal jurisdiction begins seaward of where state jurisdiction ends and extends 200–350 nautical miles seaward (the specific limits are determined in part by underwater topography). Some states, including California and Florida, have banned offshore oil drilling in state waters, but have no control over offshore oil drilling in waters that fall under federal jurisdiction.

Oil drilling is an imperfect process and spills and leaks are not uncommon. A series of major spills and accidents in 1969 and 1970 led (in the short term) to greater government regulation of this industry. The first occurred in January 1969: a blowout on an oil platform in the Santa Barbara channel (California) released an estimated 80,000 to 100,000 barrels of oil over an 11-day period. This was the largest offshore drilling accident in the United States until the Macondo blowout and drew national attention to the environmental dangers of offshore drilling. Serious accidents also occurred in February and December 1970 in the Gulf of Mexico off the Louisiana coast. Independent of major oil spills, offshore oil drilling imposes known costs on the environment from smaller leaks from platforms and oil spills from pipelines or tankers transporting oil. In addition, the process of drilling requires expulsion of large amounts of water contaminated by drilling fluids and metal cuttings

Responses to the Gulf Oil Spill

On May 2, 2010, BP began drilling a relief well intended to intersect the Macondo well at its source and allow cement to be pumped in to stop the flow of oil. On May 17, BP began drilling a back-up well, at the insistence of Ken Salazar, then–Secretary of the Interior to relieve pressure on the main well. While both were valid responses to the oil spill, neither could solve the problem quickly (e.g., the process of drilling a relief well was expected to require several months). The Macondo well was successfully capped on July 15, and was officially declared dead in September after tests confirmed that the cement seal used to cap the well was effective.

Oil drillers are required to be prepared to respond to an oil spill, and in BP's case, that was provided by contract with the Marine Spill Response Corporation, a nonprofit company created by the oil industry following the *Exxon Valdez* disaster. However, the corporation was unable to handle the problem, and eventually many different entities, from the U.S. military to local boat owners, became involved in the effort to contain the damage by methods including skimming and use of containment booms (floating artificial barriers).

Dispersants, which do not remove oil but are intended to accelerate the process by which it mixes with water and is thus diluted, were first used on the surface oil slick on April 22. Dispersants can lessen the amount of oil that reaches shorelines, reduce the amount of oil that comes in contact with wildlife, and may also accelerate the rate of biodegradation of the oil (although this point is disputed). However, use of dispersants also means more oil will move below the surface of the water and will also be spread over a wider area. In addition, some responders exposed to the dispersant have reported ill affects, such as headaches and nausea, and the refusal of Nalco (the manufacturer of the dispersant Corexit used in the Gulf oil spill) to make its formula public did not inspire confidence. Additional concerns are related to the fact that the volume of dispersant used in response to the Gulf oil spill was unprecedented: for instance, in the week from May 4 to May 10 alone, 168,988 gallons of dispersant were used in the Gulf as compared to about 5,500 gallons during the *Exxon Valdez* spill.

On May 27, Secretary Salazar ordered a six-month moratorium of drilling in the Gulf of Mexico and Pacific Ocean at a water depth greater than 500 feet, ceasing work on 33 offshore deepwater rigs in the Gulf. This decision was immediately criticized as creating a second economic disaster due to loss of revenues to the industry and those employed in it. The moratorium was modified on July 12 and lifted on October 12. On September 30, the Department of the Interior issued a new set of regulations regarding safety, emergency response, and worker training.

Impact of the Gulf Oil Spill

It may be several years before the full economic, ecological, and social impact of the Gulf oil spill can be determined. The economic impact is most obvious on two major industries in the Gulf region: tourism and fishing. Independent of actual pollution from the oil spill, both tourist areas and fisheries have suffered from association with the spill (e.g., fear that beaches might become contaminated with oil caused some people to schedule their vacation elsewhere, and Gulf seafood was perceived as unsafe even after federal and state agencies determined that it did not constitute a health risk). The oil industry also suffered financially due to the temporary moratorium imposed on deepwater drilling.

An indication of the amount of economic damage caused by the Gulf oil spill may be seen in the fact that in the first eight weeks of operation, the Gulf Coast Claims Facility, which is responsible for processing claims for compensation from the $20 billion fund established by BP for this purpose, paid out more than $2 billion to about 127,000 claimants. In comparison, the 9/11 fund paid out about $7 billion to 5,560 claimants over a period of seven years.

The cost to human health from the oil spill includes not only the 11 deaths and 17 injuries from the explosion and fire, but also the future effects that may be suffered by responders and residents exposed to the oil and dispersants. It also includes the effects of the spill and related disruption of economic activity and traditional ways of life on the mental health of individuals living in affected areas. For instance, a Gallup survey conducted in August 2010 found that diagnoses of clinical depression increased 25 percent from April to August in Gulf-facing counties, while they declined in inland counties over the same period. Residents of Gulf-facing counties also reported a 15 percent increase in stress during that period, and similar increases in sadness and worry, while all categories declined in inland counties. Calls from Gulf coast states to the National Domestic Violence Hotline also spiked between April and June 2010.

Minerals Management Service (MMS) Reorganization

At the time of the Gulf oil spill the Minerals Management Service (MMS), an agency of the U.S. Department of the Interior, was responsible for managing mineral resources on the outer continental shelf. The MMS was created in 1982 by Secretary of the Interior James Watt, and from the start it was saddled with multiple responsibilities that sometimes conflicted with each other. The primary reason the MMS was created was to promote energy independence through increased offshore drilling, but Watt also gave it the responsibility for regulatory oversight and collection of revenues from royalty payments from offshore wells. This created an obvious conflict of interest: regulations aimed at environmental protection and worker safety could easily interfere with expansion of drilling, increased production, and increased revenues.

The MMS was severely criticized after the Gulf oil spill for exerting insufficient oversight, a choice attributed to the MMS emphasizing expansion of drilling and revenue generation over safety. For instance, the MMS did not require offshore drilling operations to have approved plans to safeguard the environment and human life, a requirement imposed in other oil-producing countries such as Norway and the United Kingdom. The MMS also had insufficient staff to review billing applications or to conduct inspections, and in some cases engaged in corrupt activities. For example, a 2010 inspector general's investigation uncovered clear conflicts of interest in some MMS offices, from employees accepting gifts from oil companies to an employee who conducted inspections on a company's oil platforms while at the same time negotiating terms of employment with that company.

In June 2010, the MMS was renamed the Bureau of Ocean Energy Management, Regulation and Enforcement (BOEMRE) and split into three entities: the Bureau of Ocean Energy Management, the Bureau of Safety and Environmental Enforcement, and the Office of Natural Revenue. The creation of three separate entities to carry out the different functions handled by the old MMS was intended to clarify the purpose of each branch and remove conflicts of interest: the Bureau of Ocean Energy Management is charged with developing energy resources, the Bureau of Safety and Environmental Enforcement with inspecting offshore drilling operations and protecting the environment, and the Office of Natural Resources Revenue with revenue collection.

Investigation and Future Recommendations

On May 22, 2010, President Barack Obama created an independent nonpartisan committee—the National Commission on the BP Deepwater Horizon Oil Spill and Offshore Drilling—to analyze the causes of the spill and make recommendations to increase the safety of offshore energy production and improve the nation's ability to respond to future spills. The final report of the commission, issued in January 2011, included the following conclusions:

- The explosion was preventable and can be traced to a series of mistakes by BP, Halliburton, and Transocean that reveal serious systemic failures in risk management by the offshore drilling industry

- Fundamental reform is required in the regulatory oversight of deepwater energy exploration and production to protect human safety and the environment
- The oil and gas industry needs to institute self-policing mechanisms and other reforms to supplement governmental oversight
- The real risks of deepwater drilling and the methods of containing and cleaning up oil spills are not adequately covered by current laws and regulations
- Current scientific understanding of the environmental impact of oil drilling is inadequate, as is understanding of the environmental and human impact of oil spills

The commission's report includes suggestions in seven areas: improving the safety of offshore operations; safeguarding the environment; strengthening oil spill response, planning, and capacity; advancing well-containment capabilities; overcoming the impacts of the Deepwater Horizon spill and restoring the Gulf; ensuring financial responsibility; and promoting congressional engagement to ensure responsible offshore drilling. The commission also noted that the complicated nature of energy extraction in the United States, which is performed by private industry and often combines the efforts of a number of private service contractors (such as Halliburton in the case of the BP well), increases the complexity of regulating the process.

The commission's suggestion to ensure financial responsibility is notable. At the time of the Gulf Oil spill, liability for spills from offshore facilities was capped at $75 million, with some exceptions, such as negligence or willful misconduct. This amount is clearly inadequate to cover expenses resulting from a major spill and also reduces the incentive for a company to operate safely. BP did establish a fund of $20 billion to pay compensation to those who livelihoods were damaged by the oil spill, but it was a voluntary gesture, and there is no assurance that other companies would do likewise.

See Also: Environmental Science; Science and Technology Policy; Unintended Consequences.

Further Readings

Bolstad, Erika. "Gulf Oil Spill Fallout: MMS to Be Divided Into Three New Branches." *McClatchy Newspapers* (May 19, 2010). http://www.mcclatchydc.com/2010/05/19/94465/agency-that-regulates-offshore.html (Accessed January 2011).

Bourne, Joel K. "Gulf Oil Spill: Is Another Deepwater Disaster Inevitable?" *National Geographic* (October 10, 2010). http://ngm.nationalgeographic.com/2010/10/gulf-oil-spill/bourne-text (Accessed January 2010).

Bureau of Ocean Energy Management, Regulation and Enforcement. http://www.boemre.gov (Accessed January 2011).

Corn, M. Lynne and Claudia Copeland. "The Deepwater Horizon Oil Spill: Coastal Wetland and Wildlife Impacts and Response." *Congressional Research Service* (July 7, 2010).

National Commission on the BP Deepwater Horizon Oil Spill and Offshore Drilling. "Deep Water: The Gulf Oil Disaster and the Future of Offshore Drilling: Report to the President." https://s3.amazonaws.com/pdf_final/DEEPWATER_ReporttothePresident_FINAL.pdf (Accessed January 2011).

"Times Topics: Gulf of Mexico Oil Spill (2010)." *New York Times.* http://topics.nytimes.com/top/reference/timestopics/subjects/o/oil_spills/gulf_of_mexico_2010/index.html?scp=1-spot&sq=deepwater%20spill&st=cse (Accessed January 2011).

Sarah Boslaugh
Washington University in St. Louis

Participatory Technology Development

Participatory technology development (PTD) is an approach to development that involves collaboration between experts and local citizens and practitioners to analyze problems and find solutions that are appropriate for the specific community in question. PTD was developed in response to low rates of adoption of new agricultural technologies in developing countries and has been applied most often to agricultural development, but there is no reason the same approach cannot be applied to other questions that involve the use of outside technology to solve local problems. In fact, PTD has already been applied to other issues including natural resource management and the development of rural livelihoods. The key point with PTD is that local practitioners and citizens, for instance, farmers and other village members, participate actively in the decision-making process in all stages of the development and implementation of the technology that they will be using. This is a marked departure from the top-down, researcher-driven process that was the norm in agricultural research and development work a few decades ago.

The Green Revolution of the 1960s and 1970s greatly improved the agricultural yield in many developing countries and helped save many people from malnutrition and starvation. Great as those gains were, however, there remain new challenges for agriculture and development. Among these challenges are the need to promote equitable distribution of the benefits of increased agricultural production, to better manage the natural resources that support agriculture, to create interventions for less-favorable environments, to strengthen the ability of local farming communities to continuously improve their methods, and to build synergy between technological change and the political, cultural, and socioeconomic dimensions of agricultural innovation.

Addressing these questions requires a shift in emphasis away from simply increasing agricultural production to broader considerations of how communities function and how people best respond to change. In fact, a paradigm shift is required so that research and development is seen as an ongoing learning process involving the end users of new technology rather than a top-down system in which modern technology is developed in one location (often in the industrialized world) and then simply transferred to the end users (often in the developing world).

Participatory Technology Development With Smallholder Farmers

PTD is a general approach to development rather than a checklist or recipe to be applied in every instance, so it is useful to consider a specific example. Peter Horne and Werner Stur discuss their experience with the participatory approach to development in their work with smallholder farmers in Southeast Asia. They adopted PTD as an approach to improving agricultural production because the more typical approach of using government research and extension agencies to promote adoption of new technologies (e.g., planting new varieties of rice) and government programs (e.g., livestock credit schemes), although successful in some areas, did not work in others. For example, farmers in lowland areas were quick to adopt the use of improved fertilizers and rice varieties, but upland smallholder farmers were not. Horne and Stur decided it was necessary to involve farmers in the decision-making process because they, as outside experts, did not always understand the needs and desires of these farmers (e.g., increased productivity alone might not motivate them to adopt new technologies) and did not sufficiently understand the resources and constraints operating in upland areas. In addition, they realized that it might be unreasonable to expect these farmers to adopt a technology wholesale; rather, the farmers might be more interested in adopting certain parts of it that fit their specific needs.

Horne and Stur note that PTD is not the quickest approach to technological development but that it may be the most effective, at least in certain cases, in the long run. The process of PTD begins with two to four months of preliminary work in which the development workers choose one or more villages to work with. This choice should be made based on the best match between their particular skills and the needs of the community so that the workers' skill can most benefit the lives of individuals in the communities selected. Of course, this means that the people most affected in the village must be interested in working with the development workers because any solutions will require their active participation. Additional considerations may be any social goals specific to their organization (such as alleviating poverty, the opportunity to diffuse technology from the chosen villages to others nearby, and the presence of local partners and organizations to support the work).

Once a village is selected, the next step is participatory diagnosis (PD), a crucial but less time-consuming step that can be completed in a day. During a PD the farmers, in conjunction with the development workers, identify and prioritize the problems they would like to solve, identify who is most affected by these problems, and nominate the individuals from the village who will work with the development workers to solve the problems. There is an aspect of a screening process in a PD: at this stage, the development workers may learn that there is not a good match between the priorities of the farmers and what the development workers have to offer, and hence a particular village will not be a prime candidate for their efforts.

The next step is to explore technology options with the farmers in the selected villages, including a broad range of options that may be appropriate for different farms. This can be visualized as a multistep process beginning with having the farmers describe the causes of their main problems, listing approaches used in the past, and identifying opportunities to introduce new solutions. The next step is to search for potential solutions that may be suggested by the farmers themselves, by farmers from other villages, or may be discovered in the research literature or elsewhere. Then several options should be selected for testing on a small scale; for instance, several new types of seed could be planted in small plots to see which thrive best given the climate and soil type. During the trial period, the development worker can aid in the process by providing monitoring and measurement assistance and by interviewing the farmers involved to discover which solutions they think are the most successful and why (e.g., a high yield of a crop that their animals refuse to eat cannot be judged a success). At the end of the trial period, the farmers involved and the development workers should meet to discuss their experiences, analyze problems, and decide what to do next, and a meeting should also be held with the entire village to report on the results of the experiments.

One of the assumptions of PTD is that when farmers have seen that new technologies can provide answers to problems they care about solving, they will be interested in expanding the use of this technology and integrating it into their current farming practices. Development workers can provide support during this process by providing technical information and suggestions as to how the new technologies may be best adapted to the needs of an individual farm. They can also facilitate the exchange of information between farmers and help overcome bottlenecks such as limited availability of a particular type of seed. Evaluation and communication of results should be carried out during this phase as they were during the testing phase.

If a new technology has proved to be beneficial for one farmer, then other farmers within the village will very likely be interested in adopting it, making diffusion of the technology a natural process. The first adopters can be helpful in sharing their experiences, and the development workers can also create more formal ways to exchange information such as organizing field days to demonstrate technology options and discuss the advantages and disadvantages of each. It is likely that neighboring villages will also hear about successful technologies and may be interested in adapting them, a process that may also be facilitated by promoting the sharing of experiences between villages. Of course, technologies can seldom be transferred from one situation to another intact so a similar learning process will probably be required in new villages as in the first where the technology was tried, including starting small and seeing what works in the specific situation.

Participatory Approaches to Natural Resource Management

Traditional natural resource management programs in the past often have been largely top-down affairs in which most decision-making power lay with centralized agencies and staff. In this approach knowledge is assumed to lie with the researchers and officials who then transfer it to local people, while the latter are not involved in establishing priorities or conducting research except perhaps in a lower-skill capacity such as collecting data. In contrast, a participatory approach involves the local practitioners at all levels, including setting the research agenda, trying out different methods, and evaluating the results.

As with agricultural development, the participatory approach when applied to natural resource management may seem more complex than the research transfer approach. However, it is believed to produce better results when dealing with complex, nonlinear systems such as ecosystems that interact with other complex systems such as social, economic, policy, and institutional systems.

The participatory approach acknowledges the importance of the specific facts of every resource management situation and requires the active participation and collaboration of local leaders and resource users as well as outside authorities and experts. Part of the goal of this type of work is to develop new capacities among those involved, for instance, training local people to plan and carry out research. In addition to looking at facts specific to resource management, the participatory approach requires examination of the local social framework with consideration of how people use the resources, how they regulate resource use and plan for the future, how they resolve conflicts about resource use (e.g., resolving competing claims on the same resources), and how they relate to the authorities and to other communities.

As with PTD applied to agriculture, participatory approaches to natural resource management will only succeed if the local community has interests compatible with the expertise offered by the outside experts or researchers involved. This would generally include a long-term interest in preserving the natural resource in question and a perception that adjustments in current habits may be necessary to ensure its availability in the future. The participatory approach to resource management requires that the approach used be centered on the community and the specific location, not the resource, and that it be flexible and based on a learning process, rather than prescriptive and based on applying preexisting rules.

See Also: Appropriate Technology; Authoritarianism and Technology; Faustian Bargain; Marxism and Technology; Technology and Social Change; Unintended Consequences.

Further Readings

Economic Commission for Africa, United Nations Economic and Social Council. "Principles, Methodology and Strategy for Promoting the African Green Revolution: A Design and Training Manual." (November 2004). http://www.uneca.org/sdd/documents/agree.PDF (Accessed September 2010).

Gonsalves, Julian, Thomas Becker, Ann Braun, Dindo Campilan, Hidelisa de Chavez, et al. "Participatory Research and Development for Sustainable Agriculture and Natural Resource Management: A Sourcebook. Volume 1: Understanding Participatory Research and Development." Ottawa: International Development Research Centre, 2005. http://www.idrc.ca/openebooks/181-7 (Accessed September 2010).

Horne, Peter M. and Werner W. Stur. "Developing Agricultural Solutions With Smallholder Farmers: How to Get Started With Participatory Approaches." ACIAR Monograph No. 99. Canberra: Australian Center for International Agricultural Research, 2003. http://www.smallstock.info/reference/ACIAR/mono/99/aciar_monograph99_contents.pdf (Accessed August 2010).

Mason, Kelvin. *Brick by Brick: Participatory Technology Development in Brickmaking.* London: ITDG Publishing, 2001.

Michael, Yohannes Gebre and Karl Herweg. *From Indigenous Knowledge to Participatory Technology Development: Soil and Water Conservation.* Berne, Switzerland: Centre for Development and Environment, University of Berne, 2000.

Pichon, Francisco J., John Frechione, and Jorge E. Uquillas, eds. *Traditional and Modern Natural Resource Management in Latin America.* Pittsburgh, PA: University of Pittsburgh Press, 1999.

Probst, Kirsten and Jurgen Hagmann. "Understanding Participatory Research in the Context of Resource Management: Paradigms, Approaches and Typologies." AgREN Paper #130 (July 2003). http://www.odi.org.uk/resources/download/4267.pdf (Accessed August 2010).

Sarah Boslaugh
Washington University in St. Louis

Passive Solar

Passive solar technologies harness heat energy absorbed from the sun. Heat flows from warmer materials to cooler materials until there is no temperature difference between them. Passive solar technologies are based on this law and can be used in various applications using the heat flow methods to transfer the heat to the desired activity or space. These activities may be cooking, heating and purifying water, solar lighting, causing air movement for ventilation, and other such passive space heating and cooling techniques for solar architecture without the use of mechanical services.

Methods of Using Natural Heat Flow

Conduction conveys heat from molecule to molecule from the hot side of a material to the cooler side or to a separate cooler object in contact with it. This principle explains the reason why in hot climates, heat from the external walls and roof of a building travels to their inside surfaces, which are at cooler temperatures than the external surfaces. Similarly, in cold climates, heated homes are built of external walls that have insulation-filled cavities so that heat from the warmer interior surfaces of the walls does not travel to the exterior cold surfaces by conduction.

The ability by which heat is stored or contained in a material to enable its flow by conduction is called its thermal capacitance. The heat from solar radiation increases the temperature of the material or of its thermal mass. As a general observation, higher-density materials may have more thermal mass, like concrete, stone, tiles, and constructed masonry walls, but a large store of water with high specific heat capacity is also used effectively as a thermal store. Other high specific heat capacity phase change materials such as paraffin wax and molten salt can effectively deliver heat at high temperatures.

As one side of the thermal mass is heated by the radiation, the remaining parts receive this heat by conduction, convection, or radiation over a period of time. Thermal mass can be used in passive solar buildings by correctly positioning it to receive heat from solar radiation, which can then be transferred by the heat flow methods to heat the living and working spaces within the building.

Convection is the way heat circulates through the movement of liquids and gases, where warmer air or liquid rises and the colder part sinks. This gives rise to convection currents, which can be used in natural ventilation strategies. Solar chimneys work on convective air flows and have been used as passive ventilation techniques by civilizations that are centuries old.

Radiation is the transfer of heat from a hot body to the air surrounding it and then to a cooler body at some distance away from it. The transfer of solar energy from the sun to the Earth's surface takes place by radiation. Generally, darker colors absorb more heat energy from the sun than do lighter colors. The amount of heat absorbed by a surface also depends on its absorption coefficient, which is the property of the material that it is made of, in addition to its color. In the same way, the amount of absorbed heat that a surface radiates back to the cooler air surrounding it depends on its emissivity. A perfect emitter of heat energy is considered to have an emissivity of "1," and similarly, a hypothetical material with an emissivity of "0" would not emit any heat energy.

The greenhouse effect is now a well-known phenomenon. Heat energy from the sun is re-emitted by the heated Earth as long-wave infrared radiation. Greenhouse gases present in the atmosphere absorb this long-wave radiation and reradiate it back to the surface of the Earth in a continuous loop, thereby keeping temperatures high. This principle is used extensively in cold climates for passive heating of conservatories and sunrooms, while keeping occupants sheltered from wind, rain, and snow.

Passive Solar Design Principles

Effective passive solar design depends on five aspects or design elements, namely, the collector, absorber, quantity of thermal mass, methods of heat flow used, and the control and distribution of heat gain. These five aspects differ in the way they are used depending on whether direct or indirect solar gain is to be used in the application. Passive design techniques are also used to deflect the sunlight from overheating buildings by the use of solar shades, awnings, and louvers as well as high-albedo roof coatings and light-colored external wall paints to prevent heat entering the buildings through roofs and walls.

A solar collector may be described in general terms as the aperture or glazing that allows in solar radiation, although its type may vary from application to application. Types of solar collectors include windows, glazed facades, collectors attached to solar water heaters, or solar ovens and cookers. Solar radiation, after entering through the aperture or collector, will strike the solar absorber, which is ideally dark or black in color to absorb as much heat as possible from the sunlight striking it. The amount of the material that retains the absorbed heat is the thermal mass. It is not harmful but neither particularly advantageous to undersize or oversize thermal mass in order that optimum use be made of the heat absorbed by it, and care must be taken to ensure that it is insulated from the cold so that the heat can be used for its intended purpose for the required length of time. Heat exchange from an absorber to the thermal mass is by conduction, convection, or radiation, and the control mechanism also applies to the use of heat flow methods when required and stopping them when the heat needs to be retained or conserved.

An example of the application of direct solar gain is the principle of the greenhouse effect that is applied for passive heating of conservatories, sunrooms, or solar atria. It uses large glass windows as the aperture that allows solar radiation to pass through, but does not allow long-wave infrared radiation to escape, and keeps the heat trapped inside. Hence, the glass also serves as the control mechanism to trap the heat within the conservatory or sunspace. The absorber is the heated air inside the conservatory; the thermal mass is the floor and other surfaces that retain the heat gained through radiation.

A solar water heater is an example of how water can be effectively used as thermal mass to retain the heat by the indirect solar gain method. In this case, the absorber is the black material that receives the heat from the sunlight and passes it by a heat exchange mechanism to the thermal mass, which is the water that is inside the tank. The heated water can be used for a variety of domestic purposes as well as to contribute to some part of the water heating required for domestic space heating. However, this needs careful design of the solar thermal system, working temperatures, and the indoor heating system and must be carried out by qualified engineers and design professionals. In addition, correct positioning of the solar collector and the capacity required of the heating system as well as a backup system are some of the factors that will influence the solar thermal system design.

Economics and Solar Fraction

The solar fraction is the amount of energy supplied, out of the total, by the passive solar technology used. It is considered to be "0" for 0 percent solar energy production and up to "1.0" for 100 percent provision of all required energy supply through solar technologies. A solar fraction of a particular system is dependent on the weather data at its installation site and other site factors like orientation, size, and operation of the solar collector as well as load or use factors and storage sizes. In a hypothetical comparison, a solar system installed in an arid zone may have a higher solar fraction than a system with the same specification in a colder or wet climate. The solar fraction is an important factor in the calculation of the economies and payback of a solar energy installation, and must be given careful consideration at the design stage.

See Also: Earthships; Solar Hot Water Heaters; Sustainable Design; White Rooftops; Zero-Energy Building.

Further Readings

Baker, Nick. *Passive and Low Energy Building Design for Tropical Island Climates.* London: Commonwealth Secretariat Publications, 2010.

Haggard, Ken, David Bainbridge, and Rachel Aljilani. *Passive Solar Architecture Pocket Reference Book.* London: Earthscan, 2010.

Mazria, Edward. *The Passive Solar Energy Book.* Emmaus, PA: Rodale Press, 1979.

U.S. Department of Energy. "The History of Solar." http://www1.eere.energy.gov/solar/pdfs/solar_timeline.pdf (Accessed July 2010).

Swati Ogale
Independent Scholar

PLASMA ARC GASIFICATION TECHNOLOGY

Plasma arc gasification is an emerging waste treatment technology that uses electrical energy and high temperatures to turn garbage into usable by-products without burning it. An electrical arc gasifier passes very high voltage electrical current through two electrodes, creating an arc between them. This arc breaks down most of the waste into gas and solid waste, or slag, in a device called a plasma converter. The process can reduce the volume of waste sent to landfills and also be a net generator of electricity.

Plasma arc gasification appears to offer significant potential for reducing landfill waste and converting garbage into useful products. However, the likely costs of such facilities and the uncertain environmental impacts have complicated efforts to build them. Small facilities operate in several countries, and several large-scale facilities have been proposed in the United States and a number of other countries. Even if large-scale facilities are not constructed, advocates say the technology can be particularly effective for handling medical and hazardous waste.

One obstacle to constructing plasma arc gasification facilities has to do with basic economics: burying garbage—even though landfill space in many areas is limited and leaking waste from landfills can contaminate water—is more cost efficient than applying enormous heat to it. Whereas municipalities in 2007 were charging an average of $35 or so for dropping off a ton of garbage at a landfill, some studies claim that a ton of garbage at a plasma gasification

Slag like this is created when the plasma arc breaks down waste into gas and solids. After being cleaned, slag can be processed into bricks or synthetic gravel.

Source: iStockphoto

plant could exceed $170, although costs would likely come down as the technology becomes more established. A key technological challenge is the construction of liners to separate the high interior temperatures of the plasma system from the metal shell of the plasma container. Liners can degrade from chlorine as well as from significant variations in temperature—both of which would be present at a plasma arc gasification facility. As the technology develops, engineering designs are emerging to deal with this issue, such as the use of brick liners and slightly lower temperatures.

Plasma arc technology has also drawn some questions from environmentalists because of the contaminants that could be left in the by-products. They contend that syngas produced by plasma arc gasification, if burned for energy without proper treatment, could emit toxic acids, dioxin, and other pollutants, and the slag could retain high levels of mercury and other hazardous materials that can create challenges for solid waste disposal. Environmentalists also worry that people may become complacent about reducing the waste stream if they believe the waste can be recycled.

The technology is sometimes referred to as incinerating or burning trash, but that is inaccurate. Plasma gasification does not combust the waste as incinerators do. Instead, it converts the organic waste into a gas that still contains all the chemical and heat energy from the waste. It converts the inorganic waste into an inert vitrified glass.

The process works by passing inert gas, which is under high pressure, through the electrical arc into a sealed container of waste materials. Temperatures in the arc column can reach as high as 25,000 degrees Fahrenheit, which is hotter than the surface of the sun. Exposed to such temperatures, most types of waste are transformed into gas consisting of basic elements, while complex molecules are torn apart into individual atoms.

The by-products of plasma arc gasification consist of the following:

- Syngas, which is a mixture of hydrogen and carbon monoxide. Many waste materials, including plastics, contain high amounts of these elements, and the conversion rate into syngas can exceed 99 percent. Before the syngas can be used for power, it must be cleansed of harmful materials such as hydrogen chloride. Once cleaned, the syngas can be burned like natural gas, with a portion going to power the plant and the remainder sold to utility companies.
- Slag, which is a solid residue resembling obsidian. After being cleaned of contaminants, including heavy metals such as mercury and cadmium, slag can be processed into bricks, synthetic gravel, or materials that are mined from underground.
- Heat from the process. Depending on the individual configuration of a plant, some experts have speculated that the waste heat could be used to power steam turbines and generate more electricity.

The composition of the waste stream can affect the effectiveness of the gasification procedure. Garbage that is high in inorganic materials, such as metals and construction waste, will yield less syngas, which is the most valuable by-product, and more slag. For that reason, it may be worthwhile in certain settings to presort the waste stream. If waste can be shredded before it enters the main chamber, it will improve efficiency and help ensure that all materials are broken down.

The Sun Energy Group, which has proposed building a large-scale plasma gasification facility in New Orleans, estimates that a ton of trash will yield 55.2 kilowatt-hours of power. Sun and other companies also claim that plasma gasification plants would emit relatively small amounts of carbon dioxide because they would rely on syngas, which is similar to natural gas in that it emits less carbon dioxide than coal or oil.

Several plasma gasification plants are under consideration for permits in the United States, but none have been built yet. Plans for the nation's first plasma-based waste disposal system were announced in 2006 in St. Lucie County, Florida. County officials at the time hoped not only to reduce their waste flow but to completely empty the existing landfill within 19 years. The initial plan was for a plant with a capacity to process about 3,000 short tons of waste each day. Questions about the safety of such a facility, as well as whether the community could produce those amounts of waste, led to a revised proposal for a smaller facility that would handle about 200 short tons per day. Plasma arc gasification plants have also been proposed for Los Angeles, New Orleans, and Tallahassee, Florida.

In the past decade, several small-scale plasma gasification plants have begun operations overseas, and a number of larger-scale facilities are in the planning stages. A facility at National Cheng Kung University in Taiwan City, Taiwan, processes 3.5–5.5 short tons daily from a number of waste streams. It opened in 2005 as part of a comprehensive resource recovery facility funded by the Taiwanese government. Several comparably sized plants are in operation in Japan, while plants in England and Canada that use slightly different technologies also process relatively small amounts of waste.

See Also: Science and Technology Policy; Solid Waste Treatment; Waste-to-Energy Technology.

Further Readings

Earthanet. "Plasma Arc Gasification: Turning Garbage Into Energy." http://earthanet.com/2008/01/28/plasma-arc-gasification-turning-garbage-into-gas (Accessed September 2010).

"Florida County Plans to Vaporize Landfill Trash." *USA Today* (September 9, 2006). http://www.usatoday.com/news/nation/2006-09-09-fla-county-trash_x.htm (Accessed September 2010).

Koerner, Brendan J. "Can We Turn Garbage Into Energy?" *Slate.* http://www.slate.com/id/2181083 (Accessed September 2010).

David Hosansky
Independent Scholar

POSTINDUSTRIALISM

A postindustrial society is marked by a transition from a manufacturing-based economy to a service-based economy, usually by means of a diffusion of national capital, and mass privatization. These economic transitions are connected with a subsequent societal restructuring.

Postindustrialization is the next evolutionary step from an industrialized society and, therefore, is most evident in places such as the United States, Western Europe, and Japan. Daniel Bell describes six features of a postindustrial society. First, there is a transition from the production of goods to the production of services, with very few firms directly manufacturing any goods. Second, blue-collar manual laborers are replaced by technical and professional workers, such as computer engineers, doctors, and bankers. As the direct production of goods is moved elsewhere, those sectors of society that rely heavily on production, such as mining or automobile assembly, face large-scale unemployment. Third, an increased reliance on theoretical knowledge comes to replace practical knowledge. Fourth, postindustrial societies pay greater attention to the theoretical and ethical implications of new technologies. By giving strict attention to the impact of new technologies, environmental incidents such as Three Mile Island or massive widespread power outages can be avoided. Fifth, in order to assess the theoretical and ethical implications of new technologies, newer scientific disciplines are developed. Postindustrial societies therefore rely heavily on new forms of information technology, cybernetics, or artificial intelligence. Finally, the reliance on new technologies and scientific discipline places a stronger emphasis on the university and polytechnic institutes. Since these institutions produce the people who will create and guide new technologies, universities are crucial to a postindustrial society.

In addition to the direct economic characteristics of a postindustrial society, changing values and norms reflect the changing influences on the society. The result of the outsourcing of manufactured goods, for example, leads to changes in the way that members of a society see and treat foreigners or immigrants. Further, those individuals previously occupied in the manufacturing sector find themselves with no clearly defined social role.

There are a number of direct effects of postindustrialism on the community. For the first time, the term *community* is associated less with geographical proximity, and more with scattered, but like-minded individuals. Advances in telecommunications and the Internet mean that telecommuting becomes more common, placing people farther away from their place of work and their coworkers.

Further criticisms take the form of the relationship between manufacturing and service. Moving to a service-based economy means that manufacturing must occur elsewhere, and is often outsourced to industrial economies. While this gives the illusion that the postindustrial society is merely service based, it is still highly connected with those industrial economies to which the manufacturing is outsourced.

See Also: Futurology; Information Technology; Technological Determinism.

Further Readings

Bell, Daniel. *The Coming of Post-Industrial Society*. New York: HarperColophon, 1974.
Fisher, Dana R. and William R. Freudenburg. "Postindustrialization and Environmental Quality: An Empirical Analysis of the Environmental State." *Social Forces*, 83/1 (2004).
Milani, Brian. *Designing the Green Economy: The Postindustrial Alternative to Corporate Globalization*. Lanham, MD: Rowman & Littlefield, 2000.

Robert C. Robinson
University of Georgia

POZZOLAN

The most common modern pozzolan is fly ash, like this split fly ash being collected in the Emory River. The spherical particles in fly ash make the cement made from it especially workable and reduce water demand, and the use of fly ash does not add to any carbon impact.

Source: Steven Alexander/U.S. Fish and Wildlife Service

Pozzolan is a category of materials that exhibit cement-like properties when combined with calcium hydroxide (slaked lime) and that are typically added as a cement extender to Portland cement concrete mixtures. The category takes its name from *pozzolana*, a volcanic ash discovered in Pozzuoli, Italy, in the Mt. Vesuvius region. Siliceous and aluminous (containing silicon dioxide and aluminum oxide), *pozzolana* reacts with calcium hydroxide when mixed with it and water, forming a cement-like substance at room temperature that can set underwater. Ancient Romans mixed *pozzolana* two to one with lime before adding the solid mixture to water to create what we now call "Roman concrete," and then poured it through a long tube in order to construct underwater foundations for ports (the tube kept the cement from mixing with seawater). The piers at Cosa have survived more than 21 centuries since their original construction by this method.

Roman concrete was a predecessor of Portland cement, the most common cement in modern use. From the late Republic era through the end of the Roman Empire, Roman concrete was one of the keys to Rome's technological dominance of its world. Different *pozzolanas* from different regions—identified by color—were mixed in different ratios, established by 25 B.C.E., and led to extensive underwater construction. The rebuilding of Rome after the fire of 64 C.E. established the brick and concrete industries, as Emperor Nero demanded that most of the buildings be built of brick-faced concrete (opus latericium or opus spicatum; because the concrete surface was considered crude and ugly, it was usually covered with some sort of facing).

The formation of the cement-like substance is caused by what is called the Pozzolanic reaction:

$$Ca(OH)_2 + H_4SiO_4 \rightarrow Ca^{2+} + H_2SiO_4^{2-} + 2\ H_2O \rightarrow CaH_2SiO_4 \cdot 2\ H_2O$$

The acid–base reaction between calcium hydroxide and the silicic acid in the pozzolan results in a calcium silicate hydrate, which is responsible for the strength of cement mixtures.

The most common modern pozzolan is fly ash, a residue produced during the combustion of coal and containing large proportions of silicon dioxide. Fly ash's suitability as a pozzolan was noted around the time of World War I, but it did not come into use until the 1930s.

Until recently, fly ash was used to replace up to 30 percent of the mass of Portland cement in a concrete mixture; new mixture designs replace as much as half of the Portland cement with high-volume fly ash, and a dam project in India used roller-compacted concrete made of 70 percent fly ash. The spherical shape of fly ash's clay-like particles makes the cement made from it especially workable and reduces water demand. Because fly ash is a by-product of coal burning, the use of fly ash in cement adds nothing to its carbon impact. Portland cement, in contrast, produces about one ton of carbon dioxide for every ton of cement produced. Of course, fly ash is only carbon-neutral when it is available as a by-product: burning coal specifically to produce fly ash results in 30 tons of carbon dioxide for every ton of fly ash. Fly ash is therefore not a long-term green cement solution, presuming a global green goal of eliminating or drastically reducing the burning of coal. About 131 million tons of fly ash are produced annually in the United States alone; less than half of it is reused for any purpose. Fly ash does have environmental drawbacks. Although the U.S. Environmental Protection Agency (EPA) has ruled that it does not need to be regulated as a hazardous waste, it contains trace elements of various substances detrimental to human health, including arsenic, mercury, and beryllium, and cannot be stored (by itself, not mixed into concrete) in places where rainwater could reach it and leach the metals into the groundwater supply.

Other pozzolans include silica fume, a by-product of producing silicon and ferrosilicon alloys; ground granulated blast furnace slag, a glassy substance created by quenching with water the iron slag produced as a by-product of iron and steel making; and high-reactivity metakaolin, formed by calcining the clay mineral kaolinite in a high-heat kiln. Even apart from the increase in long-term strength when pozzolans are used in concrete, one of the most significant benefits of their use is that most of them exist as waste products of industrial processes that are unlikely to cease any time in the near future, waste products that have few or no other uses and, when not used as pozzolans, end up in landfills. Green building in the future can expect to rely heavily on pozzolan mixtures, whether in concretes or in geopolymers.

See Also: Cradle-to-Cradle Design; Green Building Materials; Green Markets; Unintended Consequences.

Further Readings

Cunningham, William and Mary Ann Cunningham. *Environmental Science: A Global Concern*. New York: McGraw-Hill, 2009.
Miller, G. Tyler and Scott Spoolman. *Environmental Science*. Florence, KY: Brooks/Cole, 2010.
Vig, Norman J. and Michael E. Kraft. *Environmental Policy: New Directions for the Twenty-First Century*. Washington, DC: CQ Press, 2009.

Bill Kte'pi
Independent Scholar

PRODUCT STEWARDSHIP

Product stewardship is a practice in which the fiscal and physical responsibility for the disposal of a product and other associated environmental harms of product production are designated as the responsibility of all parties involved in the product supply chain. Parties in the product supply chain include manufacturers, retailers, and consumers. Levels of responsibility for product stewardship practices generally increase depending upon the degree to which each responsible party can reduce environmental harm. Product stewardship recognizes that the reduction of environmental impacts of a product requires cooperation and action from all parties along the supply chain through the product's end of life.

Product stewardship is often described in conjunction with—or as a form of—extended producer responsibility. Extended producer responsibility is a practice or policy approach that shifts responsibility for product disposal from the government to the producer of the product while encouraging environmentally friendly product design. While similar in motivation, product stewardship differs from extended producer responsibility in that it is an attempt to designate responsibility for disposal and related environmental impacts to all parties in the supply chain, not only the producer of the product. Product stewardship practices, unlike extended producer responsibility, also tend to be voluntary rather than mandated.

Responsibility for product disposal and environmental harms may be designated in several ways, including assigning greater responsibility to the organization with the most control in terms of product design, equal responsibility among all parties involved, or additional responsibility to the party most easily able to pay for costs associated with recycling or changes in product design. Generally, manufacturers or producers of products are considered to have the greatest ability and, therefore, responsibility to effect product design, thus determining environmental impact. Recyclability, for example, must be designed into a product in the production stage, a responsibility that must be designated to the producer.

Product stewardship uses a life-cycle approach to reduce the environmental impacts of manufactured products. A product's life cycle includes all impacts associated with the product from initial manufacturing, or the "cradle," to disposal, or the "grave." Product stewardship, therefore, addresses both the individual environmental impacts of each stage in a product's "life" and the interactions among the stages. Although product stewardship practices focus on environmental impacts of the production, use, and disposal of a product generally, disposal of a product at end of life, or when it is no longer useful to a consumer in the original form, tends to be a more prominent focus of product stewardship practices.

A number of benefits and motivations exist for the undertaking of product stewardship. Producers may undertake practices for purposes of cost reduction. Costs may be reduced through the development of more efficient production processes or the decrease of liability by using more environmentally friendly procedures. Producers may also emphasize the use of product stewardship as a marketing strategy targeting environmentally conscious consumers. Finally, employing product stewardship practices may be used as an attempt to avoid future governmental regulation.

Although less in comparison to producers, product stewardship practices can benefit retailers. Several examples of initiatives in place in the United States use retailers as the point of collection for products at end of life. Retailers may benefit from increased sales when customers visit their store to dispose of products.

Government benefits include reduced waste disposal costs and potentially the elimination of the need for regulation. The cost for and management of solid waste is generally the responsibility of the government. Designating responsibility to all parties in the supply chain reduces the burden on the government in terms of product disposal. This reduction of burden may take a variety of forms, including the designing of recyclability into a product by the producer, collection for disposal of a product by retailers, or the reduction of harmful components in products, therefore reducing harm caused through general disposal.

Product stewardship efforts are an attempt to both encourage environmentally friendly product design and increase recycling by designating this responsibility to all parties involved in the product supply chain. Although there are many benefits to product stewardship, drawbacks include increased costs for producers and retailers, difficulties in enforcing efforts that are nonmandatory, and limitations of practices in that effectiveness is dependent upon the level of interest and involvement of all parties.

See Also: Cradle-to-Cradle Design; Design for Recycling; Extended Producer Responsibility; Sustainable Design.

Further Readings

Hickle, G. T. and D. Stitzhal. "Apportioning Responsibility for Product Stewardship: A New Case for a New Federal Role." *Environmental Quality Management* (Spring 2003).
Palmer, K. and M. Walls. "Economic Analysis of the Extended Producer Responsibility Movement: Understanding Costs, Effectiveness, and the Role for Policy." International Forum on the Environment: Resources for the Future (2002). http://weber.ucsd.edu/~carsonvs/papers/4002.doc (Accessed January 2011).

Sarah M. Surak
Virginia Tech

Pyrolysis

Pyrolysis is a type of thermolysis or chemical decomposition of organic (carbon-based) materials through the application of heat in the absence or near-absence of oxygen. Pyrolysis is thus distinct from burning, which can only take place if sufficient oxygen is present. The rate of pyrolysis increases with temperature, and in industrial applications the temperatures used are often 800 degrees Fahrenheit or higher, while in small operations the temperature may be much lower. Two well-known products created by pyrolysis are charcoal, created by heating wood in an oxygen-poor environment, and coke (which is used as an industrial fuel as well as for heat shielding), created by heating coal in an airless furnace. Pyrolysis has numerous applications of interest to green technology; for instance, as a method for removing organic contaminants from soils and oily sludges, to extract usable materials from used goods such as vehicle tires, and to create biofuel from crops and waste products.

In the process of pyrolysis, organic materials are transformed into their gaseous components plus a solid residue of carbon and ash. Pyrolysis has two primary methods for removing contaminants from a substance: destruction and removal. In destruction, the organic contaminants are broken down into compounds with lower molecular weight, while in the removal process, they are not destroyed but are desorbed from the contaminated material. Pyrolysis is a useful process for treating organic materials that "crack" or decompose under the presence of heat; examples include polychlorinated biphenyls (PCBs), dioxins, and polycyclic aromatic hydrocarbons. However, it is not useful for removing or destroying inorganic materials such as metals. According to the U.S. Environmental Protection Agency (EPA), pyrolysis is an emerging technology whose basic concepts have been validated but have not yet been evaluated according to EPA quality control standards due to insufficient data.

Recycling, Decontamination, and Fuel Production

Technology is already in use to reclaim useful materials from products that would otherwise be discarded. For instance, it is possible to use pyrolysis to break down vehicle tires into useful components, thus reducing the environmental burden of discarding the tires (a significant landfill component in many areas and a product that releases many contaminants into the air when burned) while also providing recycled materials that can replace the use of new materials in industrial processes. When treated at approximately 800 degrees Fahrenheit, the tires break down into gas and oil (usable for fuel) and carbon black (usable as filler in rubber products, including new tires and as activated charcoal in filters and fuel cells). The process has not been widely adopted due to the high capital investment and operating costs required, but improvements in technology and changes in the costs of disposal and competing materials may change in the future.

Another use of pyrolysis is to remove organic contaminants from compounds such as sewage sludge (semi-solid materials that remain after wastewater is treated and the water content reduced). Currently, about 40 percent of sewage sludge is used as fertilizer and the remainder incinerated, used as landfill, or dumped into the ocean. However, concerns about toxic components of sludge, including synthetic hormones, have led to many countries banning the use of sludge as fertilizer and have also brought into question other methods of disposal that can pollute the air, water, and land. However, if pyrolysis is used to remove the contaminants, then sludge can safely be used as a fertilizer. This not only eliminates much of the pollution and expense involved in disposing of the sludge but also creates an alternative to chemical fertilizers that would otherwise be used in agriculture.

Pyrolytic processing of biomass (biological materials such as wood and sugarcane) holds great promise for producing energy sources that could supplement or replace petroleum-based products. The process of pyrolysis causes the cellulose, hemicellulose, and part of the lignin in the biomass to disintegrate to smaller molecules in gaseous form. When cooled, these gases condense to the liquid state and become bio-oil while the remainder of the original mass (mainly the remaining lignin) is left as solid charcoal and noncondensable gases. There are a number of demonstration projects that have shown that pyrolytic biofuel production is feasible on a small scale, but the technology has not yet been tested on a small scale. Currently, bio-oil is more expensive and has a lower caloric value (i.e., it produces less heat during combustion) than conventional oil, making it more expensive and also raising storage and transportation costs. However, this may change in the future as technologies improve and if the price of petroleum-based fuel increases. The processing of biomass for fuel also holds the promise of finding a useful purpose for solid waste that would otherwise have to be disposed of in landfills or through burning.

Biochar

Charcoal has long been created from wood through pyrolysis and remains an important fuel in much of the world. Charcoal is often considered a harmful product because burning charcoal contributes to air pollution, and often human demands for charcoal have led to deforestation and environmental degradation. However, new methods for creating and using a high-grade type of charcoal created from organic materials, called "biochar," holds the promise to reduce pollution by reducing greenhouse gas emissions as well as finding useful purposes for waste (including chicken manure and scraps from corn harvesting) that would otherwise need to be disposed of. Biochar can also reduce the need for chemical fertilizers in agriculture and, when used as a substitute for wood-based charcoal, can also combat deforestation.

Biochar can be created on an industrial scale (i.e., in large factories) but also in small stoves, making it an accessible technology for rural areas and developing countries. It is also possible to harness the energy from the sun to perform pyrolysis, a process known as "solar pyrolysis," which eliminates the need for human-created sources of power in the process.

Interest in the agricultural uses of biochar was sparked by the discovery of an unusually productive soil called *terra preta* located around the Amazon River. This soil is rich in charcoal and, although there is not consensus on whether the charcoal was added by farmers or occurred naturally (as it does in some parts of Australia, for instance), it is clear that the charcoal itself is thousands of years old (according to carbon dating), meaning that it did not break down or degrade during that time. This is in distinction to compost or manure, which are often used to supplement soil but which quickly degrade in the soil. Research projects are currently under way to test the ability of charcoal added to the soil to improve productivity; for instance, depleted soils in Kenya have doubled their yields after being treated with biochar.

The mechanism by which biochar improves the agricultural productivity of soil is not entirely understood because biochar does not contain mineral nutrients as do fertilizers. One theory is that because biochar is extremely porous, it is effective in trapping water and also offers a large surface area to capture bacteria and nutrients such as nitrogen, thus improving the soil's nutrient efficiency. It can also balance soil acidity and also carries a negative electrical charge, which enables it to attract positively charged nutrients like potassium and calcium that might otherwise be washed away. In regions where it has proved successful, biochar holds the potential to drastically reduce the amount of chemical fertilizers used and also to reduce water pollution due to runoff from the fertilizers that remain in use.

Biochar has not been effective everywhere it has been tried and further research is under way to try to pinpoint where it will be most useful, as well as to isolate the mechanisms by which it improves soil productivity. In addition, the qualities of biochar depend in part on the feedstock used to create it, and if the feedstock is contaminated by heavy metals, for instance, that could pose additional concerns (although such contaminants might also be removed through the process of pyrolysis).

See Also: Biochar; Cellulosic Biofuels; Clean Energy; Recycling; Solid Waste Treatment.

Further Readings

Coleman, Mark. "Portable Pyrolysis Unit for Bioenergy." http://yosemite.epa.gov/R10/airpage
    .nsf/c36cf12146018ddc882569e5005f1951/75f93e40b30302fc88257408008069e9/$FILE/
    Coleman.pdf (Accessed September 2010).
European Biomass Industry Association. "Pyrolysis." http://www.eubia.org/211.0.html
    (Accessed September 2010).
Krietmeyer, Sharon and Richard Gardern. "Pyrolysis Treatment." In *EPA Environmental
    Engineering Sourcebook*, J. Russell Boulding, ed. Ann Arbor, MI: Ann Arbor Press, 1996.
Shah, Jasmin, M. Rasul Jan, Fazal Mabood, and M. Shahid. "Conversion of Waste Tyres Into
    Carbon Black and Their Utilization as Adsorbent." *Journal of the Chinese Chemical
    Society*, 53 (2006).
Voosen, Paul. "Once-Lowly Charcoal Emerges as 'Major Tool' for Curbing Carbon."
    New York Times (September 7, 2010).

Sarah Boslaugh
Washington University in St. Louis

R

Rainwater Harvesting Systems

Rainwater collected from unclean surface runoffs is not suitable for drinking or cooking, so this roof has a layer of ethylene propylene diene monomer (M-class), or EPDM foil, which doesn't pollute rainwater and allows the water to be used for any requirements.

Source: Wikimedia

Rainwater harvesting systems collect and store rainwater for human use—a highly advantageous system in areas facing water scarcity. Water scarcity is a pressing problem for many densely populated regions in the world. The demand for freshwater for consumption exceeds the amount of freshwater that is naturally replenished by the Earth's yearly climatic cycle. Water is demanded for human consumption and food production, for irrigation of crops as well as to take care of the needs of livestock, and in food manufacturing processes. There is growing concern as to whether we are at the point of exhausting the levels of naturally available freshwater, akin to the "peak oil" phenomenon. The United Nations (UN) defines "moderate to high water stress" as water consumption in a region that exceeds renewable freshwater resources by 10 percent. By this measure, some 80 countries, constituting 40 percent of the world's population, were already suffering from water shortages by the mid-1990s. According to another UN estimate, by 2025, 1.8 billion people will be living in regions with absolute water scarcity, and two out of three people in the world could be living under conditions of water stress.

Precipitation does not fall evenly. Recent extreme weather conditions in many countries have caused droughts in some regions and severe flooding in others, creating an imbalance in the availability of freshwater even in adjoining regions. Although these facts and statistics of water shortages are stark, it is necessary to understand that a large portion of precipitation falls in inaccessible regions or on the oceans and seas, leaving only a small proportion of the total precipitation falling on land easily available for human use. Of this, freshwater can be captured for human consumption only from containment areas that can be easily accessed, or where natural lakes and reservoirs are harnessed to supply this need. Most of the rain falling on buildings, roofs, roads, and other hard landscaping is directed away from the built environment into storm sewers for discharge and disposal. Impermeable hard surfaces are a cause of urban flooding because they do not allow rainwater to percolate into the soil below, but cause it to accumulate on the surfaces. This water then is unusable due to contamination and has to be discharged away from potable water resources through storm sewers. Rainwater that does percolate into the soil gets naturally filtered by the layers of rock and soil until it reaches and charges the groundwater table that is diminished in summer months, or that is depleted by the demand on water for human consumption. It is now being realized that rainwater harvesting or capture at the point of use for nonpotable functions, like gardening or washing clothes, significantly reduces the amount of the total freshwater that is demanded in an urban or rural setting. This saving in the demand and supply of potable freshwater is significant in large urban conglomerations with considerable economies of scale. The issue is then to design rainwater harvesting systems that fulfill the conditions of health (acceptable water quality), safety (design and installation), and serviceability (cleaning and maintenance) to be able to make use of this resource in day-to-day activities.

Design Recommendations

Rainwater harvesting (storage) systems may be of the dry or the wet type, and pressurized or nonpressurized. The simplest systems are nonpressurized dry systems where the pipes are designed to run at a slope directly from the gutters into the tank. They do not hold any water after it stops raining, and are considered to be the best option because they do not create breeding grounds for mosquitoes and other insects. However, not all configurations can enable direct runs of pipes into the tanks. In places where the tanks are located at some distance away from the collection surfaces, or where there are a series of tanks that serve a number of buildings or an area, pipes from the gutter will go down the wall and underground and then up through a riser into the tank. Thus, there may be long runs of pipes underground that may need to be in a pressurized system so that they do not constantly hold water in the wet seasons and provide a stagnant potential breeding habitat for mosquitoes. Even so, pipes in a wet system that are laid in the soil will hold an amount of water in them as long as the rain falls.

Good design features should ensure that the pipes and all other openings are insect proof, particularly in wet systems. Wire mesh screen covers on all tank inlets will also help prevent debris from entering the tank. Collection surfaces (mainly roofs) should be made of nontoxic materials, particularly avoiding lead-based paints and membranes or flashings, and tanks should be made of nontoxic and noncorrosive material. Care should be taken to ensure that the tank taps or draw-off pipes are at least 100 millimeters (mm) above the tank floor to avoid drawing out the sludge in the water supply, or be fitted with a floating arm draw-off valve. If the tank floor slopes toward the sump and washout pipe, the sludge will automatically be removed periodically. However, regular cleaning of the sump and inside surfaces of the tank is recommended, and the top of the tank should have a well-covered manhole for easy access and inspection.

Quality

Rainwater falling on surfaces mixes with the soluble and insoluble materials present on those surfaces and also collects dust, debris, and other materials as it flows down from higher elevations. Contaminants may be plants, fungi, and other organic materials, as well as inorganic substances like dissolved minerals, metals, chemicals, or water-soluble paints. As a result, rainwater collected from unclean surface runoffs is not suitable for drinking or cooking. Separation of the first flush of rainwater from the roof, gutters, and other collection surfaces improves water quality in the rainwater storage tank. Solar water disinfection may also be used as a method of rendering rainwater safe for general use by pasteurizing water and removing legionella, although this may not be possible in the monsoon season.

Alternate methods of purification include flocculation, settlement, and biofilm skimming, where bacteria, organics, and chemicals form flocs that become films on surfaces or settle to the bottom of the tanks as sludge. The pre-filtered water may then be treated further with chlorine, alum, or other chemicals if the supply is intended for potable uses. Three types of materials are commonly used as a source of chlorine: gaseous chlorine, calcium hypochlorite tablets, and sodium hypochlorite solution. Chlorination can kill all types of bacteria and make water safe for drinking. Liquid alum solution is also added to the incoming raw water. It acts as a coagulant and binds together very fine suspended particles into larger particles, which can be removed by settling and filtration. This removes objectionable color, turbidity (cloudiness), and aluminum from the drinking water. Other chemicals used for water purification are potassium permanganate, hydrated lime or calcium hydroxide, and fluoride. Each treatment requires different strengths of the chemicals applied, and the water quality must adhere to the standards specified for drinking water by local authorities.

The best filtration treatment is by the natural method through the layers of rocks and soil, but it is only useable when the rainwater is allowed to percolate into the ground to recharge the natural groundwater table. An example of a city-wide scheme is observed for the Indian city of Chennai, where building regulations have been modified for a number of years to include mandatory rainwater harvesting systems that are required to direct the flow of rainwater to charge the natural groundwater table. In addition, natural storage systems may be used in landscaping like subsurface dykes, swamps, channels, and man-made aquifers and wells.

Maintenance

Regular maintenance of rainwater harvesting systems is important and should include keeping catchments clean of accumulations of dirt, moss, lichen, and debris, and cutting back trees and branches that overhang these catchment surfaces. Regular cleaning of gutters and tank inlets and screens, and annual tank inspection and cleaning are necessary. The water should ideally be tested periodically to monitor its quality, and steps should be taken to remedy any sign of degradation.

See Also: Desalination Plants; Greywater; Solar Hot Water Heaters; Sustainable Design; Wastewater Treatment; Water Purification.

Further Readings

Commission on Sustainable Development (CSD). "Comprehensive Assessment of the Freshwater Resources of the World—Report of the Secretary-General." New York: United Nations Economic and Social Council, 1997.

Gould, John and Erik Nissen-Petersen. *Rainwater Catchment Systems for Domestic Supply.* London: IT Publications, 1999.

Jebamalar, A. "The Impact of Rain Water Harvesting in Chennai City." http://www.igcp-grownet.org/collaborators/rain-harvest.pdf (Accessed August 2010).

Kinkade-Levario, Heather. *Design for Water: Rainwater Harvesting, Stormwater Catchment, and Alternate Water Reuse.* Gabriola Island, British Columbia, Canada: New Society Publishers, 2007.

United Nations World Water Assessment Program (UN/WWAP). "World Water Development Report: Water for People, Water for Life." (2003). http://www.unesco.org/water/wwap/wwdr/index.shtml (Accessed July 2010).

Swati Ogale
Independent Scholar

RECYCLING

The modern waste management hierarchy, which is reduce, reuse, and recycle, is used as a means of addressing the growing problem of waste. Reusing and recycling both offer ways to extend the usable life of waste materials, and present opportunities to reduce the use of virgin raw materials in manufacturing. Recycling involves further processing of salvaged materials, often using different recycled materials to form a new product, whereas the reuse strategy encourages the use of the components in their original form without further processing. Recycling old materials to make new products needs a considerable amount of energy for the process, but even so, it is a fraction of the energy that may be needed for manufacturing from virgin materials.

Waste materials and substances originate from natural or manufactured products that are discarded as a result of being unusable or unwanted. Of the classifications of waste, household, construction and demolition, institutional, commercial, electronic, industrial, and biodegradable waste streams, including municipal solid waste, all provide

It is easier to recycle pure products like metal cans, and metal can also be recycled infinitely without losing any of its property. Recycling a ton of steel scrap can save 80 percent of carbon dioxide emissions compared to making steel from iron ore.

Source: iStockphoto

great opportunities for recycling and reusing waste materials. In places where recycling and reusing facilities are not available, all waste, including recyclates, are usually disposed of in landfill sites. Medical (also known as clinical), hazardous, and radioactive wastes are usually disposed of in special systems of waste management operated by administrations.

The United Nations' statistics for waste production and methods of disposal show that the largest producers of all types of waste are most commonly the richest and the most industrialized nations of the world. They also show that countries that are not able to afford alternative ways of waste management are using landfill for disposing of the majority of the waste they produce. Landfill has to date been the most commonly used method of waste disposal, but this is quickly changing as land is under pressure for development. The system of collecting waste and transporting it to the landfill site also consumes a lot of energy and produces carbon emissions. Landfill also creates other environmental hazards as toxic materials from waste leaches into the ground and pollutes the soil as well as the ground and surface water, and toxic gases are emitted from the landfill site, causing air pollution. The cost of remediation of a landfill site to be made fit for human occupation is enormous, and prone to reduce the land value to lower than its normal market price.

With growing populations in towns and cities, volumes of waste going to landfills are becoming unsustainable because existing finite landfill sites cannot accommodate the volumes of waste brought to them. This has brought about increases in landfill taxes in many places, which are ultimately paid by citizens through municipal taxation. Even so, city administrations are finding it more and more expensive to pay landfill taxes and are therefore making attempts to divert waste away from landfills by using alternative means of waste management.

Modern consumer products are mostly made from some form of plastics, textiles, metals, glass, paper, or wood. The processes to extract virgin materials and shape them into finished products uses a great deal of energy and manpower. Recycling has many benefits because it prevents the use of new or virgin raw materials, cuts down on waste disposal to landfill sites, and uses a fraction of the energy to recycle unwanted material into a new product, thus reducing carbon emissions.

The process from waste collection to manufacturing new products is a long supply and demand mechanism that involves many agencies and companies. Recycling household electrical and electronic goods is regulated in many places; requirements are often imposed on the reprocessing companies. Resources are needed to set up collection and reprocessing facilities with adequate holding areas and trained manpower, and in many countries, this initiative is being taken by city administrations in partnership with the private sector, thus creating market activity and jobs. In the past, a great deal of waste was sent by wealthier nations for disposal to developing countries with abundant cheap labor, and there were fewer environmental regulations to conform to. Handling large quantities of electronic equipment in order to recover precious metals in the circuitry led to exporting equipment to places where little regard is paid to the health of workers and the environmental consequences of poor treatment of this waste stream.

Sending waste to adequately controlled and supervised reprocessing facilities abroad is a method of waste disposal that may benefit both partner countries by providing an ethical and acceptable solution for waste disposal for one country, and creating businesses and employment in the other. Information collected for the Department of Food, Environment and Rural Affairs (DEFRA) in the United Kingdom (UK) by the Environment Agency on packaging waste shows how much material is exported and how much is recycled in the United Kingdom.

Material	Reprocessed in the United Kingdom	Reprocessed Abroad
Paper	49%	51%
Glass	81%	19%
Aluminum	66%	34%
Plastic	33%	67%
Wood	100%	0%

Even if there are benefits to partner countries for collaborating on processing waste, the sustainability of such practices, particularly relating to carbon emissions through the transport of such waste from the producer to the processor country, is frequently under scrutiny. Arguments for the practice point out the cost benefits of utilizing facilities in developing countries, while opponents are concerned about the loose environmental regulations and checks on the recycling and reprocessing facilities and practices that lead to environmental degradation in those countries. Another point raised, particularly in the case of China, is that due to the huge supply of manufactured goods, there is already considerable shipping traffic from China to developed and other countries. On the return journey for these ships, the waste to be reprocessed is sent in empty cargo containers, thus utilizing an existing service without causing additional traffic. Third, waste materials that are sent for reprocessing may themselves be the raw materials that are needed and used to manufacture new products in China, particularly plastics and PVC products. Therefore, there may be interdependencies between commercial manufacturing, trade, and waste processing that can be mutually beneficial on a case-by-case basis.

The quality of recyclables depends on the method of their collection. In a mixed method of collection, all recyclates are mixed with the rest of the waste and then sorted out and cleaned at a central sorting facility. This results in a lot of recyclates being rejected because they are too soiled to reprocess. In a commingled system, recyclables for collection are mixed, but kept separate from other landfill waste. Different materials like glass, paper, plastics, and metals are then sorted out using a combination of mechanized sorting technology and manpower, which requires up-front setup and running costs. The cleanest method of collecting recyclable materials is by source separation, where each material is cleaned and sorted prior to collection. In some countries, households collect recyclates that have a market value and sell them to small local businesses that operate as collection points. This is a source separation method incentivized by the market, and provides the incentive to keep recyclates clean and to recycle as much as possible. This system is used for glass, metal, paper, recyclable plastic including packets, clothes and textiles, leather, and other products that are valuable in the used goods or recycling markets. In richer countries that do not have many buyers of secondhand materials, recyclates are collected as part of normal waste collection by city administrations. In those cases, extensive public education programs are needed to avoid recyclate contamination. Commingled collection and mechanical sorting facilities are used in such situations.

It is evidently easier to separate pure products like metal cans, but most consumer products give rise to composite wastes that need to be reprocessed at the end of their life in order to separate different materials, which can then be sold on to recycled product manufacturers; for example, an end-of-life washing machine is collected and brought to a reprocessing facility to be broken down into its constituent materials like metal, plastic, glass, rubber, and so on. Each of these materials is then fed into separate supply chains that feed demand for those recycled materials. This type of waste disposal is also subject to rules for disposing and recycling electrical goods.

Paper is made from cellulose fiber, and an average fiber length of 3 mm is used to make good quality, strong paper. Paper can only be recycled four to six times, as the fibers get shorter and weaker each time, and some virgin pulp is added to maintain the strength and quality of the fiber. For this reason, torn or shredded paper is not suitable for recycling as the fibers may not be strong enough to produce good quality paper. "Mill broke" is virgin paper that is discarded from use due to damage or off cuts. Recycled paper is made of at least 75 percent used (de-inked) paper known as postconsumer waste and must not contain more than 25 percent mill broke and/or virgin wood pulp. More than 70 percent energy savings is made during recycling paper compared with making it from virgin wood pulp. This is because most of the energy used in papermaking is spent for pulping wood. For every ton of paper recycled, the savings of at least 30,000 liters of water, 3,000–4,000 kWh electricity (enough for an average three-bedroom house in the UK for one year), and 95 percent of air pollution can be achieved.

New glass is made from a mixture of sand, soda ash, limestone, and additives. These additives provide color or alter properties of the glass. Generally, an iron additive gives brown or green tint, and chromium/cobalt give a green/blue tint respectively, to the glass. Other metals like lead are used to alter the refractive index, alumina for durability, and boron to improve the thermal options. Therefore, while making glass out of recyclates, completely mixed glass (green, brown, blue, white) cannot be used where color purity and properties are critical. Therefore, source separation is the best method of collecting different types of glass for recycling. Color-separated glass "cullet" is used in making bottles and jars. Mixed broken glass is sent to alternative uses such as aggregates, grit blasting, and use in road surfaces and water filtration. Glass can be remelted and reused any number of times, and less energy is needed for reuse every time. After accounting for the transport and processing need, 315 kilograms of carbon dioxide (CO_2) is saved per ton of glass melted.

Metals can also be infinitely recycled without losing any of their properties. Scrap metal can be ferrous (iron/steel) or nonferrous, which includes aluminum (including foil and cans), copper, lead, zinc, nickel, titanium, cobalt, chromium, and precious metals. Valuable metals like copper, gold, silver, lead, and brass are recycled through their own well-developed recycling infrastructure as their value is recognized by the market. Aluminum is only found as a compound called alumina, which is a hard material consisting of aluminum combined with oxygen. Virgin alumina is dissolved in molten salt at a reduction plant, and a powerful electric current is run though the liquid to separate the aluminum from the oxygen. This process uses large quantities of energy. Recycling aluminum requires only 5 percent of the energy and produces only 5 percent of the CO_2 emissions compared with primary production, and reprocessing does not damage its structure. It is also the most cost-effective material to recycle. Steel is also made from iron ore, which is combined with oxygen or sometimes carbon or sulfur. The iron ore is stripped in a blast furnace to reduce it to pig iron that can then be used in steel production. Recycling one ton of steel scrap saves 80 percent of the CO_2 emissions produced when making steel from iron ore. Steel scrap is essential in the process of making new steel and can be recycled infinitely without losing its quality.

Plastics are of different types and grades, and different reprocessing techniques are required during recycling. The different types of plastic therefore need to be collected separately or sorted after collection, as reprocessors will specify which type of plastic they will accept, such as the following:

- *PET* (*polyethylene terephthalate*): Soft-drink bottles, water bottles
- *HDPE* (*high-density polyethylene*): Detergent bottles, milk jugs
- *PVC* (*polyvinyl chloride*): Plastic pipes, outdoor furniture, flooring
- *LDPE* (*low-density polyethylene*): Dry-cleaning bags, trash can liners
- *PP* (*polypropylene*): Bottle caps, drinking straws, yogurt containers
- *PS* (*polystyrene*): Cups, tableware, meat trays, packaging materials
- *OTHER* (*all other resins*): Tupperware, Nalgene bottles

Waste wood is derived from construction activity and demolition, and from commercial, industrial, and household sources for nonconstruction-related wood waste. All softwood and hardwood materials, including pallets and plywood, are recycled. The main use of clean, recycled woodchip has been in the production of chipboard, but animal bedding, equine surfacing, and garden mulches are also common uses for recycled wood. Dedicated biomass plants are also a growing market for recycled woodchip that is from lower-grade wood and is not suitable for recycling into commercial products. Medium-density fiberboard (MDF), laminated material, and railroad ties cannot be recycled and are generally reused through the secondhand market or sent to landfill. There are restrictions imposed with regard to contaminants. Ferrous contamination such as nails and screws can be readily removed by magnetic extraction, but nonferrous materials like aluminum window catches and other fittings are not so easily removed and should be removed in advance.

Waste also contains materials that cannot be stored due to their rapid decomposition, like organic waste and municipal solid waste. When these types of waste are sent to landfill, air cannot get to the organic waste. Therefore, as the waste breaks down, it creates a harmful greenhouse gas, methane, which is a cause of ozone depletion and global warming. There are different methods of disposing these waste streams, and attempts can be made to use the by-products of the disposal process for making other usable products.

Kitchen and organic garden waste is recycled in open-window composting (above ground) where oxygen helps it to decompose aerobically, and the process produces very small amounts of methane. This is suitable for organic waste that does not contain animal by-products. In-vessel composting is a method by which organic waste is treated in enclosed vessels that are generally made of metal or concrete and that allow airflow and temperature to be controlled. This process is suitable for waste containing animal by-products. Composters who treat food waste that includes animal-derived products must comply with animal by-product regulations (ABPR; e.g., Defra in the U.K.) to ensure the safety of the compost they make, in order to kill any animal pathogens that could remain in the compost and potentially spread disease. Composters must follow ABPR guidelines to make it legal for their outputs to be spread on land. Anaerobic digestion is a method by which organic waste breaks down in the absence of oxygen to produce digestate and biogas, which can used to generate energy. This is used for treating municipal solid waste and sewage.

The end product from composting is the soil improver or compost that is reused for growing crops and plants in farming and gardening. Diverting kitchen and garden waste for composting greatly reduces the volumes of household waste going to landfill. However, cooked foods or animal products must not be added to the composting mix. Types of kitchen waste that are composted include used tea bags, coffee grounds and filter paper, vegetable peelings, salad leaves and fruit scraps, cardboard, shredded or torn paper, untreated sawdust, grass and tree cuttings, fallen leaves, bark, and similar organic materials. Of these, brown materials like cardboard, sawdust, branches, bark, and fallen leaves are slower to rot and provide carbon in the mixture. Essentially, good compost is formed out of a mixture of green and brown materials. Bacteria present in the softer green materials facilitate the release of nitrogen and start the decomposition process. Nitrogen is important for forming good-quality compost.

Alcohol fuels can be produced from any waste sugar or starch like bagasse, potato, and fruit waste, and so on. The ethanol production methods used are enzyme digestion (to release sugars from stored starches), fermentation of the sugars, distillation, and drying. The distillation process requires significant energy input and hence cellulosic biomass such as bagasse, which is the waste left after sugar cane is pressed to extract its juice, can be recycled sustainably to produce biofuels.

It is important to know that biofuels are used as fuel replacements because they are more environmentally friendly than traditional fuels like gasoline and diesel in terms of their carbon emissions, and they have a high commercial value. The basic feedstocks for the production of first-generation biofuels are often important constituents of the food chain like sunflower seeds, wheat, and corn/maize, and there is criticism that the biofuel demand is diverting food away from the human food chain in times of food shortages and price rises. Therefore, these first-generation biofuels cannot be considered as recycled products.

Energy-from-waste (EfW) is carried out by pyrolysis in which waste is incinerated under pressure to generate heat and produce steam. Energy captured from the heat produced by the burning waste and the steam is also used to drive turbines to produce electricity. The residue biochar is safe to use as a fertilizer as it is disinfected at high temperature. Biochar is also being considered for carbon sequestration, with the aim of mitigation of global warming. Energy-from-waste techniques are highly advanced technological processes that require a very high level of investment and trained personnel to operate. It is not surprising that most countries have not yet adopted these technologies widely in their strategies for waste management.

See Also: Cradle-to-Cradle Design; Design for Recycling; European Union Waste Electrical and Electronic Equipment (WEEE) Directive; Materials Recovery Facilities; Resource Conservation and Recovery Act (RCRA); Solid Waste Treatment; Waste-to-Energy Technology.

Further Readings

"Automotive Plastics Recycling." *Recycling World Magazine*, 307 (December 7, 1999).
Defra: Department for Environment, Food and Rural Affairs, UK. http://www.defra.gov.uk/evidence/statistics/environment/waste/allrefs.htm (Accessed August 2010).
Goodship, Vanessa, ed. *Management, Recycling and Reuse of Waste Composite*. Cambridge, UK: Woodhead, 2009.
Polprasert, Chongrak. *Organic Waste Recycling: Technology and Management*. London: IWA Publishers, 2007.

Vollrath, K. "Battery Recycling in Europe: Confusion and High Costs." *Recycling International* (November 1999).

Waite, R. "How Recyclable Materials Are Reprocessed." In *Household Waste Recycling.* London: Earthscan, 1995.

Williams, P. *Waste Treatment and Disposal.* Hoboken, NJ: Wiley, 1998.

Swati Ogale
Independent Scholar

REFLEXIVE MODERNIZATION

Reflexive modernization is a key concept associated with the "risk society" perspective developed by sociologist Ulrich Beck and his collaborators and commentators. Beck argues that risk has become a general organizing principle for contemporary society, spanning the domains of personal identity, interpersonal and family relationships, work, organizations, institutions, law, and politics. Technological change has been one important factor in this development, along with other aspects of the industrialization and modernization of society. Applied to concepts of green technology, Beck's ideas address questions of technology assessment, development, and adoption, technological impacts and consequences (including the problem of unintended consequences), policy, and regulation. Since their introduction in the 1980s, these ideas have been widely cited, applied, expanded, critiqued, and contested in the research literatures on risk analysis, risk communication, and social studies of science and technology.

Beck argues that industrial society was originally organized around problems associated with the production of wealth and the distribution of resources. Now, however, those problems are accompanied by a parallel set of problems involving the production and distribution of risks. Of particular significance are what Beck calls "modernization risks," such as those associated with hazardous technological activities and toxic products of those activities, which have become focal points for societal concern. In this state of reflexive modernization, the risks at issue are self-generated, returning like boomerangs to impact the society that has produced them. At the same time, traditional social structures, institutions such as work and family, and cultural premises and expectations are evolving rapidly due to industrialization and rationalization.

Another implication of reflexive modernization is that the recognition of modernization risks, and of the changing nature of society more broadly, call for a fundamental reexamination of politics and social practices. Society is now challenged to look more carefully at key notions of progress, sustainability, social values, ethics, deliberation, and government. Individual, group, and organizational decision making, technology assessment and regulation, and responsibility for the consequences of technological activities all appear different in the light of such "self-confrontation."

Beck regards modernization risks as inherently "democratic": toxic materials, radiation releases, anthropogenic climate change, and other contemporary hazards span social, political, class, geographic, and temporal boundaries. However, not all groups and individuals are affected equally; instead, a range of "social risk positions" exists with implications for social and environmental justice, policy, law, and regulation. Accordingly, modernization risks must be identified and managed through more effective and inclusive democratic processes. Science has multiple roles in the era of reflexive modernization as a source of new risks, a tool for identifying and characterizing risks, and a partial source of solutions. At the same time, the increasing specialization of scientific knowledge is itself a source of risks, as narrow technical solutions to existing problems generate unforeseen new problems. Thus a need exists, as well, for reflexive examination of institutionalized science and the "politics of knowledge." Without adequate attention to these aspects of modernization, society remains mired in a state of "organized irresponsibility" in which cultural premises, institutional structures, laws, policies, and practices facilitate blindness to the full range of risks and the possibilities for managing them more democratically.

A number of commentators and critics, such as sociologist Anthony Giddens, have challenged, adapted, and further developed Beck's original concept of reflexive modernization. The resulting body of work has wide-ranging implications for notions of green technology. For example, Beck and others claim that some technologies, such as nuclear energy, are inherently problematic due to the long-term commitments they entail, the persistence over generations of their toxic waste products and environmental and health consequences, and the scale and opacity of the institutions required for their operation. In choosing such technologies, society makes long-term decisions that "close off the future" by generating irreversible and unpredictable effects. Scholars of reflexive modernization tend to advocate, instead, for more flexible and adaptable technologies that can more easily be abandoned or adapted in light of new knowledge.

Reflexive modernization scholars argue that institutions such as science, law, insurance, finance, work, and regulation have generally developed in ways that promote the expansion and sustenance of large technological and industrial systems. These institutions are grounded in modernist premises of calculable risk, rational management, and predictable and containable impacts. As such, they are increasingly unsuited for addressing modernization risks, which defy prediction and control and disseminate across boundaries. Accordingly, concepts of green technology must be linked to concepts for new decision-making processes, new models of economics and finance, and new institutional structures for regulation and governance.

Another concept associated with reflexive modernization is "individualization," the trend toward increased freedom of choice on the part of individuals. New technologies open up a broader range of possibilities for personal identity, expression, and organization of individual, family, and collective life. That freedom is accompanied by increased expectations for individual decision making and responsibility, a paradox in a context of large and inflexible institutional structures. Existing institutions, which evolved to meet the demands of early industrial society, now inhibit the new possibilities for individual choice. In many cases these institutions ascribe responsibilities for risk management to individuals, despite the systemic origins and distributed impacts of modernization risks. For example, in cases of contested illnesses in communities near hazardous industrial sites, the burden of proof typically falls upon local residents rather than the industries that generate the hazards. Scientific standards of epidemiology, legal frameworks, and insurance and financing arrangements all contribute to this shifting of responsibility. Green technologies, together with new communication technologies that have the potential to promote awareness, collaboration, and deliberation, offer new possibilities for reflexive examination and democratic revision of social institutions and practices.

Concepts of reflexive modernization have provided a critical framework for identifying and analyzing fundamental problems in late-industrial or postindustrial society. In that framework, contemporary institutions and practices evolved in tandem with technologies for energy production, manufacturing, transportation, and communication. If a new generation of green technologies is now emerging, then theories of reflexive modernization suggest that a new generation of institutions and practices will emerge as well. As contemporary society confronts itself, a new set of risks surrounds the many choices that must be made, deliberately or by default, as that historical process unfolds.

See Also: Ecological Modernization; Postindustrialism; Science and Technology Studies; Sociology of Technology; Unintended Consequences.

Further Readings

Adam, B., U. Beck, and J. Van Loon, eds. *The Risk Society and Beyond: Critical Issues for Social Theory*. Thousand Oaks, CA: Sage, 2000.

Beck, U. *Ecological Enlightenment: Essays on the Politics of the Risk Society*. Atlantic Highlands, NJ: Humanities Press, 1995.

Beck, U. *Ecological Politics in an Age of Risk*. London: Polity Press, 1995.

Beck, U. *The Reinvention of Politics*. Cambridge, UK: Polity Press, 1997.

Beck, U. *Risk Society: Towards a New Modernity*. Newbury Park, CA: Sage, 1992.

Beck, U. *World at Risk*. Cambridge, UK: Polity Press, 2008.

Beck, U., A. Giddens, and S. Lash. *Reflexive Modernization: Politics, Tradition and Aesthetics in the Modern Social Order*. Stanford, CA: Stanford University Press, 1994.

Elliott, A. "Beck's Sociology of Risk: A Critical Assessment." *Sociology*, 36/2 (2002).

Franklin, J., ed. *The Politics of Risk Society*. Cambridge, UK: Polity, 1998.

Giddens, A. *The Consequences of Modernity*. Stanford, CA: Stanford University Press, 1990.

William J. Kinsella
North Carolina State University

RESOURCE CONSERVATION AND RECOVERY ACT (RCRA)

The Resource Conservation and Recovery Act (RCRA), which is the central U.S. law governing the disposal of solid wastes, including hazardous solid wastes, was enacted by Congress in 1976. RCRA gives authority to the U.S. Environmental Protection Agency (EPA) to regulate hazardous waste from the point of generation to final disposal or storage, which is often referred to as the "cradle-to-grave" movement of waste. The law also provides a regulatory framework for nonhazardous solid wastes. Its approach is generally designed to be proactive, providing for the safe handling and containment of wastes as they are generated. RCRA has been amended on several occasions, most notably in 1984 when lawmakers expanded the law's coverage of hazardous industrial wastes, including underground storage tanks.

RCRA hazardous waste provisions regulate private businesses as well as federal, state, and local government facilities that generate, transport, treat, store, or dispose of hazardous waste. Each of these entities must comply with requirements governing the proper management of hazardous waste from the moment it is generated until it is ultimately disposed or destroyed. The law's municipal solid waste provisions regulate owners and operators of municipal landfills, setting out minimum criteria that each landfill must meet in order to operate.

Congress enacted RCRA to address the nation's growing volume of municipal and industrial waste and mounting public concerns about the unregulated dumping of toxic materials. The law, which amended the Solid Waste Disposal Act of 1965, set national goals for the following:

- Protecting human health and the environment from the potential hazards of waste disposal
- Conserving energy and natural resources
- Reducing the amount of waste generated
- Ensuring that wastes are managed in an environmentally sound manner

Prior to 1965, state and local governments were responsible for the disposal of solid waste. Garbage could be dumped haphazardly, either at the site where it was generated or transported to poorly maintained waste disposal areas. A lack of recordkeeping meant that the people who transported the waste or operated the landfills had no information about the presence of hazardous materials, and toxins sometimes built up in soils or by waterways. Dumpsites could be sold and developed, sometimes with disastrous consequences. Perhaps most notoriously, hundreds of families living in a middle-class subdivision known as Love Canal near Niagara Falls, New York, began suffering from unusually high rates of birth defects, cancer, and other health related problems. The cause, which became widely reported in the late 1970s, was that their homes and an elementary school had been built next to a former waste dump for hazardous chemicals.

Dealing With Hazardous Waste

Congress first dealt with the issue of solid waste with the 1965 Solid Waste Disposal Act. The law aimed to promote better waste disposal methods, primarily by providing research grants to state and local governments. With RCRA, lawmakers took a more comprehensive approach. The law focused on the need to establish a system to manage solid wastes, with special provisions for dealing with hazardous wastes. To reduce the stream of waste entering landfills, many of which were nearing capacity, it also encouraged more recycling.

Congress passed the law with little dissent, seeking to alleviate growing concerns about the public health impacts of hazardous waste. It banned open dumping, established a system for tracking waste, and required waste disposal facilities to take steps to prevent hazardous substances from entering the environment. RCRA did not address the cleanup of abandoned hazardous waste sites, such as Love Canal. Congress in 1980 passed separate, more controversial legislation for such situations: the Comprehensive Environmental Response, Compensation and Liability Act (CERCLA), commonly known as Superfund, which is also administered by EPA.

Under RCRA, EPA classifies waste as hazardous if it meets certain criteria such as toxicity, flammability, corrosiveness, and other hazardous characteristics. The agency uses detailed recordkeeping, known as a manifest system, to track the generation, transportation, treatment, storage, and disposal of hazardous waste. It has the authority to require businesses to clean up hazardous wastes that they have released into the environment and to require landfills to meet certain safety standards, such as the use of liners to prevent waste from leaking into groundwater. Owners and operators of hazardous waste, storage, or disposal facilities must obtain permits and meet performance standards.

RCRA allows states to administer their own hazardous waste programs as long as they meet or exceed federal standards. States, which may receive federal grants to develop solid waste management programs, must maintain an inventory of all sites where hazardous wastes have been stored. States have struggled to handle the increasing stream of solid waste, especially with landfills rapidly filling up in some areas and communities protesting plans to locate new landfills near their boundaries. Congress, in the 1984 Hazardous and Solid Waste Amendments, authorized EPA to take a more active role in assisting the states in handling nonhazardous waste landfills. To address apparent loopholes that had enabled as much as half of the hazardous waste in the nation to avoid regulation, the 1984 amendments also extended RCRA regulations to small businesses, such as dry cleaners and service stations, that produced hazardous waste.

In addition, the 1984 amendments broadened the law to regulate underground storage tanks that contained petroleum and other hazardous substances. They required states to inventory all such underground storage tanks, testing them and phasing out tanks that were subject to leaking. Owners and operators of tanks had to pay for cleanup and contamination caused by those tanks. Lawmakers in 1986 established a $500 million Leaking Underground Storage Tank Trust Fund, supported by federal taxes on motor fuels, to pay for cleaning up underground tanks if owners or operators could not be identified or were insolvent. Nevertheless, some states have run out of funds before testing and cleaning up all the leaking tanks within their borders.

Congress added a new section to RCRA in 1988 amid reports of medical wastes washing up on beaches in the United States and elsewhere. EPA conducted a two-year demonstration project to track medical waste from generation to disposal, based on the system for tracking hazardous waste. Although a nationwide program was not adopted, some states began to take steps to regulate medical waste.

Congress passed a major set of amendments in 1992, enabling states, EPA, and the U.S. Department of Justice to enforce RCRA provisions at federal facilities. Lawmakers stipulated that federal employees could face prosecution under any federal or state solid or hazardous waste law. The amendments also dealt with mixtures of radioactive and hazardous wastes at U.S. Department of Energy facilities, as well as hazardous wastes at munitions and other military sites.

Lawmakers in 1996 provided more flexibility, amending RCRA to exempt hazardous waste from regulation if it was treated to the point that it no longer was hazardous and was subsequently disposed of in certain facilities. They also exempted small landfills located in arid or remote areas from groundwater monitoring requirements, provided there was no evidence of groundwater contamination.

Recordkeeping

Although less controversial than some other environmental laws, RCRA imposes strict standards and recordkeeping requirements that have provoked some criticism and led to concerns about companies evading the law. Under RCRA, a transporter must meet EPA requirements in order to get a license for each kind of waste to be hauled. Similarly, any facility accepting hazardous wastes for disposal must obtain an EPA license specifying which kinds of hazardous waste it can accept. Regulators must approve the location and construction of a hazardous waste site as well as its ongoing operation. It can take up to four years to obtain a license, which then expires after 10 years, requiring some operators to devote a large amount of time to the licensing process.

To track hazardous wastes, regulators use extensive paperwork to accompany each batch of hazardous waste from the generator to transporters and finally to the disposal site. The paperwork, or manifest, contains about a half-dozen copies for the several entities that handle the waste, resulting in a complete paper record of the waste's movement.

Any business that generates at least one-half of a 50-gallon barrel of hazardous waste or more per month must comply with the manifest system.

The question of how to define solid waste that is regulated under RCRA has spurred numerous legal battles. Under EPA regulations, solid waste includes any garbage or refuse; sludge from a wastewater treatment plant, water supply treatment plant, or air pollution control facility; and other discarded materials, including solid, liquid, semi-solid, or contained gaseous materials resulting from industrial, commercial, mining, and agricultural operations as well as from community activities. Hazardous wastes that are generated in the home, like mineral spirits and old paint, are not covered. RCRA also addresses nonhazardous solid wastes, including garbage, nonrecycled household appliances, the residue from incinerated automobile tires, refuse such as metal scrap, wallboard, and empty containers, and sludge from industrial and municipal wastewater and water treatment.

A key legal question that has resisted easy solution is when to classify hazardous secondary materials as discarded, and therefore regulated under RCRA, if they are used or recycled in certain ways. The U.S. Court of Appeals for the District of Columbia Circuit ruled in 2000 that hazardous secondary materials that are generated and reclaimed in a continuous process within the same industry are not being discarded and therefore are not solid wastes under RCRA (*American Association of Battery Recyclers v. EPA*). In 2008, EPA issued a rule exempting certain hazardous materials from RCRA regulation if they were recycled.

People found in violation of various RCRA requirements can face fines of up to $25,000 per day per violation, or up to a year in prison.

The amount of paperwork and the minute details of operations covered by RCRA regulations have frustrated company officials. Critics have alleged that the burdensome requirements, coupled with uneven enforcement, can inadvertently encourage some companies to dispose of hazardous wastes illegally, potentially putting the public and the environment at risk.

As the economy becomes globalized, environmentalists also raise concerns about the overseas dumping of hazardous waste. With limited sites and stringent requirements for the dumping of hazardous waste in the United States, companies sometimes ship waste to overseas sites, including the Caribbean, Latin America, and Africa. A similar pattern has developed in other industrialized nations that also lack much space for solid waste. Some experts fear that developing countries are being used as dumping grounds for the hazardous wastes of wealthier nations. International agreements such as the 1989 Basel Convention on the Control of Transboundary Movements of Hazardous Wastes and Their Disposal seek to control the amount of hazardous wastes that is shipped internationally. The Basel Convention focuses in particular on reducing international movement of hazardous wastes, although it does not cover radioactive wastes.

Other Laws That Regulate Waste

The United States has several other laws regulating waste, although they are not as comprehensive as RCRA:

- *Sanitary Food Transportation Act*: This 1990 law requires the regulation of trucks and rail cars that haul both food and solid waste, which is known as "backhauling of garbage." Three agencies—the Departments of Agriculture, Health and Human Services, and Transportation—issue regulatory recordkeeping and identification requirements, decontamination procedures for refrigerated trucks and rail cars, and materials for the construction of tank trucks, cargo tanks, and ancillary equipment.

- *Clean Air Act*: The Clean Air Act Amendments of 1990 mandate more stringent federal standards for solid waste incinerators. They require EPA to issue standards for emissions from municipal, hospital, and other commercial and industrial incinerators.
- *Pollution Prevention Act*: Passed in 1990, the law declared pollution prevention to be the national policy. It directs EPA to undertake a series of activities aimed at preventing the generation of pollutants, rather than controlling pollutants after they are created. The law imposes new reporting requirements on industry, requiring certain companies to detail their efforts to reduce pollution and increase recycling. It also authorizes a program of matching grants with states to establish technical assistance programs for businesses.
- *Indian Lands Open Dump Cleanup Act*: This 1994 law requires the Indian Health Service to provide technical and financial support to inventory and shut down open dumps on Indian lands and to maintain the sites after closure. According to the Indian Health Service, only two of more than 600 waste dumps on Indian lands met EPA regulations prior to the law's enactment.
- *Mercury-Containing and Rechargeable Battery Management Act*: Passed in 1996, this law exempts battery collection and recycling programs from certain RCRA requirements. The goal is to stimulate new recycling programs. The law prohibits the use of mercury in batteries and requires labels on batteries to encourage proper disposal and recycling.

See Also: Materials Recovery Facilities; Recycling; Science and Technology Policy; Sociology of Technology.

Further Readings

Biello, David. "Trashed Tech Dumped Overseas: Does the U.S. Care?" *Scientific American*. http://www.scientificamerican.com/article.cfm?id=trashed-tech-dumped-overseas (Accessed April 2011).

Reference for Business. "Resource Recovery and Conservation Act." http://www.referenceforbusiness.com/encyclopedia/Res-Sec/Resource-Conservation-and-Recovery-Act.html (Accessed September 2010).

Wei, Norman S. "Environmental Management: Re-re-re-Defining RCRA Solid Wastes." *Pollution Engineering*. http://www.pollutionengineering.com/Articles/Column/BNP_GUID_9-5-2006_A_10000000000000148936 (Accessed April 2011).

David Hosansky
Independent Scholar

S

Science and Technology Policy

Tanks like this six-ton Special Tractor (circa 1918) were first used in World War I, when innovative science for military purposes became more important.

Source: Miranda Myrick/U.S. Army

In the contemporary world, power rests in the hands of science and scientific policy managers to a far greater extent than in the formal institutions of power. Technological expertise translates into superior industrial and military power and is perceived as the key to national dominance. Scientific decisions on matters of energy, environment, or health have a far greater impact on the daily lives of citizens than ephemeral political power. Moreover, global challenges like climate change are in essence scientific challenges that expand the power of scientists and challenge policymakers to come up with solutions that are based in science, green technology being just one these. Emerging challenges with regard to the formulation of science and technology policy revolve around the fact that economic expansion has been accompanied by growing global environmental concerns, such as climate change, energy security, and increasing scarcity of resources.

Subordinating science to the service of the state has been a key feature of the modern nation. For example, Revolutionary France, by state decree, introduced the metric system. In the United States, science and technology have always been important. Ever since the 17th century, any vision of a nation's future has incorporated a technological and scientific dimension, closely connecting science to development. Technological and scientific expertise has been key to the emergence and success of the modern state, and has been essential to U.S. emergence as a global power.

It was only in the 20th century that the organized management of advanced science and technology for national security became a key element of governance. An important reason for this was the two world wars, which saw the complete marshalling of the resources of the state—military, civilian, and intellectual—in service of the war effort. While World War I, especially in its early years, saw scientists fight and die in the trenches, as the war went on the innovative application of science for military purposes became more important, and victory increasingly became dependent on new innovations on the front, such as tanks, first deployed in the Battle of the Somme (1916). World War II saw the culmination of this trend—the Manhattan Project that developed the atomic bomb, which brought together scientists from some of America's most prestigious universities, is the best example for this. German scientists who developed the V2 rocket, first successfully tested in 1942 at the German Rocket Test Centre at Peenemunde under the direction of Wernher von Braun (1912–1977) who later designed the U.S. space program. In all countries, but especially in the United States, the war effort saw science and technology policy subordinated to and directed by the state.

The postwar period saw an enduring commitment to support science through state policy. In July 1945, Vannevar Bush (1890–1974), director of the Office of Scientific Research and Development and the key organizer of the Manhattan Project, submitted a report created at the request of President Franklin D. Roosevelt (1882–1945), *Science—The Endless Frontier*, to President Harry Truman (1884–1972) that laid the foundations of modern American science policy and inspired Truman to create the National Science Foundation, the crucial instrument of science and technology policy in the United States. The late 1940s saw the creation in the United States of the Office of Naval Research and the Atomic Energy Commission (both 1946) that would develop into the Department of Energy, research centers for the Army (1951) and the Air Force (1952), and the National Institutes of Health (NIH) in 1948. Faced with the challenge of global warming and environmental degradation, it is expected that these institutes will be in the forefront of green technology innovations, especially through the sponsoring of university research and in the series of national laboratories that are connected to them. The United States is also marked by a close alliance between the government, corporate, military, scientific, and academic interests popularly termed the *military-industrial complex*, which has an enormous influence on science and technology policymaking, not always benign. Though this alliance can be traced back to the end of the 19th century, the post-1945 decades saw an intensification of the links, which was helped by interconnected technological systems that had multiplied and grown more sophisticated over the first half of the 20th century. Green technology will also have to tread this fine balance between competing interests.

The 1960s saw the zenith of the American application of technology. Paradoxically, it also saw the first widespread questioning of the underlying premise of Western science that had been a constant since the Enlightenment—that the lot of humankind could be improved through science. Rachel Carson's *Silent Spring* (1962) revealed the ecological devastation that was a feature of pesticide-intensive agriculture; the French philosopher Jacques Ellul (1912–1994), in 1964 coined the term *technological society* to capture the mixed character of modern technology, particularly science-based technologies, arguing that humankind had made a Faustian bargain with technology; technology had fulfilled humankind's deepest desire, but it had also enslaved it. Combined with increasing disillusionment with the high-technology war in Vietnam, the impact of the energy crisis, concerns over acid rain, industrial pollution, global warming, depletion of the ozone layer, the loss of biodiversity, and the birth of the modern environmental movement, the environment became a serious factor in any national planning.

Technological failures such as the Three Mile Island nuclear accident in 1979, the explosion of the nuclear reactor at Chernobyl in 1986, the Fukushima Daiichi nuclear power plant post-earthquake/tsunami issues in Japan in 2011, and new unanticipated consequences of scientific inventions ranging from DDT to asbestos have also led to an increasing public interest in ecology. The back-to-nature aspect of the hippie movement was a popular expression of this, and the rise of the Green Party in Europe was a political expression. It also saw the breakdown of the top-down quest for modernity approach that had marked national planning since the 17th century, and especially in the early part of the 20th century. The work of Sheila Jasanoff explores the role of science and technology in the law, politics, and policy of modern democracies, paying particular attention to the nature of public reason. She questions the growing tendency throughout the world to equate human development with technological advancement, insisting that this seemingly rational strategy does not make sense without accompanying politics that probes the social foundations, presuppositions, and purposes of innovation. Policymakers now have to pay attention to ecological concerns in science and technology policy. Green technology is the latest manifestation of this trend.

The Challenge of Sputnik

It is often a perceived technological challenge that crystallizes a nation's commitment to science and technology policy. For the United States, this occurred with the Soviet launch of the first artificial satellite, Sputnik, on October 4, 1957, which shook America's perception of itself as the world's leading scientific and technological power. Spurred by this challenge, the United States made a series of decisions that laid the basis of its technological dominance. These include the passage of the National Aeronautics and Space Act (Space Act) of 1958 that created the U.S. National Aeronautics and Space Administration (NASA), the National Defense Education Act (1958) that encouraged students to pursue education in science and engineering, and the creation of the Advanced Research Projects Agency (ARPA), now known as DARPA, under the U.S. Department of Defense, which has been at the forefront of technological innovation for military purposes, the most prominent example being the Internet. Funding for the National Science Foundation (NSF) grew from $3.5 million in 1952 to $500 million in 1968. Military funding for corporations and universities has also been a key component of U.S. science and technology policy. U.S. science and technology policy has responded best when faced with a crisis. In the 1960s, the main challenge was to put a man on the moon; in the 1970s and the 1980s, the energy crisis; and in the 1990s, challenges like global warming.

The key instrument for science and technology policymaking in the United States is the U.S. Office of Science and Technology Policy (OSTP) that was created by Congress in 1976. The organization has a broad mandate to advise the president and others within the Executive Office of the President on the effects of science and technology on domestic and international affairs. The 1976 act also authorizes the OSTP to lead interagency efforts to develop and implement sound science and technology policies and budgets and to work with the private sector, state and local governments, the science and higher education communities, and other nations toward this end.

Directing science for the purposes of the state is not unique to the United States. Other nations have similar polices. In the case of China, since 1978, the State Council and the Central Committee of the Communist Party of China (CPC) have put investment in science and technology at the fore of its reform process, particularly the March 1985 Decision on the Reform of the Science and Technology Management System and the May 1995

Decision on Accelerating Scientific and Technological Progress. India, after independence in 1947, invested heavily in a series of superb engineering colleges, the Indian Institutes of Technology (IITs), modeled on the Massachusetts Institute of Technology (MIT) and charged with the training of an elite core of engineers. In fact, the training at these institutes was so good that many of these engineers later became prominent in Silicon Valley. It also, paradoxically, gave India a key advantage in a globalized service economy after liberalization in the 1990s, and is at the forefront of green technology in that country. For both China and India, reconciling economic growth with the environment is the key challenge.

Globalization

Globalization has given rise to new challenges. In a networked society, the role of the nation-state in sponsoring research to address local needs is increasingly questioned. For example, legislation favoring universities, such as the U.S. Morrill Land-Grant Acts statutes that allowed the creation of land-grant colleges, and including the Morrill Act of 1862 (7 U.S.C. § 301 et seq.) and the Morrill Act of 1890 (the Agricultural College Act of 1890 [26 Stat. 417, 7 U.S.C. § 321 et seq.]) put American needs, especially local needs such as agriculture, as the top priority. This trend, evident until the 1980s, has reversed. Federal funding for science and technology has fallen, though it still remains the largest contributor. Fears have been expressed that falling investment in science and technology will lead to a decline in the technological power of the United States. Some have even gone to the extent of arguing that unless there is a climatic crisis, spurring research into green technology and alternative energy sources, the United States will continue to lag behind or be subordinate to opposing interests.

Globalization has also changed the face of policymaking. In the contemporary world, national forging of science and technology policy is confronted by global challenges, such as worldwide regulatory frameworks, differing ideas of intellectual property rights, a technologically networked world where the imposition of national policy and national laws is extremely difficult, and challenges like climate change where, for meaningful success, the needs of the nation should be subordinate to those of the world. Thus, for example, even if one country tries to confine information and knowledge within its boundaries, the globalized nature of technology, especially the Internet, makes this difficult. Moreover, national governments find it increasingly difficult to implement restrictive policies, given the ubiquity of technology. For example, national Internet restrictions can be bypassed using proxy servers or Virtual Private Networks, and proprietary intellectual property can be easily transferred and duplicated or reverse engineered. The latter has particular implications for green technology. While the United States has been in the forefront of the demand that green technology be subject to intellectual property rights, countries in the developing world have demanded that such innovations be cheap and, in the absence of that, are likely to "steal" the technology. Finally, there has been a marked reluctance on the part of newly emerging economic powers such as China and India to subordinate economic growth to environmental concerns.

Moreover, technology has made innovation global. Transnational knowledge networks, mediated by technology, whether a call center, an industry-university collaboration, or diasporic networks of scientists who have their roots in a certain mode of training, as, for example, the alumni of MIT, all illustrate how a space of places (the territorial nation-state) has been replaced by a space of flows. Communicative networks and a transnational elite that circulate among knowledge centers such as Seattle, Silicon Valley, Bangalore, and

Shanghai often meet across boundaries using technology, and the transference of skills and even jobs across boundaries has created new identities that have little to do with the traditional boundaries of the nation-state. This global elite is extremely influential in the establishing of national science and technology policy. A good example for this is outsourcing, where the needs of citizens, including environmental regulations, are subordinate to the interests of multinational corporations. Another key input into the forging of science and technology policy comes from international organizations such as the United Nations and movements such as Greenpeace.

This merging of various concerns and actors has been termed the *network society*. In such a society, hierarchies, so essential to the top-down model that has been characteristic of national science and technology, have morphed into networks, which means that centers of knowledge production must channelize and produce knowledge, and that this knowledge cannot be subordinate to the interests of a single nation-state. Thus, nodes in the network such as Silicon Valley are extremely important, leading to close alliances between industry and academe, but these networks are not confined to the United States alone. From this perspective, higher education appears as a capital investment in the production of knowledge, just like machines used to be the capital of industrial production, with the important difference that industry was physically bound, while knowledge is not.

However, though the world moves toward a globalized society, it remains in thrall to the charm of borders and lines drawn on maps. Technological cooperation and rivalry reflect new conceptions of borders and new political entities. Similar rivalry exists in the pursuit of green technology, with fears emerging over a "green tech" race. By 2010, China had become the world's largest exporter of solar panels and wind turbines, and some American commentators warned that China would take the lead in the quest for clean technology. In 2009, Chinese investment in clean energy reached $34.6 billion, far more than any other country in the Group of 20 major economies.

Implicit in the forging of modern scientific policy is the belief that the key marker of modernity is technology. After the Industrial Revolution, the rhetoric of technology became the primary measure of intelligence, rationality, and the good society, supplanting Christianity for Europe. For the West, modernity became the core of its self-definition. Weapons, mass production, and communication networks were the new symbols of power. Colonial empires meant that this view spread throughout the world and science and technology policy in postcolonial states, buttressed by an unshakable belief in scientific objectivity, epitomized by the technological planners of modern states, mirrored the 17th-century Enlightenment project to better the lot of humankind through direct and deliberate action. Thus, environmentally destructive scientific innovations such as green revolutions, nuclear power plants, jet fighters, and great dams reflected Western ideas of technological progress and expanded the power and prestige of local elites. Crucially, environmentally friendly traditions, especially of indigenous people, such as reverence for nature, were seen as nonprogressive or romantic utopias. Ironically, green technology, at least in its principles, goes back to respect for the environment.

See Also: Faustian Bargain; Intellectual Property Rights; Science and Technology Studies.

Further Readings

Adas, Michael. *Dominance by Design: Technological Imperatives and America's Civilizing Mission.* Cambridge, MA: Belknap Press of Harvard University Press, 2005.

Adas, Michael. *Machines as the Measure of Men: Science, Technology, and Ideologies of Western Dominance*. Ithaca, NY: Cornell University Press, 1989.

Freenburg, Andrew. *Between Reason and Experience: Essays in Technology and Modernity*. Cambridge, MA: MIT Press, 2010.

Harrison, Carol E. and Ann Johnson. "Introduction: Science and National Identity." *Osiris*, 2/24, special issue, National Identity: The Role of Science and Technology (2009).

McClellan, James E. and Harold Dorn. *Science and Technology in World History: An Introduction*. Baltimore, MD: Johns Hopkins University Press, 2006.

Neal, Homer A., Tobin L. Smith, and Jennifer B. McCormick. *Beyond Sputnik: U.S. Science Policy in the Twenty-First Century*. Ann Arbor: University of Michigan Press, 2008.

Pursell, Carroll W. *The Machine in America: A Social History of Technology*. Baltimore, MD: Johns Hopkins University Press, 2007.

U.S. Office of Science and Technology Policy. http://www.whitehouse.gov/administration/eop/ostp (Accessed August 2010).

Sabil Francis
University of Leipzig, Germany,
École Normale Supérieure, Paris

SCIENCE AND TECHNOLOGY STUDIES

Science and technology studies (STS) is a new and dynamic interdisciplinary field that tries to understand how science and technology shape human lives and livelihoods and how society and culture, in turn, shape the development of science and technology. Instead of looking at science and technology as value-neutral, function-oriented concepts, it examines how science and technology are human institutions, situated in wider historical, social, and political contexts. It seeks to understand the origins, dynamics, and consequences of science and technology. The discipline tries to develop increasingly sophisticated understandings of scientific and technical knowledge and of the processes and resources that contribute to that knowledge. STS scholars also look at the directions and risks of emerging technology, including green technology. Increasingly, in a bid to avoid the earlier pitfall of deploying new technology such as insecticides or asbestos only to realize their negative consequences a few years later, a new field of anticipatory governance has emerged. In green technology, green nanotechnology, which tries to incorporate environmentally safe features at the design stage, is a good example of this. STS, by not treating science and technology as independent of the political process, contributes to the study of modern-day knowledge societies.

The word *science* comes from the Latin *scientia*, meaning knowledge, and it is generally accepted to mean "knowledge attained through study or practice." Implicit in this definition is the idea that science is objective, universal, based on empirically valid data, neutral, and above politics, ideology, or culture. The meaning of the word *technology* has shifted over time. A combination of the Greek *techne*, "art, craft," with logos, "word, speech," when it was first used in ancient Greece it meant a discourse on the arts, both fine and applied. The term also refers primarily to Western technology that became dominant after the 15th-century European Renaissance. In 17th-century England, it referred to the applied arts. By the early 20th century, the word incorporated means, processes, and ideas as well as tools and machines. And at the height of confidence in man's scientific power that marked the mid-20th century, it meant the power to "correct," as one Soviet scientist put it, "nature's mistakes." Such unquestioning confidence in the powers of science has waned, and as greater awareness of the consequences of unbridled scientific and industrial progress, such as climate change, becomes evident, the new quest is to find environmentally sound technologies, including green technology. Implicit in the idea of technology is progress, and the contradictions between progress and technology, for instance, the potential of advanced technology such as nuclear weapons to destroy life, are a fruitful subject for science and technology studies.

In line with broader trends in the academic world, STS brings together insights from several disciplines, including sociology, history, philosophy, and anthropology, and it is diverse and dynamic in its methodological approaches. STS incorporates several fields that emerged more or less independently well into the 1980s, including science studies, a branch of the sociology of scientific knowledge that studies scientific controversies in their social context; the history of technology; the history and philosophy of science; science, technology, and society studies that paid special attention to groups that were marginalized in and by technology such as women; and finally science, engineering, and public policy studies that heralded the emergence of such concepts as anticipatory governance that are of extreme importance in the nano-cogno-bio-info knowledge society.

Conceptually, the field was marked by several key turning points. These include the realization that technology was not value neutral but was influenced by the society in which it emerged, and that technology did not emerge independently but at a conflux of social need, social resources, and a sympathetic social ethos. One of the key divisions in the discipline is between the "High Church" and "Low Church" of STS. While the former tries to critically examine traditional perspectives in philosophy, sociology, and history of science and technology, the latter focuses on reform or activism, critically addressing policy, governance, and funding issues as well as individual pieces of publicly relevant science and technology; it tries to reform science and technology in the name of equality, welfare, and environment. Thus, the field is not confined to academics, and STS scholars engage activists, scientists, doctors, decision makers, engineers, and other stakeholders on matters of equity, policy, politics, social change, national development, and economic transformation. In the United States, the National Science Foundation (NSF) has been active in funding research into the social implications of emerging technology. In a trend that is increasingly evident worldwide, several governments have sponsored programs that study the ethics, history, philosophy, and social study of science and technology. Policy-wise, STS has been able to challenge the post–World War II ideal of an autonomous science and independent elite decision makers and to push for the democratization of science and technology by allowing input from the public and from activists. For instance, the interaction of American AIDS activists and AIDS researchers in the 1980s and early 1990s influenced research into the disease.

Elements of STS

One of the key elements of STS is a focus on the relationship between science and technology and extant categories of social thought such as race, gender, poverty, development, trust and credibility, participation and democracy, health and pathology, risk and uncertainty, globalization, and environmental protection. Emerging subfields include the social study of medicine, STS and the environment, reproductive technologies, science and the military, and science and public policy. The field is also closely linked with other emerging areas of scholarly endeavor such as global studies, and there is a growing amount of research concerned with science and technology in comparative and global perspectives, done by an international community of scholars.

Although STS was institutionalized only in the second half of the 20th century, glimmers of its basic concepts were evident long before that. Ludwik Fleck (1896–1961), a Polish medical doctor and biologist, came up with the idea of a "Denkkollektiv" (thought collective), an idea that is echoed in Thomas Kuhn's concepts of paradigm shifts, and French philosopher Michel Foucault's (1926–1984) episteme. Fleck also spoke of the "tenacity of closed systems of opinion" and introduced the idea of comparative epistemology, which was to become important in the history of science. By interrogating whether there could be an absolute truth in scientific research and pointing to sudden shifts in science, he anticipated later developments in social constructionism, and especially the development of critical science and technology studies.

However, it was Thomas Kuhn in his *The Structure of Scientific Revolutions* (1962) who laid some of the foundations of the discipline, including the idea that (1) a scientific community cannot practice its trade without some set of received beliefs, (2) normal science is based on the idea that scientists know what the world is like, (3) research is an attempt to force nature into conceptual boxes that have been provided by professional education, and (4) scientific revolutions take place when the existing paradigms of scientific knowledge are radically challenged. Most importantly, rather than see science as a given, Kuhn's work opened the way to look at science as a social activity.

Other pioneering efforts were David Bloor's (1976) and Barry Barnes's (1974) articulation of the "strong program" that looks deeper into the "causes of knowledge." At the start, this scrutinized the interests that went into the making of science, interests that deeply influenced scientific knowledge. One example of this is feminist work that reveals how technology has almost always been constructed as masculine and how particular scientific claims contribute to the construction of gender. Thus, ideology rather than objective scientific knowledge plays an important role in the construction of knowledge. The 1970s saw a major thrust on empirical research with a number of researchers—most prominently, Harry Collins, Karin Knorr Cetina, Bruno Latour, Michael Lynch, and Sharon Traweek—going to laboratories to watch and participate in the work of experimentation, the collection and analysis of data, and the refinement of claims—to study the culture of science. Other leading scholars in STS are Steve Woolgar, who conceptualized the term *turn to technology*; Trevor Pinch and Wiebe Bijker, who emphasized the social construction of technology; Ruth Schwartz Cowan, who has looked at the role of women in technology; Michael Adas, who looks at the roots of Western technological dominance; and Thomas Hughes and Carroll W. Pursell, who have done sterling work on the social history of technology in the United States, along with Adas.

The processes that technology incorporates, the social practices that constitute it, and the myriad ways we interact with it are at the heart of STS. The terms *actor-network*, which emphasizes the importance of interactions and network formation in the development of technology, and *sociotechnical ensemble*, which has been defined as hybrid networks that enclose human as well as technological components and relations between the components, are significant to its specific function, are used in STS to suggest that objects and humans not be seen as separate entities but that material objects and humans mutually constitute each other. Thus, instead of seeing the two as separate analytical categories, any analysis must start from the fact objects and people are always entangled to various degrees, and that social, political, cultural, and economic values affect innovation and that innovation in turn affects them. In fact, the more complex the technology, the greater the entanglement.

STS recognizes the fact that technologies are central to identity—the key difference between the contemporary world and past worlds is the kind of technology that is used and the way in which it is used in daily life. STS also intensively looks at the relationships of power that are at the heart of technology and development. These relationships of power have the force of common sense and therefore are hegemonic, or in other words, they are so deeply rooted in social life that they seem natural to those they dominate. A good example of this would be gender or masculine and feminine roles in many sciences, but especially the life sciences. Feminist scholars have shown how the "biologization" of gender differences or the argument that males and females have innate natural differences ignores the social contexts that can produce gendered behavior, and challenges the ideological framework that supports sex difference research.

While outstanding work is being done in Eastern S&T and challenging the explicit or implicit assumptions of the superiority and universality of Western science is a flourishing feature of the discipline, STS at present predominantly looks at Western science and technology, a reflection of the ascendancy of Western science and technology around 1500, an ongoing incorporation of much of the world into its network, and a powerful cultural assumption, implicit or explicit, that the West has a mission to spread its technology. The constructivist strand of STS has also looked at the connections between science, technology, and development. One key strand of this type of literature is how science and technology, rather than being universal, are local, and how the introduction of Western science in the colonies, or by modernizing governments in postcolonial contexts, led to conflict and cooperation with localized science and technology. In the same context, scrutiny of Eastern medicine or of medical practices like shamanism or faith healing in modern clinical laboratories fails because, decontextualized, they are reduced to their mere ingredients. In the same way, Western dairy farming practices often translate poorly in Asia and Africa, but locally designed technology such as the bush pump in Zimbabwe, which is assembled by the villagers themselves, often succeeds.

A lively new field, it is marked by an increasing number of undergraduate and postgraduate degrees and publications such as *Social Studies of Science; Science, Technology and Human Values; Research Policy; Science as Culture;* and *Science and Public Policy.* Professional associations include the Society for the History of Technology (SHOT, founded 1958), Society for Social Studies of Science (4S, founded 1975), the European Association for the Study of Science and Technology (EASST, founded 1981), and specialized associations such as the American Association for the History of Medicine or the Special Interest Group on Computers, Information and Society under the auspices of SHOT. Generally, while the history of medicine has received enormous attention in recent years, several aspects of science and technology studies, especially histories of universities from an STS viewpoint or the social history of the personal computer—why people bought computers and how they used them—are unexplored.

Leaders in STS

Among other pioneers of STS was British social scientist Michael Mulkay, who in 1979 conceptualized the idea of a "richer notion of materiality," arguing that objects were not merely objects but that a true history of any object encompasses technology, the social practices that constitute it, and the myriad ways in which people interact with it. Thus, for example, a broken television set is not the same as a working one, which interacts with many aspects of life, in a complex socio-legal-technical framework, or what in STS is termed a *sociotechnical ensemble*. A television set interacts with electricity, the television industry, the government, media companies, and the communications network, and so on. For some programs such as *American Idol* or *Big Brother*, viewers are more intensely involved. From an STS perspective, this interaction would also include how people used the telephones that are used to call in votes. Another example of an STS perspective would be Diane Vaughan's 1996 sociological analysis of why the space shuttle *Challenger* crashed, which narrates how decisions were made by managers and engineers who gradually, despite several warnings, showed poor judgment, buttressed by a culture of high-risk technology. The study looks at how the testing of the O-rings of the solid-fuel booster rockets was carried out. In making a social analysis of the accident, Vaughan tries to understand how technical uncertainty was dealt with by different groups of engineers working within different organizational contexts.

A rich vein of STS literature deals with how and why new standards were adopted. In essence, any technological artifact has a history of its own. In other words, rather than reduce any technology to merely its function, STS tries to study the social role of technical objects and the lifestyles that they make possible. While it is implicitly assumed that the extant nature of things is the only way in which they could have been created, STS recognizes that technological development is not unilinear but can veer off in many directions and could reach generally higher levels along several different paths. Further, STS argues that it is not technology that determines society but that any technology is subject to technical and social factors. STS stands in contrast to technological determinism, the idea that it is technology that determines culture and society, an idea that is still dominant in policymaking, especially corporate policymaking, and is so deeply rooted that it is considered common sense.

One of the best examples of this is the analysis of the bicycle by Pinch and Bijker. When the bicycle was first conceived of as a human-powered mode of transport in the 19th century, designers had to make a choice between one that had large front wheels, enthusiastically advocated by ardent racing fans, and one that had equally sized wheels—the bicycle as we know it now—which was much safer. Pinch and Bijker term this variability of goals the *interpretative flexibility* of technologies. What a technology is depends on what it is for, and that is often in dispute at the early stages of design, when several alternative designs exist. Over time, and over the process of invention, this flexibility of technologies declines and ends with the consolidation of a standard design capable of prevailing for an extended period. This process can be seen in the story of the bicycle, the automobile, and the Internet. Constructivism, which is a major strand of thought in STS, brings together new concepts such as the theory of large-scale technical systems, social constructivism, and actor-network theory.

See Also: Engineering Studies; Participatory Technology Development; Science and Technology Policy; Social Construction of Technology.

Further Readings

Adas, Michael. *Dominance by Design: Technological Imperatives and America's Civilizing Mission*. Cambridge, MA: Belknap Press of Harvard University Press, 2005.

Adas, Michael. *Machines as the Measure of Men: Science, Technology, and Ideologies of Western Dominance*. Ithaca NY: Cornell University Press, 1989.

Bijker, Wiebe, Thomas Hughes, and Trevor Pinch, eds. *The Social Construction of Technological Systems: New Directions in the Sociology and History of Technology*. Cambridge, MA: MIT Press, 1987.

Bijker, Wiebe E. and John Law. *Shaping Technology/Building Society: Studies in Sociotechnical Change*. Cambridge, MA: MIT Press, 1995.

Cowan, Ruth Schwartz. *More Work for Mother: The Ironies of Household Technology From the Open Hearth to the Microwave*. New York: Basic Books, 1983.

Feenberg, Andrew. *Between Reason and Experience: Essays in Technology and Modernity*. Cambridge, MA: MIT Press, 2010.

Fleck, Ludwik. *Genesis and Development of a Scientific Fact*. Chicago, IL: University of Chicago Press, 1981.

Fuller, Steve. *Philosophy, Rhetoric, and the End of Knowledge: The Coming of Science and Technology Studies*. Madison: University of Wisconsin Press, 2004.

Hackett, Edward J. and Society for Social Studies of Science. *The Handbook of Science and Technology Studies*. Cambridge, MA: MIT Press, 2008. Published in cooperation with the Society for the Social Studies of Science.

Kuhn, Thomas. *The Structure of Scientific Revolutions*. Chicago, IL: University of Chicago Press, 1962.

Latour, Bruno. *Science in Action: How to Follow Scientists and Engineers Through Society*. Cambridge, MA: Harvard University Press, 1987.

Latour, Bruno and Steve Woolgar. *Laboratory Life: The Construction of Scientific Facts*. Princeton, NJ: Princeton University Press, 1986.

Pursell, Carroll W. *The Machine in America: A Social History of Technology*. Baltimore, MD: Johns Hopkins University Press, 2007.

Sismondo, Sergio. *An Introduction to Science and Technology Studies*. Chichester, UK; Wiley-Blackwell, 2010.

University of Wisconsin–Madison, Science and Technology Studies. http://www.sts.wisc.edu (Accessed August 2010).

Sabil Francis
University of Leipzig, Germany,
École Normale Supérieure, Paris

Smart Grid

A smart grid is a modern solution to the demands of electricity distribution, and overlays the existing electricity distribution grid with digital technology to interact with the appliances plugged into the grid at the end-user level in order to increase efficiency. Smart grids and smart meters are among the common suggested remedies to modern energy problems, including energy inefficiency, overconsumption, and grid outages as the result of weather, disaster, or attack. In the United States, the 2007 Energy Independence and Security Act set aside funding for the 2008–2012 period to support smart grid development; established a

Most electricity usage meters still need to be manually read by an in-person visit to the home, whereas smart meters communicating via a wireless network would track electricity flow over time.

Source: iStockphoto

funds-matching program for states, utilities, and consumers; and created the Grid Modernization Commission to recommend protocol standards. The act initially allocated $100 million per fiscal year in funding; the American Recovery and Reinvestment Act of 2009 further allocated $11 billion for the creation of a national smart grid.

Until the smart grid, power grids have been broadly the same since the 1950s. They have kept up neither with increasing consumer demand nor with alternative energy sources, the possibility of security threats, the need to conserve electricity, or the increasing demand for uninterruptible electricity supply for certain businesses, requiring hospitals and other buildings to use emergency generators that are not part of the local electricity distribution system. At the same time, they have failed to take advantage of many advances in technology that would enable a grid that can respond to circumstances and conditions, a grid that is reactive rather than a passive channel through which electricity passes. In an age in which cities are adopting free wireless Internet initiatives and mobile phones have laptop-equivalent computing power, it is somewhat remarkable that most electricity usage meters still need to be manually read by an in-person visit to the home (and make usage reporting errors with nontrivial frequency)—such is the degree to which power grids have lagged behind other areas of technology.

The primary goals of a smart grid project are to modernize the power grid in terms of its reliability, security, and efficiency; to decentralize power generation so that energy customers can also be energy suppliers; to encourage the use of renewable energy and make more affordable and appealing such technologies as solar energy, wind turbines, and electric vehicles; and to make power consumption more flexible and less biased in favor of traditional nonrenewable sources. Furthermore, the electricity market will be broadened because a nationwide smart grid will make it easier for producers of alternative energy sources to sell electricity to customers anywhere in the country.

A smart grid consists of an intelligent monitoring system that tracks the electricity flowing through the grid and communicates with the production, transmission, distribution, and consumption parts of the grid in order to respond dynamically to grid conditions. Grid upgrades included with the smart grid include better potential integration with renewable energy sources and a net metering system that credits end users who feed electricity back into the system (e.g., with the use of solar panels or wind turbines). Power grids until recently were unidirectional: electricity flowed from a source to the end user. Though current grids can be used bidirectionally, with power fed back into them, the smart grid is bidirectional by design.

Using the smart grid, electricity will cost more during peak hours and less during off-peak hours, and end users will have the option of taking advantage of this. Appliances that can be run at any hour, such as a washing machine in a residence or factory processes at a manufacturing plant, can be set to be automatically activated at times when the demand on the grid is low and prices are at their lowest. Other appliances can be set to be automatically shut off during peak usage hours, to keep the electric bill low and reduce unnecessary demand on the grid. These decisions are made by the customer: the grid simply makes the option available and responds according to the settings the customer selects.

The grid ideally can respond to grid events anywhere in the distribution chain, even increasing air conditioning on hot days. The amount of bandwidth required for the grid to operate is very small, as is its cost of operation once put in place. One of the grid's strengths is its ability to respond to crisis. Smart grid projects are designed so that if a national grid—the electricity distribution interconnection backbone—fails, local grids can continue to operate, and vice versa. Until the 2007–2008 initiative to begin serious development of a national smart grid, most efforts had transpired at the municipal level, just as it has done with net metering. Austin, Texas, for instance, replaced one-third of its meters with smart meters communicating via a wireless network in 2003 and has been upgrading its grid gradually since, including the use of sensors and smart thermostats. Boulder, Colorado, put the first components of its smart grid online in the summer of 2008. The threat of terrorist attacks on the national grid and power distribution problems like the rolling blackouts in California or the long-term outage following a 2008 ice storm in central New England have increased the desire for a grid that is less vulnerable to outages.

One of the elements of the smart grid frequently mentioned in public discussion is the capacity to self-heal, which is sometimes misunderstood. Because of the amount of embedded sensors and monitoring equipment, and the grid's flexibility and ability to respond to events as they occur, some damage can be prevented before it occurs by better managing power distribution, avoiding overloads, and so on. The grid cannot self-repair. But if damage occurs, it may be able to continue to function despite that damage, or to function better than existing power grids would be able to. It may be able to avoid cascading power failures, for instance. In areas where underground cables are the bulk of the distribution network, outages can be avoided almost entirely—when there is a failure in one part of the distribution network, power can still reach end users through other cables. This is not the case with above-ground power lines because of the difference in the way they are networked, but situations like the 2008 ice storm—in which much of the delay to repairs was caused by the utility companies simply not knowing the extent or nature of the damage until making a physical visit to each site where service was out—can be avoided because when something goes wrong, the monitoring system reports it. This helps the grid persist despite natural disaster as well as man-made attacks.

The objections raised to smart meters, which have limited their adoption, include health concerns. In Maine, for instance, consumers have voiced concerns over the potential health problems caused by the meters' constant radio wave emissions, as well as the possibility of fire hazard when installing the meters in old colonial homes. In California, health concerns motivated the ban of smart meters in the town of Fairfax. Furthermore, there are privacy concerns related to the behavioral data that can be observed through electricity usage, particularly in a day and age when consumer behavior is so closely scrutinized by marketers and corporations looking to target their efforts.

See Also: Appliances, Energy Efficient; Appropriate Technology; Cogeneration; Microgeneration.

Further Readings

Flick, Tony and Justin Morehouse. *Securing the Smart Grid: Next Generation Power Grid Security*. Burlington, MA: Syngress, 2010.

Fox-Penner, Peter S. *Smart Power: Climate Change, the Smart Grid, and the Future of Electric Utilities*. Washington, DC: Island Press, 2010.

Gellings, Clark W. *The Smart Grid: Enabling Energy Efficiency and Demand Response*. Boca Raton, FL: CRC Press, 2009.

U.S. Department of Energy. "Smart Grid." http://www.oe.energy.gov/smartgrid.htm (Accessed January 2011).

Vig, Norman J. and Michael E. Kraft. *Environmental Policy: New Directions for the Twenty-First Century*. Washington, DC: CQ Press, 2009.

Bill Kte'pi
Independent Scholar

SOCIAL AGENCY

Social agency refers to the capacity of individuals to act independently and to make their own free choices. Closely related to the sociological concept of structure, social agency is a central ontological issue in environmental studies, political science, sociology, and a variety of other disciplines. Often framed as the battle of socialization versus autonomy, social agency colors theories and arguments related to the causes and effects of many environmental issues, including pollution, sustainability, authoritarianism, and community. Social agency has developed and changed as a theory and has subtly different uses and meanings in different contexts and when used by different theorists. Despite variations in the use of the term, *social agency* broadly encompasses discussions regarding humans' ability to act independently. Within the context of environmentalism, the sustainability movement, and green business ventures, social agency has several ramifications. It suggests that individuals may be capable of making their own choices without external limitations. Social agency theory also supports the theory that individuals construct, and reconstruct, their understanding of the world in which they live. Finally, social agency implies that individual choices will affect the overall health of society, yet attempts to ensure that these choices will not be oppressed, judged, or limited.

Structure and Agency

Classical and contemporary sociology have been shaped by an ongoing debate over the primacy between structure and agency. Within the context of environmentalism, these constructs affect perspectives related to the role individuals have in shaping the world around them. Individuals' roles in making choices that allow sustainable practices are dependent upon how one views structure and agency. Debates regarding the nature of social structure and agency reflect a philosophical debate displaying two contrasting visions of reality, how one observes groups and their interrelations.

"Social structure" refers to patterned social arrangements that form the society as a whole. Social structure determines, to some degree, the actions of individuals socialized into that structure. Social structure is based on the notion that predetermined roles influence or limit the opportunities and choices that individuals can make. Examples of social structure include class, gender, occupation, race, and age. These structures are immobile and create constant pressure on the individual. This pressure affects and limits choices that an individual can make, theoretically limiting potential action that might be taken to assist the environment. Social structure consists of a variety of norms that shape behaviors within the social group, such as how members of a group value or have access to certain goods and services.

"Social agency," a related concept, resists the pressure of social structures, allowing individuals to determine their own roles and self-made choices. A person's agency is the independent capability or ability to act on his or her will. This ability is affected by the cognitive belief structure one has formed through prior experiences. Society's and the individual's perceptions also shape choices, and the structure and environment one is in affects decisions and actions. Social agency supports the concept that individuals or organizations make choices and decisions that are sometimes very different from what one would expect from their upbringing, history, or circumstances, such as when a giant retailer embraces green packaging or a suburban couple installs solar panels on their home even though such choices cause them increased out-of-pocket expenses compared to other options. Social agency supports the mobility of individuals to explore a free range of roles and believes that individuals shape their own behaviors, often ignoring society's norms.

The Environment and Sustainability

Social agency affects an individual's response to the environment and sustainability. The environment encompasses one's attitude toward all natural things, living and nonliving, that occur on Earth, while sustainability emphasizes the capacity to endure and to maintain the planet in such a manner that future generations will have access to a high quality of life and natural resources. Social agency emphasizes that individuals have the ability to make choices that allow them to positively affect the environment and sustainability decisions.

Social agency supports the proposition that individuals can resist the pressure of societal norms to examine, explore, and make decisions regarding the environment. Ultimately, individuals can make decisions to protect elements of the environment, even when these decisions run counter to beliefs and values with which they have been inculcated. This hypothesis stresses that individuals act independently at times, not as part of their social stratum. Individuals operating with social agency are able to react to changes in the environment, observing potential problems, identifying causes of these, and then appreciating a need to overcome the problem. Social agency thus allows individuals to act to assist the environment and produce solutions that assist in more sustainable actions. Research can be conducted to determine those natural essentials of humans' interaction with the environment and also to distinguish which elements are most harmful. Individuals exhibiting social agency take the steps to fully educate themselves regarding the environment and then take actions that support that which they have discovered. Other members of their social stratum might view this type of action as eccentric or unusual, since it goes against the consensus of the social structure.

After learning about threats to the environment, individuals exhibiting social agency will conduct research on the most efficacious behaviors that can be taken to protect it. These behaviors are then either taken, thus adding them to the individual's situation, or eliminated, consequently removing a strain on the surrounding environment. This approach may well be different from the way others in the same social stratum choose to proceed, especially those who do not exhibit a large degree of social agency. Those who embrace more of a social structure approach view those who exhibit social agency as separating from the accepted or "normal" way of life and as making a choice to live in an unfamiliar and unacceptable manner. Indeed, those taking a social structure approach tend to view the choice of those proceeding under a social agency approach as seeking extra work or spending time on irrelevant matters. Social agency manifests itself in a variety of ways, such as a high school student who is interested in protecting the environment. This student might be viewed by others in his or her social stratum as using the local public library in "abnormal" ways, such as using the facility to do research on environmental issues rather than for homework or pleasure reading. Social agency might also cause an individual to begin to view certain household objects or materials differently than the norm of his or her social group. For example, a suburban resident might begin to consider the environmental effects of common household products used around the house rather than merely focusing on their cleaning properties.

Most importantly, social agency affects an individual's willingness and ability to take certain steps to remedy environmental risks and harms that he or she identifies. A high degree of social agency greatly increases the likelihood that an individual may take positive action to address those risks or harms. Such positive action might include changing personal habits so that less environmentally harmful choices are made with regard to household purchases, selecting alternative energy sources for home heating and cooling, or joining ecological groups to advocate change on a broader level. The willingness to go against the grain is one of the more significant manifestations of social agency.

Theory Behind Change

Debates regarding the primacy of structure or agency relate to central questions that bedevil sociologists. Focusing on the influence, meaning, and nature of the social world, theories concerning structure and agency examine the social world, what it is composed of, its causes, and its effects. Functionalists, or those who interpret society as relatively static arrangement with interrelated parts, believe structure and hierarchy are stabilizing forces that allow societies to thrive and prosper. Opponents of structure, such as Marxists, emphasize that social structure can act to the detriment of many members of a society. Both groups include such concepts that are material (e.g., economic factors) and cultural (e.g., customs, ideologies, norms, and traditions) within the definition of "structure."

Traditionally, discussions about social agency focused on the attributes, worth, and significance of structure and on its impact on independent action. As a result, much of classical sociology advocated the importance of structure. Theorists believed that structure accounted for many unique aspects of the social world that could not be explained simply by the sum of the individuals present. Proponents of this viewpoint developed schools of thought including structuralism, functionalism, and Marxism. A second wave of sociologists eschewed this approach, instead positing that the aggregation of decisions by individuals are the dominant force in shaping society, thus advocating strongly for the significance of social agency. This position was advocated by such theoretical systems as methodological individualism, social phenomenology, interactionism, and ethnomethodology.

Many modern social theorists take a middle road, attempting to balance the two previous positions. Those holding this perspective see structure and agency as complementary forces rather than polar opposites. Modern social theorists believe that while structure influences human behavior, humans are also capable of changing the social structures that they inhabit. Since all human action exists within the context of a preexisting social structure that is governed by a set of norms that in turn are distinct from those of other social structures, all human action is at least partly predetermined based on the varying contextual rules under which it occurs. Reflexive feedback, however, sustains and modifies the structure and rules so that they are not permanent and external, but instead influenced through social agency. Social agency thus becomes the avenue for advocacy and change.

Advocating Change

Within the context of environmentalism, social agency permits those interested in environmental issues to examine, advocate, and change practices that impede sustainable development and green growth. Specifically, social agency allows members of communities to identify problems that affect them and to strive for changes that will improve their lives and those of others. For example, individuals noted that degradation of the Earth's atmosphere had occurred. The Earth's atmosphere is composed of a layer of gases containing chiefly nitrogen, oxygen, argon, and carbon dioxide. This protective layer regulates the Earth's climate, moderating otherwise frigid night temperatures and inferno-like daytime conditions. The carbon dioxide serves as a means to hold an efficient amount of heat on Earth, acting like a filter that traps radiation from the sun and allows a habitable temperature. Too much carbon dioxide, however, produces a greenhouse effect, altering temperatures on Earth and causing global warming and deterioration of the atmosphere. Social agency allowed the identification of this environmental problem and spurred individuals willing to question the social structure to seek a response.

Since social agency affects individual responses to the environment and sustainability, it allows individuals to choose to engage in behaviors, such as reducing greenhouse gas emissions, that make a difference. Social agency permits an individual to fully educate himself or herself about the best methods to conserve energy and to take action regardless of social norms that might otherwise limit action. Multiple pathways exist that allow the conservation of energy or the reduction of harmful emissions. For example, those focused on reducing greenhouse gas emissions might select to concentrate on the transportation industry, but even there find multiple avenues to pursue in a quest for more sustainable alternatives. The extraction of ore and minerals from the Earth causes harmful emissions of greenhouse gases, as does the manufacturing processes by which automobiles, airplanes, motorcycles, bicycles, trains, and other modes of transportation are made. Social agency allows individuals to make choices that enable them to reduce their contribution to greenhouse gas emissions, decisions that require careful thought to the consequences that stem from any possible alternative.

Social agency allows individuals to overcome limitations placed upon them by their social stratum and other forces to engage in actions that protect and improve the environment. The majority of Americans use cars or buses for transportation daily, with little thought about harmful emissions released as a result. Individuals ignoring social norms, those with the capacity to act independently, ride bicycles in a move to reduce greenhouse gases. Although millions of cars are driven daily, the choice to use alternative modes of transportation provides one example of social agency. Small actions such as this, when taken cumulatively, can make a tremendous difference over time. Social agency allows individuals to make choices that defy social norms to protect future generations and to further a cause in which they believe.

Social agency can also affect decisions made in the home. Although few realize it, private homes can be a major source of harmful emissions. Greenhouse gases are emitted through many common household tasks. Individual choices can greatly reduce greenhouse gas emissions caused by household behaviors. If a person throws away several pounds of garbage per day, for example, this refuse will eventually be sent to the landfill, where it will release harmful emissions as it decomposes. Choosing to recycle, compost, or otherwise reduce personal garbage production reduces the need to transport refuse to landfills. Such choices may also increase enrichment of the soil in home gardens, which in turn will reduce carbon dioxide in the atmosphere. Small choices such as the decision to compost are based on social agency that spurs the creation of a better environment. Social agency allows one's experiences to alter decisions that are accepted as the norm by society. The circumstances and situations one engages in affect the choices made. Although as a society the United States may accept many environmentally destructive practices, social agency permits individuals to recognize the harm in the status quo and to advocate change. Social agency also spurs those concerned with the environment to engage in more sustainable behavior and allows trial and error to suggest which pathways are the most efficacious in achieving desired goals.

Conclusion

Acting independently and gaining confidence and knowledge from the process allows individuals to make informed decisions that positively affect the environment in which they live. Social agency affects an individual's responses to the environment and sustainability insofar as it allows the choice of actions that circumvent and avoid societal behaviors. Individuals may demonstrate social agency by considering the need to engage in environmentally friendly behaviors as well as ways of doing so. Individuals can overcome the limits pressed upon them to take actions that protect the environment. Societal norms and mores have limited effect on their choices. Social class, gender, ethnicity, income levels, or occupational status do not impede decisions of those who exhibit social agency.

See Also: Actor-Network Theory; Authoritarianism and Technology; Democratic Rationalization; Frankfurt School; Sociology of Technology.

Further Readings

Archer, M. *Realist Social Theory: The Morphogenetic Approach.* New York: Cambridge University Press, 1995.
Barker, C. *Cultural Studies: Theory and Practice.* London: Sage, 2005.
Guha, R. *Environmentalism: A Global History.* New York: Longman, 1999.
McNeill, J. R. *Something New Under the Sun: An Environmental History of the Twentieth-Century World.* New York: W. W. Norton, 2001.
Ritzer, G. *Sociological Theory.* New York: McGraw-Hill, 2010.

Stephen T. Schroth
Jason A. Helfer
Cassandra R. Harrell
Knox College

SOCIAL CONSTRUCTION OF TECHNOLOGY

The standard view of technology, popularly known as "technological determinism" and "technological constructivism," was largely motivated by a reductionist approach. According to this approach, it is possible to analyze technological change along a fixed unidirectional path with reference to economic laws or some inner technological logic. Under the framework of this approach, respective societies adapt to newly introduced technologies. However, the constructivist approach began to take hold in the mid-1980s. This new stance toward technology shaped social models that provided room for social factors that could affect the growth of technology as opposed to technology following its own momentum. Accordingly, technology and technological change cannot be understood without recourse to the social context. As a supporting example, it has quite often been shown that in the Indian context, farmers successfully experiment and introduce more technological innovations compared to the big agricultural companies, possibly because of their direct experience with agricultural practices. The most succinct description is captured in one phrase: "technology as a social product."

Social construction of technology, or SCOT, an example of strong social constructivism, draws inspiration from sociology of scientific knowledge (science as a social activity). It is best studied within the broader field of science and technology studies. SCOT does not treat the success or failure of technology in isolation but strongly advocates an intimate relationship between technology, human action, and society. Wiebe Bijker and Trevor Pinch are the two famous adherents of SCOT. According to SCOT, success and failure should not and cannot be treated as intrinsic technological variables (as exemplified by engineering prowess) but are determined with respect to the needs, problems, and solutions that accrue from the various interactions of the concerned social groups (as exemplified by existing markets, action groups, producers, etc.).

Bijker and Pinch advance the principle of symmetry or methodological relativism for assessing the success and failure of technological artifacts. The principle holds that the same explanations should be used for both successes and failures, consequently giving way to a neutral position with respect to social factors that could affect the final acceptance and/or rejection of technology. This principle was put forth for doing away with the common observations that success of technologies is often attributed to technological advancements while failure of technologies was usually attributed to sociological factors. Therefore, SCOT should not be merely understood as a "theory" but rather as a "methodology" that helps one to understand steps that go into gauging technology.

Five Components

The conceptual framework supporting SCOT is composed of the following five basic components:

1. *Interpretive flexibility*: This idea basically views the technology design process as an open scenario that could yield varying outcomes dependent upon unique social circumstances. In a similar fashion within the SCOT framework, technological artifacts could take on different designs, but the final design is often an outcome of intergroup (interested parties) negotiations.

2. *Social groups*: These groups act as agents of action and are capable of imparting different meanings to artifacts through their actions. During the course of technology development, different groups embody different interpretations of the artifacts and negotiate in the design process for the final outcome.

3. *Closure and stabilization*: The technology development process is often laden with controversies owing to different interpretations of the proposed artifact. The negotiations continue unless the artifact becomes appealing to all concerned. At some point in time, the process achieves closure where no further revisions in decisions are possible and ultimately the artifact stabilizes in nature.

4. *Wider context*: The wider context subsumes the sociocultural and political contexts within which the artifact takes shape. These contexts affect the artifact both directly and indirectly and could affect contexts that bear factors affecting intergroup interactions, their power relations, and the rules that govern these interactions. Most critiques often base their criticisms of SCOT on this fourth factor.

5. *Technological frame*: This component is the latest to have been added in SCOT. It stands for an amalgam of goals, problems, testing procedures, and so forth, and helps in imparting structure to the social group's thinking, problem-solving strategies, and other design activities with respect to technological artifacts.

Criticism

Toward the end it is also important to consider the specific criticisms of the SCOT approach. Langdon Winner is considered the most important critic of the SCOT approach (other noted criticisms appear in the writings of Stewart Russell) after his publication titled "Upon Opening the Black Box and Finding It Empty: Social Constructivism and the Philosophy of Technology" where several important points were raised:

- SCOT only deals with explaining the origins of technology but ignores the effects of technology beyond the emergence (like the consequences of choosing a particular artifact)
- Social groups that play a role in technology development are often given importance but other groups like those that are excluded from the process or even suppressed by a particular artifact choice are often ignored
- There are no considerations for moral principles
- Some of the dynamics are not covered within the intergroup relations and negotiations such as intellectual, cultural, or social origins of social choices about technology are not given due consideration

In defense of some of the conceptual underpinnings, Pinch (1996) points out that SCOT is endowed with the capacity to understand power relations and social structures and hence some of the criticisms that are focused on it rather pertain to the early work done within SCOT. The earlier work had more to do with the designing of technology than with examining the broader contextual issues that could shape technology. Bijker (1995) also points out there is a dearth of detailed technical analysis in the larger sociological literature on power.

See Also: Actor-Network Theory; Science and Technology Studies; Sociology of Technology; Technological Determinism; Technology and Social Change.

Further Readings

Bijker, W. E. "Do Not Despair: There Is Life After Constructivism." *Science, Technology and Human Values*, 18/4 (1993).

Bijker, W. E. *Of Bicycles, Bakelites, and Bulbs: Toward a Theory of Sociotechnical Change.* Cambridge, MA: MIT Press, 1995.

Bijker, W. E. "Sustainable Policy? A Public Debate About Nature Development in the Netherlands." *History and Technology*, 20/4 (2004).

Bijker, W. E., T. P. Hughes, and T. J. Pinch. *The Social Construction of Technological Systems: New Directions in the Sociology and History of Technology.* Cambridge, MA: MIT Press, 1987.

Bijker, W. E. and J. Law. *Shaping Technology/Building Society: Studies in Socio-Technical Change* Cambridge, MA: MIT Press, 1992.

Bijker, W. E. and T. J. Pinch. "SCOT Answers, Other Questions: A Reply to Nick Clayton." *Technology and Culture*, 43 (2002).

Klein, H. K. and D. L. Kleinman. "The Social Construction of Technology: Structural Considerations." *Science, Technology & Human Values*, 27/1 (2002).

Pinch, T. "The Social Construction of Technology: A Review." In *Technological Change: Methods and Themes in the History of Technology*, R. Fox, ed. Oxfordshire, UK: Harwood Academic Publishers, 1996.

Pinch, T. J. and W. E. Bijker. "The Social Construction of Facts and Artefacts: Or How the Sociology of Science and the Sociology of Technology Might Benefit Each Other." *Social Studies of Science*, 14 (1984).

Russell, S. and R. Williams. "Opening the Black Box and Closing It Behind You: On Micro-Sociology in the Social Analysis of Technology." Edinburgh PICT Working Paper No. 3. Edinburgh, UK: Edinburgh University, 1988.

Winner, L. "Upon Opening the Black Box and Finding It Empty: Social Constructivism and the Philosophy of Technology." *Science, Technology, and Human Values*, 18/3 (1993).

Neha Khetrapal
Indian Institute of Information Technology

SOCIOLOGY OF TECHNOLOGY

The most basic questions in the sociology of technology can be broadly formulated as, "How does technology influence and shape human society and how does society shape the development and use of technology?" The term *technology* in its broadest sense refers both to tools and to a system of knowledge with practical uses; thus, the history of technology could be dated back to prehistoric humans who used tools such as clubs, hammers, and fire to attempt to control and shape the world.

Technology is often contrasted with science: technology is more focused on developing and using knowledge and tools that have applications for practical problems (e.g., the best way to build a bridge in a specific geographical context) rather than accumulation of knowledge for its own sake. However, the boundaries are fluid; for instance, scientific innovations are often put to use by technology, and the need for knowledge to solve a practical problem may influence the direction of scientific research. Technology has been a major force in social change throughout much of human history, particularly since the Industrial Revolution, but the rate and scope of change is arguably greater today. For this reason, discussions of the sociology of technology often center on relatively recent technological developments, such as space travel and nuclear power plants.

The influence of a given technology on society may be far-reaching, influencing the organization of human relationships, the organization of society, and the ways in which societal resources are invested. For instance, the invention of the automobile created a new technology, which offered a novel means of transportation, which, in turn, greatly facilitated individual travel. However, the success of the automobile as a technological innovation was partly due to the willingness of society to make many changes to facilitate its use. For instance, in order for cars to become a significant means of transportation, an infrastructure and support system had to be developed, including paved roads, a system for training and licensing drivers, traffic laws and a means to enforce them (hence the expansion of police departments and the court system), and a geographically dispersed system of repair shops and gas stations. If these social changes had not been made, the automobile might have remained a curiosity or rich person's plaything rather than an influential means of transportation.

Theories of the Relationship Between Society and Technology

One of the earliest formulations of the relationship between society and technology was that of *technological determinism*. This school of thought views technology as an autonomous force that shapes society, which can cause social change without being influenced in turn by society. This is an extreme position that discounts the role of humans in creating and adopting technology, but is useful for focusing attention on how the development of specific technologies (which may seem relatively minor when viewed in isolation) may influence major changes in human society. One famous example is the theory of the historian Lynn White, Jr. that the introduction of the stirrup caused the development of the feudal system in medieval France. White's argument is that adoption of the stirrup made armed cavalry the greatest military force of medieval Europe. In the face of this unprecedented capacity for war the peasants, who could not afford to own horses, were forced to seek protection from aristocratic lords who could afford to maintain cavalries and thus offer them protection. This theory is no longer accepted due to criticisms on several grounds, among them that use of the stirrup did not in fact confer so great an advantage and that cavalry were successfully used in France before the introduction of the stirrup.

Several other arguments make technological determinism an uncommon theory today. One is that the same technologies have had different effects in different societies (such as the ways gunpowder was used in China versus in the United States) suggesting that the technology is not an independent or unstoppable force, but merely offers opportunities to people who choose how to use it depending on their own needs and desires. Equally important is the fact that the historical record shows that societies choose whether to adopt technologies or not and how to use them. To take the example of power generation, the technology of the photovoltaic cell was discovered in the United States in the 1950s, but the country chose not to invest significant resources in developing the potential of solar power. At the same time, the United States did choose to invest in the development of nuclear power, and that technology has become an important supplier of electricity in some areas of the country while solar power plays a much smaller and more specialized role.

A different theory of the relationship between society and technology is that of social construction of technology (SCOT), which argues that human action shapes technology and that the use of a technology is embedded in a social context. In this school of thought, a society (or influential stakeholders within a society) plays the key role in deciding which of several competing technologies will be adopted and how they will be used. For instance, when the bicycle was a relatively new technology, numerous alternative designs were initially available, but the familiar design of the modern bicycle eventually won out while the others were discarded because they did not serve the needs of contemporary society. Although SCOT may be seen as a correction to the excessive power granted technology in technological determinist theory, it has been criticized for ignoring the influence technologies may exert on societies after they are adopted.

A third theory is that of *technological momentum*, a term coined by Thomas P. Hughes that incorporates elements of both technological determinism and social construction of technology but views the relationship between society and technology as both reciprocal and time-dependent. In this formulation, society may exert influence on technology, technology may influence society, and their relationship may change over time. Typically, when a technology is relatively new and undeveloped, society may hold more power to shape it while the technology must demonstrate that it can fulfill important needs in that society. Once a technology becomes more developed and the society has more invested more in it, power may shift so that the technology may be able to exert power to shape the society. For instance, the Muscle Shoals Dam in the United States was built during World War I to supply energy for a nitrogen fixation facility that was needed for the war. After the war it was no longer needed for this purpose, but since the dam represented a large societal investment, other uses were sought for it, and eventually it became part of an electrification and flood control project that brought major changes to the region.

Diffusion of Technology

Interest in the diffusion of technology has its roots in the studies of anthropologists, historians, and sociologists interested in how new knowledge and practices spread and were adopted or rejected by different human societies. Early studies of diffusion include the adoption of the horse among Native Americans, and the spread of corn (maize) cultivation from North America to Europe. Studying the diffusion of technology provides an excellent opportunity to see the reciprocal influence between societies, and technology and diffusion theory provides an analytical framework for the study of social change. Diffusion theory is also used today to inform programs intended to deliberately spread knowledge or technologies (e.g., beneficial health practices or improved agricultural techniques) among human societies.

Although every case is different, a few commonalities exist that can predict or explain how different societies will react to and shape a technology. One relates to how they view the technology: if it is perceived as a good fit with the current practices and values of a society, if adoption of the technology presents low social and economic costs, and if it is relatively simple, it is more likely to be adopted. In contrast, technologies that are perceived to be at odds with current values and practices, that present high social and economic costs, and that are relatively complex, are more likely to meet with resistance.

Technologies that are flexible enough to be adapted to the conditions of a specific society also have a better chance of being adopted, as are those that can be observed or tried without a large commitment on the part of the potential adopter.

Methods of communication also play a role in whether technologies are adapted by a society or not. Mass communications (e.g., advertising on television or radio) may reach many people and inform them of the existence of a new technology but face-to-face communications (e.g., if a trusted neighbor is using the technology) are more likely to result in adoption. Technologies adopted by high-status individuals who are perceived as being members of the social system are more likely to be adopted by others than those used only by outsiders or marginal members of the society.

People and societies differ in their eagerness, or lack thereof, to adopt new technologies. For purposes of analysis, individuals in a society may be divided into the following classifications: innovators, early adopters, early majority, late majority, and late adopters. An innovator is an individual who is among the first to adopt a new technology, but has less influence because he or she may be viewed as somewhat marginal to his or her own society. Early adopters are usually highly integrated into the social system and serve as role models for others: when the early adopters take up and adopt a technology, it becomes far more likely that others in a society will do likewise. From them, an innovative technology will spread to the early majority (generally, individuals who are exposed to many sources of information and interact frequently with their peers), then to the late majority (generally those farther from sources of new information, who lack the resources to take chances and thus may adopt new technologies only due to economic necessity or peer pressure). Late adopters are the last group to take up a new technology (by which time it may no longer be considered new by most members of the society).

Social structure—and an individual's place within it—frequently exerts a strong influence on how quickly a new technology spreads through a society. Although the choice to adopt or not may be individual, extra-individual factors such as the political and economic structure of the society and the geographic location of the different individuals may also exert strong influence. Social norms can exert as strong an influence as formal structures: for instance, if a technology is viewed as appropriate for males but not for females—or is associated with a less powerful minority group—that may override many other considerations.

Diego A. Comin and Bart Hobijn looked at the diffusion of technology in 166 countries over almost 200 years (1820–2003) to see which factors influenced the adoption of technology at the country level. They looked at a number of different technologies in six different fields: transportation, information technology, healthcare, telecommunications, steel production, and electricity. Overall, Comin and Hobijn found that the average adoption lag (the time between the invention of a technology and its adoption by a particular country) was 47 years, but that there were major differences between countries in the speed by with which new technologies were adopted. They also found the speed of technology adoption is accelerating and that more variance in the speed of adoption is explained by the technology itself rather than by country characteristics.

Many differences in the adoption rate among countries and over time were observed by Comin and Hobijn, underlining both the influence individual societies may exert on technology and the ways that adoptions of technology may transform societies by facilitating increased prosperity. They found that the United States and the United Kingdom have been leaders in adopting new technology over the entire time period, factors they attribute in part to the high income and productivity levels of these two countries. They see a different pattern with Japan; until 1867, Japan was a very slow adopter of most technologies, but the industrialization process that began with the Meiji restoration accelerated the adoption of technologies and by the 20th century, Japan adopted most technologies at a speed comparable to the United States and faster than most Organisation for Economic Co-operation and Development (OECD) countries. Comin and Hobijn note that increased living standards and per capita income proceeded parallel with increasingly rapid adoption of technology. Similar results were observed with the "East Asian Tigers" (Hong Kong, Singapore, South Korea, and Taiwan), which experienced both rapid growth in per capita GDP and increased speed of technology adoption in the second half of the 20th century, a major change from the previous 150 years in which their technology adoption gaps were often longer than those in sub-Saharan Africa and Latin America. Finally, countries in sub-Saharan Africa were very slow to adopt new technologies, and their economies have also grown more slowly than average while Latin American countries were quick adopters before 1950, but slow thereafter, a fact reflected in their declining prosperities.

The Sociology of Risk Perception

New technologies often bring with them new risks: for instance, the large-scale use of chemicals in industry brings the risk of chemical spills and other forms of pollution while the adoption of nuclear technology in power plants brought with it the risk of a nuclear accident. Sociology has focused on two particular questions with regard to risks posed by technology: "What influences how people perceive risk?" and "How does the interaction between technology and social organizations influence the creation of technological risk?" Perceived risks are important because they can strongly influence public opinion and thus may be a greater factor in which technologies are adopted or not adopted than actual mathematical calculations of risk.

There is a large literature on the human perception of risk, what factors influence it, and how it may diverge from the realities of risk from a statistical or empirical point of view. For instance, a study by the U.S. Centers for Disease Control and Prevention determined that the five things parents were most concerned about with regard to their children (under age 18) were kidnapping, school snipers, terrorists, dangerous strangers, and drugs. In reality, the greatest dangers to children in the United States are automobile accidents, homicide (usually by someone known to the child), child abuse, suicide, and drowning. Both psychological and sociological explanations have been offered for this difference. Psychologically, most people are not comfortable with statistical concepts and are not adept at abstract reasoning but tend to react emotionally to things that seem strange or different. So the fact that the automobile is a familiar technology tends to override awareness that it is also a dangerous one (over 33,000 Americans died in car accidents in 2009) because the very familiarity causes people to discount the danger. Child abduction by a stranger is a very rare occurrence and seems to be decreasing (about 100 took place in the United States in 1999 versus 200–300 in 1988) but the perception of danger may be magnified by its very rarity. The same is true of adult perceptions of risk: people tend to believe that the risk to life by spectacular but rare events such as plane crashes or nuclear accidents are much greater than they really are, while failing to recognize the dangers to life posed by common chronic diseases, such as asthma and diabetes.

Social communications also play a role in how risks are perceived: for instance, in the United States when a potential child abduction is reported, this information is broadcast repeatedly on the radio and television (an "AMBER alert," named for Amber Hagerman, who was abducted and murdered in 1996). Although most abducted children are returned unharmed, the perceived threat reinforced by the repetitive broadcasts may leave a strong emotional impression, making the event seem more common and dangerous than it really is. Automobile accidents are so common that they may not be reported by the news media, and even if reported are simply one story on the evening news, not a special broadcast repeatedly interrupting other programming. Social stigma can also play a role: for instance, the risk of driving drunk has been widely publicized and the hazards posed by drunk drivers are well understood, while research indicates that cell phone use while driving poses a similar risk, but because it has not been given the same publicity it is not perceived by the public as a major source of risk.

These considerations are important because new technologies are often seen as more dangerous than they really are, and reports or discussions in the news media may play on the natural human tendency to fear the unfamiliar. In addition, toxic threats, such as those posed by chemical pollution (like the Love Canal) or nuclear accidents (like Three Mile Island and Fukushima Daiichi nuclear power plant), may seem more threatening than natural disasters (like tornadoes or floods) because they have no definable end. While a tornado can cause great destruction quickly, it eventually moves on, whereas health risks from contamination could become evident years after the initial exposure, so a person never knows when he or she can feel safe again. In addition, the physicist Chauncey Starr demonstrated that people had greater acceptance of risk if they felt they were in control of the risk (e.g., by driving a car) as opposed to someone else being in control (e.g., by flying as a passenger in an airplane). Starr was a proponent of the adoption of nuclear energy and his studies of risk were motivated by public attitudes toward safety as applied to environmental risks, finding that people would accept high levels of risk from personal behaviors such as smoking but were averse to even small risks if imposed on them, such as construction of a nuclear power plant, even for a desired good such as generation of electricity.

The Sociology of Risk

Another branch of the sociology of technology looks at how interactions between technology and different modes of social organization may create risk. For instance, a society that is very hierarchical in nature, or that grants a great amount of power to people considered to be experts in some matter, may be at greater risk while using certain technologies than a more egalitarian society. It is also concerned with ways organizations take symbolic actions in order to appear in control of risks: for instance, petroleum companies have plans on file to contain oil spills even though no oil spill has ever been successfully contained—so the creation of the plan may be seen as an attempt to bring order to something that is in fact beyond human control.

The sociology of risk has been used to analyze technological disasters and to examine how sociological and technological factors combined to cause the disaster. One example is the explosion of the Space Shuttle *Challenger* in 1986, which failed shortly after takeoff, killing all crew members. Diane Vaughan analyzed the sociological context of this disaster, in particular the social organization of U.S. National Aeronautics and Space Administration (NASA). Clearly, the general public was shocked by the disaster, both because of the loss of life and also because it seemed inconceivable that an agency with access to so many resources and technical experts could have made such a terrible mistake. Vaughan found the root cause in the organization of the agency rather than any particular decision or the action of any one individual. Despite ample evidence that the shuttle launch should be delayed, the culture of NASA was focused on moving forward with the event so the definition of acceptable danger was regularly revised and thus a level of risk that would have drawn the attention of someone outside the system was entirely acceptable to those working inside it.

To use another example, the accidental release of toxic chemicals at the Union Carbide plant in Bhopal, India, in 1984 may have caused over 10,000 deaths and 500,000 injuries (estimates differ widely), making it one of the worst industrial accidents in history. Examination of the circumstances surrounding this accident suggest that a high degree of risk was accepted by plant management due to the plant being located in a poor, developing nation. These risks include locating a dangerous chemical facility near a densely populated area and failing to impose safety standards that would have been expected were the plant located in, say, Europe or the United States. People working within the plant also became accustomed to the high level of risk, both because it was familiar and because their options for other employment were limited. The normalizing of a high level of risk meant the plant operated under conditions that made occurrence of the resulting disaster less shocking that it would seem to someone who assumed the plant was operated with a more typical attitude toward danger.

Normal accident theory, a concept developed by Charles Perrow, suggests that some technologies are too complex to ever be operated by fallible humans without significant risk. In fact, according to this theory, in sufficiently complex technologies the occurrence of accidents is a normal event that should be expected rather than an unexpected deviation from the norm. Perrow argues that in very complex technologies it is impossible to calculate all the ways in which an accident could occur: a failure in one part of the system, accompanied by a failure in an entirely separate and unrelated part, could lead to consequences that could never have been predicted. He argues that efforts to ensure the safe operations of such complex technologies, for instance by building in layers of redundancies, are more useful in creating the illusion or appearance of safety than its reality.

High-reliability theory, a competing formulation of the sociology of risk, argues that although failure is inevitable in complex systems, it is possible to organize the systems so that failure will not be catastrophic. In addition to building redundancy into the system (in the hopes of limiting the damage of failures), social organization can help limit the probability of disasters. For instance, hierarchical methods of organization may contribute to the probability of an accident because people low in the hierarchy may be most likely to know when a hazard occurs. If these people were empowered to intervene if they see a hazard and to report mistakes, and if those mistakes are treated as a learning experience for the entire organization rather than as an occasion for blame and punishment, the probability of a serious disaster can be significantly lessened.

See Also: Actor-Network Theory; Information Technology; Marxism and Technology; Participatory Technology Development; Social Construction of Technology; Technological Determinism; Technological Momentum; Technology and Social Change; Technology Transfer; Unintended Consequences.

Further Readings

Belkin, Lisa. "Keeping Kids Safe From the Wrong Dangers." *New York Times* (September 18, 2010). http://www.nytimes.com/2010/09/19/weekinreview/19belkin.html?ref=weekinreview (Accessed September 2010).

Comin, Diego A. and Bart Hobijn. "An Exploration of Technology Diffusion" (April 2008). http://hbswk.hbs.edu/item/5918.html (Accessed September 2010).

Haywood, Trevor. *Info-Rich—Info-Poor: Access and Exchange in the Global Information Society.* London: Bowker Saur, 1995.

Hill, Michael W. *The Impact of Information on Society: An Examination of Its Nature, Value and Usage.* London: Bowker Saur, 1999.

Lyon, David. *The Information Society: Issues and Illusions.* Cambridge, UK: Polity Press/Blackwell, 1988.

Merritt, Roe Smith and Leo Marx, eds. *Does Technology Drive History? The Dilemma of Technological Determinism.* Cambridge, MA: MIT Press, 1994.

Perrow, Charles. *Normal Accidents: Living With High-Risk Technologies.* Princeton, NJ: Princeton University Press, 1984.

Starr, Chauncey. "Social Benefit Versus Technological Risk." *Science*, 165 (1969).

Vaughan, Diane. *The* Challenger *Launch Decision: Risky Technology, Culture and Deviance at NASA.* Chicago, IL: University of Chicago Press, 1996.

Weinstein, Jay. *Sociology/Technology: Foundations of Postacademic Social Science.* New Brunswick, NJ: Transaction Books, 1982.

Sarah Boslaugh
Washington University in St. Louis

Solar Cells

Solar cells use a semiconductor to convert the solar energy of sunlight into electricity via the photovoltaic effect. Individual cells are assembled in order to form solar panels and arrays. First observed by French physicist Alexandre-Edmond Becquerel, son of electricity pioneer Antoine Cesar Becquerel, as a teenager in 1839, the photovoltaic effect is the process by which voltage or electric current is created in a material that has been exposed to electromagnetic radiation (such as light). The first solar cell was built in 1883 by American inventor Charles Fritts, who coated selenium (a semiconductor) with a thin layer of gold. The solar cell worked but had an efficiency of only about 1 percent, making it impossible to generate enough electricity to recover the cost of the materials except over a very long period of time. Improvements to solar cells in the mid-20th century brought efficiency up to 5 percent and gradually climbed, with many cells on the 21st-century market exceeding 20 percent efficiency (the most efficient are over 40 percent but are used for satellites). Modern thin-film solar cells can be expected to last at least 20 years and are half the thickness of solar cells used in the 1990s, making them significantly cheaper. The energy payback time of solar cells—the time it takes for an energy system to pay back the energy used to manufacture it—is hard to normalize because location and other factors have such a large impact, but in some cases it has fallen to less than one year, and rarely exceeds four years.

Solar cells are connected in modules, which are connected in an array. There are a number of types of solar cells, often referred to by the material used to make their thin film. Thin film solar cells include thin-film silicon (the same material, just thinner, used in the 1990s), cadmium telluride (CdTe), and copper-indium selenide (CIGS), which vary in efficiency and cost. The most efficient cells are multi-junction devices that use multiple thin films, each of which is most efficient at a certain portion of the electromagnetic spectrum. The cost of gallium and germanium has risen as demand has increased for multi-junction solar cells using gallium arsenide (GaAs), gallium indium phosphide (GaInP), and germanium thin films.

Solar cells can power a specific device—there are, for instance, solar-powered cell phone chargers, lights and lanterns (which charge during the day to be used at night), security cameras, and MP3 players. The U.S. National Aeronautics and Space Administration (NASA) has even developed a solar-powered aircraft. But a more flexible use of solar cells is to feed the collected energy into batteries, or to use an inverter to feed the electricity back into the grid—either way, the electricity is then free to be used by anything that can be plugged in.

Solar cells work because as the photons in sunlight strike the cell, some of them are absorbed by the semiconducting material. As each photon is absorbed, its energy is passed on to an electron. The extra energy frees the electron and allows it to move freely through the semiconductor, while also producing heat from any energy that is not successfully absorbed. This is one of the factors affecting the efficiency of a solar cell, but efficiency is not as critical to practicality in solar energy as it is in the use of nonrenewable fuels, because solar energy itself is free and inexhaustible. Efficiency is primarily of interest in determining how much electricity can be generated by a given number of solar cells, especially in areas with cloud cover where solar collection is not optimal. Furthermore, a more energy-efficient solar cell is not necessarily more cost-efficient if it costs significantly more money to buy or make; in the end, what matters is the cost to the consumer for the kilowatt-hour of energy produced by the cell.

The most significant advance in photovoltaics (the science of solar cells) in recent years has been the thin-film solar cell, which is manufactured by laying thin layers ("thin film," with a thickness ranging from under a nanometer to several micrometers) of photovoltaic material on a substrate. In 2009, thin-film solar cell sales overtook the older bulk crystalline designs. One of the advantages of thin-film cells is that they can be used to create semitransparent solar cells that can be applied to windows, generating electricity while still allowing light into the house and appearing no different than ordinary window tinting. This greatly increases the available installation area, especially for homeowners, but also opens up the possibility of skyscrapers coated in semitransparent solar cells. Most thin-film cells are not semitransparent, and those that are cost more.

Organic solar cells use conductive organic polymers to absorb light and transport the charge; they have a low production cost when dealing with high volumes but are also less efficient. One class of organic solar cell is the dye-sensitized solar cell, for which French physicist Michael Grätzel won the 2010 Millennium Technology Prize. Though dye-sensitized cells, which use a porous layer of titanium dioxide nanoparticles (the semiconductor) covered in a layer of photosensitive dye (the light absorber), have a lower conversion efficiency than many other thin-film cells, the cost to manufacture is very low, and it is expected that when manufactured in volume (to benefit from economies of scale), the cost per kilowatt-hour will be low, a rate competitive with fossil fuels. Significant development remains to be done in dye-sensitized solar cells, which have only entered the commercial market in the past few years.

Whatever sort of cell is used, interconnected solar cells make up solar panels; multiple panels are joined in a photovoltaic array, and the installation is completed by adding an inverter and batteries. The inverter is necessary because the photovoltaic produces, and batteries store, direct current electricity, while household electrical systems use alternating current. The solar array may be fixed in a specific position, whether on a rooftop, on windows as discussed above, or on a fixed rack mount. When mounted, the array is typically fixed at a tilt angle equivalent to the latitude of the location for maximum solar collection. Moving mounts are also available and may either be manually adjusted (typically along only one axis, adjusted seasonally to maximize sun exposure) or adjusted automatically (often along both axes, to follow the sun's movements over the course of the day). Because solar power is an intermittent source, unavailable at night and low-yield during cloudy conditions, batteries are necessary to store power for times when demand exceeds the solar panels' supply. Alternately, surplus electricity can be fed into the municipal grid and will be credited to the consumer in areas where utility companies offer net metering.

See Also: Appropriate Technology; Best Available Technology; Carbon Market; Clean Energy; Cogeneration; Concentrating Solar Technology; Green Roofing; Microgeneration; Passive Solar; Smart Grid; Solar Hot Water Heaters; Solar Ovens; Sustainable Design; Zero-Energy Building.

Further Readings

Miller, G. Tyler and Scott Spoolman. *Environmental Science*. Florence, KY: Brooks/Cole, 2010.

Nelson, Jenny. *The Physics of Solar Cells (Properties of Semiconductor Materials)*. London: Imperial College Press, 2003.

Sahin, Abban and Hakim Kaya. *Thin-Film Solar Cells (Energy Science, Engineering and Technology)*. Hauppauge, NY: Nova Science Publishers, 2010.

Vig, Norman J. and Michael E. Kraft. *Environmental Policy: New Directions for the Twenty-First Century*. Washington, DC: CQ Press, 2009.

Bill Kte'pi
Independent Scholar

Solar Hot Water Heaters

Solar energy is harnessed in the form of heat energy and is used to produce hot water. The practice of using the sun for heating water for domestic use was attempted informally as far back as the 19th century, and the first patented solar hot water system was sold commercially in 1891, in Baltimore, Maryland. Invented by Clarence Kemp, the system was called the "Climax," and was popular in California and other states in the United States. The incentive for householders to invest in this heating system was the comparatively high cost of using conventional fuels for heating and the inconvenience of having to use fires and stoves to have hot water. Solar thermal system designs were experimented with in the following years, and in 1909, William J. Bailey patented a system that separated the water storage tank from the heating element that absorbed the energy from the sun. This is essentially the basis of the design of solar hot water heaters used today, even though the current technology and efficiency are considerable improvements over past heating systems.

Integrated collector storage systems, like that shown here, integrate the water tank with the solar collector and heat water directly in the tank.

Source: iStockphoto

A typical solar water heater consists of a solar collector mounted on the roof of a building or in any appropriate place where it receives direct sunlight. The solar collector is connected to a water storage tank, which may be located differently according to the type of solar heating system installed. The unheated water is circulated through the collector by tubing that absorbs the heat from the collector and transfers it to the circulating fluid. This fluid may be the water that is to be heated or a high-capacity heat exchange fluid that carries the heat to the water in the tank. These are passive methods of facilitating the exchange of heat energy from a solar collector to the unheated water in a storage tank or cylinder, without the use of mechanical services. However, a small amount of conventional electricity may be used in some active solar hot water systems to circulate the heat exchange fluid around the system, but this energy used to operate mechanical pumps and controllers is very little compared to the energy that might have been used to heat the water in the whole tank. A small scale domestic solar hot water system is different from the electricity-producing solar photovoltaic (PV) systems as it does not convert the sun's energy into electricity, but stores it in the form of heat energy in water.

Active and Passive Systems

Active solar hot water systems use mechanical pumps and differential controllers to regulate and direct the flow of the heat transfer fluid or water from the solar collector to the tank or vice versa. The controller senses the temperature difference between the water in the tank and the temperature in the solar collector, and the pump is switched on when the water in the tank cools below the temperature of the collector, for example, if the temperature differential setting is 0 degrees to 10 degrees Celsius, the pumps are switched off automatically by the controller when the differential is 0 degrees Celsius to prevent additional heating and switched on when the differential is 10 degrees Celsius or above to boost heating. Some pumps run on mains electricity and others operate on electricity generated by a solar PV panel connected to it. The solar PV–dependent systems will circulate the fluid only when the sun is shining and store the heated water in well-insulated tanks for nighttime space heating. They may use mains electricity as a backup for days when the sun is not strong enough to generate electricity for running the pumps. In active solar hot water systems, the water storage tanks can be located inside the roof space or in any other location that will minimize heat loss to the cold atmosphere, as the flow of water does not depend on a gravity system. These tanks can therefore be combined with the hot water cylinders in domestic space heating systems, and the solar hot water system can be used to preheat water in the cylinder in winter for space heating.

A type of passive solar hot water system is that in which the water tank is integrated with the solar collector, called the integrated collector storage system. The collector forms the top of the water storage tank and heats the water directly in the tank. In addition, a thermosyphon passive system can also be configured with the tank separate from the collector, but located at a higher elevation than the collector. Cold water from a height flows down to the solar collector by gravity, and as the water heats up while passing through the collector, it rises to reach the storage tank again. There is no need for mechanical pumps to circulate fluid around the system, as this works on convective heat flows and gravity. Convection is the way heat circulates through the movement of liquids and gases, where warmer air or liquid rises and the colder part sinks down. Thus, phase change (liquid–vapor) can be used to cause the heated fluid in the collector to evaporate and rise to transport heat from the collector to the storage. These types of passive systems can deliver expected results in hot climates where night or wintertime freezing is not present, or a good amount of sunlight is received even in colder but not freezing temperatures.

Solar Collectors

Solar collectors are devices that absorb heat energy from the sun. They are of three main types, namely, flat plate, integrated collector, and evacuated tube. A flat plate collector consists of a solar absorber plate (usually made of copper and painted black), a transparent glass cover to let in sunlight but prevent heat loss, tubing attached to the absorber plate that circulates the water (or heat exchange fluid) to and from the storage tank, and insulation backing to prevent heat loss from the back and sides of the box. Large-scale solar hot water systems are used in nondomestic settings to heat swimming pools, and they may be unglazed or without glass covering the collector.

One of the problems in cold climates is that of damage to the tubing due to freezing of the water in them. For this reason, sealed collector units are used with a heat exchange fluid that has an anti-freezing liquid mixture running through the tubing. This liquid mixture collects the heat and then transfers it to the water in the tank by another simple heat exchange process. After the heat is withdrawn from the liquid it returns to the collector for reheating in a continuous loop. The advantage in this is that the water in the tubing does not need to be drained on severely cold days to prevent it from freezing and bursting the tubes. Integrated collectors are similar to flat plate collectors, only the collector forms the top surface of the tank in which heated water is to be stored. The remaining walls of the tank are insulated to prevent heat loss. Evacuated tube collectors consist of multiple glass tubes parallel to each other mounted on a plate or in a box. Each tube is actually made of two concentric glass tubes with a vacuum between them. The inner tube is also coated with a black or dark thermal absorbent. Inside it runs U-shaped copper piping containing the liquid, which may be water or the heat transfer fluid. The sun's energy falls on the glass tubes and is absorbed as heat energy by the thermal absorbent on the inner glass tube. Any heat likely to be lost by re-radiation is not able to escape because of the vacuum between the glass tubes. The heat is thus retained in the inner glass tube and absorbed by the copper piping and the liquid inside it and carried to the water in the storage tank. Very high temperatures can be achieved by evacuated tubes because heat losses are minimized by the vacuum, and the collector gross area is larger than a single flat plate collector of the same size due to multiple evacuated tubes. Temperature limitation and overheating mitigation measures need to be built into all solar water heaters, especially when installed in hot climates.

Design, Efficiency, and Annual Costs

The output of a solar hot water system generally depends on the efficiency of the collector and the effectiveness of the whole system design. Individual collectors and whole systems are rated separately for their efficiencies because the collector efficiency depends on the performance of one component (solar absorber), while the whole system efficiency depends on a number factors (water and ambient temperatures, system configurations, insulation, water volume, the type of collector, heat exchange mechanism efficiencies, the location and local weather at the installation, and the amount of sunlight received by the collector, etc).

A method of quantifying the performance of a solar hot water system is by calculating its annual running cost from its efficiency or solar energy factor (SEF). This is defined as the energy delivered by the system divided by the electrical or gas energy put into the system. SEFs range from 1.0 to 11, with efficiency increasing with higher numbers.

Solar fraction (SF) is also used to calculate the efficiency of a system. If the temperature of the unheated water in the tank is 10 degrees Celsius and the required temperature is 60 degrees Celsius, then the amount of solar energy that is used to raise the temperature from 10 to 60 degrees Celsius is the SF for the day. If it delivers all the energy toward this requirement, the SF is 100 percent or 1.0 (maximum). For no solar heating (on rainy or cloudy days), the SF will be "0." The monthly SF is the average of the values calculated throughout the month.

However, there are some losses that need to be accounted for in the calculation of SF. These are the heat exchanger losses or the difference in the output temperature of the heat exchanger from the storage tank temperature, and the energy used for running the pumps and controllers if conventional electricity is used. For solar PV–dependent pumps, this heat loss is zero. If the SF value is calculated without taking into account these losses, the value of the monthly average SF will be higher. Therefore it is necessary when comparing the SF for specific systems to ensure that the same calculation procedure for SF has been used.

Solar hot water collectors should be oriented considering local landscape features that shade the collector daily or seasonally (foggy mornings or cloudy afternoons), as these factors may affect the collector's optimal efficiency. Even so, performance and efficiency are not the only criteria in choosing a solar energy collector. Designing an efficient solar hot water system will need appropriate sizing of the collector and storage tank according to the use requirements for hot water. Up-front material and installation cost, payback or saving in energy costs over the short and long term, quality of installers, availability of service and parts, and the expected life of the equipment are also important points to consider. The system may be well designed, but may not perform according to the predicted efficiencies if not properly installed and operated.

See Also: Passive Solar; Solar Ovens; Sustainable Design; White Rooftops; Zero-Energy Building.

Further Readings

Bainbridge, David. "The Integral Passive Solar Water Heater Book." The Passive Solar Institute. http://www.mangus.ro/pdf/Manual%20incalzire%20solara%20pasiva.pdf (Accessed July 2010).

Baker, Nick V. *Passive and Low Energy Building Design for Tropical Island Climates.* London: Commonwealth Secretariat Publications, 2010.

Chiras, Dan. *The Homeowner's Guide to Renewable Energy: Achieving Energy Independence Through Solar, Wind, Biomass and Hydropower.* Gabriola Island, British Columbia, Canada: New Society Publishers, 2006.

Perlin, John. "Solar Evolution: The History of Solar Energy." California Solar Center. http://www.californiasolarcenter.org/history_solarthermal.html (Accessed December 2010).

Ramlow, Bob and Benjamin Nusz. *Solar Water Heating: A Comprehensive Guide to Solar Water and Space Heating Systems.* Gabriola Island, British Columbia, Canada: New Society Publishers, 2006.

Swati Ogale
Independent Scholar

SOLAR OVENS

A solar oven, also sometimes known as a solar cooker, is a device that harnesses the sunlight as a source of energy to produce heat for cooking foodstuffs. Solar ovens have the advantage of requiring no fuel, thus removing any ongoing costs associated with cooking; furthermore, they present a lower fire risk than traditional open fires.

In the developing world, wood is a common fuel used to cook foodstuffs. The collection of wood and unsustainable forestry practices driven by the necessity of living with scarce resources leads to deforestation. This in turn has a number of environmental impacts—including the loss of forests as a "carbon sink" and the desertification of land previously occupied by trees.

Types of Solar Cookers

There are a great many designs of solar cooker, employing an array of different materials and approaches, deployed as the solution for a common problem. There are a number of design factors that any successful solar cooker must satisfy. The design must be capable of concentrating sunlight from over a wide area to a central point. At this central point, a black surface helps to convert the sunlight into heat, which is used to heat the food. Once the heat is generated, it must be trapped and insulated from the air outside the cooker. Solar ovens can be categorized into a number of different design configurations.

A box cooker consists of a number of mirrored panels that focus their beams toward an insulated box structure. This has a transparent top used to admit the solar radiation. Cooking pots and pans are painted black and placed "inside the box."

A hybrid cooker is a box cooker that is equipped with a supplementary electrical heating system, which can be used to provide additional heat when it is overcast, cloudy, or at night. These tend to be larger, fixed installations for use by a community or group.

Parabolic cookers are capable of generating high temperatures, but present challenges in construction. A parabolic mirror is used to focus the sunlight to a central point. The cooking container is placed at this point.

Panel cookers are an incredibly cheap design of solar cooker, where panels, often made of corrugated cardboard and covered with a cheap reflector such as tin foil or Mylar, focus the sun's rays onto a black cooking pot, which is kept inside a temperature-resistant plastic bag.

Solar kettles are devices used to generate boiling water and steam using solar power. This may be done to produce hot water to cook and make drinks with or to sterilize water. Such devices may rely on parabolic reflectors, or on evacuated tubes to generate the high temperatures. Such devices may also be used as improved autoclaves delivering dry heat to sterilize surgical instruments. Vacuum tubes have the advantage that they can function with diffuse sunlight and can also store heat.

Cooking With a Solar Cooker

Solar cooking requires a slightly different approach to food preparation. Furthermore, the constraints imposed by the sun mean there will be an optimum window of sunlight in which food cooks best. Food cooks faster when it is in smaller pieces, therefore, those using solar cookers often chop food into smaller parts to accelerate the cooking process. To heat up the solar cooker, it is turned to face the sun. The solar cooker may require regular realignment to ensure that it receives the optimum solar gain.

Food in a solar cooker is not stirred or agitated as a rule; this is partly because the food cooks much slower, and partly because when the solar cooker is opened, there is the potential for heat to escape.

Solar Cookers in Context

Solar cookers are an accessible technology that has the potential to deliver cooked food and the potential for sterilizing water for the great many at the "bottom of the pyramid" in the developing world. In terms of human health benefits, solar cookers offer many advantages over the smoky indoor stoves or open fires that are commonly accessible in the developing world. Although they do not require fuel to be gathered or bought, they do require patience and education of communities in order for them to gain social acceptance.

See Also: Appliances, Energy Efficient; Passive Solar; Solar Hot Water Heaters.

Further Readings

Algifri, Abdulla H. and Hussain A. Al Towaie. "Efficient Orientation Impacts of Box-Type Solar Cooker on the Cooker Performance." *Solar Energy*, 70/2 (2001).

Halacy, D. S. and Beth Halacy. *Cooking With the Sun*. Lafayette, CA: Morning Sun Press, 1992.

Lof, G. O. G. "Recent Investigations in the Use of Solar Energy for Cooking." *Solar Energy*, 7/3 (1963).

Mullick, S. C., T. C. Kandpal, and A. K. Saxena. "Thermal Test Procedure for Box-Type Solar Cookers." *Solar Energy*, 39/4 (1987).

Gavin D. J. Harper
Cardiff University

SOLID WASTE TREATMENT

Solid waste is leftover solid material from industrial, commercial, domestic, and other human activities. Solid waste management is the collection, source separation, storage, transportation, transfer, processing, treatment, and disposal of waste materials. Different management and treatment processes are employed to manage the solid wastes based on the location and setting in which they are generated, such as developed or developing countries, rural or urban areas, and residential or industrial setting. To the extent possible, management of solid wastes also involves deriving benefits from them; otherwise the materials within the waste would be lost, and simultaneously pollute the environment.

In the United States and other developed countries, conventional practices for mixed municipal solid waste (MMSW) management are as follows. First, the MMSW is collected by on-site or curbside collection methods where they are already separated into reusable and nonreusable wastes. The nonreusable wastes are treated in the landfill by conventional waste-to-energy technology for energy production. After energy production, the rest of the material is processed through anaerobic digestion to produce manure or compost. The reusable waste materials are recycled. The MMSW management is an orderly seven-step process that includes reduction, reuse, recycle, compost, incineration (with energy as a result), transfer to landfills, and incineration (with no energy).

Waste management technologies (WMT) reduce the amount of waste and unsalvageable materials and provide proper, economical solid waste disposal practices. Common established WMTs followed in the United States are composting, incineration, landfill, recycling, and windrow composting.

Composting is a process that aerobically decomposes organic materials derived from plant and animal matter. Composting is simple and can be done in any setting. It is part of source reduction techniques. People compost their organic wastes in their homes and on their farm land. Industrially, cities and factories compost their solid waste using more advanced aerobic decomposition techniques. Composting produces new fertile soil rich in nutrients, and is used in gardens, landscaping, horticulture, and agriculture. It reduces the amount of fertilizer that needs to be applied to the crops, thus further reducing environmental pollution. Other engineering applications of composting that help environmental sustainability include: its use as a soil conditioner; its use as a natural pesticide for soil; its ability to aid erosion control, land and stream reclamation, and wetland construction; and its use as a landfill cover.

Windrow composting is large-scale composting. Organic matter or biodegradable waste, such as animal manure and crop residues, are piled in long rows called windrows. These rows are generally turned from bottom to top to improve porosity and oxygen content. The turning helps remove moisture and redistributes cooler and hotter portions of the pile. Machines are used to turn the piles to enhance the decomposition process. Temperature records are maintained for the windrows to determine the optimum time to turn them for quicker compost production. Automated temperature data collection is used to save the workers from hazardous gas inhalation. This process produces hazardous gases, but the advantages obtained, due to the process's solid waste source reduction ability, minimize the environmental impact.

Incineration is the combustion of organic substances contained in waste materials or thermal destruction of solid wastes. The end products of incineration are ash, gas, and heat. Inorganic constituents of the solid waste are burned to produce flue gases. Organic wastes become ash after the thermal treatment, and this is then buried in landfill. Modern incineration systems change the chemical, physical, or biological composition of solid wastes through the use of high temperatures, controlled air flow in the incinerator, and excellent mixing of the wastes. State-of-the-art air pollution control devices capture particulate and gaseous emission contaminates from the incinerator to reduce air pollution. The heat generated by the incineration process is sometimes used to generate electric power. Incineration with energy recovery is one of several waste-to-energy technologies that are part of the solid waste treatment. Others are gasification, plasma arc gasification, pyrolysis, and anaerobic digestion. If the generated gas and ash are not properly handled, they create hazardous environmental conditions. Therefore, this is regulated by the U.S. government.

The landfill process is the burying of waste. Generally, nonusable inorganic wastes are landfilled. Single producers bury their own waste at the place it is produced, known as internal waste disposal landfilling. Landfills in an urban are used by many producers and receive solid wastes from households or commercial settings as well as nonhazardous sludge, industrial solid waste, and construction or demolition debris. Municipal solid waste landfills are carefully designed. These structures are built into or on top of the ground for solid waste disposal so that the wastes can be isolated from the surrounding environment, including groundwater, air, rain, and scavenging animals. This isolation is accomplished by having an impervious bottom-liner in the landfill, and by covering the wastes daily with soil. A typical landfill is constructed with cells to store solid compacted wastes, a storm water drainage facility to collect rainwater that falls into the landfill, a leaching collection system to collect percolated contaminated water to prevent flow into groundwater, a methane collection system, and a covering or cap to seal the top of the landfill. Landfills are generally constructed away from populated areas because of their foul smells and are often constructed in ditches so as to create a level surface when filled.

Recycling is another source reduction technique. Usable materials are separated from the solid wastes beginning at the waste-generating locations, such as households, or commercial or industrial locations. In most cities, recycling centers separate materials, including glass, paper, cardboard, rubber materials, plastics, iron or other metals, textiles, and electronics, using mechanical means. These separated, reusable materials reduce the following: consumption of new raw materials, energy use, air pollution (from incineration), water pollution (from landfilling), and greenhouse gas emissions.

Cities are constantly modernizing their solid waste disposal treatment processes. In 2004, New York City evaluated new and emerging solid waste management technologies and found a few technologies to be more efficient than the conventional processes. ArrowBio Process, for example, is a unique, patented, anaerobic digestion process in which the incoming MSW is deposited into a water bath that separates the components by density in order to remove the inorganic portion of the MSW (including recyclables) and prepares the organic fraction for digestion. Then, the saturated organic material or organic slurry is pumped to an anaerobic digester, an upflow anaerobic sludge blanket (UASB) type, to produce biogas for electricity production; the digested solids are then dewatered and aerobically composted to produce a soil amendment. The 2004 New York City report, presented to the city's Economic Development Corporation and the New York City Department of Sanitation, also evaluated other anaerobic digestion technologies including the following: BTA Process Technology, a three-stage anaerobic digestion process (acidification, solids hydrolysis, and methanization); BIOCEL process, an anaerobic digestion process that is intended for the source-separated fraction of MSW; and the Valorga process. The report also evaluated a hydrolysis process (the CES OxyNol hydrolysis process, which includes preparation, acid hydrolysis, acid recovery, fermentation, and distillation); a thermal process from Dynecology (gasification with briquetting of RDF/Coal/Sewage Sludge); and a gasification process from Twin-Rec fluid bed gasification technology. Details about the emerging solid waste treatment technologies can be obtained from the report.

See Also: Anaerobic Digestion; Design for Recycling; Recycling; Waste-to-Energy Technology.

Further Readings

The City of New York. "Evaluation of New and Emerging Solid Waste Management Technologies." http://www.nyc.gov/html/dsny/downloads/pdf/swmp_implement/otherinit/wmtech/phase1.pdf (Accessed September 2010).

Coufal, C. "In-House Windrow Composting Q and A." http://www.thepoultrysite.com/articles/1269/inhouse-windrow-composting-q-and-a (Accessed September 2010).

Epstein, E. *"Industrial Composting: Environmental Engineering and Facilities Management*, 2nd ed. Boca Raton, FL: CRC Press, 2010.

Freudenrich, C. 2009. "How Landfill Works." http://science.howstuffworks.com/landfill6.htm (Accessed September 2010).

Kolenbrander, A., J. Todd, M. S. Zarske, and J. Yowell. "Lesson: 3RC (Reduce, Reuse, Recycle and Compost." http://www.teachengineering.org/view_lesson.php?url=http://www.teachengineering.com/collection/cub_/lessons/cub_environ/cub_environ_lesson05.xml#objectives (Accessed September 2010).

Oppelt, E. T. "Incineration of Hazardous Waste—A Critical Review." *Journal of the Air Pollution Control Association*, 37/5 (1987).

Panda, S. S. "Landfills." *Green Cities: An A-to-Z Guide*, Paul Robbins and Nevin Cohen, eds. Thousand Oaks, CA: Sage, 2011.

Sudhanshu Sekhar Panda
Gainesville State College

STRUCTURATION THEORY

Structuration theory is a sociological concept that offers perspectives on human behavior based on a synthesis of structure and agency effects (termed *duality of structure*). Instead of describing the capacity of human action as being constrained by powerful stable societal structures, or as a function of the individual expression of will, structuration theory acknowledges the interaction of meaning, standards and values, and power. By taking account of the dynamic relationship between these different facets of society, a more comprehensive method of examining both the "greening" of society and the introduction of new technologies can take place.

Theories of Structure and Agency

The nexus of structure and agency has been a central tenet in the field of sociology since its inception. Theories that argue for the preeminence of structure (also called the objectivist view in this context) resolve that the behavior of individuals is largely determined by their socialization into that structure. Structures operate at varying levels, with the research lens focused at the level appropriate to the question at hand. At its highest level, society can be thought to consist of mass socioeconomic stratifications (e.g., gender or class structures). On a mid-range scale, institutions and social networks (such as religious or familial structures) might form the focus of study, and at the microscale one might consider how community or professional norms constrain agency. Structuralists describe the effect of structure in contrasting ways. Émile Durkheim highlighted the positive role of stability and permanence, whereas Karl Marx described structures as protecting the few, doing little to meet the needs of the many.

In contrast, proponents of agency theory (also called the subjective view in this context) consider that individuals possess the ability to exercise their own free will and make their own choices. Here, social structures are viewed as products of individual action that are sustained or discarded, rather than as incommensurable forces.

The Particular View of Structuration Theory

More recently, sociologists have questioned the polarized nature of the structure–agency debate, highlighting the synthesis of these two influences on human behavior. A prominent scholar in this respect is British sociologist Anthony Giddens, who developed the concept of structuration. Giddens argues that just as an individual's autonomy is influenced by structure, structures are maintained and adapted through the exercise of agency. The interface at which an actor meets a structure is termed *structuration*.

Thus, structuration theory attempts to understand human social behavior by resolving these competing views of structure–agency and macro–micro perspectives. This is achieved by studying not the actor or the structure but the processes that take place at this interface. Structuration theory takes the position that social action cannot be fully explained by the structure or agency theories alone. Instead, it recognizes that actors operate within the context of rules produced by social structures, and only by acting in a compliant manner are these structures reinforced (i.e., they have no inherent stability outside human action—they are socially constructed). Alternatively, through the exercise of reflexivity, actors evolve structures by acting outside the constraints placed on them. Giddens's framework of structure differs from that in the classic theory. He proposes three types of structure. The first is signification, where meaning is coded in the practice of language and discourse. The second is legitimation, consisting of the normative perspectives embedded as norms and values. Giddens's final structural element is domination, concerned with how power is applied, particularly in the control of resource.

The concept of anthropocentric "climate change denial" is an interesting example. As an element of discourse, this relatively recent phrase has superseded that of "climate skeptic," introducing a new normative element that suggests something inarguably true is being denied (rather than challenged, or doubted in the more traditional skeptical fashion). This effectively brackets those with an opposing view into a disempowered position of ignorance and further legitimizes the climate change policy agenda to release resources that further strengthen the paradigm. Regardless of one's personal opinion on the veracity of climate change claims, this exemplifies how the signification of the notion of "denial" ripples through Giddens's layers of structure.

Applications in Green Technology

Structuration theory suggests that all social change emerges from a combination of macro- and micro-level influences, and the application of structuration theory to technology is established notably in the field of information technology in organizations. Notable researchers include Wanda Orlikowski, who uses the theory to examine how technology is used, what meanings are ascribed to it, how norms are affected, what the power response would be, and what the implications might be. This is in contrast to a deterministic approach, where technology is placed firmly in the setting of social structure, restricting human agency. Instead, a degree of interpretative flexibility is recognized, where actors take on different meanings for the same technologies.

Green technologies offer fertile soil in which to undertake the structuration approach. Innovations in environmental technology cannot be understood purely in terms of the actions of individuals, nor in terms of class or of institutional or community structures. Instead, the level of green technology observed now can be viewed as a result of the synergy between Giddens's structural parameters. The signification of the concept of environmental responsibility takes place in a diffusion of individual environmental conscientiousness and the meaning ascribed to scientific information and climate models in the 1980s onward. These quantitative models facilitated dominant forces to develop the national targets agreed in the Kyoto Protocol (1997) that would become actionable in the prioritization of green technologies, the release of resources for investment and capacity building, and eventual environmental impact legislation applicable at the organizational level. As companies develop and communicate the green metrics required to reduce their environmental impact, the pursuit of greener operational routines becomes legitimized through an expectation of—for example—recycling at the individual and organizational levels. Developing a new language of environmental impact and a greater availability of benchmark data establishes the informed, motivated investor who goes on to further evolve the structural elements of society in new ways.

As well as examining contributory factors to these broader trends, structuration theory offers a viewpoint from which to think about how new technologies are introduced and used. This iterative description of the interrelationship of different levels of structure alongside individual agency identifies the complexity and interdependence involved in making such technologies prevalent, the nuances of which would be lost in a more static structure versus agency debate.

See Also: Actor-Network Theory; Authoritarianism and Technology; Marxism and Technology; Social Agency; Sociology of Technology.

Further Readings

Bryant, C. and A. Jarey. *Giddens' Theory of Structuration: A Critical Appreciation.* London: Routledge, 1991.

Giddens, Anthony. *The Constitution of Society: Outline of the Theory of Structuration.* Queensland, AU: Polity Press, 1984.

Orlikowski, Wanda. "The Duality of Technology: Rethinking the Concept of Technology in Organizations." *Organization Science*, 3/3 (1992).

Walley, Liz and M. Stubbs. "Termites and Champions: Case Comparisons by Metaphor." *Greener Management International*, 29 (Spring 2000).

Beverley J. Gibbs
University of Nottingham

SUSTAINABLE DESIGN

The term *sustainable design* is used in this entry to refer to design that is motivated by humanity's capacity to endure. The most commonly cited definition of sustainability comes from the Brundtland Commission (the World Commission on Environment and Development), which stated in its 1987 report, "Humanity has the ability to make development sustainable—to ensure that it meets the needs of the present without compromising the ability of future generations to meet their own needs." Although the statement is vague, it discusses the needs of both future and present generations. In the context of design, described above as inclusively anything involved in shaping environments, sustainable design must contribute to balance the natural cycles and resources of the planet.

Construction wastes constitute the largest fraction of landfills; that's why sustainable approaches, such as using recycled materials like straw bales to build houses, are promoted.

Source: Colin Rose/Wikimedia

Sustainable design was born out of the environmental movement in the United States in the 1950s that pointed out the environmental costs associated with many technological innovations of the time. Today, however, we see almost an inverted perspective in that it is often optimistically understood that technological innovations can be utilized to foster sustainable design. Technological development from within the fields of design can be focused to transcend ecological equilibrium, to have positive embodied energy assessments, material reductions, and even change living pattern trends.

The general term *design* encompasses the description of a field composed of numerous disciplines, such as planning, architecture, interior, landscape, graphic, fashion, automotive, and product design. Typically, design, which is equally used as a verb, involves the events and processes that take something from an idea to an actual built item, either an object or an environment. It is also important to understand that design could include anything that is done to meet a perceived need or desire, and therefore is not limited to the disciplines listed above. It could be defined inclusively as any process involved in shaping the environments of daily experience.

This entry elaborates on the above definition of sustainable design through the lens of some of the critical issues of design related to global ecological issues. Further, the following section summarizes some principles of sustainable design. Through following these simple principles of design, it is possible to lessen the negative impact of the designed world on the planet. Sustainable principles of design allow for the design for the needs, wants, and desires of today without compromising the future.

Sustainable Design Principles

The goal of sustainable design should be reached by working equally toward environment, society, and economy. These three areas are often referred to as the "three pillars" of sustainability. These three areas are not mutually exclusive and can be mutually reinforcing; a principal goal of sustainability is to achieve a balance of all systems that will last over time. In other words, a sustainable design should (1) benefit the environment, (2) improve the lives of humans, and (3) be cost effective. In addition to such considerations, sustainable designs should adhere to the following principles.

Energy Efficiency, Carbon Footprint, and Life-Cycle Assessment

The primary principle of sustainable design involves the reduction of energy use. Good design should produce products, environments, and lifestyles that require less energy. After all, one of the largest contributors to awareness of sustainability is climate change, which is spurred by carbon emissions and other greenhouse gases. These result from the burning of fossil fuels, which in turn result from nonrenewable resources. Energy use starts with the manufacturing of a product and continues until its eventual deconstruction. Therefore, designs should evaluate the energy used from the very beginning of the process in manufacturing, to the transportation of goods, to the energy expended during operation, and even the energy required for demolition. This inclusive assessment of design should also assess the "carbon footprint," or the carbon dioxide emissions of a design. This can be achieved by multiplying the anticipated energy usage of a project by pollution emissions provided by the local utility companies for various greenhouse gases.

An important strategy is to use less energy-intensive means to cut down on life-cycle energy costs of the designed object or environment. In some disciplines, such as product or fashion design, the greatest amount of energy is used during the manufacturing and transportation stages, and therefore efforts should be made to reduce the energy use during that phase. Designers can choose materials that require less energy to manufacture, select manufacturing plants that use renewable energy sources, and design packaging to optimize transportation efficiencies.

In other disciplines, such as automotive, appliance, and architectural design, where the end product has a longer life span and requires energy to operate, the greatest amount of energy is used during operation. It is therefore most critical to consider efforts to reduce energy used during operation, in addition to efforts to reduce energy use during the other stages. In buildings, heating and cooling comprise the majority of energy costs, both environmentally and economically. Designers can combat this by using passive design measures, such as passive solar heating, thermal mass, shading, and/or natural ventilation in their building designs so that less mechanical heating and cooling are required, thus reducing energy flows. In addition, renewable energy can be harnessed on-site to offset the power used by appliances or to provide energy for mechanical systems. Similarly, where possible, planners can design developments so that the majority of buildings are oriented east–west because this orientation reduces unwanted heat gains from east and west façades and greatly reduces the amount of energy required for heating and cooling. Planners and policymakers can work to encourage the use of public transportation in order to reduce the energy used by single-occupancy vehicles. Automotive designers should strive to design vehicles that require less fossil fuel usage to run and create innovative solutions that use alternative sources of energy. Appliance designers can similarly work to design products that use less energy and water over the life span of the product.

In addition to the operational costs, it is also important to consider the role that materials play in sustainable design. An important strategy therefore is to choose materials that are locally manufactured and produced. The use of local materials cuts down on materials transportation costs and all of the associated environmental costs and also educates the community about construction and manufacturing methods that are readily available.

Conservation of Natural Resources and Waste Reduction

Sustainable design should produce products, environments, and lifestyles that require fewer natural resources from the planet and produce less waste. Therefore, it is preferable to select materials that are recycled, salvaged, and/or sustainably produced. For example, rapidly renewable materials, those typically harvested within a 10-year cycle, such as bamboo, strawboard, or cork, have a smaller environmental impact on the land. Recycled content, both pre- and postconsumer types, reduces the amount of virgin materials required for production while simultaneously reducing waste. Examples include fleece jackets made of recycled water bottles or houses made of straw-bales. How long the material will last greatly contributes to its sustainability. In terms of longevity, designers should design products with a long life and low maintenance. This reduces the amount of materials that will end up in landfills and have to be replaced with new materials.

Designers should also consider the "afterlife" of a product or building and should design for deconstruction, which is selective dismantlement of building components, specifically for reuse, recycling, and waste management. It is not necessary that each item be put in the landfill once its useful life has ended. Products should be designed to facilitate disassembly and reuse or recycling. In addition, much energy is expended and waste created during the demolition of buildings. Constructing a building in a way that it can be taken apart and separated into various materials is helpful. Architects should always ensure that adequate measures have been taken to reduce construction waste because construction waste constitutes the largest fraction of landfills. A sustainable approach that further reduces waste is to design objects, buildings, and environments for multiple or flexible uses. For example, a parking lot can be for parking cars at night and a market during the day. Alternatively, a conference bag can turn into a shopping tote. Finally, when it is unavoidable to send a product to the landfill, then it is possible to design objects so that they are biodegradable. The number of products that are biodegradable has increased greatly in recent years, such as diapers, plastic cutlery, and plastic bags.

Biomimicry

Sustainable design is design that is in balance with the natural cycles and resources of the planet. One way to approach this is to look at how nature has designed the world. Biomimetics is a scientific approach that studies systems, processes, and models in nature and then imitates them to solve human problems. Put simply, nature is the largest laboratory that ever existed and ever will. In addressing its challenges through evolution, nature tested every field of science and engineering, leading to inventions that work well and last. Nature has found what works, what lasts, what is necessary and not necessary, and, consequently, what is appropriate for this planet. In the future, architects might design buildings to mimic the sophisticated energy management systems of nature and allow the building to change its enclosure and ventilation systems as required to respond to variations in temperature, wind, daylight, and moisture conditions.

Reduce Environmental Impact

Another important principle involves design that minimizes impact on the environment. Human beings account for a small percentage of the occupants of the planet, but their environmental impact is proportionally huge. Designers should attempt to tread lightly on the landscape. This can be accomplished in many ways. For example, in selecting sites for development, greenfield sites (properties not previously developed) should be left alone and efforts should be made to develop brownfield sites (properties previously used for industrial purposes, which may or may not be contaminated). In landscape design, storm water management measures such as bioretention strategies should be used to control water naturally to allow water to recharge the groundwater systems for future generations. In addition, reducing the use of impervious surfaces in construction helps to allow water to penetrate the ground as well as reduces the urban heat-island effect, which can cause higher temperatures in urban areas, further exacerbating summer cooling loads.

Environmental impact can also be reduced by choosing materials that use less water and have less of a negative impact on the planet. The manufacturing process pollutes the Earth, uses nonrenewable resources, and can contribute to poor human health. Life-cycle assessment tools can analyze these in great detail. For example, it is possible to evaluate how a particular material contributes to acid rain, global warming, and eutrophication, negatively affects human health, and water usage, and then make material specifications based on this information.

Encourage Healthy Lifestyles

Good sustainable design is that which contributes to healthy people, and at the very least, buildings and products should not be harmful to occupants and users. Product manufacturers should use materials that are nontoxic, such as being free of formaldehyde or volatile organic compounds (VOCs). In the developed world, people spend 80–90 percent of the time inside buildings. Therefore, it is important to carefully research and select products and architectural finishes that are used indoors. In addition, proper ventilation, good views, and proper daylighting can all contribute to healthier and happier occupants. Building design can integrate readily accessible and visually appealing stairs to encourage more people to walk rather than take elevators. Healthy lifestyles can be encouraged by the development of walking environments, mixed-use neighborhoods, bicycle lanes, and design that encourages pedestrian activity. This is in contrast to the suburban sprawl condition that currently afflicts society.

Desirable Design

Finally, sustainable design should be appealing and desirable to a consumer. Sustainable design must also enhance human lifestyles. Good sustainable design must be in tune with the people for whom it was designed. In simple terms, the better the design, the longer it will be in use. In addition, good design should respect local traditions, such as local materials, culture, and existing vernacular designs. Sustainable design does not necessarily have to look sustainable, but it does need to perform sustainably.

Role of Design in the Environmental Crisis

It can be said that the environmental crisis is a crisis that stems from design. It is a crisis that has resulted from a long period of irresponsible design that spans the manufacturing of products to land use patterns. It is a crisis that stems from design that has for a long time been too focused on the short-term specific interests of humans at the expense of the other creatures and the limited resources of the planet.

To some extent, design is in large part to blame for climate change. This is due to the fact that design allows people to live more luxuriously and often encourages wasteful behavior. Designed products appeal to consumers, encouraging them to purchase items that they do not need. For example, patrons wait in long lines to buy the latest mobile telephone, even when their old one is working fine. In architecture, vernacular buildings were designed to take into account local climatic conditions because there were no heating and cooling systems available. Currently, in architectural design, mechanical systems have become the norm and cheap energy has lessened the need to use passive strategies for heating and cooling buildings.

On the other hand, there is great potential in the realm of design to overcome the mistakes of the past. Often, it is the case that good design must be encouraged through legislative incentives. While policy changes are improving standards in all areas of design, these changes in standards are very slow acting. It is the case, therefore, that sustainable design must be desired by the consumer in order to create a demand that is served by design. It is also important for designers to take it upon themselves to make improvements for the greater good of the environment. Designers need to understand that their design decisions can often have implications for a very long time and need to see beyond the immediate short-term interests in order to benefit the future.

Some designers are taking it upon themselves to make a difference. One example of this is the Architecture 2030 Challenge, which asks that the global architecture and building community develop designs for new buildings, developments, and major renovations that reduce fossil-fuel, greenhouse-gas-emitting energy consumption by setting targets until reaching a goal of carbon neutrality (using no fossil-fuel, GHG-emitting energy to operate) by 2030. The challenge recommends that these targets be accomplished by implementing innovative sustainable design strategies and/or generating on-site renewable power and/or purchasing renewable energy and/or certified renewable energy credits. This effort can have a great impact because approximately 40 percent of energy usage can be attributed to buildings. On the other hand, improvements in other sectors can also make a large difference; for example, the transportation sector accounts for 32 percent of energy use, and the industrial sector accounts for the remaining 28 percent.

Conclusion

For improvements to occur in the environment, the design world needs to understand the profound impact of its work. Sustainable design must satisfy the immediate needs, wants, and desires of today without compromising the future. Design should attempt to meet the criteria discussed above by following these simple principles. There is a need for good design, and a need for a future in which to enjoy it. Designers need to understand that their design decisions can often have implications for a very long time and that they must contribute to balance the natural cycles and resources of the planet.

See Also: Cradle-to-Cradle Design; Green Building Materials; Passive Solar; Zero-Energy Building.

Further Readings

"EIA—Greenhouse Gases, Climate Change, and Energy." U.S. Energy Information Administration, Independent Statistics and Analysis. http://www.eia.doe.gov/bookshelf/brochures/greenhouse/Chapter1.htm (Accessed July 2010).

Fox, Michael and Miles Kemp. *Interactive Architecture*. New York: Princeton Architectural, 2009.

Glicksman, Leon R. and Juintow Lin. *Sustainable Urban Housing in China: Principles and Case Studies for Low-Energy Design*. Dordrecht, Netherlands: Springer, 2006.

Lechner, Norbert. *Heating, Cooling, Lighting: Sustainable Design Methods for Architects*. Hoboken, NJ: Wiley, 2009.

McDonough, William and Michael Braungart. *Cradle to Cradle: Remaking the Way We Make Things*. New York: North Point, 2002.

McHarg, Ian L. *Design With Nature*. Hoboken, NJ: Wiley, 1992.

Stitt, Fred A. *Ecological Design Handbook: Sustainable Strategies for Architecture, Landscape Architecture, Interior Design, and Planning*. New York: McGraw-Hill, 1999.

Van der Ryn, Sim and Stuart Cowan. *Ecological Design*. Washington, DC: Island Press, 2007.

Williams, Daniel Edward. *Sustainable Design: Ecology, Architecture, and Planning*. Hoboken, NJ: Wiley, 2007.

Yeang, Ken. *Ecodesign: A Manual for Ecological Design*. Hoboken, NJ: Wiley, 2008.

Juintow Lin
California State Polytechnic University, Pomona

SYSTEMS THEORY

Systems theory or the systems approach is a way of looking at things as parts of an organized and complex whole rather than as a collection of isolated phenomena. The term *systems theory* was coined by Austrian biologist Ludwig von Bertalanffy in 1928 as an alternative to the assumptions (which can be traced back to Descartes) that a system could always be broken down into individual components that could be analyzed separately and then added back together in a linear fashion to describe the total entity. The latter type of thinking dominated scientific research at the time but Bertalanffy found it inadequate and proposed an alternative view in which a system was characterized by nonlinear interaction among components that were nonlinear so that the functioning of the individual parts could not simply be added together to explain the functioning of the system.

Systems theory is a general approach to analysis that has been applied to many fields of study, including ecology, engineering, industrial management, mathematics, philosophy, computing, and human relationships. Regardless of the field of study, systems theory offers a holistic approach to looking at any system that emphasizes the arrangement and interaction of various elements in a system (e.g., the human body) rather than focusing exclusively on the discrete properties of specific elements (e.g., cells). Systems theory has been highly influential in environmental studies, and the concept of the ecosystem (defined below) is now accepted as a key level of environmental organization that is often the organizing focus of conservation efforts.

Although systems theory is a general approach that can be used to understand many natural phenomena, the description of any particular system will be highly specific to that system. To begin with, the definition of a system is in part conceptual in that two observers might define the boundaries of a system differently. In part, this depends on the aims of those doing the defining and the context in which they are working, so that what is a system in one view could be a component in another. For instance, an engineer or industrial designer might conceptualize an automobile as a system with components such as the battery and the engine, with the main interest being how those parts interact. From the point of view of an urban planner or sociologist, the automobile might be considered a component in a larger system of regional transportation. In the case of an ecosystem, one biologist might define a stand of trees and their immediate environment as an ecosystem while another might consider them a component in a much larger forest system.

Disputes often arise over defining the boundaries of the system because this decision will in part determine whose interests should be considered and who gets to make decisions about the system and its functioning. For instance, is a factory considered part of a system that ends with the corporation that owns it, or is it part of a larger system that includes the surrounding community (and ultimately, the entire world)? Is a natural area the property of only the person or entity that holds title to it, or does it somehow belong to a larger community? These decisions are important: in the case of the factory, if the system ends with the corporation that owns it, then the factory owner may feel free to release pollutants into the surrounding community and has no incentive to install equipment or change operating practices to reduce harm to that community. In the case of a natural area, bitter disputes have been fought over whether the need to preserve a habitat for a species (a choice that presumably benefits the world at large) is sufficient to override the choices of the owner or surrounding residents who may wish to use the land for some other purpose.

A system generally includes some sort of controls or value system: for instance, mammals have a system of homoeostasis that can cause different components in the body to perform different behaviors (e.g., sweating to lower the temperature and shivering to increase it) to keep body temperature within an appropriate range. In the case of a corporation, the values of the governing body (often a board of directors) will inform the implementation of specific controls (such as installing scrubbers to remove pollutants from smoke discharged into the surrounding air). However, in a complex system there may be a discrepancy between the control behaviors that optimize survival or production for one component and those that would be better for the health of the whole. For example, in an ecosystem some species may be better at reproducing and acquiring food or territory to the point that they cause the extinction of other species and may threaten the viability of the ecosystem as a whole. This is often a problem when a new species is introduced to an area where it was previously unknown and where it may have no natural predators, as was the case when rabbits were introduced to Australia.

Ecosystem Theory

The term *ecosystem* as used in biology refers to a specific level of organization that is composed of organisms and their interactions with each other and their environment and in which energy is exchanged and materials are cycled through the system. Ecosystems embody the concept of a complex system in which the sum is greater than the whole of the parts because of the constant interaction among living organisms, dead organic matter, and the environment.

The term *ecosystem* was first used in 1935 by the British ecologist A. G. Tansley, but the concept of an ecological system dates back much further; for instance, it is implicit in discussions of the food chain or of population regulation. However, until the 1960s, the concept of the ecosystem was seldom applied in analysis; instead, different components of the environment were managed as distinct units (e.g., forest management and pest control).

In 1964, British-American zoologist G. Evelyn Hutchinson contrasted two ways of studying large ecosystems: the holistic approach in which the system is treated as a black box that does not require specification of its internal components and the merological approach in which the focus is on individual parts of the system. To put it another way, in the holistic approach a scientist begins with the entire system and "models down" (an approach typical of engineers and physicists interested in the system as a whole), while in the merological approach one begins with the individual components and "models up" (an approach more typical of biologists and others concerned with a particular species). The holistic approach, incorporating the concept of the ecosystem, is the dominant approach today, aided in part by development of new mathematical tools to model populations.

The specific components of the system will vary from one ecosystem to another but will always include some biotic (living) and abiotic (nonliving) components, their interactions, and a source of energy. At one extreme would be a very simple ecosystem consisting of a single plant in a terrarium, including soil, water, and nutrients and exposed to sunlight: the plant is a biotic element, the air and soil are abiotic components, and the sun is a source of energy. At the other extreme, you could consider the entire biosphere of Earth to be an ecosystem.

Materials cycle through an ecosystem but energy does not; hence, an external source of energy (such as the sun) is necessary. In the process of photosynthesis, plants (primary producers) use the sun's energy to convert inorganic materials such as carbon dioxide into organic materials (e.g., edible leaves, grasses, grains) that are consumed by animals (primary consumers) who themselves are consumed by other animals (secondary consumers) in a food chain or food web. Decomposers such as bacteria break organic matter down to its inorganic constituents so that it can be used again for primary production. Defining how an ecosystem functions and how it reacts to changes (e.g., if components are added or subtracted to the system) is the focus of much environmental research.

See Also: Environmental Remediation; Geoengineering; Science and Technology Policy.

Further Readings

Bates, Frederick L. *Sociopolitical Ecology: Human Systems and Ecological Fields.* New York: Plenum Press, 1977.

Grant, W. E. "Ecology and Natural Resource Management: Reflections From a Systems Perspective." *Ecological Modelling*, 108 (May 1998).

Hagen, Joel Bartholemew. *An Entangled Bank: The Origins of Ecosystem Ecology.* New Brunswick, NJ: Rutgers University Press, 1992.

Jorgensen, Sven Erik, Brian D. Fath, Simone Bastianoni, et al. *A New Ecology: Systems Perspective.* Oxford, UK: Elsevier, 2007.

Meadows, Donella H. *Thinking in Systems: A Primer.* White River Junction, VT: Chelsea Green Publishing, 2008.

New Hampshire Public Television. "NatureWorks." http://www.nhptv.org/natureworks/nwepecosystems.htm (Accessed September 2010).

Rooney, N., K. S. McCann, and D. L. G. Noakes, eds. *From Energetics to Ecosystems: The Dynamics and Structure of Ecological Systems.* Dordrecht, Netherlands: Springer, 2007.

Sarah Boslaugh
Washington University in St. Louis

TECHNOLOGICAL AUTONOMY

The central role and influence of technology in society has been considerably disputed and is still one of the prime debates in science and technology studies. Within this context, the notion of technological autonomy is particularly controversial due to its portrayal of a self-directing phenomenon that prevails over other factors. Rather than a mere product of human action, technology is presented as a self-generating, self-controlling, self-propelling, self-perpetuating, and self-expanding force. It is seen for the most part as independent from, or out of, human control, and conducting its operations with regard to its own internal needs and dynamics. A large number of social thinkers, writers, and philosophers have worked on this concept of a self-governed technology, usually criticizing its devastating effects as well as its deviation from human purpose. At times with strong aversions, and others with reinterpretations, the debate on technological autonomy revolves around the existence of essential characteristics that may constitute a technological imperative and our possibilities to master or control them.

There is a widespread belief on technology's power to shape modern life, or more emphatically, on its driving force as an independent entity. This idea of self-government appears in many instances as a feature of technological determinism. But it does not necessarily entail a cause-and-effect association between technology and social change, as it refers primarily to the intrinsic characteristics that compose the existence of technology's agency. It is in this sense that technology is often described as having a "will of its own," as, for example, in discourses about the all-encompassing impact of computers and networks that rush us into "information societies." In another sense, technology is also depicted as inanimate objects "coming to life" in anthropomorphisms or technological animisms. This notion is represented through narratives of scientists and digital enthusiasts like Hans Moravec or Kevin Kelly, and also in stories from more skeptical writers such as E. M. Forster, Thomas Carlyle, George Orwell, or Kurt Vonnegut.

In fact, many thinkers of technological society have questioned the notion of technology as an autonomous and positive force of progress. This severe critique refers not only to its human and environmental costs, but also to Western civilization's fundamental relationship with nature. For example, the writings of Oswald Spengler and Martin Heidegger denounce technique's manifestation as a nihilistic will to power, or as an irrepressible destiny advancing through the manipulation and conquest of the material world. Moreover, framed in ontological terms, Heidegger claims that technology is relentlessly overtaking us as the modern mode of revealing. In the ongoing transformation of our environment, everything and everyone is degraded to raw resources to be used in technical processes that are instrumental in their essence.

From a sociological perspective, one of the most striking analyses comes from Jacques Ellul, who asserts explicitly that Technique, with a capital T, has become autonomous. In his macroscopic "characterology," this author offers a diagnostic of a self-sufficient and self-justifying technical system that follows its own criteria of efficiency, overpowering any other economic, political, cultural, or moral factor. There is a sense of self-fulfillment in this system, understood as a hegemonic social fact that prevails over and annuls human autonomy. Thus, in the face of a unified, universal, and self-enhancing Technique, for Ellul, the individual and collective abilities to choose the direction and the circumstances of social life become severely compromised.

We have also witnessed a visible concern in other authors over the conditions and consequences of technology, namely, over its imperative influence and destructive effects. It is worth mentioning Bernard Charbonneau, who shared Ellul's concerns over technological autonomy. To Charbonneau, the technical imperative since industrialization translates into the pursuit of efficiency and power, with its own ends and effects that suspend human judgment. Lewis Mumford, though not fully committed to a notion of autonomy, chooses to deplore the historical existence of "megamachines" and their mechanical orders, without human guidance and hostile to life. For Marshall McLuhan, from light bulbs to television, the technical characteristics of means elicit certain effects on human experience, thus containing and discovering their own goals. And in a more direct approach, Neil Postman states that technique tends to become autonomous by itself, like a robot that has escaped its master's grip.

There has been, however, a fierce backlash against these and similar views. They have been criticized as "essentialists" or "substantivists" because of their perspectives on technology's intrinsic properties of rationality and efficiency. Their visions have also often been dismissed for defining technology as an abstract and metaphysical entity immune to human reconstruction, while focusing on its negative effects of domination or even totalitarianism. Instead, scholars such as Trevor Pinch, Wiebe Bijker, John Law, Michel Callon, or Bruno Latour prefer to argue that technologies are subject to "social construction" in a "seamless web" of technology and society. In their works, an "interpretative flexibility" in artifacts poses different ways to interpret and design them, depending on social interests and resources, as in Bijker's study of the bicycle.

Nevertheless, if most contemporary perspectives have rejected the autonomy of technology, others have responded to its flaws and incorporated some of its elements. Several philosophers of technology have transformed previous criticisms through an "empirical turn," more concerned with concrete existences and manifestations of diverse technologies. For instance, phenomenologist Albert Borgmann compliments "substantivists" for their clarifications of our technical choices and commitments, while he reprimands them for a sort of obscure principle of technology. As an alternative, Borgmann not only focuses on the properties of specific devices like the CD player or central heating, but equally presents a severe diagnostic of their bodily and social disengagement. Also from a phenomenological framework, Don Ihde acknowledges the works of previous philosophers and offers a different understanding of artifacts with intentionality, not in the sense of autonomous agency, but as world mediators, like we see in his example of eyeglasses and their sensory configurations of space and time.

In similar ways, thinkers such as Andrew Feenberg and Langdon Winner have also redefined the debate, although in more political grounds. For Feenberg, technology in itself is not responsible for the overall degradation of social life, but the antidemocratic values that govern it and prevent its radical reconstruction. This author recaptures Herbert Marcuse and his notion of technique not as an essence but as a means of social domination menacing human autonomy. But Winner is more uneasy with the decline of our ability to know or control technical means as a symptom of autonomous technology. For him, technical systems impose norms and forms of power, as in Robert Moses's bridges in New York and their dissuasion of public transportation. Plus, technologies are inaccessible to most people and even experts, and extremely complex with ever-accelerating uncontrolled consequences. Winner calls ultimately for new technologies with an adequate scale and structure, with a high degree of flexibility and mutability, and with the direct participation of those who experience them.

See Also: Science and Technology Studies; Social Construction of Technology; Sociology of Technology; Technological Determinism; Technology and Social Change.

Further Readings

Ellul, Jacques. *The Technological System,* Joachim Neugroschel, trans. New York: Continuum, 1980 [1977].
Feenberg, Andrew. *Questioning Technology*. London: Routledge, 1999.
Winner, Langdon. *Autonomous Technology: Technics-out-of Control as a Theme in Political Thought*. Cambridge, MA: MIT Press, 1977

Susana Nascimento
Lisbon University Institute

TECHNOLOGICAL DETERMINISM

A term originally coined by sociologist Thorstein Veblen in the early 20th century and later expanded on and developed by his student Clarence Ayres, the theory of technological determinism states that a society's technology shapes—determines—its social and cultural institutions. It must be understood as part of the broader philosophical view of determinism—the idea that causality is so specific, the universe so orderly, that present and future outcomes of events are determined by prior states—that was so prevalent in the West at the time.

By the end of the 19th century, there was a sense of intellectual completion in the Western world. Modern psychiatry was in its infancy but implied the possibility of an understanding of the human mind. The theory of evolution, though it would be contested on religious grounds by the growing number of religious conservative reactionaries unhappy with the modern world, suggested an understanding of the human species. The world had been mapped and explored and even seen from the skies. Newtonian physics, which seemed perfect and incontestable, described a physical world that acted predictably, according to unchanging, knowable laws, many of which had already been discovered. Billiard balls struck on a table would move and roll in a predetermined fashion according to the way they

The printing press (and subsequently a linotype machine like this) has been named as a cause of the creation of the nation-state.

Source: iStockphoto

were struck—the properties of the universe, the properties of the balls and the table, and the properties of the strike were the prior conditions that made those movements inevitable. The world was considered "mechanical": extremely complex, extremely detailed, but with no more randomness than the gears that move within a clock. Even Mark Twain—best remembered now as a humorist but a staunch modern man in his day, a supporter of inventors and progressives and a critic of charlatans and superstition—devoted much of his nonfiction writings to explaining determinism. A century later, we know that the world is not as orderly as it seemed, that the makeup of the human mind is more complex, that Newtonian physics was incomplete, and even the Einsteinian physics that followed it had to be augmented by quantum physics. There are aspects of the physical world that act in ways we cannot predict, even when modern physics says we possess all the relevant information—events that seem to be not determined but random, such as why one atom of a radioactive element decays at a given moment when identical atoms in the same mass do not. But determinism has remained a popular notion, in part because it is a compelling idea that is easy to explain and easy to apply to the explanation of phenomena in the world.

Technological determinism bridges the gap between the physical and social worlds to some degree and builds on the notion in evolutionary biology that an organism's environment will impact its evolution because of the process of adaptation over many generations. Under technological determinism, introduced technologies bring about social changes. Specific variants of this generalization have certainly been put forth throughout history, from the worry that widespread books and literacy would ruin peoples' memories to various editorials on the social impact of the Internet. Technological determinism goes further than simply saying that technology allows for social change, or can bring it about: it positions technology as the primary causal element in social change. Media theorist Marshall McLuhan has credited the printing press with the creation of the nation-state, for instance: the press allowed for the production of pamphlets and newspapers that spread ideas faster (and with less signal loss) than word of mouth or face-to-face communication could have, and that furthermore made more important the role of ideas in politics rather than factions or aggregations of power. The printing press, like the Internet and all the developments between the two, changed the form in which information could be presented, which from a technological determinist view means it resulted in changing the nature of information that was presented and the way information was received.

Cultural critic Neil Postman, who wrote primarily in the last decades of the 20th century, described the technological determinist view of technology as the idea that "uses made of technology are largely determined by the structure of the technology itself, that is, that its functions follow from its form." Because radios are designed to receive a range of radio signals rather than specific preset signals, for instance, pirate radio stations—in many cases, a force for social change—could develop, not only without being intended by the manufacturers of radios, but against their interests. "The printing press, the computer, and the television are not therefore simply machines which convey information," Postman writes. "They are metaphors through which we conceptualize reality in one way or another. They will classify the world for us, sequence it, frame it, enlarge it, reduce it, argue a case for what it is like. Through these media metaphors, we do not see the world as it is. We see it as our coding systems are. Such is the power of the form of information."

Technological determinism is, on the face of it, compatible with the inevitability thesis, which says that once a technology is introduced, the use to which it is put is the inevitable use given the nature of the technology and of the culture possessing it. This is, after all, a determinist view. But proponents of the inevitability thesis are not necessarily technological determinists, and many articulations of technological determinism skate around the inevitability thesis by describing the development of technology as following its own path, primarily outside cultural influence.

Just as determinism itself can be articulated in different degrees—for instance, there is the compatibilist versus incompatibilist disagreement on whether or not free will is compatible with determinism—so too can technological determinism. Soft determinists, analogous to the compatibilists, give society some agency in deciding how technology will impact its institutions. Hard determinists do not, believing society organizes itself to meet the needs of its technology, not the other way around.

In the century since its introduction, technological determinism has repeatedly been challenged. Work in the philosophy of technology, particularly on the social construction of technology and cross-cultural science and technology studies, has painted a picture of the world that resists causal models and seems to contradict the basic precepts and implications of technological determinism, at least in its early articulations. On the other hand, social scientists who study the social construction of technology sometimes advance a view of technology that is nearly a mirror image of technological determinism, one in which social and economic institutions strongly determine the shape taken by technology, rather than the other way around.

See Also: Democratic Rationalization; Innovation; Technological Autonomy; Technological Momentum; Technology and Social Change.

Further Readings

Feenberg, Andrew. *Transforming Technology: A Critical Theory Revisited.* New York: Oxford University Press, 2002.

Kaplan, David M., ed. *Readings in the Philosophy of Technology.* Lanham, MD: Rowman & Littlefield, 2004.

Postman, Neil. *Technopoly: The Surrender of Culture to Technology.* New York: Vintage, 1992.

Scharff, Robert C. and Val Dusek, eds. *Philosophy of Technology.* Hoboken, NJ: Wiley-Blackwell, 2003.

Bill Kte'pi
Independent Scholar

TECHNOLOGICAL MOMENTUM

Historian of technology Thomas P. Hughes coined the term *technological momentum* to describe a theory of the relationship between society and technology that incorporates constructs from two other theories: technological determinism and social construction. Hughes views the relationship between technology and society as reciprocal and time dependent, so that technology neither determines changes in society nor society changes in technology, but that both may influence each other, and their relationship can change over time. Hughes first applied this concept to hydrogenation in Germany in the early 20th century, but has since used it to explain such varied social and historical phenomena as the development of light and power companies in the United States and the survival of work begun by the Manhattan Project after the conclusion of World War II. The theory of technological momentum has also been used by other authors, such as David Nye, who used it to explain how different energy creation and consumption systems have risen and fallen in importance over the years in the United States.

Technological momentum theory is differentiated from two alternative theories of the relationship between society and technology. Theories of technological determinism posit that technical forces determine changes in society and culture: for instance, that the creation of the stirrup set off a chain of causes and effects that led to feudalism or that invention of the steam engine led to the development of factories and bourgeois society. Theories of social construction argue that it is society that shapes technology because social interest groups determine which technologies are developed and prosper. For example, in the early history of the bicycle there were many alternative designs, but eventually most were eliminated and the modern design of the bicycle was adopted and sustained because it met the needs of powerful interest groups in Western society.

Technological momentum takes more of a reciprocal view of the relationship between society and technology, so that each can influence the other depending on specific historical and cultural circumstances. Technological momentum also considers the relationship between society and technology to be time dependent and thus expected to vary, depending on the state of the development of the technology. For instance, for a new technology, the power may lie more in the hands of society and the technology must demonstrate that it can meet social needs and desires, while with a more mature technology, greater power may lie with the technology and society may be shaped in order to accommodate or perpetuate it.

An Example: The Electric Bond and Share Company

Technological momentum theory may be used to explain the history of large technological systems such as hydroelectric power systems or the extensive research laboratories and attendant systems developed in the United States during World War II. For instance, Hughes used the theory of technological momentum to examine the history of the Electric Bond and Share Company (EBASCO), a U.S. electric utility holding company (a company that exists primarily or entirely for the purpose of owning or holding shares of other companies). EBASCO was established in 1905 by the General Electric Company, and both controlled various electrical utility companies and technical subsystems (e.g., electric light and power networks) and provided financial, management, and engineering construction services for them. While a technological determinist might emphasize how the availability of electric lighting affected patterns of work and home life and the availability of electric appliances changed women's roles and a social constructivist might emphasize how rapid urbanization and population increased and shaped the development of electrical supply systems, from the point of view of technological momentum theory, EBASCO both shaped and was shaped by society.

In addition, technological momentum theory emphasizes how the relationship between EBASCO and American society changed over time: in particular, how, as the company became larger and more complex, it was more able to shape society to meet its own needs while earlier in its history it was more shaped by the needs of society. By the 1920s, EBASCO had become quite powerful due to its size and the importance of the good that it supplied: it was comparable to a large railroad company in terms of capital investment, number of customers, and influence on government at all levels. This influence was seen in many different areas: for instance, the company employed many electrical engineers who were economically committed to the well-being of EBASCO and to the perpetuation of methods of doing business that favored their specialized knowledge. They exerted influence on the development of the engineering profession through professional organizations, engineering schools, and so on. EBASCO also influenced the industries and communities to which it supplied power because of their shared economic interests.

Technological Systems and Technological Momentum

Technological momentum does not apply only to specific companies but can also be used to understand the history of entities such as bodies of skill and knowledge, specialized machines and processes, organizational bureaucracies, and physical structures. The "momentum" in technological momentum refers to the fact that, as a large physical object in motion is difficult to stop, so a technology that has become sufficiently large and powerful may also be difficult to stop and may in fact find ways to perpetuate its own existence. For instance, the discipline of civil engineering greatly developed in the 19th century in response to the needs of railroad construction. This knowledge was transmitted to the next generation through engineering schools, and graduates of these schools as well as experienced engineers sought opportunities to apply this knowledge outside the field of railroad building, both for their own careers' sake and to solve societal problems. With the growth of cities in the late 19th century and the ensuing traffic problems, many individuals who had worked on building the cross-country rail network found outlets for their skills in the creation of urban transit systems, for instance, using their knowledge of building railroad tunnels to construct subway systems and their knowledge of constructing railroad bridges to create urban elevated rail systems.

The same principle applies to large physical structures. For instance, the Muscle Shoals Dam (Wilson Dam) was built in the United States during World War I in order to supply electricity for a nitrogen fixation facility. With the end of the war, there was no longer high demand for this project, but the dam and power station remained. They represented a large investment so various interest groups sought to find a use for these facilities. After a failed attempt to create an industrial complex to take advantage of these facilities, they found a new purpose under the Tennessee Valley Authority as part of a development project to bring electricity and flood control to the region.

See Also: Marxism and Technology; Technological Determinism; Technology and Social Change; Technology Transfer; Unintended Consequences.

Further Readings

Feenberg, Andrew F. "Momentum: A Concept in Technology Studies." http://www.sfu
.ca/~andrewf/momentum.htm (Accessed September 2010).

Hughes, Thomas P. "Technological Momentum." In *Does Technology Drive History? The
Dilemma of Technological Determinism*, Merritt Roe Smith and Leo Marx, eds.
Cambridge, MA: MIT Press, 1994.

Hughes, Thomas P. "Technological Momentum in History: Hydrogenation in Germany
1898–1933." *Past and Present*, 44/1 (1969).

Nye, David. *Consuming Power: A Social History of American Energies*. Cambridge, MA:
MIT Press, 1999.

Nye, David. *Technology Matters: Questions to Live With*. Cambridge, MA: MIT Press, 2006.

Sarah Boslaugh
Washington University in St. Louis

TECHNOLOGICAL UTOPIAS

The term *technological utopia* refers in a broad sense to the use of machines and civil engineering to create an ideal society. In ancient Greek, the term *utopia* means "no-place" but is pronounced as "good place," making a utopia a nonrealized good place.

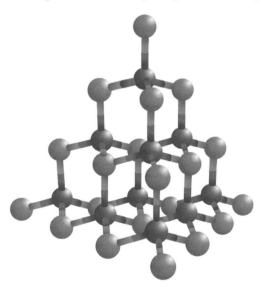

This cadmium sulfide molecule can be prepared by traditional chemical methods, but Eric Drexler's utopian vision included assemblers that would be able to fabricate larger nanoparticles.

Source: U.S. National Aeronautics and Space Administration/Marshall Space Flight Center

Utopias are always presented in a time frame; they are set in the past as a golden age narrative, in the present as a travel narrative, or in the future as a science fiction narrative. Technological utopias are predominantly set in the future. Despite not being in the "here and now," utopian visions are set in the real world, with ordinary people, and the same natural laws. The potential for utopian visions to become reality is what gives them power. In this sense, utopias contrast with purely fantasy worlds in that they are places that should exist or at least be pursued, according to the narrator. The visions are to be realized by organizing society and the environment. The means of organization in technological utopias are science and engineering.

Literary technological utopias become more common around the period of the scientific revolution. In the 1600s, some scientists began to assert that there was a need for a complete new system of knowledge, leading to a new golden era. They also argued that humanity had a religious duty to technically control nature, an ability lost with the expulsion from the Garden of Eden. Others envisioned that if the true causal structure of nature were known, humans would be able to fly and ordinary metals could be turned into gold.

Technological utopianism in the present era has arguably declined due to negative consequences of advanced technology such as environmental pollution and the nuclear threat. There are, however, still active techno-utopian visionaries and followers. Influential contemporary technological utopias have been proposed by nanotechnology advocate Eric Drexler and singularity spokesperson Raymond Kurzweil.

K. Eric Drexler (1986) described how nanotechnology, the manipulation of single atoms and molecules at the nanometer level, will create a new era of prosperity and affluence. He envisioned programmable molecular machines, called "assemblers," that would build perfect molecular structures in virtually any form. These assemblers would be used to construct basically any material possible. Some of the claimed benefits of the technology included the production of synthetic meat and the molecular disassembly of pollution created by the Industrial Revolution. Humans would live in ecological balance. There would also be social harmony because there would be no more conflict over resources. Alternative lifestyles would be possible with the help of assemblers. For example, one could live a Stone Age life but without the diseases and starvation.

Raymond Kurzweil (2005) envisioned an exponential acceleration of technology in which humans increasingly became enmeshed with machines. Genetics, Drexlerian nanotechnology, artificial intelligence, and robots all become one in what Kurzweil called a "technological singularity." In the singularity, new types of hybrid organic/machine intelligences will be formed that are far superior to existing organisms and machines. The mix of machine and biology would end death and starvation, crafting a world beyond our human comprehension. Nature would eventually become absorbed by the new intelligences and, ultimately, all material in the universe would be converted into sentient substrates.

Common themes in contemporary technological utopias are the notion of a purely material universe and the idea that there is no fundamental difference between nature and technology. A human cell and a nano-machine are held to be equally alive, and human thoughts are to be downloadable into computers. Nature, technology, and humans merge to become one. Concepts such as war, death, birth, and technology cease to exist as we understand them. It is no surprise that technological utopias have inspired science-fiction writers, while tech visionaries are often accused of writing science fiction. This is to the dismay of the narrators, who want their foresights to be perceived as realistic scenarios, not pure fantasies.

See Also: Green Nanotechnology; Marxism and Technology; Technology and Social Change.

Further Readings

Drexler, K. Eric. *Engines of Creation: The Coming Era of Nanotechnology*. New York: Anchor Books, 1986.

Kurzweil, Raymond. *The Singularity Is Near: When Humans Transcend Biology*. New York: Viking Press, 2005.

Shapin, Steven. *The Scientific Revolution*. Chicago, IL: University of Chicago Press, 1996.

Mikael Johansson
University of California, Santa Barbara

TECHNOLOGY AND SOCIAL CHANGE

Every historical epoch has its defining ethos in technology. For the Roman Empire, it was an extensive network of estates powered by slaves; for medieval Europe, feudalism and the church were the most important bulwarks of life on Earth. However, for the modern period, technology and the changes it brings about are the defining feature of the epoch. Technological innovation, though, is not a contemporary phenomenon. Innovation, especially in military and agricultural technology, such as the crossbow that was superior to the longbow or the heavy-wheeled plow that made cultivation of the rich, heavy, and wet soils of northern Europe possible, is a feature of history. However, technological innovations in the past were often localized, spread slowly, and took years to manifest in social change. For instance, the plow made its first appearance in the 5th century in Slavic lands and appeared only in the 8th century in the Rhineland. Such technology led to social change over long periods of time.

French historians such as Fernand Braudel (1902–1985) have pointed to the existence of long cycles of time over which the social change that technology unleashed become evident. Braudel's ideas of long cycles of time have inspired many other works—Jacques Le Goff famously looked at how the invention of mechanical clocks changed European conceptions of time, and popular authors such as Jared Diamond have used the idea of long cycles of time to argue that the seeds of European dominance in the 15th century were laid when horses were domesticated and steel was invented. Thus, the Spanish conquistador Francisco Pizarro (1476–1541) was able to lead a vastly outnumbered Spanish force to victory over the Inca Empire because the European system had enormous biological, technological, institutional, and cultural advantages over the native Inca systems.

One of the best examples of technology and social change is the invention of the printing press by German printer Johannes Gutenberg in 1440. The easy availability of books, considered a luxury in medieval times, unleashed tremendous social change such as the 16th-century Reformation that challenged the dominance of the Catholic Church, a direct result of the translation of the Bible into vernacular languages, and political revolutions all over Europe in the next centuries. Social change that is created by technological innovation may be fast or slow. The social change that movable-type printing brought about was not a sudden phenomenon. It took nearly 100 years from the translation of the Bible to the creation of the modern state system, enshrined in the Treaty of Westphalia (1648) that made religion a private matter. On the other hand, in the 20th century, the pace of change due to innovations—such as the automobile that led to the replacement of animal power with mechanical power for transport, transatlantic ocean liners that gave way to airplanes, increasing urbanization as a result of technological innovation that favored the cities—can all be measured in decades.

Tremendous social change, driven by technology, became evident only after the 19th-century Industrial Revolution. What changed in the mid-19th century was an increase in receptivity to innovation and a faster breaking down of the barriers of scientific knowledge. This resulted in a series of "ages" that can be described by the predominant technology of the era, thus successively electrical, chemical, aeronautical, electronic, nuclear, and space ages of technology. This kind of increasing competence ran parallel to the competence of the age, which is a key element of the technology an age adapts. For example, it was only in the 18th century that the steam engine, an idea that had been known since the time of ancient Greece, was applied in a practical way. The elastic properties of steam had been known for a long time, but it was only the ability to cast iron cylinders of considerable dimensions with accuracy that converted it into a machine that could produce power economically and effectively. Similarly, the development of photography was based on knowledge of the different effects light has on various substances, something known since the Middle Ages. It was only in the late 19th century that the intellectual milieu became receptive to photography.

Technology and Social Change in the Contemporary Era

The contemporary era is marked by technology that is increasingly sophisticated and the defining feature of each decade is new technology. The history of our era can be described in terms of innovation—the automobile, the airplane, radar, radio, television, computers, the Internet—the list of innovations is endless. In the contemporary era, the term *technology* has been closely linked with invention (the creation of a new idea) and innovation (the first use of a new idea). The history of the 20th century is a history of social change that new inventions have brought about, and green technology might be the latest in a long series of innovations that include flight (1903), nuclear power (1945), contraception (1955), and the Internet (1965). Social change can also take place over decades. The invention of the contraceptive pill, for example, over decades led to major social changes such as the liberation of women from the drudgery of domestic life and decreasing populations, especially in the West, which faces demographic change, the social impact of which is still unclear.

The 20th century has also seen tremendous environmental impact as a result of technological innovation. Of special relevance to green technology is how a society's responses to its environmental problems always prove significant in averting or leading to collapse. Studying seven cases of collapse—Easter Island, Pitcairn and Henderson Islands, the Anasazi tribe of North America, the Maya in Central America, the Vikings in Iceland, and the Norse in Greenland, author Jared Diamond in his book *Collapse* (2005) lists eight factors that have historically contributed to the collapse of past societies: deforestation and habitat destruction, poor soil and water management, overhunting, overfishing, the effects of introduced species on native species, overpopulation and the increased per capita impact of people. All of these show how relevant green technology can be to contemporary problems. Diamond also argues that four new factors mark our current era of technologically induced growth: anthropogenic climate change, the buildup of toxins in the environment, energy shortages, and the buildup of greenhouse gases in the atmosphere.

In the field of green technology, one major innovation in the 20th century that led to the end of the biblical plague of famine in most parts of the world was the Green Revolution. Historically, stabilizing a population's food supply has meant improving agricultural methods, harvesting tools, equipment, transportation, and infrastructure, all of which inevitably involve technology. The Green Revolution, which swept the world in the 1950s and 1960s, applied agricultural innovations such as hybrid seeds, chemical fertilizers, pesticides, and irrigation methods. To many—including Norman Borlaug, the American scientist known as the father of the Green Revolution—these technologies were essential to prosperity, though critics charged they were environmentally devastating.

Ironically, green technology is also the answer to one enduring problem that has bedeviled new technology—the rise of new problems when older ones have been solved. To take a specific environmental example, early-19th-century American farmers used a variety of methods to control insects, with biological methods predominating. However, in the

20th century, chemical methods predominated until the publication of Rachel Carson's *Silent Spring* (1962), which revealed the havoc that chemical methods could wreak on the countryside. Carson's major contribution was the recognition that humans were an integral part of the natural world. In a world that since the 17th century had been used to the Promethean notion that science and technological development would give man ultimate control of nature, this was radical. Since the 1960s, there has been intense interest in technology that is not intrusive and that tries to work with nature rather than against it. This is reflected in interest in technologies that address questions of environmental pollution and public health, community right-to-know legislation, and the personal values embedded in scientific practice. Events in the 1970s, the sharpening of the nuclear arms race, the 1979 near-meltdown of the nuclear reactor at Three Mile Island, and increasing pessimism led to a questioning of the ability of technology and technocrats to control things and an idea that society was not efficiently administered or organized at all that marked a key shift from the belief that science could conquer nature to the realization that science could only work in harmony with nature. Thus, in the 20th century, the growing awareness of the dangers that unbridled technology can cause is one the key changes in social attitudes that directly influenced the quest for green technology, including renewable sources to replace nuclear power plants, green nanotechnology to ensure that nanotechnology will not have an adverse impact on the environment, and technology that is designed to reduce global warming.

Another change in attitudes that is significant to green technology is the growing recognition that there are technologies of the poor that have as much relevance as the technologies of the rich. Earlier notions of scientific progress, epitomized by the modernization paradigm that guided development in most postcolonial states and in Europe after the 18th century, favored heavy resource-intensive and environmentally unfriendly technology such as dams, nuclear power, and huge manufacturing plants. Where social attitudes toward the environment are still subordinate to the lure of economic progress, there is still a preference for such technologies, in spite of the harm that they can cause the environment. Examples of these include the Three Gorges Dam in China, which is the world's largest hydropower project, and the Narmada Dam in India.

Global Changes

Another key element of contemporary social change is that it is global. Manuel Castells, for example, has theorized the existence of a "space of flows," which Castells uses to explain how technology has enabled a separation of the production process from nation-states, enabling a disjointed patchwork of command and control centers to coexist with geographically diverse centers of production. Applying this concept to modern industrial production, one can delineate a nonhierarchical system that is not geographically bound. This has challenged ideas of the developed West and the underdeveloped East, or such territorially bound definitions as the first world and the third world. Thus, the contemporary world is marked by research and development (R&D) in the highly innovative industrial core areas, skilled fabrication in the core and semi-periphery, semiskilled large-scale assembly and testing work that has shifted to the periphery, and finally the customizing of devices and after-sales maintenance and technical support that is organized in regional centers throughout the globe. However, rather than being territorially defined, the core and the periphery can exist in the same country, for example, R&D for Microsoft that takes place in Bangalore, India, which also hosts a burgeoning service industry that serves the need of the core that is the advanced, industrialized West.

Technological change can be spurred by war, though war is not an essential condition to scientific progress. Two of the greatest periods of human innovation, from the late 19th century to the eve of World War I and the post-1945 period, have seen enormous technological and scientific progress. Wartime innovations have led to major changes in food storage and agriculture, with considerable impact on the environment. For example, the need to supply enormous numbers of troops in Europe and the Far East led to massive U.S. government investment in new preservation techniques such as spray-drying and dehydration. In fact, during the war, while the average American male civilian ate 125 pounds of meat in 1942, a typical soldier was allotted 360 pounds. Through the power of processed foods and new cooking innovations such as the microwave oven, a meal that took one hour to produce in 1965 was shaved down to only 35 minutes by the mid-1990s. Other key innovations were the washing machine, the domestic electric oven, and oral contraception. The greatest impact of such technology that cut down the chores of housekeeping was that it freed women from the household and led to a massive influx of women into the workforce. The social changes that this brought about, most prominently the easing of the economic dependence of women on men, were tremendous.

Technological innovations have traditionally been held to cause sudden and dramatic changes, though explanations for economic growth that favored technological change have now given way to more nuanced studies of the impact of technology on society, especially in the emerging field of science and technology studies. However, while the speed of social change can be debated, the fact that technological innovations starting in the 19th-century Industrial Revolution dramatically altered societies cannot be denied. Thus, coal and coal-using technology so dramatically transformed industry toward the end of the 18th century, transforming Britain from an iron-importing country to one that dominated the world market in wrought iron. The combination of coal technology, steam engines, and heavy industry dramatically transformed Britain. Writers such as Charles Dickens (1812–1870) gave life to the dramatic inequalities that the Industrial Revolution caused in works like *Hard Times* (1854). William Blake coined the expression *dark Satanic Mills* to refer to the destruction of nature and human relationships that followed the early Industrial Revolution, which threw traditional and independent craftsmen out of work, especially in the textile industry. They, and many writers like them, gave fictional expression to the tremendous social changes that new technology brought in Britain.

More important was the social thinking that the new technology and its changes gave rise to. They included socialism, which advocated public ownership of the means of production; Marxism, which called for a takeover of the state by the working class; and anarchism, which considered the state superfluous and called for a stateless society, or anarchy. All of these inspired revolutionary movements throughout Europe. There were others, such as industrialist Robert Owen (1771–1858), who experimented with self-sustaining communes, an idea that environmentalists would adopt later, and Henri de Saint Simon (1765–1825), who coined the term *socialisme* and argued for technocracy and industrial planning.

In the 20th century, these philosophies would lead to social experimentation on a grand scale, but an experimentation that at its core was based on control of technology and the means of production. These include Marxist revolutions, most prominently in Russia (1917) and China (1949), the early-20th-century rise of fascism and Nazism in Italy and Germany, and the emergence of the social welfare state, in varying forms, in postwar Europe. What is important is that all of them advocated some kind of social order that was in the final analysis a product of technology. Thus, Marxist states advocated state-driven industrial planning, and fascism saw the rise of the corporatist state that called for management of the economy through the state or through privately controlled organizations (corporations). With reference to green technology, fears have been expressed regarding both the power of corporations, especially in the oil, gas, and automobile sectors, to block sustainable but expensive green technology and the tremendous concentration of power in the hands of corporations in areas such as biotechnology.

Sometimes the Unexpected Happens

Often, technology has not brought change in the way that was expected. The history of science and technology is replete with instances where technology did not bring about the social change that it was expected to. When electricity was first introduced, it was predicted that it would make cities green, an outcome that did not occur. The information technology revolution of the last decades of the 20th century also did not bring about expected positive social outcomes. It was hoped that the infrastructure of information and communication technology (ICT) would decentralize production and bring about tremendous social change by enabling work to be done from anywhere in the world. In reality, local, national, and increasingly globalized information and communication networks have broadly inflated a range of social and spatial boundaries, that is, the reach of active social connectivity, yet they remain unmistakably centered in the metropolitan core. Thus, while in theory the Internet is borderless, its core elements are designed and situated in well-defined spaces, where large research, development, software production, mass media, commodity manufacturing, sales and advertising, finance, commerce, state intelligence, and other commanding interests usually gather—and where most strategic decision making and revenue ultimately reside. The first-tier information cities are New York, Tokyo, and London, and secondary nodes can be found in cities such as Bangalore and Shanghai. Socially, this has meant the transfer of jobs such as back-office operations from the West to countries in the East. As a result of the ICT revolution, U.S. firms began to outsource computer and software development to India, exploiting the differential rates of income as a source of profit. Thus, in 2003, while an American engineer earned $50,000 to $60,000, the comparable figure for an Indian engineer was $3,000. Such outsourcing affected customer services, back-office accounting, and even secretarial work with enormous consequences for the poor and working class in advanced nations, especially the United States.

Thus, one of the less salubrious effects of the ICT revolution was the breaking down of the geographical barriers of social inequality. ICT has led to the creation of islands of plenty among poverty, a phenomenon that is as evident in the inner cities of the United States as it is in newly emerging regions such as Shanghai and Bangalore. The Internet, which traces its roots to military research in the United States, ironically broke down the centralization of information in the hands of corporate, government, media, and university authorities. Activists, the most prominent among them the website wikileaks.org, have enabled people to bypass electronic communications media that broadcast centrally determined messages to mass passive audiences and communicate with each other quickly and directly. Information technology also has led to radical changes in the way people communicate in society. Face-to-face interaction, long a staple of human culture, is increasingly replaced by "interface" with technological "terminals" of communication, electronic devices that acquire a life of their own. Games like Second Life enable people to interact virtually and to fulfill various roles. In particular, television has had tremendous social impact, with critics charging that it disrupts social and family life, allows TV news to exercise great agenda-setting power, and trivializes complex political and social issues.

A key component of technology and social change is its international and transnational nature. While invention, especially in the past, was localized, its spread across borders has led to complex changes. Whether openly or by stealth, nations have always tried to acquire the latest technology. The Venetians tried to obtain the secret of Greek fire from the Byzantine navy during the late Middles Ages and the British, in a precursor of contemporary struggles over intellectual property rights and green technology, tried to prevent the export of their steam engines and textile machinery to rival powers.

The time gap between new technological innovations and the social change they bring about has narrowed from centuries to mere decades, and the coming nano-cogno-bio-info knowledge economy is likely to accelerate this trend. Some futurists argue that new technology will so shape the channels of our experience, transforming our conception of the "real," redefining what we mean by "community," and, some would maintain, what we mean by our "selves." It is likely that in the future, human beings will spend as much time in virtual electronic space as they do in reality. It is likely that humankind is now on the threshold of a posthuman era in which persons merge with intelligent machines. In such a posthuman state, there would be no borders between bodily existence and computer simulation, between cybernetic mechanism and biological organism. One of the key components of this world will be ubiquitous special-purpose chips, "smart" devices, and agents that interact with human beings constantly. Most "computers" will be embedded in wearable micro-devices and implants, blurring the divisions between man and machine.

See Also: Intellectual Property Rights; Marxism and Technology; Science and Technology Policy; Science and Technology Studies.

Further Readings

Castells, Manuel. *The Network Society: A Cross-Cultural Perspective*. Cheltenham, UK: Edward Elgar, 2004.

Diamond, Jared M. *Collapse: How Societies Choose to Fail or Succeed*. New York: Penguin Books, 2006.

Diamond, Jared. *Guns, Germs and Steel: The Fates of Human Societies*. New York: W. W. Norton, 1997.

Feenberg, Andrew. *Between Reason and Experience: Essays in Technology and Modernity*. Cambridge, MA: MIT, 2010.

Goff, Jacques. *Time, Work, & Culture in the Middle Ages*. Trans. Arthur Goldhammer. Chicago, IL: University of Chicago Press, 1980.

Heilbroner, Robert L. *The Limits of American Capitalism*. New York: Harper & Row, 1966.

King, Steven and Geoffrey Timmins. *Making Sense of the Industrial Revolution: English Economy and Society 1700–1850*. Manchester, UK: Manchester University Press, 2001.

Lenoir, Tim. "All but War Is Simulation: The Military-Entertainment Complex." *Configurations*, 8/3 (2003).

McClellan, James E. and Harold Dorn. *Science and Technology in World History: An Introduction*. Baltimore, MD: Johns Hopkins University Press, 2006.

Nowak, Peter. *Sex, Bombs and Burgers: How War, Porn and Fast Food Created Technology as We Know It*. London: Allen & Unwin, 2010.

Pursell, Carroll W. *The Machine in America: A Social History of Technology*. Baltimore, MD: Johns Hopkins University Press, 2007.

Sassen, Saskia. *Cities in a World Economy*. New York: Colombia University Press, 2006.

Sabil Francis
University of Leipzig, Germany,
École Normale Supérieure, Paris

TECHNOLOGY TRANSFER

Technology transfer refers to the transfer of knowledge, not just to the transfer of goods or service. For example, giving a country a wind turbine like this would not be technology transfer, whereas teaching it the mechanisms behind its power or use would be.

Source: Argonne National Laboratory, U.S. Department of Energy

Technology transfer (TT) is a complex, multidimensional concept. Its definition varies widely depending on the subject discipline, context, and perspective from which it is conceptualized. In the climate change context, the Intergovernmental Panel on Climate Change (IPCC) defines TT as a broad set of processes covering the flows of know-how, experience, and equipment for mitigating and adapting to climate change among different stakeholders, such as governments, private sector entities, financial institutions, nongovernmental organizations (NGOs), and research/education institutions. Here TT has been conceptualized in a very broad and inclusive manner that encompasses diffusion of technologies and technology cooperation not only across countries but also within countries and among various categories of entities. However, it is international TT, and more specifically North-South TT that has turned out to be the most widely articulated and rather controversial dimension of the concept.

As the lion's share of the generation of commercially significant technology still takes place in developed countries, particularly within the transnational corporations (TNCs) originating from these countries, TT in adequate scale and at affordable prices from developed to developing countries plays a significant role in the economic and technological development of developing countries. According to the 1985 Draft International Code of Conduct on the Transfer of Technology negotiated under the aegis of the United Nations Conference on Trade and Development (UNCTAD), TT refers to the transfer of "systematic knowledge" for the manufacture of a product, or the application of a process or for the rendering of a service and does not extend to the transactions involving the mere sale or mere lease of goods. This definition, developed specifically in the context of international TT, emphasizes the knowledge element of TT—clearly indicating that mere import of goods (say, equipment) cannot be regarded as TT. A successful TT is a complex process that goes through several phases of consolidation: (1) absorption and learning, (2) adaptation to the local environment and needs, (3) assimilation of subsequent improvements, and finally (4) generalization. TT in this sense underscores the importance of technology learning and technological capability building in developing countries.

Neoclassical Versus Evolutionary Approaches

The aforesaid conceptualization of TT is quite in line with the evolutionary theory of economic growth, which marks a significant departure from the way TT is articulated by the traditional neoclassical economic theory. In the highly simplified neoclassical world, technology is codified information that can be transmitted fully between firms. TT, in this world, is like the sale of a product. There are no tacit elements in the transfer, no learning costs, and no need to make adaptations. Developing countries simply import and apply technologies, picking them in line with their factor prices. More importantly, once they have imported a technology, they can use it efficiently from the very beginning, again without extra cost or effort. The neoclassical framework does not quite depict the ground realities of technology transfer and technology development in developing countries, however. Instead, the evolutionary theory appears to be a more apt framework for explaining these complex processes. Technology, according to this school of thought, is not sold in embodied forms, nor is it fully codifiable. Technology has important "tacit" elements that need effort and time to master. Hence, transfer necessarily requires learning. Technological mastery entails building costly new capabilities, which not only requires time, effort, and investment, but is also fraught with risks and uncertainties. Efficient use of an imported technology cannot, therefore, be assumed to be an automatic or simple process.

Modes

The mode or vehicle of TT influences to a great extent the terms and conditions on which a technology is transferred. Modes of TT can broadly be classified into two categories: (1) internalized, namely, from a TNC to foreign affiliates under its control (i.e., foreign direct investment), and (2) externalized, namely, between independent firms. Internalized modes can take place either through the establishment of a new affiliate in a host country or through the acquisition of a local firm that can be turned into a suitable recipient. Externalized modes of transfer take a variety of forms, such as minority joint ventures, franchising, capital goods sales, licenses, technical assistance, and subcontracting or original equipment manufacturing arrangements. While internalized modes necessarily involve TNCs, externalized ones may also involve TNCs selling technologies on contract.

Not all modes of technology import are equally conducive to indigenous learning. Internalized transfers come highly packaged with complementary factors to ensure their efficient deployment. The retention of technology and skills within the network of a TNC may hold back deeper learning processes and spillovers into the local economy, especially where the local affiliate is not developing research and development (R&D) capabilities. Externalized transfers, generally, tend to call for greater learning effort by the recipient. Technology licensing—the most common form of externalized transfer—typically involves the purchase of production or distribution rights, protected by some intellectual property rights (IPRs; e.g., patents, trade secrets, copyrights, and trademarks) and ideally the provision of technical assistance and know-how required to make effective exploitation of those rights. The transfer of know-how or tacit knowledge and the provision of technical services are essential to ensure that the licensor secures the capabilities to use the technology in an effective way and succeeds in adopting and adapting the technology. Given the potential for greater learning, deepening, and externalities, externalized modes are often regarded as superior to the internalized mode, as developing countries go higher up the ladder of technological capability building. Thus, where a choice exists between internal transfers to foreign affiliates or external transfers to local technology recipients, developing-country governments may wish to intervene to affect the terms of transfer associated with each modality, for example, where incentives are offered to TNCs for the transfer of advanced technical functions. Another approach is to upgrade the capacity of the host economy to receive and benefit from TT. However, it is argued by some commentators that in practice, the externalized modes also often suffer from serious flaws owing to various imperfections in the technology market and the dominance exercised by the technology-owning firms, particularly in the context of the strengthened intellectual property (IP) protection under the World Trade Organization Agreement on Trade-Related Aspects of Intellectual Property Rights (WTO TRIPS) and beyond.

IP and TT

Commercial technology is usually exploited through the application of intellectual property rights (IPRs), which helps to increase the value of the technology to its owner by creating relative scarcity through legally restricted access to it. According to proponents of strong intellectual property (IP) regimes, IPRs can serve as an important support for markets in technology characterized by the problems of appropriability and asymmetric information. Without adequate protection from leakage of new technical information, firms would be less willing to provide it on open technology markets. In an environment of weak IPRs, they may choose not to transact at all or may opt for other strategies that may have adverse implications for TT, such as offering older-generation technologies or keeping the information within the firm by dealing only with subsidiaries, and so on. According to the opponents, however, the availability (and enforceability) of IPRs will, by no means, create a sufficient incentive for TT to occur and instead can sometimes make TT more problematic. A weak IP regime may offer local firms in the recipient countries some scope for imitating foreign technologies and reverse engineering products—options that would be rather limited if IP protection were strengthened. Moreover, inventors by virtue of their lead times and IP protection may be expected to sell technologies at a price higher than marginal cost, which is socially less than optimal for the recipient country, at least in a static sense. High levels of IP protection can also deepen negotiating imbalances in IP licensing in favor of the licensor, often resulting in the imposition of abusive practices that restrain competition. They criticize the one-size-fits-all approach (as adopted by the WTO TRIPS regime) as far as the role of IP in technology development and transfer is concerned, particularly when technological capabilities are asymmetrically distributed across countries and the need for IP protection also varies from one technology to another. Although there exists a large body of empirical literature on the relationship between IP and TT, the findings turn out to be diverse and often ambiguous. A few studies that attempt to assess the weight of IPRs on TT decisions indicate that they generally are of medium importance, among various other factors.

Barriers

Barriers to TT may vary from one technology to another and may manifest themselves differently in different countries. In the climate change context, the Technology Needs Assessments (TNAs) carried out by 56 non–Annex I Parties of the UNFCCC (United Nations Framework Convention on Climate Change), economic and market barriers were found to be the most frequently identified barriers, followed by barriers relating to human capacity, information and awareness, institutional, regulatory, policy related, and technical. Most of the TNA reports underlined the role of governments in helping to remove such barriers through the formulation of effective policies, regulations, standards, codes, and other measures. According to the IPCC, there is no preset answer to enhancing TT. The identification, analysis, and prioritization of barriers should be country specific. It is important to tailor action to the specific barriers, interests, and influences of different stakeholders in order to develop effective policy tools.

See Also: Innovation; Intellectual Property Rights; Science and Technology Policy.

Further Readings

Foray, D. "Technology Transfer in the TRIPS Age: The Need for New Types of Partnerships Between the Least Developed and Most Advanced Economies." Issue Paper No. 23. Geneva: International Centre for Trade and Sustainable Development, 2009.

Intergovernmental Panel on Climate Change (IPCC). "IPCC Special Report: Methodological and Technological Issues in Technology Transfer." (2000). http://www.ipcc.ch/ipccreports/sres/tectran/index.php?idp=2 (Accessed September 2010).

Lall, S. "Technological Capabilities and Industrialization." *World Development*, 20/2 (1992).

United Nations Conference on Trade and Development (UNCTAD). "Transfer of Technology." UNCTAD Series on Issues in International Investment Agreements. New York: United Nations, 2001.

United Nations Framework Convention on Climate Change (UNFCCC). "Second Synthesis Report on Technology Needs Identified by Parties Not Included in Annex I to the Convention." FCCC/SBSTA/2009/INF.1. May 2009.

Kasturi Das
Research and Information
System for Developing Countries

THERMAL DEPOLYMERIZATION

Thermal depolymerization (TDP) is a multistage thermochemical conversion and separation process for converting biomass into usable fuels, oils, gases, and other hydrocarbon-based energy products. The technique relies on chemical decomposition of biomass feedstock inside a chemical reactor under high pressure and temperature conditions that mimic those used by nature in the production of conventional fossil fuels. Although many similar processes have been around for hundreds of years, only recently have advances in chemical engineering technology demonstrated the TDP process to be cost effective and energy efficient.

Typical biomass feedstocks used in TDP processes include turkey offal, manure, animal waste, sewage, restaurant grease, plastic bottles, cornstalks, and paper pulp effluent. These feedstocks provide the source of carbon-containing organic material that is chemically decomposed through the high pressure and temperature conditions inside the reactor. Biopolymers comprising repeating chemical units of carbon-containing molecules strung together in a chain are broken down into smaller units. For example, triglyceride fat molecules are broken down into short hydrocarbon chains during TDP through a process that is analogous to "thermal coking" or "cracking" technology used in the petroleum-processing industry. Amino acids can be recovered from TDP processes, producing nutrient-rich materials that can be used as fertilizer. An activated carbon adsorbent material can also be recovered from TDP processing. The quality and range of recoverable products obtained through TDP will depend on the specific feedstock, the temperature and pressure conditions chosen, and the purification processes used.

Political History

Thermal depolymerization has recently received increased attention due to the rapidly growing biodiesel and renewable diesel energy markets. These markets have been supported through a series of tax incentives that promote the biodiesel industry in the United States. Tax incentives that were passed as part of the Energy Policy Act of 2005 provided subsidies to companies using thermal depolymerization to produce renewable diesel fuel from biological materials.

On April 3, 2007, the Internal Revenue Service (IRS) issued regulatory guidance to clarify specifically which types of fuel would qualify for the tax credit. The 2007 statement indicated that thermal depolymerization was to be construed most broadly to include any process that reduces complex organic materials through the use of pressure and heat to decompose long-chain polymers into short-chain hydrocarbons with a maximum chain length of around 18 carbon atoms. The statement also indicated that TDP processes that use a catalyst are to be included under the tax subsidy laws. This last provision made it possible for large, integrated oil companies to add small amounts of animal or vegetable oils into conventional diesel oil refining processes and claim a tax credit of $1 per barrel of diesel produced.

In 2007, the National Biodiesel Board argued that these types of coprocessed diesel, or "blender" products, should not qualify for the tax incentive. In August 2009, the Biodiesel Tax Incentive Reform and Extension Act of 2009 sponsored by Senator Maria Cantwell (D-WA) was introduced. This bill, in committee in the Senate at the time of this writing, redefines renewable diesel to exclude coprocessed biodiesel from the tax subsidy program. The 2009 bill simplifies the definition of "biodiesel" and changes the incentive from a blender credit to a production tax credit.

Future Outlook

Although thermal depolymerization offers a direct route for reprocessing of biomass waste and decreasing landfill use, the technique does not offer a clean energy source free of carbon dioxide emissions. The fuel produced using TDP undergoes combustion like conventional fossil fuels, and so production of greenhouse gases is not avoided. TDP proponents, meanwhile, argue that the building of biomass reprocessing infrastructure will boost industry and job growth while decreasing the nation's dependence on imported oil and gas.

Technical barriers have partially limited the extent to which TDP has been adopted in the 21st century. For example, variability in the composition of the biomass feedstock over time can lead to difficulty in achieving controlled processing conditions, resulting in inconsistent fuels being produced. There currently exists, therefore, a demonstrated need for technological innovations capable of making TDP processing as economically and energetically efficient and chemically consistent as possible. Energy researchers and industry analysts will undoubtedly closely follow scientific and commercial developments in this arena in search of a clear route to simple, sustainable, and profitable TDP production technologies for the future.

See Also: Biochemical Processes; Cellulosic Biofuels; Pyrolysis; Thermochemical Processes.

Further Readings

Duane, D. "Turning Garbage Into Oil." *New York Times Magazine* (December 14, 2003).
Lemley, B. "Anything Into Oil." *Discover Magazine* (May 2003).
Ragauskas, A. J., et al. "The Path Forward for Biofuels and Biomaterials." *Science*, 311/5760 (2006).

Michael A. Nash
University of Washington, Seattle

THERMAL HEAT RECOVERY

Heat recovery is a general term used to denote the process of utilizing some of the heat energy contained in the output of a process or an effluent and generally indicates that the heat would be otherwise unused and dissipated into the immediate environment. The prevalence of heat processes in energy systems, such as for space heating in the domestic sector, for generating motive power in the transport sector, and for producing electricity for use in all sectors, means that heat recovery has a wide area of potential application. In general, energy savings of up to 30 percent can be possible through heat recovery measures. Sources of waste heat are ubiquitous because the heat loss is inevitable, but waste heat is not always suitable for heat recovery. There are various technological options available for heat recovery, although economic or technical constraints sometimes preclude their successful application.

In many heat- and/or electricity-generating processes, the heat available for recovery is a by-product of the process itself; the heat demand of the process is met, and any excess or waste heat is exhausted from the process. The laws of thermodynamics indicate that heat transfer will naturally occur from higher to lower temperatures. The temperature of a process's waste heat is thus inevitably lower than the temperature of the process itself.

The tendency of systems toward a thermodynamic equilibrium (i.e., reaching the same temperature) is the ultimate reason for heat loss and means that in all heat-generating processes there is some waste heat. Not all of the lost heat can be recovered, however, and

A researcher measures the air flow distribution through heat exchangers, which transfer heat between fluid streams and are essential to heat recovery technology.

Source: National Institute of Standards and Technology, U.S. Department of Commerce

its suitability for recovery depends upon several factors, including the amount of heat being lost (in terms of power) and its heat-flux density (the rate of heat flow per cross-sectional area), the nature of the environment, the temperature of the heat, and other process-specific considerations—such as the rate of cooling, which must be controllable in some industrial processes such as glass manufacture.

In fact, the temperature of the waste heat is a crucial factor, along with the physical quantity, in determining the feasibility for heat recovery. Generally speaking, the higher the temperature, the more useful the heat is and the more suitable it is for generating electricity (as opposed to being used directly). This is because of the temperature difference required to drive thermodynamic processes, as mentioned above; with only a small temperature difference between source and sink, the forcing mechanism for heat transfer is weak.

Heat loss from a process occurs through three main mechanisms: radiation as electromagnetic waves; convection, which is the transmission of energy through thermal currents (in fluids); and conduction, which is the direct transmission of heat through a substance. Heat recovery technologies employ one or a combination of these mechanisms in order to recover waste heat.

Heat exchangers are devices used to transfer heat between fluid streams. They are a crucial and ubiquitous technology for heat recovery, as they enable the transfer of heat energy between hot and cold streams. Heat exchangers can be classified into three main types: recuperators, regenerators, and evaporative heat exchangers. Recuperators enable the transfer of heat between fluids on either side of a dividing wall, and they operate continuously. Regenerators and heat wheels allow the transfer of heat to and from an absorbent medium, such as heat-conducting bricks, and operate periodically, with a loading phase during which hot fluid charges the device, and an unloading phase during which the heat is transferred to a cooler fluid. A regenerator is stationary with respect to moving streams, whereas a heat wheel rotates between ducts containing hot and cold streams. Evaporative heat exchangers continuously and evaporatively cool a liquid in the same space as the coolant, such as occurs in power-station cooling towers. There are various types of heat exchangers in each of these categories, each with specific advantages and disadvantages.

Heat exchangers are widely employed to recover waste heat from processes, either to be used directly as heat or to generate electricity. In industry, recovered heat may be used for preheating raw materials, drying operations, and/or raising process steam, and in other sectors it may be used for space and water heating. The scope for heat recovery from conventional domestic heating systems is limited, however, given that typical gas-fired boilers have efficiencies of around 90 percent. The generation of electricity from waste heat is often more favorable than using recovered heat directly due to the versatility and relatively high value of electricity compared to heat. This is because electricity can be used for power as well as heat applications and because it is highly transportable (heat can also be transported, but only for relatively short distances without large losses). Electricity generation with conventional thermodynamic cycles (i.e., such as in large power plant) is possible only with high-temperature sources of waste heat. At lower temperatures, nonconventional cycles such as the organic Rankine cycle (ORC) are employed. This cycle uses an organic working fluid with a lower boiling point than water so that the evaporation occurs at a much lower temperature, and the vapor can be used to drive a turbine and generate electricity.

Other technologies relevant to heat recovery include heat pumps and heat pipes. Heat pumps are simple thermodynamic machines in which low-temperature heat from a source is transferred to a higher-temperature sink using mechanical or high-temperature heat energy. In industry, there are a number of possible applications wherever it is desirable to pump low-temperature waste heat into a higher-temperature environment. In the domestic sector, ground or air source heat pumps upgrade ambient sources of heat to temperatures suitable for domestic heating. Heat pipes enable the transfer of heat over moderate distances with a very low heat loss and without the need for mechanical pumping. These may be used in combination with combined heat and power (CHP) systems in order to transport the heat to district heating schemes or adjacent industrial facilities.

In practice, limitations on heat recovery include requiring a use for the recovered energy; unless electricity can be generated, the heat's transportability is limited. Furthermore, corrosive gases in exhaust streams can lead to fouling and corrosion of heat transfer surfaces (in heat exchangers), making regular cleaning and maintenance necessary, which can render the plant uneconomic. In the case of heat recovery at high temperatures, the materials employed clearly need to be able to withstand such temperatures. Finally, the inflexibility that can result from recovering heat, especially where one is contractually bound to supply it to a third party, can also be a key reason for not exploiting apparent opportunities.

See Also: Clean Energy; Ecological Modernization; Green Building Materials; Solar Hot Water Heaters.

Further Readings

Drapcho, Caye, John Nghiem, and Terry Walker. *Biofuels Engineering Process Technology.* New York: McGraw-Hill Professional, 2008.

Ganapathy, V. *Industrial Boilers and Heat Recovery Steam Generators: Design, Applications, and Calculations.* Washington, DC: CRC Press, 2002.

Mousdale, David M. *Biofuels: Biotechnology, Chemistry, and Sustainable Development.* Washington, DC: CRC Press, 2008.

Russell McKenna
Karlsruhe Institute of Technology

THERMOCHEMICAL PROCESSES

Thermochemistry is the branch of physical chemistry that deals with the ways in which chemical reactions either absorb or produce heat. Researchers in the field are playing a significant role in promoting sustainability by discovering ways in which thermochemical processes can be used to deal with waste products that threaten the environment. Such processes are being used around the world to convert trees, grasses, agricultural crops, agricultural residues, forest residues, animal waste, and municipal solid waste into ethanol, renewable diesel, gasoline, and hydrogen that are used for fuel. Other uses for waste products include the generation of electricity and the chemical production of plastics, adhesives, paints, dyes, detergents, and a score of other products. The fact that since the 1970s, researchers have been experimenting with ways to employ thermochemical processes to use water to generate hydrogen for use as an alternative source of power has enormous implications for a world in which natural gas resources are constantly being depleted. In the 21st century, thermochemical process research is being conducted in the United States, Switzerland, Turkey, Russia, and many other countries around the world.

Both developed and developing nations have begun using the thermochemical process known as biomass conversion to produce energy that is carbon dioxide neutral. This conversion generally takes place through gasification, carbonization, or direct liquefaction. The advantages inherent in using gasification are that the process produces little sulfur or nitrogen oxide; its particulate emissions are lower than those of traditional fuels; and it can help to stabilize existing carbon dioxide concentrations in the atmosphere by reducing the need for fossil fuels and promoting greater carbon dioxide absorption in vegetation and soil. The end products of gasification are suitable for use as an energy resource for both homes and industries. In 2008, such conversions represented 35 percent of total energy consumption in developing countries and 14 percent of worldwide consumption. Internationally, 64 percent of biomass resources are derived from wood and wood wastes. Municipal solid waste provides 24 percent of such resources, and agricultural waste and landfill gases furnish another 5 percent each.

Practical Applications

In the United States, poultry farms produce some 5.6 million tons of litter, consisting of bedding such as wood shavings and peanut hulls, manure, and food that has been spilled. Environmentalists are concerned because this litter often leads to both land and water pollution, and health officials and the general public are concerned about its impact on heightened vulnerability to mad cow disease or avian influenza, both of which could potentially be spread through litter transference. A team of researchers at Virginia Tech's College of Agriculture and Life Sciences discovered a means of turning poultry litter into bio-oil using a transportable pyrolysis unit that vaporizes poultry litter through heat, condensing it to bio-oil, and producing fertilizer via slow release from the reactor. In turn, the gas produced by the unit is used as fuel to operate the system. Since the unit can be transported to the site of the litter, it reduces the chance of transferring diseases or infections. The thermochemical process used to convert the litter has the added advantage of destroying any microorganisms contained in the litter. The quality of bio-oil produced is dependent on conditions such as age and content of the litter. Because of its importance to both the environment and the financial health of the Shenandoah Valley, the support team for this project included specialists from the Virginia Cooperative Extension, state officials, conservation organizations, and representatives from private industry. Funding was generated through a million-dollar grant from the National Fish and Wildlife Foundation's Chesapeake Bay Targeted Watershed Program. Similarly, researchers at Iowa State University developed a corn stover model that uses waste from corn to produce bio-oil, flammable gas, and charcoal as well as a number of by-products.

Around the world, government officials are working with private industry to promote sustainability through the use of thermochemical processes that convert the mountains of solid waste that threaten the environment into other products. In Storey County, Nevada, for instance, officials have partnered with Fulcrum Sierra Bio Fuels, LLC, to construct the Waste-to-Ethanol Production Facility in the Tahoe–Reno area. The conversion takes place through a two-step process in which a downdraft, partial oxidation gasifier converts waste into synthesis gas. The second step uses a patented process to convert the gas into ethanol.

Many environmentalists see coal gasification as the wave of the future in the field of using environmentally motivated thermochemical processes. The gasifier offers a low-impact method of breaking coal down into its original chemical components. In most cases, the process uses heat to force steam and oxygen or air to transform coal into carbon monoxide, hydrogen, and other gases without harming the environment. The U.S. Department of Energy's Office of Fossil Energy is experimenting with a new coal gasifier that is capable of processing biomass and municipal and industrial waste as well as coal with considerably more efficiency and reliability, and less environmental impact than gasifiers currently in use.

In Russia, thermochemical processes are being used to break down toxic and radioactive wastes. This is accomplished by causing the combustion of powder metal constituents along with hazardous waste components. The end result is that organics are completely decomposed during the process, while toxic and radioactive components are retained as mineral-like or glass components that can be stored, transported, or disposed of without threatening the environment.

See Also: Biochemical Processes; Science and Technology Policy; Thermal Heat Recovery; Waste-to-Energy Technology.

Further Readings

Balat, Mustafa. "Mechanisms of Thermochemical Processes, Part 1: Reactions of Pyrolysis." *Energy Source*, 30 (2008).

Funk, Ted L. "The Global Demand for Biofuels: Thermochemical Processing and Utilization of Biomass." http://cgs.illinois.edu/powerpoint/the-global-demand-biofuels-thermochemical-processing-and-utilization-biomass (Accessed September 2010).

Greiner, Lori. "Thermochemical Process Converts Poultry Litter Into Bio-Oil." http://www.eurekalert.org/pub_releases/2007-08/vt-tpc081307.php (Accessed September 2010).

L'vov, Boris V. "Thermochemical Approach to Solid-State Decomposition Reactions Against the Background of Traditional Theories." *Journal of Thermal Analysis and Calorimetry*, 96/2 (2009).

Ojovan, M. I., et al. "Thermochemical Processes Using Powder Metal Fuels of Radioactive and Hazardous Waste." *Journal of Process Mechanical Engineering*, 218 (2004).

Soetaert, Wim and Erick J. Vandamme. *Biofuels*. Chichester, UK: Wiley, 2009.

U.S. Department of Energy. "Gasification Technology R&D." http://www.fossil.energy.gov/programs/powersystems/gasification/index.html (Accessed September 2010).

Elizabeth Rholetter Purdy
Independent Scholar

U

Unintended Consequences

When some cities banned plastic bags, stores switched to sturdier paper and reusable plastic bags. However, consumers still disposed of the thicker-walled bags, leading to greater stress on city waste facilities than before the ban.

Source: iStockphoto

Intentional human actions cause multiple effects. Some of these effects are planned, and others occur unexpectedly. Unintended consequences are unplanned outcomes that occur due to the implementation of a technology, policy, or other initiative. Social scientists typically categorize them as beneficial, detrimental, or controversial. Unanticipated consequences follow directly or indirectly from human activities but occur at a future time and possibly in a different location. Therefore, they can be difficult to identify or directly link to a triggering activity. As a result, journalists, technologists, and policymakers sometimes overlook the impacts of unintended consequences. Negative unanticipated consequences can be challenging to evaluate and remedy because they arise within complex ecological interactions and social conditions. Green technologies (e.g., wind turbines, solar cells, and biofuels) and initiatives (e.g., efficiency, recycling, and organics) yield distinct unanticipated consequences that can partially or fully offset intended environmental benefits.

Theorists of economics, political science, history, and sociology have long evoked the concept of unintended consequences, sometimes called "the law of unintended consequences." The notion is embedded in other common concepts such as snafu, Murphy's Law, serendipity, windfall, the butterfly effect, and perverse incentive. The concept of unintended consequences is central to moral philosophies of consequentialism, which hold that people should judge actions based on the outcomes they create. For instance, in 1848, French economic journalist Frédéric Bastiat wrote: "In the economic sphere an act, a habit, an institution, a law produces not only one effect, but a series of effects. Of these effects, the first alone is immediate; it appears simultaneously with its cause; *it is seen*. The other effects emerge only subsequently; *they are not seen*." He reasoned that social scientists should recognize and account for these unseen effects.

In 1936, Robert K. Merton advanced a definition of unintended consequences that would go on to inform much contemporary thought on the subject. He pointed out two methodological pitfalls that arise when putting the term to work. First, social scientists must determine how much of an observed consequence can be rightly attributed to a purposive action. To what extent, for instance, can the rise of organized crime be blamed on Prohibition? The second challenge for social scientists is to determine the intended purpose of an action in the first place. Consequences of actions can be rationalized after the fact, as exemplified by the horseman who, after being thrown from his horse, declared that he was "simply dismounting."

Merton and other theorists have identified numerous factors that lead to unanticipated consequences: ignorance, error, greed, shortsightedness, cognitive processes, emotional bias, and even the world's inherent complexity as elaborated in chaos theory. Merton argued that people may occasionally be so eager to realize the immediate effects of an act that they give no consideration to other potential consequences. Similarly, people may overlook further consequences when their fundamental values oblige them to pursue an action. The resulting unintended consequences may actually change basic values over time.

Types of Unintended Consequences

People commonly consider unintended consequences to be negative or positive, but they may also be perverse, neutral, or even controversial. The actual categorization may depend on the observer's perspective. For instance, a medical drug produces many effects. Some are intended while others are not. The unintended consequences can be the following:

- *Positive*: The drug yields a beneficial side effect in addition to the intended effect. Aspirin is a pain reliever but also acts as an anticoagulant, which can help prevent heart attacks and reduce damage caused by thrombotic strokes.
- *Negative*: The drug produces a detrimental side effect in addition to the intended effect. Human immunodeficiency virus (HIV) medications save lives but they can reduce a user's appetite and even trigger nightmares.
- *Perverse*: The drug produces exactly the opposite of the intended result. Antibiotics can induce antibiotic-resistant strains of bacteria. Also, doctors have discovered that some drugs intended to prevent heart arrhythmias actually turned out to be pro-arrhythmic in practice.
- *Controversial*: The drug creates an effect that some view as detrimental but others view as beneficial. Some heart medications induce hair growth, and some pain relievers produce euphoric sensations.

A drug may produce any combination of these effects on the body. Public policies, environmental initiatives, business dealings, and other human undertakings regularly produce unplanned outcomes as well. Therefore, unanticipated consequences are a topic of concern and study across a wide spectrum of disciplines. For instance, developmental economists have shown that simplistic food aid can worsen long-term food security of a target region if international organizations deploy the aid without accounting for local economic conditions. If a community is flooded with free food from abroad, local farmers cannot compete and may subsequently earn too little to plant their fields the following season. In this case, the food aid induces the perverse unanticipated consequence of worsening food security by putting local farmers out of business. Developmental economists have developed strategies for avoiding these consequences. For instance, a charity might secure funds for local farmers or introduce the food aid at market prices so local farmers can compete with the imported food.

When San Francisco and other cities banned plastic bags, stores switched to sturdier paper and reusable plastic bags. However, consumers still disposed of the thicker-walled bags, leading to greater stress on city waste facilities than before the plastic bag ban was implemented. In contrast, Seattle stores charged a small fee for each bag. Shoppers brought their own reusable bags to avoid the small charge. This policy yielded the intended effect of decreasing waste without the perverse unintended consequence. However, critics point out that although bag charges are successful from a waste and carbon perspective, bag fees place a disproportionate burden on poor residents.

Unintended Consequences of Alternative Energy

Fossil fuel energy yields many benefits but the associated extraction operations, distribution networks, and combustion practices yield a host of negative unintended consequences. Environmental groups, politicians, and businesses frame green energy technologies as clean alternatives to fossil fuels. However, green energy alternatives generate unanticipated consequences of their own.

As with traditional energy production, the unanticipated consequences arising from green technologies can generate political tensions. Once a government or organization backs a certain green technology, it risks losing credibility if detrimental consequences are exposed. For instance, in 2008, riots broke out around the world in response to rising corn prices. Some blamed the increase on weather conditions, others claimed that demand from India and China was to blame. The World Bank studied the price jump but kept its findings secret, presumably because they might have upset the bank's major donor, the United States. However, *The Guardian* obtained a leaked copy of the report and published its findings. The World Bank study group had determined that the rise in corn prices was an unintended consequence of biofuel production. The report concluded that biofuel producers' demand for corn pushed prices higher for everyone, including those who needed corn for food.

Economists and ecologists have identified numerous other unanticipated consequences of biofuel production. Biofuel proponents maintain that their fuel cycles net no additional carbon dioxide (CO_2). In theory, biofuel feedstock plants absorb and offset combustion-related CO_2 emissions. However, when Indonesian swamps were drained in order to grow palm oil crops, soil decomposition accelerated, unexpectedly releasing large quantities of greenhouse gases into the atmosphere. France, Germany, and other European nations withdrew support for palm oil when they discovered that these rogue emissions were more than 10 times greater than the potential savings afforded by converting from petroleum to palm-based biofuels.

Biofuel producers can refine fuel from sugarcane but critics maintain that sugarcane crop practices endanger rainforests and biodiversity. Authors of an article published in the journal *Science* argue that the benefits of producing biofuels from sugarcane are greatly diminished if the unanticipated consequences of sugarcane production are taken into account. They argue that carbon-rich rainforests are frequently leveled to make room for sugarcane plantations. This not only interrupts the carbon cycle but also endangers local biodiversity, hydrological functioning, and soil stability. Ideally, farmers would plant bio-fuel crops exclusively on abandoned farmland, but such land is relatively rare and usually less fertile. Even on suitable sites, crop residues left behind from farming activities release methane, a greenhouse gas with 23 times the warming potential of CO_2. Furthermore, fertilizing fields of sugar, corn, rapeseed, and other biofuel feedstocks with nitrogen-rich fertilizers yields nitrous oxide. Nitrous oxide has a global warming potential 296 times greater than CO_2 and additionally damages stratospheric ozone.

Many people admire solar photovoltaic cells for silently extracting clean energy from the sun's rays but the panels contain heavy metals that can leach into groundwater when disposed at the end of their life cycle, according to the Silicon Valley Toxics Coalition. Photovoltaic manufacturers employ toxic and explosive compounds that can lead to unintended health risks for workers and local residents. While solar cells do not produce CO_2, the photovoltaic manufacturing industry emits nitrogen trifluoride (NF_3) and sulfur hexafluoride (SF_6)—greenhouse gases that are 10,000 to 25,000 times more harmful than CO_2, according to the Intergovernmental Panel on Climate Change. The unintended consequences of photovoltaic production offset at least part of the carbon and environmental benefits of solar cells.

Wind turbines generate energy from a freely available and renewable resource. However, large turbines can disturb residents and therefore regularly generate not-in-my-backyard (NIMBY) resistance when sited near residential communities. If sited in remote regions, associated maintenance roads can inadvertently afford poachers and loggers access to ecologically sensitive areas.

Alternative energy generation may also instigate unintended macroeconomic consequences. Alternative energy promoters aim to reduce dirty fossil fuel use by expanding clean energy production. However, increasing any form of energy supply can exert downward pressure on energy prices, thereby stimulating overall demand for energy services. Economists warn that without appropriate countermeasures, any increase in energy production, alternative or conventional, may unintentionally perpetuate energy-intensive modes of living. Also, when energy consumers believe their energy is derived from clean sources, they may be less concerned about conserving it.

Unintended Consequences of Energy Efficiency

Instituting energy-efficiency measures can lead to both beneficial and detrimental unintended effects. According to behavioral psychologists, when energy users employ more efficient energy technologies, they may in turn increase their frequency of use. In one study, participants who purchased energy-efficient washing machines subsequently started doing more loads of laundry. When individual or organizational energy consumers institute energy-efficiency measures, such as using more efficient light bulbs or machinery, they also save money on energy. However, consumers may choose to spend these savings on other products or endeavors that still lead to energy consumption. In this case, money-saving energy-efficiency measures can unintentionally stimulate other forms of consumption, leaving overall energy footprints unchanged. Energy-efficiency measures can spur similar unintended effects on a macroeconomic scale. Efficiency measures frequently lead to larger profits, which can spur more growth and higher energy consumption overall. This unintended consequence of energy efficiency is termed the *Jevons paradox*. It is named after William Stanley Jevons, who in 1865 explained how James Watt's introduction of the steam engine greatly improved efficiency, which in turn made steam engines more popular and subsequently drove the use of coal ever higher.

Energy efficiency advocates argue that instituting energy taxes or other incentives designed to thwart energy demand can block some of these unintended consequences. They point to California, which instituted a system called "decoupling" three decades ago. Decoupling is a financial arrangement that rewards energy companies for selling less of their energy services, rather than more. Since its introduction, decoupling has stabilized per capita electricity consumption in California even though national per capita electricity consumption surged 50 percent higher over the same period.

Energy reduction endeavors can clearly spur positive unintended consequences as well. For instance, when cities started to shift to energy-efficient LED municipal lighting, they also realized maintenance savings and traffic safety improvements because the new bulbs failed less frequently than the bulbs they replaced. In older cities, builders often constructed dwellings shoulder-to-shoulder in order to efficiently utilize urban space and save energy (heat transfers from one flat are absorbed by others, reducing everyone's energy bills). Physical proximity brought people closer together in novel ways, allowing for the efficient, walkable neighborhoods and cosmopolitan exuberance now taken for granted in cities such as Paris, Tokyo, New York, and London. Downshifters (people who choose to greatly reduce their material consumption) often unexpectedly discover new interests and report higher satisfaction with their low-consumption lifestyles.

Unintended Consequences of Organic, Fair Trade, and Local Food

Numerous mainstream environmental organizations and concerned citizens throughout the world support organic, fair-trade, and local food initiatives. These movements aim to bring agriculture and food processing and distribution activities into line with ecological justice and sustainability principles. These initiatives yield many intended benefits, but their successes are at least partly offset by detrimental unintended consequences. For instance, fair-trade programs aim to assist small farmers by guaranteeing that buyers will purchase their commodities, such as coffee and sugar, at a price above market value. This system produces two distinct negative unanticipated consequences. First, guaranteeing an elevated price leaves producers with no incentive to maintain or improve quality. Second, fair-trade subsidies may block market signals by subsidizing goods that are being overproduced. Typically, overproduction drives prices lower, signaling producers to switch to other crops. Fair-trade subsidies can prevent this signal from getting through and may even attract more producers to market. Intensified overproduction shoves market prices even lower, risking leaving all non–fair-trade producers poorer—unless program directors institute measures to counteract this unintended consequence.

Local foods often require little energy to distribute. However, if local farmers employ heated greenhouses or inefficient transport methods, locavores may unintentionally expand their energy footprints when prioritizing local fruits and vegetables over those shipped from warmer climates via efficiently packed containers. While locavorism clearly benefits local farmers and communities, it can unintentionally hurt export farmers in the global south.

Organic farmers reduce environmental harms stemming from pesticides and fertilizers. However, organic farming techniques require extensive plowing in order to control weeds. This in turn requires more petroleum. Additionally, Nobel Peace Prize laureate Norman Borlaug has argued that in order to supply organics to everyone on the planet, cultivated land would have to be tripled because organic farming is more land intensive than conventional farming. Therefore, he claimed that demand for organic foods unintentionally places rainforests and other sensitive areas at risk. Here, the intended benefits (reducing fertilizer and pesticide contamination) are difficult to weigh against the unintended consequences (increased petroleum use and rainforest endangerment).

Critiques of Unintended Consequences

Critics of the concept of unintended consequences point out that the concept can obscure deeper structural problems that should be addressed. For instance, journalists, corporations, and politicians frequently frame oil spills as accidents, or unintended consequences of resource extraction. However, they could alternately frame spills as the inevitable and expected outcome of an undertaking with extreme environmental risks.

Some political and economic theorists stress the many negative unintended consequences of government spending and regulation in order to argue for limiting the government's reach. Others claim that this use of the concept of unintended consequences is politically motivated and suspect. Presumably, if legislators suspend an activity in order to eliminate its unintended consequences, the intended benefits of the activity will also be lost. These theorists maintain that all human actions yield unanticipated consequences and strong governance, even if imperfect, is required to prevent even greater injustices from harming people and their ecosystems.

See Also: Appropriate Technology; E-Waste; Luddism; Reflexive Modernization; Science and Technology Studies.

Further Readings

Bastiat, Frédéric. "What Is Seen and What Is Not Seen." *Ideas on Liberty*, 51 (2001).
Beck, Ulrich. *Risk Society: Towards a New Modernity*. Newbury Park, CA: Sage, 1992.
Chakrabortty, Aditya. "Secret Report: Biofuel Caused Food Crisis." *The Guardian* (July 3, 2008).
Herring, Horace and Steve Sorrell, eds. *Energy Efficiency and Sustainable Consumption: The Rebound Effect*. New York: Palgrave Macmillan, 2009.
Merton, Robert K. "The Unintended Consequences of Purposive Action." *American Sociological Review*, 1/6 (1936).
Scharlemann, Jörn P. W. and William F. Laurance. "How Green Are Biofuels?" *Science*, 319 (2008).
Zehner, Ozzie. *Coming Clean: The Dirty Truth About Clean Energy and the Real Future of Environmentalism*. Lincoln: University of Nebraska Press, 2011.

Ozzie Zehner
University of California, Berkeley

University-Industrial Complex

The term *university-industrial complex* refers to the close connections between universities and industry that are a feature of the intellectual landscape in many countries, but especially in the United States. The 19th-century Industrial Revolution saw the emergence of new civic universities in Britain and Germany that were to train workers for the new age of manufacturing. This marked a shift from the craft traditions that had dominated the intellectual landscape of medieval Europe into the systematic application of scientific advance for industrial purposes. State support for the university often took the form of technical universities that were meant to train workers. University–industry linkages in this period focused on certain key departments such as chemistry, and the university laboratory was the major source of new inventions.

However, what makes the modern university-industrial complex significant is the overwhelming role that technology plays in contemporary life. New advances in the nano-cogno-bio-info knowledge society are crucially dependent on research in university labs, and as more nations move toward a knowledge society and a knowledge economy, the alliances between the state, industry, and universities have become much closer and have been christened the "triple helix" of innovation.

The rise of the university-industrial complex was the result of a series of events that took place in 1980 that have been referred to as the second transformation of the university. These included a steady decline in federal funding for universities, which in real terms fell by 9.4 percent per full-time academic researcher between 1979 and 1991; the passage of the Bayh-Dole Act in 1980 (the Patent and Trademarks Laws Amendments Act of 1980) that aimed at making it easier to transfer technology developed in universities to industry, crucially allowing universities, small businesses, and nonprofit institutions to hold exclusive patent rights to federal government–sponsored research; and the U.S. Supreme Court decision in *Diamond v. Chakrabarty* that ruled "a live, human-made micro-organism is patentable subject," paving the way for close collaboration between life sciences companies and university research labs and the emergence of biotechnology as a lucrative industry. Over the past two decades, universities have developed ever-closer links with industry as competition for government research funding has become more intense, institutional costs have risen, and demographic changes have adversely affected the job market for graduates in science and engineering.

The emergence of the university-industrial complex was the latest stage in the consolidation of the intellectual landscape that was one of the key features of the 20th century, which saw the birth of the modern military-university-industrial complex, most prominently in the United States. World War II had shown the value of the university-based scientist and institutionalized the bureaucratic infrastructure of science and technology policy. University scientists such as Robert Oppenheimer—the creator of the atomic bomb—and the success of the Manhattan Project that developed nuclear weapons, and other wartime initiatives such as the development of penicillin and streptomycin—cemented an alliance between the university and the state that was termed the *social contract* between science and the state. Moreover, cutting-edge applications, for example, research in high-energy physics, were astronomically expensive. The immediate postwar period saw an increase in university funding by the federal government. From 1953 to 1968, public support grew by 12 to 14 percent annually. Whereas funding for scientific research from all sources totaled $31 million in 1940, federal funding alone reached $3 billion in 1979, much of it dispensed by the National Institutes of Health and other new agencies.

However, the emergence of neoliberalism in the 1980s meant that investing in higher education, especially in the United States, was not seen as a public good anymore. Education was increasingly seen as a marketable commodity rather than something that had to be funded by taxpayers. The 1990s saw many Western governments try to achieve growth in student numbers with lower unit costs. In the United States, the change in policy climate underscored trends that had already been evident for some years in some areas. For example, in 1965, the federal government provided about $126 million for research facilities on university and college campuses; by 1979, that figure had shrunk to about $32 million. The 1980s saw the acceleration of these trends. According to figures from the National Science Foundation, while total research and development (R&D) funding in the United States has hovered around 2.5 percent of the GDP, the share of the federal government has fallen while that of industry has risen. However, the federal government still is the primary source of support for academic R&D.

It is estimated that by 1990, there were 1,056 university-industry research and development centers that spent a total of $2.9 billion on R&D, which was twice the National Science Foundation's $1.3 billion that the federal government provided to research that year. In fact, in 1990, in a trend that continues to the present, around half of industry support for academic R&D went to university-industry R&D centers. From 1980 to 1998, industry funding for academic research expanded at an annual rate of 8.1 percent, reaching $1.9 billion in 1997—nearly eight times the level of 1977. Industry supported universities to the tune of $850 million in 1985, and in 1995, that figure had risen to $4.25 billion. To take one prominent example of corporate funding, the University of California, Berkeley was given $94.7 million in 2006, compared to a figure of $22.2 million between 2002 and 2003. Examples of the commercialization of such university research are the founding of the Advanced Technology Development Center located at the Georgia Institute of Technology in Atlanta, Georgia; the Rensselaer Polytechnic Institute Incubator Center in Troy, New York, in 1980; and the Science Park near Yale University in New Haven, Connecticut, in 1983. University incubators were, in fact, established at most major universities in the United States, for example, Berkeley Business Incubator, Austin Technology Incubator, USC Columbia Technology Incubator, University of Maryland at College Park Technology Advanced Program, and Boston University Photonics Center, among others.

The University-Industry Complex in the United States

The university-industry complex is particularly strong in the United States. For example, in 1994, 15 years after the passage of Bayh-Dole, industry support accounted for 7 percent of university research; the comparable figure for Japan was 3.6 percent. Further, while industry research links in Japan are more informal, they tend to be much more formalized in the United States. The greater strength of the university-industrial complex in the United States also reflects differing institutional structures. Unlike most countries, which have a central ministry of education, the higher education sector in the United States is not the responsibility of the federal government, but of the states and of private institutions. This means that money from the federal government comes from a variety of sources, especially the Department of Defense (DOD), and the kind of projects that are funded reflect the research interests of these agencies. The most famous example for this is the Internet, which began as a DOD initiative to create a decentralized communication infrastructure that could survive a Soviet nuclear attack. In fact, most of the key innovations in the 1960s and 1970s came from the departments of space, energy, and defense. The United States also is a nation where there is considerable ideological opposition to the idea that universities should be funded by taxpayer money, in contrast to many European nations, with the exception of Britain, and especially the Scandinavian countries where university education is seen as a right rather than a privilege. The legal climate in the United States also favors university-industry alliances. In addition to the Bayh-Dole Act, 1980 also saw the passage of the Economic Recovery Tax Act, which extended industrial R&D tax breaks to support research at universities.

One of the key changes in the feature of the intellectual landscape in the United States was the passage of the Bayh-Dole Act, which was passed to "promote the utilization of inventions arising from federally supported research . . . [and] to support the commercialization and public availability of inventions" (35 U.S.C. 200). The Bayh-Dole Act was passed to stem the growing gap between the United States and other countries in the West by allowing industry to benefit from innovations that were discovered in universities. Crucially, the act allowed universities to patent the results of federally funded research. Immediately after World War II, the United States was in the forefront of technological innovation, a combination of the remarkable lead it had taken during the war and the destruction of the scientific capability of its rivals, especially Germany, which had been the leading scientific and technological power in the early decades of the 20th century. However, by the 1960s, the lead had narrowed. Between 1966 and 1976, the U.S. patent balance decreased with respect to the United Kingdom, Canada, West Germany, Japan, and the Soviet Union. By 1975, it was negative for the last three. The proportion of the world's major technological innovations produced by the United States decreased from 80 percent in 1956–1958 to 59 percent in 1971–1973. The Bayh-Dole Act was an attempt to neutralize this gap by allowing U.S. universities, still some of the best in the world, to benefit from the enormous creativity and innovation that they nurtured. It allowed the universities to gain royalties from campus based inventions by allowing the transfer of exclusive control over such inventions to other parties. It also set a uniform patent policy for federal agencies and enabled small businesses and universities to retain title to inventions made through federally funded research.

Universities and Patents

As a result of the act, the number of patents granted to universities rose sharply. Before Bayh-Dole, universities produced roughly 250 patents a year (many of which were never commercialized); in fiscal year 1998, however, universities generated more than 4,800 patent applications. There was a rapid rise in technology transfer from U.S. universities to firms through such mechanisms as patents, licensing agreements, research joint ventures, and university-based start-ups. The most famous example of university-to-industry research transfer is the search engine Google, which was developed at Stanford. One key element in such partnerships is the creation of spin-offs, where one technology or its derivative is used to start a new company.

With regard to green technology, a major point of disagreement is the extent to which laws such as the Bayh-Dole Act that were designed in the 1980s to offset decline in federal funding of U.S. universities must be made valid. Clauses in the act that allow universities to grant exclusive rights are particularly problematic. For example, universities can give one sole company worldwide exclusive licenses, which can make the technology prohibitively expensive.

Green technology challenges the fundamental assumption behind the Bayh-Dole Act that the granting of exclusive patent rights to companies by the university is the only way for companies to profit from the innovations that they commercialize. While this is true to a certain extent in the pharmaceutical industry given the high costs of clinical trials, and the long and cumbersome process for Food and Drug Administration (FDA) approval, green technology is effective in combating climate change only if it is deployed on a large scale. High licensing fees and the scattering of innovations among several universities would mean that the technology would be expensive. Environmental activists point out that green technology is an attempt to reverse environmental degradation and combat climate change, but if the technology, due to intellectual property rights, is too expensive for the developing world, then the very purpose of developing it would be defeated.

Moreover, federal funding of green technology has increased under the administration of President Barack Obama, and it is expected that university research labs will create green technology, such as ways to store solar energy as fuel. While the current provisions of the act allow the federal government to opt for compulsory licensing of technology deemed to be in the interest of the public, this clause has been restricted to exceptional circumstances. Cheaper green products would encourage the cutting of carbon emissions. One suggestion is to make patent rights in the case of green technology the opposite of those in the pharmaceutical industry. This would mean that exclusive patent rights, rather than being the norm, would be the exception. Moreover, many innovations that had their origins in university labs, such as search engines or computers, and emerged out of federally funded research, were commercialized without the need of patents. Currently, it is argued that university technology transfers remain preoccupied with the generation of revenue rather than making new technology available to the public. Ironically, of the many patents that universities file, only a few—the "blockbuster" patents—raise any significant revenue. At present, for most universities, revenue generated from patents barely covers legal expenses. And given the fact that the key impetus behind the development of green technology is to disseminate environmentally friendly technology as widely as possible, fears have been expressed that the restrictive patent regime that is at the core of the extant university-industrial complex will hinder widespread adoption of new technology so crucial to climate change combat efforts.

One area in which the university-industrial complex has been exceptionally active is the biotechnology industry. The 1980s saw the emergence of the science-based model of university–industry collaborations with this new and promising industry as the prime example. Another result was the emergence of the "triple helix" of industry-state-university collaboration. In the United States, this alliance even physically changed the landscape. Computer engineering and biotech firms began to cluster around academic research centers such as Highway 128 around Boston, Massachusetts, Silicon Valley in the San Francisco Bay Area, California, the Research Triangle in North Carolina, and Los Angeles's Aerospace Alley. Such conglomerations of scientific expertise have become a feature of the U.S. landscape. University–industry linkages have also followed the needs of the regional economy, primarily because in the United States, education is under the control of the state and not the federal government. Thus, for example, the University of Georgia in its agricultural and engineering programs addresses the needs of the peanut-farming industry, and Stanford University has close links with Silicon Valley. Moreover, industry–university linkages are heavily concentrated in those areas where industry has immediate advantage, such as engineering, biotechnology, computer science and computer engineering, molecular biology, and clinical medicine. Universities that are primarily engineering colleges such as the Massachusetts Institute of Technology (MIT) and the Rensselaer Polytechnic University in New York have natural links with industry.

Key factors contributing to the growth of the university-industrial complex are the change in the nature of innovation and the emergence of new industries where knowledge, rather than heavy industrial infrastructure, is indispensable. Biotechnology, one of the areas where the new collaboration was visible, was a key factor in the rise of the university-industrial complex. This period also coincided with the rise of the software industry, which also depended crucially on innovation within university campuses. Informal links, for example, between venture capitalists—often alumni—and the university strengthen the university-industrial complex. The strength of the venture capital industry, the decentralization of control, the diversity of institutions, and the wide range of sources of support in the United States mean that the innovation landscape in the United States, in comparison to other countries that can depend on funds from the national government, is dynamic. In fact, after the recession of the 1980s, the industrial resurgence of the United States was crucially dependent on knowledge-based industries.

Challenges and Problems

While industry funding has enabled universities to confront immediate financial challenges, it has led to its own set of problems. Corporate funding means that universities are subject to political and economic pressure from which they were traditionally immune. Fears have been expressed about the dominance of the economic agenda and its potential to inhibit features that have always been seen as essential to the university—the importance of institutional autonomy, the erosion of the university's key role in culture and society, the decline of the pursuit of knowledge for its own sake, the sidelining of subjects that are not seen as economically viable such as the humanities, in short, the slow eroding of the universities' "public" role, differing conceptions of how public and private interests should be balanced in matters of patent and copyright, how industry funding may be affected by the dissemination of research antagonistic to corporate or industry goals, divided loyalties, and accountability. Critics have charged that professors who own stock in companies would be influenced by the fact. They have also charged that the Bayh-Dole Act gave corporations the rights to commercialize publicly funded research and restrict access to such inventions in a morass of patent law. One particular point of contention has been the precise length of time for innovations to be kept secret. Industry argues for a longer period of time, asserting that early disclosure would mean that potential rivals would benefit, while universities ideally should support shorter periods of time. A survey of 210 life-science companies, conducted in 1994 by researchers at Massachusetts General Hospital, found that 58 percent of those sponsoring academic research require delays of more than six months before publication.

In 2010, a proposal was placed before the U.S. Congress: the Federal Research Public Access Act (FRPAA), H.R. 5037 and S. 1373, seeks to ensure and maximize the public's return by delivering open online access to the results of research funded through 11 federal agencies no later than six months after publication in a journal.

Universities have also, in opposition to their traditional mission of producing knowledge for its own sake, begun to operate technology licensing offices and aggressively protect their intellectual property. The federal government, still the largest sponsor of research, would prefer shorter periods so that innovations can benefit the public. In fact, the Bayh-Dole Act gives the government "march-in" rights to license the invention to a third party, without the consent of the patent holder or original licensee, where it determines the invention is not being made available to the public on a reasonable basis. Thus, if the government deems fit, it can issue a compulsory license. Environmental activists have demanded that this be made the rule, rather than the exception, in the case of green technology.

Concern has also been expressed regarding the extent to which nations, locked in technological competition, actually benefit from such collaboration. The international nature of modern corporations means that, according to figures from the Organisation for Economic Co-operation and Development (OECD), subsidiaries of foreign-owned multinational enterprises account for a rising share in national research, development, and innovative expenditure in almost all OECD countries. The presence of guest researchers from foreign firms in U.S. laboratories, close collaborations with multinational industry, and the licensing of faculty inventions to foreign firms also mean that the innovation within the framework of the nation-state, which gives some countries a crucial advantage in a knowledge economy, is increasingly challenged by the borderless nature of the global knowledge society.

Finally, corporations often insist that they have a role in deciding how research funding is to be spent. Thus, for example, an agreement between University of California, Berkeley, and the Swiss pharmaceutical company Novartis in 1998 stipulated that in return for $25 million for research in the Department of Plant and Microbial Biology, Novartis would have the first right to negotiate licenses on roughly a third of the department's discoveries. Moreover, this right was not restricted to research that was funded by Novartis; it also included the results of research funded by state and federal sources. The agreement also granted Novartis two of five seats on the department's research committee, which was in charge of how the money would be used. It was the first time that a single company had provided fully one-third of the research budget of an entire department at a public university. Industry funding of universities has also led to some incongruous situations—Freeport McMoRan, a mining firm that has been accused of environmental misconduct in Indonesia, sponsors a chair in environmental studies at Tulane. Another area of concern with corporate funding has been suspicion that funding would be naturally directed to areas that had immediate commercial potential. For example, one of the largest grants to a university was that of the oil company British Petroleum (BP) that gave £19.5 million to Cambridge University in 1998 for an institute specializing in petroleum studies.

See Also: Biotechnology; Intellectual Property Rights; Science and Technology Policy; Science and Technology Studies.

Further Readings

Branscomb, Lewis M., Richard L. Florida, and Fumio Kodama. *Industrializing Knowledge: University-Industry Linkages in Japan and the United States.* Cambridge, MA: MIT Press, 1999.

Cunningham, Stuart, et al. "New Media and Borderless Education: A Review of the Convergence of Global Media Networks and Higher Education Provision." Canberra City, Australia: Department of Employment, Education Training and Youth Affairs, 1998.

de Faria, Pedro and Wolfgang Sofka. "Knowledge Protection Strategies of Multinational Firms—a Cross-Country Comparison." *Research Policy*, 39/7 (2010).

Hall, Peter. "The University and the City." *GeoJournal*, 41/4 (1997).

Press, Eyal and Jennifer Washburn. "The Kept University." *Atlantic Monthly* (March 2000).

Renault, Catherine Searle. "Academic Capitalism and University Incentives for Faculty Entrepreneurship." *Journal of Technology Transfer* (2006).

Simons, Maarten, et al. "Introduction: The University Revisited." *Studies in Philosophy and Education*, 26/5 (2007).

Sabil Francis
University of Leipzig, Germany,
École Normale Supérieure, Paris

Waste-to-Energy Technology

The terms *waste-to-energy* (WtE) or *energy from waste* (EfW) refer to the general approach of obtaining useful energy from different waste sources. Although there are several types of waste that are suitable for this sort of approach and several different processing routes, the combustion (incineration) of municipal solid waste currently accounts for the majority of processing capacity in this area, and is therefore the focus of this entry.

While there are obvious environmental benefits resulting from the exploitation of waste streams that would otherwise be sent to landfill sites, the primary motivation for waste incineration tends to be to reduce the volume and hazard of waste, as well as to capture and/or destroy potentially harmful substances. In this regard, it is useful to consider a possible hierarchy for sustainable waste management, which places incineration as the penultimate option, second only to landfilling. The complete framework places the emphasis on reducing, reusing, and recycling, before incinerating and landfilling. Hence, the amount of waste produced in the first place should be reduced through systemic efficiency improvements, thus reducing the primary material and energy requirements of production systems.

Types and Composition of Waste

Municipal solid waste (MSW) is the term used for the mainly solid waste collected from households and businesses, generally excluding hazardous types of waste and often sent to landfill in many countries. The composition of the waste varies by country, depending largely upon the degree of customer separation in the specific country. In some countries, there is little or no separation of different waste streams, and other countries such as Germany have advanced waste separation schemes, for example, into biomass, plastics and metals, paper, glass, and other waste.

In general, the biomass fraction (including kitchen waste, paper products, and wood) of MSW accounts for around half of the energy content, and the other half comes from plastic and rubber. The average energy content (i.e., net calorific value) of MSW is around 10 gigajoules per tonne (GJ/t), compared to about 24 GJ/t in coal.

Methods of Energy Conversion (Combustion and Other Routes for Fuel Production)

The general and most ubiquitous approach for energy conversion from waste streams is combustion of the waste to produce heat and/or electricity, although it is worth noting that not all waste incineration plants capture and use the energy content of the waste.

There are alternative means of obtaining fuels from various waste streams, however, that are potentially more efficient than using (combusting) the waste directly because the produced fuel can be burned at higher temperatures or used in fuel cells. An example of an alternative process is anaerobic digestion, whereby organic waste is broken down by bacteria in the absence of oxygen to produce carbon dioxide and methane—a mixture known as biogas. This biogas can be used directly or upgraded (i.e., separated) to be used as biomethane. Alternative options for upgrading include fermentation, for example, to produce ethanol, and several thermal processes to produce fuels in different forms. The first of these is gasification, which produces a hydrogen and carbon monoxide mix known as syngas from a solid carbon-containing substance such as coal or biomass. This syngas can either be used directly or upgraded into methane or liquid fuels. Another thermal process for the conversion of waste is pyrolysis, which also produces syngas from biomass, by heating organic material to high temperatures in the absence of oxygen. The focus in this entry, however, is on incineration.

The current state of the art for waste incineration involves the burning of received waste in an "as received" state, referred to variously as mass burning and moving grate technology. The moving grate was the most significant development in waste incineration in recent times, as it allows the waste to be continually fed into the furnace, and it accounts for around 90 percent of all European MSW incineration plants. The combustion process is basically an oxidation of the combustible materials in the waste, which has two main products: the combustion gases containing the available fuel as heat, and the solid residue that accounts for about 10 percent and 25 percent of the initial waste volume and weight, respectively.

The combustion process has to be carried out at temperatures above 850 degrees Celsius in Europe, in order to ensure that the organic elements and pollutants in the waste are destroyed, which is achieved by ensuring an excess amount of air is always present. The combustion air consists of primary and secondary air flows, whereby the former is the air required for combustion itself and is therefore supplied through the moving grate mechanism. The secondary air is usually injected through inlet nozzles at high velocity, which serves to ensure thorough turbulent mixing of the combustion gases and also acts as a temperature regulator.

The use of this energy for heating applications depends on the location of the incinerator in relation to local heat demands. Without a significant heat demand in the vicinity, as well as the necessary infrastructure, such as a district heating network, the production utilization of the heat might not be technically or economically feasible. This, combined with the fact that the siting of incinerators was until relatively recently quite contentious (mainly due to emissions, see below), means that extensive heat integration has often been difficult in the past. Nevertheless, the overall efficiency of pure electricity generation from waste-to-energy facilities is around 30 percent, so that it is desirable to produce electricity and heat in combined heat and power plants. In these cogeneration plants, overall efficiencies on the order of 80 percent are achievable.

Waste Treatment

The main waste products from waste incineration are the solid ash from the furnace and the emissions of various oxides and other pollutants to air; the treatment of the flue gases also produces some solid residues, as described below.

The solid waste fraction, known as incinerator bottom ash (IBA), also includes non-combustible components such as glass, iron, and other metals—typically, this fraction is around 20 percent. Metals such as iron are generally removed from the IBA before it is further processed; processing options include aging to improve its physical and chemical integrity and thermal or chemical processing. The main application areas for the processed IBA are in the construction industry, for example, as landfill cover, road foundations, and aggregate in asphalt. Depending upon the type of flue gas treatment employed, the composition of the fly ash (i.e., solid residue from flue gas treatment) varies. It may be washed off to produce an alkaline water solution for use in wet scrubbers for the removal of acidic gases, further processed to recover metal components, and/or vitrified in order to produce a product similar to glass, although this process is not widely employed within Europe. The products from the removal of the acidic gases from the flue gases may also be sold into several markets such as for road treatment, building products, and as a water softener in the case of sodium chloride.

The emissions to air include particulate matter and dust, acidic gases, heavy metals, carbon compounds, and greenhouse gases. Generally speaking, in recent years, the emissions from waste incineration plants have been drastically reduced. The flue gas treatment is carried out by a combination of processes, depending upon the gas's composition and the specific emissions to be removed. Particulate matter can be removed by a variety of filtration devices (e.g., wet and dry electrostatic precipitators, ESPs) and acidic gases such as sulfur dioxide are removed with alkali reagents (so-called scrubbers) that react with the acid to form a pH neutral product. Volatile organic compounds (VOCs), including dioxins and mercury compounds, can be removed by activated carbon, a solid reagent. Nitrous oxide emissions can be reduced by selective catalytic reduction devices (also used to remove dioxins).

There was previously great concern about the emission of dioxins and mercury from these plants but recent developments in legislation mean that this is no longer such a concern in the public eye. In particular, regulations relating to best available technology (BAT) in Europe and maximum available control technology (MACT) in the United States have resulted in large investments for retrofitting processes with dust-bag filters, and thus drastic reductions in these emissions.

Environmental Benefits

In general, incineration plants are environmentally favorable to landfilling, because of the overall lower carbon dioxide (CO_2) emissions, but also because of the avoided methane (CH_4) emissions. This is even more the case where energy recovery is employed at the site. As landfill waste decomposes it produces methane, which is a greenhouse gas with a much larger global warming potential that CO_2. In addition, there are problems with leaching associated with landfill sites, whereby local groundwater can be contaminated, particularly if the ground is permeable. This risk can be reduced or completely avoided through incineration of the waste that is suitable. In many cases, this biogas is collected at landfill sites, but even in this case, around 25 percent of this gas ends up escaping. In addition, nonrecyclable and nonorganic materials do not decay in landfill, but can be disposed of and their volume drastically reduced through incineration. It is estimated that the incineration of one ton of MSW instead of landfilling saves around one ton of CO_2 emissions.

Waste-to-energy plants generally have lower carbon dioxide emissions than conventional fossil fuel–fired plants, and the biomass fraction in the waste means that the combustion is effectively CO_2 neutral, as the released CO_2 was captured by the organic material during its lifetime. On the other hand, the nitrous and sulphur oxide (i.e., NO_x and SO_2) emissions from waste incineration plants are on a comparable level with coal and oil plants, and the flue gases must therefore be treated as described above.

Additional environmental benefits of incineration over landfilling include the significantly reduced space requirement to dispose of the waste, as well as the raw material savings in the construction industry as the solid residues are recycled.

Economics

The economics of waste-to-energy strongly depend on the local conditions, such as the availability of waste, the degree of preprocessing required, environmental legislation, and taxes and subsidies such as gate fees for disposing of waste. For example, heat prices from incineration in Denmark are competitive with heat from other sources, but this is largely due to the fact that the energy recovery facilities operate on a cost recovery basis as well as the taxes on landfill and incineration. In general, waste has an economic advantage over alternative fuels because it is already collected and centrally handled—but it should be considered that it is only really attractive as an alternative to landfill.

The costs of an EtW plant break down roughly equally between operational and capital expenditure, and the revenues consist mainly of the gate fee, which is a charge made to take the waste in the first place. Small revenue streams come from the sale of recycled materials from the waste treatment processes as well as, in some countries, from policies that offer an economic incentive for waste combustion (such as the Renewables Obligation Certificates in the United Kingdom).

Current Status Quo in Europe and the United States

The most common grate technology is that developed by Martin GmbH (of Munich, Germany) has a global installed capacity of around 60 million tonnes per annum (Mt/a). The Von Roll (of Zurich, Switzerland) mass burning process has around 32 Mt/a installed worldwide. All other processes together have a total estimated capacity of around 40 Mt/a, which means the total global installed capacity for waste incineration is around 130 Mt/a in about 600 facilities. There have been significant capacity additions in the decade following 1995, amounting to around 150 facilities and 16 Mt/a capacity.

In the United States, the total incineration capacity is around 30 Mt/a, or 23 percent of global capacity, and over two-thirds of this capacity is concentrated in just six states on the east coast, namely Connecticut, New York, New Jersey, Pennsylvania, Virginia, and Florida.

Within Europe the largest number of waste-to-energy plants, and therefore the greatest capacity, are to be found within France, Germany, the Netherlands, Switzerland, Italy, Spain, the United Kingdom, and Scandinavian countries. The average rate of MSW incineration across Europe is about 25 percent, but there is a large variation across Europe within the approximate range of 0–60 percent.

In particular, Denmark has set an example in this regard by legally prohibiting the land-filling of waste that is suitable for incineration since 1997 (effective through the Environmental Protection Act and the Waste Order, the latest version of which is from 2000). It is also prohibited to export waste for disposal, except for specific residues from flue gas treatment at waste-to-energy facilities. This means that, while the overall waste incineration capacity in Denmark is relatively low, as a whole the waste management system is the most efficient in Europe. In fact, Denmark by far has the highest waste incineration rate per capita (600 kilograms per annum [kg/a]) in Europe; in second place is Sweden with around 350 kg/a per person.

As well as this ban on landfilling certain waste, a crucial aspect in the development of this system has been local (municipal) management, which relies on cooperation between municipalities and establishment of not-for-profit (cost-covering) companies, which operate the waste incineration facilities. This ensures that the cost of incineration is low, in fact the gate fees in Denmark are some of the lowest in Europe, and the costs for individual households are comparable with the costs of waste collection bag (around 0.33 euros per week). Such landfill bans are now a part of the EU Waste Strategy and were therefore recently introduced in Germany and Sweden.

Conclusion

In the context of a sustainable waste strategy that primarily focuses on reducing, reusing, and recycling waste streams, waste incineration generally presents an environmentally favorable alternative to landfilling. The generation of heat and electricity at such plants relies on a local demand for the heat and the necessary infrastructure to transport it.

In order to keep track of developments in the amount of waste being produced, incinerator capacities need to be planned and expanded well in advance, given that the new plants take up to five years to plan and build. The lack of incineration capacity outside Denmark is partly due to uncertainty relating to the framework conditions and partly due to the fact that in many countries the public sector has contracted out the establishment of waste to energy facilities.

Especially in Europe, the expansion of the incineration sector is foreseen over the next few decades as alternatives are sought for waste diverted from landfill by the Landfill Directive.

See Also: Best Available Technology; Best Practicable Technology; Cogeneration; Design for Recycling; Electrostatic Precipitator; Flue Gas Treatment; Materials Recovery Facilities; Recycling; Thermal Heat Recovery.

Further Readings

International Solid Waste Association. "Energy From Waste: State of the Art Report," 5th ed. Vienna, Austria: Working Group on Thermal Treatment of Waste, 2006.

Psomopoulos, C. S., A. Bourka, and N. J. Themelis. "Waste to Energy: A Review of the Status and Benefits in the U.S.A." *Waste Management*, 29 (2009).

Young, Gary C. *Municipal Solid Waste to Energy Conversion Processes: Economic, Technical and Renewable Comparisons*. Hoboken, NJ: Wiley, 2010.

Russell McKenna
Karlsruhe Institute of Technology

WASTEWATER TREATMENT

The construction of centralized wastewater collection systems and treatment plants in the United States was one of the great public health success stories of the millennium. Today, however, many question the ability of the current model to meet the challenges of climate change and urban growth demands as well as the need to replace much of the existing infrastructure. This model uses vast quantities of clean water as transport, does a poor job of reusing nutrients, is increasingly energy intensive, disperses contaminants, disrupts natural ecological cycles, and is divorced from the bigger picture of water in the urban and natural environment. The Aspen Institute characterizes the issue this way: "While the traditional definition of water infrastructure focused mainly on physical structures associated with drinking water supply and distribution and disposal of wastewater and storm water, a sustainable water infrastructure [is needed that] integrates these traditional components with the protection and restoration of natural systems, conservation and efficiency, reuse and reclamation, and the active incorporation of new decentralized technologies, green infrastructure and low impact development to ensure the reliability and resilience of our water resources."

Over the past two decades, many communities have begun the work of shifting to a new water management paradigm that is integrated into the fabric of the city and also seeks to replicate the natural water cycle. Some of the change is coming from the traditional water utilities, while other efforts have come about by activist mayors and city staff, innovative architects, developers, and dedicated nonprofit organizations. The following describes the state of the art at the water utility, the building, and the community level.

Centralized Utility Innovations

The most farseeing utilities are beginning to view wastewater as a source of valuable water, energy, and nutrient resources, and it is a rare utility that is not now trying to become carbon neutral with respect to its own operations. As such, various "green" technologies are being pioneered.

Energy Recovery

Many wastewater utilities are producing energy from biosolids and food scraps obtained from the local solid waste management agency. The East Bay Municipal Utility District in the East San Francisco Bay Region has such a program. Food waste is composted in a closed-system anaerobic digester, similar to the way sewage sludge is broken down into biosolids at wastewater treatment plants. This process captures volatile organic compounds that contribute to ground-level ozone pollution and produce methane that can be used to generate electricity, and reduces the need for landfilling the food scraps. If half the food waste in the United States went through a similar process, there would be enough power each year for 2.5 million homes.

Nutrient Recovery

A variety of green technologies are being used to recapture the nitrogen and phosphorus in wastewater to keep these chemicals from receiving water bodies where they cause "dead zones." One West Coast firm contracts with wastewater utilities to turn these substances into struvite, a crystalline substance that can be sold commercially for fertilizer. Other utilities take the effluent, dry it out, and sell the biosolids to farmers as fertilizer. There have been some concerns, however, that the biosolids may contain heavy metals. Both efforts not only reduce the amount of pollutants going into the receiving water bodies, but if adopted on a widespread basis, could also reduce the need for mining phosphorus for agricultural fertilizer, a substance that is projected to be in scarce supply in 40 years.

Recycled Water for Nonpotable Use (Purple Pipes)

In water strained areas, recycled water has been used for decades for agricultural irrigation. Utilities are now able to produce "fit for purpose" water at either the central plant or at a satellite plant with varying degrees of quality to be used for industrial, agricultural, commercial, and residential use. Purple pipes are used to transport the water to its destination to distinguish it from traditional potable water supplies. There are many new residential developments where recycled water is piped back to residences for use in outside irrigation, including Santa Rosa, California; a golfing community in North Carolina; and Civano in Tucson, Arizona. Use of recycled water from the utility inside commercial build ings for toilet flushing is generally permitted (Irvine, California, pioneered this use), and California recently passed a law allowing recycled water to be used inside residential buildings. Although recycled water is a relatively low-energy-intensity resource, the high cost of dual plumbing inside buildings is a major barrier. Some water-short areas, such as Maricopa County, Arizona, are encouraging the retrofit of existing developments for recycled water. The water provider for Oakland, California, ran a purple pipe from its central plant to retrofit its headquarters building toilets for recycled water.

Direct and Indirect Potable Reuse

Visionaries in the water reuse community look to direct and indirect reuse of potable water processed at a central treatment plant as the next frontier for sustainable water. In the case of indirect potable reuse, recycled water is pumped back into a water source, such as Lake Mead in Nevada, or into a depleted underground aquifer, such as in Orange County. Orange County takes wastewater through a fourth treatment process, reverse osmosis, which effectively filters out pathogens and also many of the emerging contaminants, such as drugs and metals. The combined fresh and recycled water is then piped back to the water treatment plant and treated again to potable water levels before being distributed to consumers. Direct potable reuse would skip this last step and put the treated recycled water directly into the water distribution pipes already in existence. Since pipes are often viewed as 80 percent of the cost of the system, direct potable reuse might seem the most advantageous way to proceed; however, community acceptance lags.

Constructed Wetlands

The use of constructed wetlands is an area of considerable interest today because of wetlands' important role in carbon sequestration and their ability to naturally treat wastewater. Constructed wetlands can be designed for storm or sanitary wastewater treatment, and for surface or subsurface flows. They can be used in rural areas where land costs are low for secondary or tertiary wastewater treatment with low energy costs. In urban areas of the United States, where land costs are higher, they are more useful for tertiary treatment and to serve as buffers for wet weather effluent flows both from storm sewers and from combined sanitary and storm water sewers. Few of the constructed wetlands in the United States produce water for reuse, although the output from some experimental sites is being pumped back into groundwater sources. A large-scale constructed wetlands project provides primary, secondary, and tertiary wastewater treatment for the 19,000 residents of Arcata, California.

Sustainable Storm Water Technologies

Storm water can cause flooding, erosion, property damage, and the destruction of wildlife and the contamination of water supplies. The latter is particularly true in urban areas where the natural cycle of water percolating into the soil or evaporating into the atmosphere is disrupted by roofs, impervious pavements, and by the channeling of storm water into underground sewers. Most "green" innovative technologies being pioneered at the beginning of the 21st century involve increasing the retention of storm water in the location where it falls—a decentralized approach that requires the involvement of local general-purpose governments and the development community.

Rainwater Harvesting

Although the cost per acre-foot of water saved by rainwater harvesting can be high, these systems are popular for buildings seeking LEED certification because local regulations generally permit the use of untreated rainwater for irrigation and, if treated, for toilet flushing. Many feel, however, that if the true costs to all agencies over the lifetime of a facility were included, the costs would be competitive with more traditional water supply methods. Santa Fe, New Mexico, now requires all new construction to provide for rainwater harvesting, and the practice is now legal in Colorado, where formerly it was prohibited because of water rights laws. The state of Texas publishes a manual on how to install rainwater harvesting systems that is the "bible" in the United States. This practice is also common in European buildings. Membrane bioreactors (MBRs) are often used to treat rainwater for use of toilet flushing due to impurities.

Blue-Green Design

Nonpoint source pollution regulations in the United States, water shortages in the United States, and lack of land in the European Union (EU) have given rise to site and building designs that attempt to mimic the natural water cycle. These designs are referred to as either low-impact development (LID), water-sensitive urban design, or "blue-green design." Efforts to make driveways and parking lots more permeable include the use of swales instead of storm sewers, and rain gardens are commonly promoted as best practices for new construction. Green roofs and walls are also used to reduce storm water flows, which also helps reduce the urban heat island effect. The Ballard Library in Seattle, the Solaire Building in New York, the San Francisco Art Museum, and others in the United States and Europe have green roofs and walls. A green wall on a Parisian building has become a favorite spot for tourist picture-taking. Chicago has developed a manual for homeowners wishing to retrofit a green roof. Many of these developments also incorporate water into the design of the neighborhood, site, or building to bring people closer to water.

Stream and Creek Restoration

In efforts to reduce the speed of storm water flows and to promote infiltration, along with increasing wildlife diversity and the amenity value of the natural waterways in communities, many cities are restoring streams, creeks, and canals to their natural conditions. Zurich has a project to open its canals, while Copenhagen also has a canal reuse project. Leicester, England, is promoting a more natural water cycle with its Riverside Park project. Berkeley, California, is above-grounding and restoring portions of Strawberry Creek, a practice shared by many other U.S. cities. In South Korea, a large freeway constructed in the bed of a former river was dismantled and the river restored, to the delight of residents when dragonflies returned to the area.

Citywide Natural Storm Water Programs

In the United States, concerns about sewer overflows during storms have led some cities to develop decentralized natural solutions instead of building new billion-dollar facilities. Some Australian cities view the entire city as a catchment or watershed. Chicago initiated efforts to retrofit all its alleys with permeable pavers, while Philadelphia, Pennsylvania; Syracuse, New York; and Cleveland, Ohio, have developed comprehensive storm water programs that involve recreating wetlands, above-grounding streams, and extensive tree-planting programs, as well as water gardens for municipal facilities.

Adaptation to Extreme Weather Events

Rising coastal waters and inland-area flooding has triggered the preparation of adaptation plans by some governments that promote green technologies for wastewater. Miami–Dade County, Florida; Seattle, Washington; and New York City's planning efforts promote blue-green designs, street design along with more traditional levee infrastructure. For example, the 35 jurisdictions in Dade County are trying to coordinate their street design regulations to ensure that all street levels are raised to prepare for higher flood levels. In the Netherlands, Rotterdam is developing new communities that place garages on the first level in case of flooding, and emergency-retreat areas in higher areas. Storage areas for extreme storm water events are located in areas underneath the built developments. Floating cities are also being proposed.

Neighborhood and Site-Level Wastewater Technologies

Some of the most exciting advances are being made in the development of neighborhood and site-level wastewater technologies that have the potential to be significantly more eco-friendly than the traditional technology. They range from greywater recycling and black-water reuse to urine diversion and full nutrient and energy recovery.

Greywater Recycling

Greywater recycling is an area that has great potential for potable water demand reduction. Greywater is usually defined as wastewater from showers, washing machines, and hand-washing facilities, but both definitions and regulations on use vary from state to state and from country to country. Some distinguish between light grey and dark grey—the latter including wastewater from the kitchen. The Centers for Disease Control in the United States has embarked on a project with the U.S. Environmental Protection Agency (EPA) to develop national guidelines for the safe use of greywater that addresses both treated and untreated greywater:

- *Untreated greywater for irrigation:* Most jurisdictions in the United States permit the use of untreated greywater at the household level for irrigation, differing on the definition of "grey" and whether it can be used above ground or not. Current California regulations now define greywater as untreated water from showers, wash basins, and laundry, and it may be used above ground. Arizona's liberal state code allows localities to develop their own greywater regulations. This autonomy has resulted in many innovative building designs and retrofits. There are a variety of design and engineering issues with reusing untreated water for the design professional because of wide variation in the composition of the effluent, especially depending on whether laundry and kitchen water is used. Household-level greywater systems are common in Europe but on the other hand, Polderdrift, Arnhem in the Netherlands uses a centralized natural greywater recycling system for irrigation.
- *Treated greywater:* Generally, local governments in the United States permit greywater reuse for toilet flushing if it is treated. For example, the National Resource Defense Council Building in Santa Monica, California, uses water treated with reverse osmosis and ultraviolet for toilet flushing. On-site greywater systems for toilet flushing are common outside the United States. For example, in Annecy, France, an apartment building for 40 households and 120 users has such a system. The Eco-House, in Vauban, Germany, as well as the Miyako Hotel in Japan, and the Rouse Hill development in Sydney, Australia (projected for 50,000 homes), will have such systems. Many athletic facilities have been using greywater for toilet flushing for years, including the Millennium Dome in London and Sydney Olympic Park in Australia.

On-Site or Neighborhood Recycling of All Wastewater (Blackwater or Brownwater)

On-site or neighborhood recycling of all wastewater is an area where the technology is advancing rapidly. There are two approaches: one is the package plant that is suitable for both new construction and retrofits, and the second is the "boutique" or engineered approach using natural systems. Estimates for potable water demand reduction of such systems range from 50 to 80 percent.

- *Package plants for individual buildings or cluster developments:* The miniaturization of MBRs and rapidly declining costs have resulted in robust small-scale package wastewater treatment plants that can operate with a minimum of maintenance. These systems promise to be the next major breakthrough for widespread use in the United States, especially as they become less energy intensive. Most are not carbon neutral unless total life-cycle costs are included. On-site MBR plants today are typically used for high-value real estate development, for municipal facilities, for demonstration projects, or for eco-tourism, primarily outside the United States. Of note are a suite of projects in New York City for large-scale commercial and residential buildings, including the Solaire, a 250-unit residential building where potable water demand was reduced 50 percent by using an on-site MBR and a rainwater collection system. These plants have the potential to be used for "eco-blocks," for the retrofit of deteriorated sanitary sewer pipes in residential neighborhoods, redevelopment areas, shopping centers, or universities where a satellite treatment plant is appropriate and where a second set of distribution pipes can be installed.

- *On-site small-scale natural systems:* Sustainable water management systems in this category include the trademarked "Living Machine," which can be designed for grey- or blackwater. This is currently being used at Oberlin College, where flows are sometimes too low to support the system. The Living Machine has been recently installed in a nature center in a San Francisco park and in many new headquarters buildings of water utilities. Natural systems also include composting toilets, which require considerable space. Obtaining permits for these systems can be a challenge, however, and therefore a variety of guerrilla systems exist in individual homes. If a reliable prepackaged natural "plant" with a small footprint were readily available, many U.S. suburban single-family homes would likely install them.

Source Separation, Urine Diversion, Full Nutrient, and Resource Recovery

The most sustainable wastewater treatment and reuse model has been developed in Europe and in the developing world. This model attempts to replicate the natural water and wastewater cycle with the goal of "zero emissions." Black, yellow, and greywaters are separated at the source, treated, and reused with nutrient recovery on site. There are no permitted systems like this in the urban United States, although one has been proposed for Palo Alto, California.

- *Urine diversion:* Urine-diverting toilets are not produced in the United States but there are several manufacturers in Europe. Pilot projects include a four-plex in Switzerland, the Solar City project in Linz, Austria, and many in urban areas of Sweden. Other examples include the communities of Lanxmeer, Netherlands; Lambertsmuhle, Germany; and Svanholm, Denmark. Generally, urine diversion systems involve a cluster development with a separate pipe system for urine collection, often using vacuum toilets and centralized storage. Clogging of the urine pipes can occur but is easily resolved. Blackwater may be processed on-site or discharged into the municipal wastewater system. Urine is transported to local farms for reuse as fertilizer, where permitted.

- *Total resource recovery:* Total resource recovery systems include rainwater harvesting; separation of grey, black, and yellow water; the use of urine for fertilizer; and the recovery of food scraps and feces into energy. Some of these systems are in peri-urban areas where on-site natural treatment can be used for greywater, including constructed wetlands. Others are institutional buildings in more urban areas where MBRs are used to process blackwater on-site. OtterWasser GmbH piloted a system for a four-unit residential building in Lubeck, Germany. A vacuum-biogas system for blackwater and biowaste was implemented in Freiburg Vauban, Germany. GTZ headquarters and Huber Technologies have a similar zero-emission, cradle-to-cradle system with source separation and biogas production. The Jenfeld District in Hamburg has implemented a nutrient and energy recovery system that uses a truck transport system for various components, and plans are under way for additional districts.

Conclusion

A variety of practices of green wastewater treatment exist. The field is in the middle of a major paradigm shift whereby the entire system, including treatment and collection and distribution systems, is being rethought. This reorganization will require not only input from the water industry and its related professional and academic institutions, but also the dedicated participation of land use planners, architects, national and local officials, naturalists, biologists, landscape architects, and the nonprofit community. The good news is that, for the most part, the technologies exist to move into an integrated urban water management. What is needed to synthesize the individual components into a set of standard practices is a sea change in the institutions that control these systems, which can only happen by including the larger community.

See Also: Composting Toilet; Greywater; Membrane Technology; Rainwater Harvesting Systems; Water Purification.

Further Readings

Asano, Tak, Franklin Burton, Harold Leverentz, Ryujiro Tsuchihashi, and George Tchobanoglous. *Water Reuse: Issues, Technologies, and Applications*. Wakefield, MA: Metcalf & Eddy, 2008.
Aspen Institute. "Sustainable Water Systems: Step One—Redefining the Nation's Infrastructure Challenge" (2009). http://www.aspeninstitute.org/sites/default/files/content/docs/pubs/water_infra_final.pdf (Accessed July 2010).
Judd, Simon. *The MBR Book: Principles and Applications of Membrane Bioreactors for Water and Wastewater Treatment*, 2nd ed. Amsterdam, Netherlands: Elsevier Science, 2010.
Steinfeld, Carol and David Del Porto. *Reusing the Resource: Adventures in Ecological Wastewater Recycling*. Concord, MA: Ecowaters, 2004.
Von Munch, Elisabeth. "Basic Overview of Urine Diversion Components" (2009). http://betuco.be/compost/Urine-diversion-ecosan.pdf (Accessed July 2010).
Water Environment Research Foundation (WERF). "Knowledge Area: Decentralized Systems." http://www.werf.org/AM/Template.cfm?Section=Decentralized_Systems (Accessed January 2011).
Water Environment Research Foundation (WERF). "Knowledge Area: Stormwater." http://www.werf.org/AM/Template.cfm?Section=Stormwater3 (Accessed January 2011).

Vicki Elmer
University of California, Berkeley

WATER PURIFICATION

Access to an adequate water supply is a requirement for sustaining human life. Although much of the water on Earth is not fit for human consumption, the process of water purification renders the water to be safe for this use. Water purification involves a process by

New water-purification standards have caused some traditionally clean water sources, such as natural springs like Silver Glen Spring in Florida (pictured), to be re-evaluated.

Source: Sandra Friend/Forest Service, USDA

which undesired chemicals, organic and inorganic materials, and biological contaminants are removed from raw water. The purification of raw water takes place in various settings, including large-scale municipal water purification, portable and emergency water purification, industrial water purification, and small-scale distillation of water. All of these methods focus on the central goal of removing undesired chemicals and contaminants from water. Different methods are favored to produce water for a variety of purposes. One of the major purposes of water purification is to provide clean drinking water for humans. Many other uses of water purification, however, are utilized for other purposes such as meeting the needs of medical, pharmacology, chemical, and industrial applications for clean and potable water. These processes reduce the concentration of things such as suspended particles, parasites, bacteria, algae, viruses, and fungi as well as other dissolved and particulate material that water may have come in contact with as a result of the water cycle.

Determining Water Quality

The quality to which water must be purified is typically set by government agencies. Whether local, national, or international, these standards will typically outline a set of minimum and maximum concentrations of harmful contaminants that are allowed to be within the water for it to still be considered safe. Multiple processes have been developed to test water's contamination levels, as it is nearly impossible to examine water simply based on appearance. Regular household methods such as boiling water or utilizing an activated carbon filter, such as the popular Brita water filters, have been developed to remove water contaminants to a certain degree. Although these methods are popular because they can be used widely and inexpensively, they are insufficient in many cases because they do not address the possibility of more dangerous contaminants. New standards of purification have also caused a reevaluation of traditional water providers, such as natural spring water. Natural spring water had historically been considered clean for all practical purposes. Recently, however, natural spring water has come under scrutiny and is subject to treatment and batteries of tests. Chemical analysis that determines the content of water as well as the concentration of contaminants contained therein has become recognized as the only reliable means of determining which methods of purification are necessary.

The purification of water has taken on a new importance as, according to a 2007 report by the United Nations World Health Organization (WHO), 1.1 billion people lack access to a safe and reliable drinking water supply. Safe and reliable water has been defined as water of a quality that can be consumed without immediate or long-term harm from consumption.

Safe and reliable water is necessary if it is to be consumed by humans or used in the preparation of food. Lack of sanitary drinking water has been attributed to 88 percent of the 4 billion annual cases of diarrheal disease reported worldwide. Unsafe water, coupled with inadequate sanitation and hygiene, leads to approximately 1.8 million deaths due to diarrheal diseases each year. As a result of these diseases, a growing consensus has developed to provide potable water to as much of the Earth's population as possible.

Sources of Water

A source of water must be secured by most locations so that water is available for human use. The supply of water used by humans comes from a multitude of sources, including groundwater, lakes and reservoirs, rivers and canals, and desalinated water. Groundwater is the water that arises naturally from deep ground and rain that has occurred during rainfall, and is often referred to as spring water or well water. Lakes and reservoirs represent that water which is stored aboveground and that stands naturally at the head of rivers. Rivers and canals account for lowland surface waters and may contain bacteria and algae. Desalinated water is that which comes from the ocean after having salts and other minerals removed so that it is fit for human consumption. After a water supply is secured, water is tested so that it can be determined if it will be safe for human consumption. If necessary, water purification can be provided to allow this to be so.

Water Purification

After a source of water is secured, it must be tested for contaminants and then purified so that it may be used by humans. A great majority of water used in the Western world is treated at water treatment plants. To ensure purification, plants utilize different methods and protocols in the pretreatment process. Although not all plants use the same methods, depending on their size and the severity of the contamination, practices have been standardized to ensure general compliance to national and international standards. The majority of water is purified after it is pumped from its natural source or directed via pipelines into holding tanks. This process is important as the pipes and pumps must be constructed in a way that cannot cause contamination. After the water has been transplanted to a central location, the process of purification may begin.

Pretreatment

A series of methods are used to assure the cleanliness of water, a process that includes removing biological contaminants, chemicals, and other materials. The first step in this process, known as pretreatment, is screening. Screening is used to remove large debris such as sticks and trash from the water to be treated. Screening is generally used when purifying surface water. Surface water presents a greater risk of having been polluted with large amounts of contaminants from daily human interaction. The second step in purification occurs through storage. Water is held in reservoirs for long periods of time in order to allow natural purification to take place in a controlled environment. Water stored in these slow sand filters can be utilized in emergency situations such as short periods of drought or pollution in larger bodies of water that will take a longer process for decontamination. The third step in the purification process is known as preconditioning. Water with high mineral content is known as hard water. Although hard water is not harmful, water treatment plants use sodium carbonate to force out calcium carbonate, which is one of the main components in shells of marine life and is an active ingredient in agricultural lime. Preconditioning ensures the same consistency as soft water, which is achieved by general water purification. The final step of the pretreatment process is known as prechlorination. Prechlorination consists of purification plants chlorinating incoming water to minimize the growth of organisms in pipes and tanks. This process, though a standard in purification in many parts of the world, has begun to be discontinued due to the potential for the quality of water to be affected and concern for the environmental ramifications of chlorine.

Other Purification Steps

Once the various pretreatment processes have removed contaminants that are visible from the water, chemical treatment and refinement can occur. This process begins with flocculation, a step that clarifies water. Flocculation removes the cloudiness or haziness from fluids that can be caused by the growth of phytoplankton or the disturbance of land by human interactions. As human interaction is common near urban areas, turbidity (i.e., cloudiness) is common in water coming from locations near municipalities. Through flocculation, solids, also known as precipitates, are formed and removed through physical methods. These precipitates arise from the coagulation of small particles present in raw water being absorbed into the particles of the precipitate. After this process is begun, they are passed through a sand filter or a mixture of sand and anthracite. After water exits the flocculation basin, the next step in water purification is called sedimentation. Water enters a sedimentation basin, which is sometimes referred to as a decant pond and which controls wastewater. Sedimentation basins move treated waters along through the purification process while allowing remaining particles to settle. Sedimentation uses gravity to allow particles to settle to the bottom of the tank. Sludge forms that appear on the floor of the tank are removed and treated. From this basin, water is moved to the next step, filtration, which removes the remaining suspended particles and unsettled floc. Filtration involves moving water through a rapid sand filter, which removes organic compounds affecting taste and odor. While some rapid sand filters use rapid gravity, others employ pressure filters to achieve the same result. In areas where sufficient space exists, slow sand filters or lava filters are also used for filtration. Disinfection is the final step in water purification. During this step, harmful microbes are filtered out through adding disinfectant chemicals. Disinfection are used to kill pathogens, such as bacteria, viruses, and protozoa. Disinfection usually involves a form of chlorine, especially chloramines or chlorine dioxide. Chlorine is a toxic gas, resulting in some danger from release associated with its use. To avoid these risks, some water treatment plants use ozone, ultraviolet, or hydrogen peroxide disinfection instead of chlorine.

Alternative Methods of Purification

Certain areas of the world do not have access to water treatment plants and must use alternative methods of purification. These methods include boiling, granular activated carbon filtering, distillation, reverse osmosis, and direct contact membrane distillation. Although often more expensive than traditional water treatment methods, these processes do have the benefit of having fewer environmental risks than large-scale water purification. As a result of the risks from mistake and human error, many municipalities have moved from using chlorine and chloramines as an agent of purification. This is due to the risks associated with chlorine leaks as well as the highly corrosive nature of chloramines, which can dissolve the film within water pipelines, releasing lead, a known neurotoxin, into the water supply. As a result of these risks, investigations into safer methods continue.

See Also: Best Available Technology; Best Practicable Technology; Desalination Plants; Rainwater Harvesting Systems; Wastewater Treatment.

Further Readings

Crittendon, J. C., R. R. Trussell, D. W. Hand, K. J. Howe, and G. Tchobanoglous. *Water Treatment: Principles and Designs.* Hoboken, NJ: Wiley, 2005.

Kinkade-Levario, H. *Design for Water: Rainwater Harvesting, Stormwater Catchment, and Alternate Water Reuse.* Gabriola Island, British Columbia, Canada: New Society Publishers, 2007.

Tchobanoglous, G., F. L. Burton, and H. D. Stensel. *Wastewater Engineering: Treatment and Reuse.* New York: McGraw-Hill, 2003.

Stephen T. Schroth
Jordan K. Lanfair
Knox College

WHITE ROOFTOPS

Solar radiation reaches the Earth's surface partly as visible light, and is absorbed as heat energy by the surfaces that it strikes. Buildings conduct and radiate this heat into the building spaces that they cover through their walls and roofs. The color of the surface influences the amount of visible light reflected by it, and hence the amount of heat absorbed; for example, white or light colored walls reflect the incident visible light, and therefore, a smaller percentage of the heat energy is absorbed by them. The principle of using white rooftops or coating surfaces in light- or nearly white-colored materials, minimizes heat gain through exposed roof surfaces of buildings.

White rooftops scatter a percentage of visible light back into the atmosphere without converting it into heat energy. Visible light is not affected by the greenhouse gases present in the atmosphere, and can escape back into outer space. For surfaces receiving visible light, the capacity to store absorbed heat depends upon their mass or density. Therefore, black or dark roofs (e.g., asphalt-covered roofs or slate roofs) readily absorb large amounts of solar radiation (visible, infrared, and UV) as heat energy and store it for long periods because of their high density or mass. Further, heat flows from the hot surfaces of roofs to their cooler internal surfaces by conduction. The heated internal surfaces then radiate heat to the indoor air, increasing indoor temperatures and internal cooling loads. White rooftops, by their reflectivity, help to reduce the heat loads inside buildings and thus reduce energy demand for air conditioning.

White rooftops can also be high albedo roofs. Albedo is the ratio of the amount of solar radiation reflected from a surface to the total amount reaching that surface. Since infrared radiation (outside of the visible spectrum) conveys the largest amount of heat energy to a material, a high albedo material must be capable of reflecting the visible spectrum and re-radiating the heat absorbed from the infrared spectrum of solar radiation effectively. The color of a material indicates its reflectivity to light or the visible spectrum only, and therefore color and composition of a material are important factors that determine how much solar radiation is absorbed and radiated by that material.

A measure of evaluating both these factors is the solar reflectance index (SRI), which incorporates solar reflectance and emissivity in a single value. A standard black material (low reflectance equals 0.05, high emittance equals 0.90) has an SRI of "0," and a standard white material (high reflectance equals 0.80, high emittance equals 0.90) has an SRI of "100." This means that external building surfaces made from materials having a higher SRI value are less likely to cause overheating of internal spaces by heat absorbed from incident solar radiation on the building.

A roof can be designed to be an inherently cool roof for new buildings, and can be built using high-SRI value materials like white vinyl. Roofs on existing buildings can be modified to receive specifically designed high-SRI value white roof coatings to make the surface of the roof highly reflective. Such white roof coatings contain transparent polymeric materials such as acrylic, and white pigments to make them opaque and reflective. These coatings typically reflect 70 to 80 percent of the sun's energy. Despite their white appearance, special pigments strongly absorb the 5 percent or so of the sun's energy that falls in the ultraviolet range and help protect the polymer material and the substrate underneath from ultraviolet damage.

When selecting and applying white roof coatings, care should be taken to ensure that the substrate is smooth, clean, dry, and not prone to water leakage or accumulation. Coatings can be selected to retard accumulation of dirt and mildew growth. It is recommended that the thickness of the applied coating is one millimeter or more to increase the reflectance of the coating.

With energy costs continuing to rise, the popularity of eco-friendly and environmentally conservative building practices is increasing. The city of Los Angeles, California, has taken proactive steps to research strategies to combat the urban heat island (UHI) syndrome, and recommends that its residents apply reflective or white coatings to rooftops. U.S. Secretary of Energy Dr. Steven Chu also points out the benefits of white rooftops, and under his mandate, U.S. Department of Energy buildings have had new white rooftops installed.

See Also: Green Building Materials; Passive Solar; Sustainable Design; Zero-Energy Building.

Further Readings

Cool Roofs Materials Database. http://eetd.lbl.gov/coolroofs (Accessed July 2010).

Gartland, Lisa. *Heat Islands: Understanding and Mitigating Heat in Urban Areas*. London: Earthscan, 2008.

Halewood, Justin and Pieter De Wilde. "Cool Roofs and Their Application in the UK." Bracknell, UK: IHE BRE Press, 2010.

Swati Ogale
Independent Scholar

WIND TURBINE

Technologies for harnessing wind energy have been used for thousands of years, with the most recent innovation being the wind turbine. This entry reviews wind turbine technology, power estimation, turbine siting factors, environmental and socioeconomic concerns with wind turbines, and the extent of current wind power generation.

Early civilizations used wind power for sailing vessels and in milling grains and pumping water through the use of windmills. Perhaps the most commonly known windmill, the traditional north European tower windmill type, used elongated rectangular sails attached to four radial arms to turn a rotor for grinding or water pumping. The radial arms rotated around a horizontal axis and were positioned in the prevailing wind direction. These windmills were a common site on the landscape in Holland and in England, where it has been estimated over 10,000 windmills were located prior to the Industrial Revolution. While windmills convert wind energy into power for milling or pumping, the term *wind turbine* has been used to refer to the technology used to convert wind energy into electricity due to the similarity to gas and steam turbine technologies. Although the two terms are often used interchangeably, strictly speaking, windmills are not wind turbines and the correct use of the *wind turbine* term is increasingly being integrated into mainstream references to renewable energy.

There are two primary types of wind turbines currently used in implementation of wind energy systems: horizontal axis wind turbines (HAWTs) and vertical axis wind turbines (VAWTs). HAWTs are the most commonly used type of wind turbine and are usually made with two or three blades per turbine or with a disc containing many blades (multi-bladed type) attached to each turbine. VAWTs are able to harness wind blowing from any direction and are usually made with blades that rotate around a vertical pole. HAWTs are characterized as either high or low solidity devices, in which solidity refers to the percentage of the swept area containing solid material. High solidity HAWTs include the multi-bladed types that cover the total area swept by the blades with solid material in order to maximize the total amount of wind coming into contact with the blades. An example of the high solidity HAWT is the multi-bladed turbine used for pumping water on farms often seen in the landscapes of the American west. Low solidity HAWTs most often use two or three long blades and resemble aircraft propellers in appearance. Low solidity HAWTs have a low proportion of material within the swept area, which is made up for by a faster rotation speed used to fill up the swept area. Low solidity HAWTs are the most commonly used commercial wind turbines as well as the type most often represented through media sources. These HAWTs offer the greatest efficiency in electricity generation and therefore are the most cost-efficient type of wind turbine available today.

The lesser-used VAWTs include designs that vary in shape and method of harnessing wind energy. The Darrieus VAWT is the most implemented of vertical axis turbines and uses curved blades in a curved arch design. H-type VAWTs use two straight blades attached to either side of a tower in an H-shape, and V-type VAWTs use straight blades attached at an angle to a shaft, forming a V-shape. Most VAWTs are currently experimental and are not economically competitive with HAWTs. However, there is continued interest in research and development of VAWTs, particularly for building integrated wind energy systems.

Estimating Power Generation

The expected power generated from a particular wind turbine is estimated from a wind speed power curve derived for each turbine, usually represented as a graph showing the relation between power generated (kilowatts) and wind speed (meters per second). The wind speed power curve varies according to variables unique to each turbine such as number of blades, blade shape, rotor swept area, and speed of rotation. In order to determine how much wind energy will be generated from a particular turbine at a specific site location, the turbine's wind speed power curve needs to be coupled with the wind speed frequency distribution for its site. The wind speed frequency distribution is a histogram representing wind speed classes and the frequency of hours per year that are expected for each wind speed class. The data for these histograms are usually provided by wind speed measurements collected at the site and used to calculate the number of hours observed for each wind speed class.

The expected wind energy is calculated for the turbine at its site location from the wind speed frequency distribution by multiplying the number of hours of wind in each class by turbine power estimate at each wind speed from the wind speed power curve. These data can then be represented in a wind energy distribution, a histogram displaying the expected annual electricity generated from each wind speed class in kilowatts per hour (kWh). The total amount of energy expected to be generated from each turbine per year can then be summed from the data within its operating range. Rough initial estimates can also be made at potential wind turbine sites by using average annual wind speeds taken from site measurements or estimated from sources such as wind speed maps. This rough estimate of annual electric production in kilowatt-hours per year at a site can be calculated from a formula multiplying average annual wind speed, swept area of the turbine, the number of turbines, and a factor estimating turbine performance at the site. However, additional factors may decrease annual energy production estimates to varying degrees, including loss of energy due to distance of transmission as well as the availability of the turbine. Availability is a measure of how reliably the turbine will produce power when the wind is blowing, with most commercial turbines functioning at over 90 percent availability.

Gathering wind speed data at a site can be costly and time consuming; particularly if a high-quality assessment is to be conducted, longer periods of measurement are needed for accurate interpolation and statistical analysis of site measurements. However, secondary data sources of wind speed are increasingly becoming available at a variety of scales of study in the form of wind speed maps. For example, an extensive set of maps has been generated for the European Union countries at the regional scale and at the national scale in the United Kingdom and the United States. These kinds of maps are generally produced from geographic information systems (GIS) analysis of digital elevation models (DEMs), wind speed data, wind speed computer models, and variations in topography such as at ridge tops or in open sea. Wind resource maps are often used for a preliminary assessment of site conditions for wind turbines and wind farms, locations with multiple wind turbines managed together. The U.S. National Renewable Energy Laboratory (NREL) produces wind resource maps at the national and state scale displaying wind power density, or mean annual power available per square meter of swept area of a turbine at an altitude of 50 meters. NREL's wind resource maps use a standardized wind power class ranking system ranging from Class 1 to Class 7, with most viable wind resources being at Class 3 (300–400 watts per sq. m at 50 m altitude).

Concerns About Wind Turbines

A major concern of wind turbine siting relates to negative environmental impacts associated with noise, visual disturbance, and impacts on wildlife. Two kinds of noise associated with turbines are mechanical noise, that produced by its equipment such as its gearbox, and aerodynamic noise, that produced from the movement of air over the blades. Mechanical noise is often readily abated by simply altering mechanical components of turbines. Aerodynamic noise, often described as a "swishing" sound, is a factor of types of blades and speed of rotation. However, wind turbine noise in decibels has been found to be no greater than that experienced by traveling in a moving car and often on par with typical rural nighttime background noise. Other concerns involve flicker zones, areas where light may be reflected off the spinning blades, and areas of electromagnetic interference affecting television and radio signals within close proximity to turbines. Wind turbines have also been associated with killing birds, largely due to reports at one wind farm at Altamont Pass, California. However, it is estimated that one or two birds per turbine per year are affected by wind turbine placement, with the majority having no impact at all. However, a much higher number of bats have been reported killed by wind turbines. While the exact cause of these fatalities is unknown, the migration and mating behavior of migratory tree bats is the most widely discussed and is currently being researched by biologists.

The largest concerns with wind turbine placement are found in public perceptions of their visual impact and concerns about the return on investment in wind developments. For example, much controversy has surrounded the 130 turbine, 420 MW Cape Wind project off the coast of Massachusetts, approved for development in April 2010 by U.S. Department of Interior Secretary Ken Salazar after an eight-year federal review. Located offshore in Massachusetts's Nantucket Sound, the project has drawn opposition centered on potential negative effects the wind farm might have on viewsheds within range of tourist destinations and second homes along Cape Cod. Opponents with the greatest impacts on the project were Massachusetts Senator Edward M. Kennedy, who had a family home within range of the wind farm, and the Wampanoag Indian tribes, who objected due to concerns of disturbance to cultural resources. Concerns also arose over potential negative impacts on birds, historic sites, and the region's fishing industry. The Cape Wind project is significant to the development of offshore wind projects in the United States, as it has led to a call for a shorter permitting process and an increase in discussion of potential wind development in states along the northeast and mid-Atlantic coast. For example, East Coast governors established the Atlantic Offshore Wind Energy Consortium, and Governor Christie signed the State of New Jersey's Offshore Wind Economic Development Act in summer 2010.

Wind power generation is increasing around the globe, but most rapidly in Europe and in the United States, which holds nearly 80 percent of global wind power installations. Wind power capacity continues to double every three years, and is expected to employ 1 million persons by 2012. Worldwide wind power capacity was estimated to be 175,000 MW as of June 2010 by the World Wind Energy Association, with the United States (over 36,000 MW) and China (over 33,000 MW) having the greatest wind power capacity. Germany, Spain, and India comprise the remainder of the top five countries in terms of wind power capacity. Denmark continues to lead in percentage of power generated from wind at nearly 25 percentage. Globally, Asia has the largest percentage of new wind turbine installations, and China remains the largest market as it has more than doubled its installations four years in a row. Texas led the United States in 2009 in annual capacity with over 2200 MW, and Iowa led in the percentage of in-state power generation with 18.8 percent. In 2009, the United States had 2.4 percent of power generated from wind, and four states had over 10 percent of their power generation from wind, including Iowa, South Dakota, North Dakota, and Minnesota.

As of 2010, offshore wind generated just over 3 MW of wind power capacity and had been primarily developing in northern Europe, where 26 offshore wind farms have been constructed as of 2009. The United Kingdom leads in the number of offshore wind developments, followed by Denmark, the Netherlands, Belgium, Sweden, and Germany. Offshore wind developments are expected to grow in Europe and the United States; however, these projects rely upon government subsidies and incentives due to the high construction costs, lengthy permitting processes, and investment capital.

Wind turbines come in sizes ranging from small-scale, used for providing electricity to rural homes or cabins, to community-scale (see photos), used for providing electricity to a small number of homes within a community, to industrial scale, where the largest of turbines are placed in the form of a wind farm with multiple turbines in rural areas or offshore.

Greater public understanding of the differences in sizes of wind turbines and the facts of their environmental and economic impacts, through resources such as websites discussing wind initiatives, is needed to assist in further developing wind energy as a viable clean energy source.

See Also: Clean Energy; Ecological Modernization; Innovation; Participatory Technology Development.

Wind turbines come in sizes ranging from small-scale, used for providing electricity to rural homes or cabins, to community-scale, such as these wind turbines in Boone, North Carolina.

Source: Jessica McLawhorn

Further Readings

Appalachian State University. "North Carolina Wind Energy." http://www.wind.appstate.edu (Accessed September 2010).

Baban, S. M. J. and T. Parry. "Developing and Applying a GIS-Assisted Approach to Locating Wind Farms in the UK." *Renewable Energy*, 24/1 (2001).

Boyle, G., ed. *Renewable Energy: Power for a Sustainable Future*, 2nd ed. Oxford, UK: Oxford University Press, 2004.

European Wind Energy Association. http://www.ewea.org/index.php (Accessed September 2010).

National Renewal Energy Laboratory. "Wind Research." http://www.nrel.gov/wind (Accessed September 2010).

Rodman, L. C. and R. K. Meentemeyer. "A Geographic Analysis of Wind Turbine Placement in Northern California." *Energy Policy*, 34 (2006).

Christopher A. Badurek
Independent Scholar

Z

ZERO-ENERGY BUILDING

Six photovoltaic roofing products are being tested, which can help ZEBs produce their energy on-site to meet their electricity, heating, or cooling needs.

Source: National Institute of Standards and Technology, U.S. Department of Commerce

Zero-energy building (ZEB) or *net zero-energy building* are general terms applied to buildings with zero net energy consumption and zero carbon emissions, calculated over a period of time. ZEBs usually use less energy than traditional buildings, as well as generating their own energy on site to use in the building; hence, many are independent from the national (electricity) grid. They are in response to ever more stringent environmental standards, both regulatory and voluntary, introduced to address increasingly significant environmental issues such as climate change, resources, pollution, ecology, and population. Many people in developing countries (and elsewhere) already live in zero-energy buildings out of necessity, including huts, tents, and caves exposed to temperature extremes and without access to electricity. The energy in the building can be measured in many ways (e.g., cost, energy, or carbon emissions) and different views exist on the relative importance of energy production and energy conservation in achieving a net energy balance.

In the United Kingdom (UK), the government set out a target for all new homes to be zero carbon (i.e., energy) from 2016, with a progressive tightening of the energy-efficiency building regulations by 25 percent in 2010 and by 44 percent in 2013. Wide debate still exists, however, regarding the definition of "zero carbon," particularly around the eligibility of "off-site renewables." The UK Code for Sustainable Homes (CfSH) was launched in 2006 and has since become the most significant policy framework for reducing energy and environmental sustainability in house building. The method is based on the Building Research Establishment (BRE) EcoHomes version of the BREEAM (BRE Environmental Assessment Model) methodology adapted to relate closely to building regulations and government policy (see www.breeam.org). The method sets mandatory minimum standards for energy, water, construction and household waste, materials, and lifetime homes that relate to key government targets and policies. It has six levels, and since 2007, Level 6 has required a net zero carbon (energy) solution. The CfSH assesses the sustainability of a home by awarding points in nine design categories, including the following:

- Energy and carbon dioxide (including insulation, electric lighting, heating, domestic appliances)
- Materials (responsible sourcing of construction and finishing elements)
- Ecology (protection or enhancement of site habitats)
- Waste (household recycling facilities, site waste management, composting facilities)
- Pollution
- Health (specific room daylight factors, sound insulation, Lifetime Homes)
- Water (internal and external potable water consumption)
- Surface water runoff
- Management (Home User Guide, site information, Considerate Constructors Scheme)

The UK government recommends a three-step approach to reaching the zero carbon homes standard, based on the following:

- A high level of energy efficiency in the fabric and design of the dwelling
- Carbon compliance—a minimum level of carbon reduction to be achieved from on-site technologies (including directly connected heat networks)
- Allowable solutions—a range of measures available for achieving zero carbon beyond the minimum carbon compliance requirements

Energy Generation for ZEBs

ZEBs need to produce their own energy on site to meet their electricity and heating or cooling needs. Various microgeneration technologies may be used to provide heat and electricity to the building, including the following:

- *Solar*: Solar hot water, photovoltaics (PV)
- *Wind*: Wind turbines
- *Biomass*: Heaters/stoves, boilers, and community heating schemes
- *Combined heat and power* (CHP) *and micro CHP for use with the following fuels*: Natural gas biomass sewerage gas and other biogases
- *Community heating*: Including utilizing waste heat from large-scale power generation
- *Heat pumps*: Air source (ASHPs) and ground source (GSHP), and geothermal heating systems
- *Water*: Small-scale hydro power
- *Other*: For example, fuel cells using hydrogen generated from any of the above "renewable" sources

Many homebuilders have serious concerns about whether microgeneration and renewable energy technologies can deliver the energy generation requirements to produce adequate working, cost-effective ZEBs. Builders fear that owners and occupiers may not accept the required new technologies and could choose to retrofit energy-intensive appliances and systems, which would ultimately undermine the zero-energy objectives. There are further concerns that failure to maintain the new systems and technologies adequately may expose owners and occupiers to health and safety risks.

Home Energy Rating

A Home Energy Rating System (or HERS) is a measurement of a home's energy efficiency used primarily in the United States. HERS Ratings make use of a relative energy-use index called the HERS Index—a HERS Index of 100 represents the energy use of the "American Standard Building" and an index of zero indicates that the proposed building uses no net purchased energy (a zero-energy building). Other countries also have similar schemes; in Australia it is known as the House Energy Rating and is based upon a five-star rating, and in the UK, the Energy Performance Certificates are rated from A to G. HERS provide a standardized evaluation of a home's energy efficiency and expected energy costs. The evaluation is conducted in accordance with uniform standards and includes a detailed home energy use assessment, conducted by a state-certified assessor, using a suite of nationally accredited procedures and software tools. The rating can be used to judge the current energy efficiency of a home or to estimate the efficiency of a home that is being built or refurbished. The U.S. Department of Energy recommended Home Energy Ratings report will typically contain the following:

- Overall rating score of the house
- Recommended cost-effective energy modifications
- Estimates of the cost, annual savings, and useful projected life of the modifications
- The potential improved rating score after the installation of recommended modifications
- The estimated projected annual energy costs for the existing home, before and after the modifications

The UK Energy Performance Certificates (EPC) Scheme

EPCs are required in England and Wales for domestic properties with three or more bedrooms. Rental properties, which have a certificate valid for 10 years, have been required on new tenancies after 2008. Display Energy Certificates (DECs) are also required for larger public buildings, thus enabling everyone to see how energy efficient the country's public buildings are. The DEC has to be displayed at all times in a prominent place clearly visible to the public—and it is accompanied by an Advisory Report that lists cost-effective measures to improve the energy rating of the building. The introduction of EPCs has been controversial, however, and has been opposed by many in the UK housing industry such as the Royal Institute of Chartered Surveyors.

See Also: Appliances, Energy Efficient; Green Building Materials, Microgeneration, Sustainable Design.

Further Readings

Department for Communities and Local Government (CLG). "Code for Sustainable Homes: Technical Guide—2010. http://www.communities.gov.uk/publications/planningand building/codeguide (Accessed April 2011).

Department for Communities and Local Government (CLG). "Homes for the Future: More Affordable, More Sustainable." (2007). http://www.communities.gov.uk/documents/ housing/pdf/439986.pdf (Accessed June 2010).

Directgov. "Energy Performance Certificates." http://epc.direct.gov.uk (Accessed June 2010).

Goodier, C. and W. Pan. *The Future of Housing*, London: RICS Book, 2010.

National Home Energy Rating/National Energy Services. http://www.nher.co.uk (Accessed April 2011).

NHBC Foundation. "Zero Carbon Compendium, Who's Doing What in Housing Worldwide." http://www.nhbcfoundation.org/Portals/0/Zero%20Carbon%20 Compendium%202010%20Web.pdf (Accessed June 2010).

Residential Energy Services Network (RESNET). http://www.resnet.us (Accessed June 2010).

Chris Goodier
Loughborough University

Green Technology Glossary

A

Actor-Network Theory: In social studies of science and technology, the term is used to describe an analytical approach that shows how networks of relations between material and discursive explain society.

Air Pollution: Contaminants or substances in the air that interfere with human health or produce other harmful environmental effects.

Alternative Energy: Usually environmentally friendly, this is energy from uncommon sources such as wind power or solar energy, not fossil fuels.

Annual Solar Savings: The annual solar savings of a solar building is the energy savings attributable to a solar feature relative to the energy requirements of a nonsolar building.

Appropriate Technology: A normative approach to designing and implementing technological systems, often simpler, less capital- and energy-intensive ones.

B

Backflow/Back Siphonage: A reverse flow condition created by a difference in water pressures that causes water to flow back into the distribution pipes of a drinking water supply from any source other than the intended one.

Batch Heater: This simple passive solar hot water system consists of one or more storage tanks placed in an insulated box that has a glazed side facing the sun.

Best Available Technology: A term used to refer to the most effective measures (according to U.S. Environmental Protection Agency guidance) for controlling small or dispersed particulates and other emissions from sources such as roadway dust, soot and ash from woodstoves, and open burning of brush, timber, grasslands, or trash.

Biogas: A combustible gas created by anaerobic decomposition of organic material, composed primarily of methane, carbon dioxide, and hydrogen sulfide.

Biomass: Any organic matter that is available on a renewable basis, including agricultural crops and agricultural wastes and residues, wood and wood wastes and residues, animal wastes, municipal wastes, and aquatic plants.

C

Cadmium Telluride: A thin-film semiconductor used in photovoltaic technologies, typically paired with cadmium sulfide.

Carbon Footprint: A popular term describing the impact a particular activity has on the environment in terms of the amount of climate-changing carbon dioxide and other greenhouse gases it produces. A person's carbon footprint is the amount of greenhouse gases that his or her way of life produces overall. It is also a colloquialism for the sum total of all environmental harm an individual or group causes over their lifetime. People, families, communities, nations, companies, and other organizations all leave a carbon footprint.

Carbon Offsets: Financial instruments, expressed in metric tons of carbon dioxide equivalent, that represent the reduction of carbon dioxide or an equivalent greenhouse gas. Carbon offsets allow corporations and other entities to comply with caps on their emissions by purchasing offsets to bring their totals down to acceptable levels. The smaller voluntary market for carbon offsets exists for individuals and companies that purchase offsets in order to mitigate their emissions by choice. There is a great deal of controversy over the efficacy and truthfulness of the offsets market, which is new enough that, in a best-case scenario, the kinks have not yet been worked out, while in the worst, it will turn out to be a dead end in the history of environmental reform.

Chemical Stressors: Chemicals released to the environment through industrial waste, auto emissions, pesticides, and other human activity that can cause illnesses or death in plants and animals.

Clean Power Generator: A company or other organizational unit that produces electricity from sources that are thought to be environmentally cleaner than traditional sources. Clean or green power is usually defined as power from renewable energy such as wind, solar, and biomass energy.

Concentrating (Solar) Collector: A solar collector that uses reflective surfaces to concentrate sunlight onto a small area, where it is absorbed and converted to heat or, in the case of solar photovoltaic (PV) devices, into electricity.

Contingency Plan: A document setting out an organized, planned, and coordinated course of action to be followed in case of a fire, explosion, or other accident that releases toxic chemicals, hazardous waste, or radioactive materials that threaten human health or the environment.

Copper Indium Gallium Diselenide: A thin-film semiconductor used in the photovoltaic industry, typically paired with cadmium sulfide or zinc compounds.

Crystalline Silicon Photovoltaic Cell: A type of photovoltaic cell made from a single crystal or a polycrystalline slice of silicon. Crystalline silicon cells can be joined together to form a module (or panel).

D

Dioxin: Any of a family of compounds known chemically as dibenzo-p-dioxins. Concerns about it arise from its potential toxicity as contaminants in commercial products. Tests on laboratory animals indicate that it is one of the more toxic anthropogenic (man-made) compounds.

E

Earthship: A type of sustainable design for housing utilizing passive solar heating and recycled materials for construction found in the U.S. Southwest desert, often associated with a particular design from Taos, New Mexico.

Energy: The capability of doing work; different forms of energy can be converted to other forms, but the total amount of energy remains the same.

Energy Star: A joint program formed between the U.S. Environmental Protection Agency and the U.S. Department of Energy to identify and label high-efficiency building products.

Epidemiology: Study of the distribution of disease or other health-related states and events in human populations, as related to age, sex, occupation, ethnicity, and economic status in order to identify and alleviate health problems and promote better health.

Episode (Pollution): An air pollution incident in a given area caused by a concentration of atmospheric pollutants under meteorological conditions that may result in a significant increase in illnesses or deaths. May also describe water pollution events or hazardous material spills.

Exposure: The amount of radiation or pollutant present in a given environment that represents a potential health threat to living organisms.

Extended Producer Responsibility: A mode of product stewardship that encourages sustainable design by making manufacturers responsible for their products at the end of their useful like.

F

Federal Implementation Plan: Under current U.S. law, a federally implemented plan to achieve attainment of air quality standards, used when a state is unable to develop an adequate plan.

Finished Water: Water is "finished" when it has passed through all the processes in a water treatment plant and is ready to be delivered to consumers.

Fluorocarbons (FCs): Organic compounds analogous to hydrocarbons in which one or more hydrogen atoms are replaced by fluorine. Originally used in the United States as a propellant for domestic aerosols, FCs are still found in coolants and some industrial processes. FCs containing chlorine are called chlorofluorocarbons (CFCs). They are thought to be allowing more harmful solar radiation to reach the Earth's surface by modifying the ozone layer.

Fuel Economy Standard: The Corporate Average Fuel Economy Standard (CAFE) made effective in 1978. It enhanced the U.S. fuel conservation effort imposing a miles-per-gallon floor for motor vehicles.

Fugitive Emissions: Emissions not caught by a capture system.

G

Geothermal Energy: Any and all energy produced by the internal heat of the Earth.

Global Warming: An increase in the near-surface temperature of the Earth. Global warming has occurred in the distant past as the result of natural influences, but the term is most often used to refer to the warming predicted to occur as a result of increased emissions of greenhouse gases. Scientists generally agree that the Earth's surface has warmed by about 1 degree Fahrenheit in the past 140 years. The Intergovernmental Panel on Climate Change (IPCC) recently concluded that increased concentrations of greenhouse gases are causing an increase in the Earth's surface temperature, and that increased concentrations of sulfate aerosols have led to relative cooling in some regions, generally over and downwind of heavily industrialized areas.

Greenhouse Effect: The warming of the Earth's atmosphere attributed to a buildup of carbon dioxide or other gases; some scientists think that this buildup allows the sun's rays to heat the Earth, while making the infrared radiation atmosphere opaque to infrared radiation, thereby preventing a counterbalancing loss of heat.

Greenhouse Gas Emissions: Any emissions that are released by humans (though naturally occurring in the environment), mainly through the combustion of fossil fuels, and that have a warming potential as they persist in the atmosphere, contributing to the greenhouse effect.

Greenwashing: A marketing ploy for businesses to jump onto the green movement bandwagon. They are not genuinely interested in sustainability, but are simply trying to improve their standing with the public by paying lip service. A company interested in "going green" for public relations reasons is greenwashing.

H

Heat Absorbing Window Glass: A type of window glass that contains special tints that cause the window to absorb as much as 45 percent of incoming solar energy, to reduce heat gain in an interior space.

HEPA: A high-efficiency particulate air filter.

High-Level Nuclear Waste Facility: Plant designed to handle disposal of used nuclear fuel, high-level radioactive waste, and plutonium waste.

Household Hazardous Waste: Hazardous products used and disposed of by residential as opposed to industrial consumers. Includes paints, stains, varnishes, solvents, pesticides, and other materials or products containing volatile chemicals that can catch fire, react or explode, or that are corrosive or toxic.

Hybrid Vehicle: Vehicle that uses both a combustible form of fuel (gasoline, ethanol, etc.) and an electric motor to power it. Hybrid vehicles use less gasoline than a traditional combustion engine, and some even have an electric plug-in to charge the battery.

I

Incident Command Post: A facility located at a safe distance from an emergency site where the incident commander, key staff, and technical representatives can make decisions and deploy emergency manpower and equipment.

Insolated Solar Gain System: A type of passive solar heating system where heat is collected in one area for use in another.

Irradiation: Exposure to radiation of wavelengths shorter than those of visible light (gamma, x-ray, or ultraviolet), for medical purposes, to sterilize milk or other foodstuffs, or to induce polymerization of monomers or vulcanization of rubber.

ISO: The International Organization for Standardization is an international organization that coordinates industrial and commercial standards for products and practices.

L

LEED (Leadership in Energy and Environmental Design): An organization that has created the Green Building Rating System that encourages and accelerates global adoption of sustainable green building and development practices through the creation and implementation of universally understood and accepted tools and performance criteria.

Life-Cycle Analysis: The assessment of the environmental impacts of a product across all stages of its development, from resource extraction through production, use, and disposal.

Lifetime Exposure: Total amount of exposure to a substance that a human would receive in a lifetime (usually assumed to be 70 years).

M

Materials Recovery Facility (MRF): A facility that processes residentially collected mixed recyclables into new products available for market.

Maximally (or Most) Exposed Individual: The person with the highest exposure in a given population.

Megawatt: One thousand kilowatts, or 1 million watts; standard measure of electric power plant generating capacity. It is assumed that 1 MW is enough to power 700 to 1,000 homes.

Module: The smallest self-contained, environmentally protected structure housing interconnected photovoltaic cells and providing a single DC electrical output; also called a panel.

N

Net Metering: A method of crediting customers for electricity that they generate on site in excess of their purchased electricity consumption. Customers with their own generation offset the electricity they would have purchased from their utility. If such customers generate more than they use in a billing period, their electric meter turns backward to indicate their net excess generation. Depending on individual state or utility rules, the net excess generation may be credited to the customer's account (in many cases at the retail price), carried over to a future billing period, or ignored.

Net-Zero Energy: Characteristic of a building that produces as much energy as it consumes on an annual basis, usually through incorporation of energy production from renewable sources such as wind or solar.

NIMBY: An acronym for "not in my backyard" that identifies the tendency for individuals and communities to oppose the siting of noxious or hazardous materials and activities in their vicinity. It implies a limited or parochial political vision of environmental justice.

Nuclear Reactors and Support Facilities: Uranium mills, commercial power reactors, fuel reprocessing plants, and uranium enrichment facilities.

P

Panel (Solar): A term generally applied to individual solar collectors, and typically to solar photovoltaic collectors or modules.

Persistent Toxic Chemicals, Persistent Pollutants: Detrimental materials, like Styrofoam or DDT, that remain active for a long time after their application and can be found in the environment years, and sometimes decades, after they were used.

Photochemical Smog: Air pollution caused by chemical reactions of various pollutants emitted from different sources.

Photovoltaic (Solar) Cell: Treated semiconductor material that converts solar irradiance to electricity. When grouped, they are called solar arrays, modules, or panels.

Planned Obsolescence: The art of making a product break/fail after a certain amount of time: not so soon that you will blame the manufacturer, but soon enough for you to buy another one and make more profit for them.

Point-of-Use Treatment Device: Treatment device applied to a single tap to reduce contaminants in the drinking water at the faucet.

Pollution: Generally, the presence of a substance in the environment that because of its chemical composition or quantity prevents the functioning of natural processes and produces undesirable environmental and health effects. Under the U.S. Clean Water Act, for example, the term has been defined as the man-made or man-induced alteration of the physical, biological, chemical, and radiological integrity of water and other media.

Pollution Prevention: Identifying areas, processes, and activities that create excessive waste products or pollutants in order to reduce or prevent them through alteration, or eliminating a process. Such activities, consistent with the U.S. Pollution Prevention Act of 1990, are conducted across all EPA programs and can involve cooperative efforts with such agencies as the U.S. Departments of Agriculture and Energy.

Polychlorinated Biphenyls: A group of toxic, persistent chemicals used in electrical transformers and capacitors for insulating purposes, and in gas pipeline systems as lubricant. The sale and new use of these chemicals, also known as PCBs, was banned in the United States by law in 1979.

Postconsumer Waste: In the recycling business, material that has already been used and discarded by consumers, as opposed to manufacturing waste. Using products with "postconsumer" recycled content actually keeps waste out of landfills and incinerators, unlike "postindustrial" recycled content, most of which would get recycled anyway.

Pyrolysis: A means of heating organic matter to decompose in the absence of oxygen. Used to make biochar.

R

Radioactive Waste: Any waste that emits energy as rays, waves, streams, or energetic particles. Radioactive materials are often mixed with hazardous waste from nuclear reactors, research institutions, or hospitals.

Refuse Reclamation: Conversion of solid waste into useful products; such as composting organic wastes to make soil conditioners, or separating aluminum and other metals for recycling.

Reverse Osmosis: A treatment process used in water systems by adding pressure to force water through a semipermeable membrane. Reverse osmosis removes most drinking water contaminants; also used in wastewater treatment. Large-scale reverse osmosis plants are being developed.

S

Science and Technology Studies: A field of inquiry that is concerned about the role of culture, politics, and society on science and technology, and vice versa.

Semiconductor: Any material that has a limited capacity for conducting an electric current.

Silicon: A chemical element, of atomic number 14, that is semi-metallic, and an excellent semiconductor material used in solar photovoltaic devices; commonly found in sand.

Single-Crystal Material: In reference to solar photovoltaic devices, a material that is composed of a single crystal or a few large crystals.

Smart Grid: An electricity generation infrastructure that utilizes information technology to improve the efficiency of the system.

Smog: Air pollution typically associated with oxidants.

Superconducting Magnetic Energy Storage (SMES): SMES technology uses the superconducting characteristics of low-temperature materials to produce intense magnetic fields to store energy.

Sustainability: Process designed to give support or relief to, carry, withstand, and meet the needs of the present without compromising the ability of the future to meet its needs.

T

Technology-Based Standards: Industry-specific effluent limitations applicable to direct and indirect sources developed using statutory factors, but not including water-quality effects.

Temperature Coefficient (of a Solar Photovoltaic Cell): The amount that the voltage, current, and/or power output of a solar cell changes due to a change in the cell temperature.

Teratogen: A substance capable of causing birth defects.

Teratogenesis: The introduction of nonhereditary birth defects in a developing fetus by exogenous factors such as physical or chemical agents acting in the womb to interfere with normal embryonic development.

Thermodynamic Cycle: An idealized process in which a working fluid successively changes its state (from a liquid to a gas and back to a liquid) for the purpose of producing useful work or energy, or transferring energy.

Thermodynamics: The study of the transformation of energy from one form to another, and its practical application.

Toxicity: The degree to which a substance or mixture of substances can harm humans or animals.

Tracking Solar Array: A solar energy array that follows the path of the sun to maximize the solar radiation incident on the cell's surface.

Turbine: A device for converting the flow of a fluid (air, steam, water, or hot gases) into mechanical motion.

U

Unglazed Solar Collector: A solar thermal collector that has an absorber without a glazed covering, like those used to heat swimming pools.

V

Variance: Government permission for a delay or exception in the application of a given law, ordinance, or regulation.

VOCs (Volatile Organic Compounds): Gases emitted from liquid or solid substances that may cause short-term and long-term harmful health effects. Examples of products containing VOCs include paints and lacquers, paint strippers, cleaning supplies, pesticides, building materials and furnishings, office equipment such as copiers and printers, correction fluids and carbonless copy paper, graphics and craft materials including glues and adhesives, permanent markers, and photographic solutions.

Vulnerability Analysis: Assessment of elements in the community that are susceptible to damage if hazardous materials are released.

W

Waste Minimization: Measures or techniques to reduce waste generated during industrial production processes; also refers to recycling and other efforts to reduce the amount of waste.

Water Pollution: Includes chemicals and debris that render water unusable for natural habitat, human consumption, and recreation.

Water Turbine: A turbine that uses water pressure to rotate its blades, usually for generating electricity.

Water Wheel: A wheel that is designed to use the weight and/or force of moving water to turn it, primarily to operate machinery or grind grain.

Dustin Mulvaney
University of California, Berkeley

Sources: U.S. Environmental Protection Agency (http://www.epa.gov/OCEPAterms), U.S. Energy Information Administration (http://www.eia.doe.gov/tools/glossary)

Green Technology
Resource Guide

Books

Allen, Edward. *How Buildings Work: The Natural Order of Architecture*. New York: Oxford University Press, 1995.

Anastas, Paul and John Warner. *Green Chemistry: Theory and Practice*. New York: Oxford University Press, 1998.

Baker, Nick. *Passive and Low Energy Building Design for Tropical Island Climates*. London: Commonwealth Secretariat Publications, 1987.

Beauchamp, T. and J. Childress, eds. *Principles of Biomedical Ethics*. Oxford, UK: Oxford University Press, 2008.

Beck, Ulrich. *Risk Society: Towards a New Modernity*. Newbury Park, CA: Sage, 1992.

Behling, Sophia and Stefan. *Sol Power: The Evolution of Solar Architecture*. Munich: Prestel, 1996.

Bell, Daniel. *The Coming of Post-Industrial Society*. New York: Harper Colophon, 1974.

Berry, R. *The Ethics of Genetic Engineering*. London: Routledge, 2007.

Bijker, W. E. *Of Bicycles, Bakelites, and Bulbs: Toward a Theory of Sociotechnical Change*. Cambridge, MA: MIT Press, 1995.

Bijker, W. E., T. P. Hughes, and T. J. Pinch. *The Social Construction of Technological Systems: New Directions in the Sociology and History of Technology*. Cambridge, MA: MIT Press, 1987.

Bijker, W. E. and J. Law. *Shaping Technology/Building Society: Studies in Socio-Technical Change*. Cambridge, MA: MIT Press, 1992.

Bookchin, Murray. *Post-Scarcity Anarchism*. Oakland, CA: AK Press, 2004.

Burley, J., ed. *The Genetic Revolution and Human Rights*. Oxford, UK: Oxford University Press, 1999.

Burley, J. and J. Harris, eds. *A Companion to Genethics*. Oxford, UK: Blackwell, 2002.

Carson, Rachel. *Silent Spring*. Boston, MA: Houghton Mifflin, 1962.

Chiras, Dan. *The Homeowner's Guide to Renewable Energy: Achieving Energy Independence Through Solar, Wind, Biomass and Hydropower*. Gabriola Island, British Columbia, Canada: New Society Publishers, 2006.

Diamond, Jared. *Guns, Germs, and Steel: The Fates of Human Societies*. New York: W. W. Norton, 1999.

Drexler, Eric K. *Engines of Creation: Challenges and Choices of the Last Technological Revolution.* New York: Doubleday, 1986.

Dyson, A. and J. Harris, eds. *Ethics and Biotechnology.* London: Routledge, 1994.

Ellul, Jacques. *The Technological Society.* New York: Vintage Books, 1964.

Ellul, Jacques. *The Technological System.* New York: Continuum, 1980.

Feenberg, Andrew. *Questioning Technology.* London: Routledge, 1999.

Finegold, D., et al. *Bioindustry Ethics.* Oxford, UK: Elsevier, 2005.

Fox, Michael and Miles Kemp. *Interactive Architecture.* New York: Princeton Architectural Press, 2009.

Glicksman, Leon R. and Juintow Lin. *Sustainable Urban Housing in China: Principles and Case Studies for Low-Energy Design.* New York: Springer, 2006.

Glover, J. *Choosing Children: Genes, Disability, and Design.* New York: Oxford University Press, 2008.

Glover, J. *What Sort of People Should There Be?* New York: Penguin, 1984.

Goodman, Paul. *New Reformation: Notes of a Neolithic Conservative.* Oakland, CA: PM Press, 2010.

Goodship, Vanessa, ed. *Management, Recycling and Reuse of Waste Composite.* Cambridge, UK: Woodhead, 2009.

Gordijn, B. and R. Chadwick, eds. *Medical Enhancement and Posthumanity.* Dordrecht, Holland: Springer, 2008.

Gordon, Richard and Joseph Seckbach, eds. *The Science of Algal Fuels: Psychology, Geology, Biophotonics, Genomics and Nanotechnology.* New York: Springer, 2010.

Harris, J. *Clones, Genes and Immortality.* Oxford, UK: Oxford University Press, 1998.

Harris, J. *Enhancing Evolution.* Princeton, NJ: Princeton University Press, 2007.

Heidegger, M. *The Question Concerning Technology and Other Essays.* New York: Harper & Row, 1977.

Herlock, J. H. Cogeneration: *Combined Heat and Power Systems: Thermodynamics and Economics.* Oxford, UK: Pergamon Press, 1987.

Herring, Horace and Steve Sorrell, eds. *Energy Efficiency and Sustainable Consumption: The Rebound Effect.* New York: Palgrave Macmillan, 2009.

Hodge, R. *Genetic Engineering.* New York: Facts on File, 2009.

Hubbard, R. and E. Wald. *Exploding the Gene Myth.* Boston, MA: Beacon Press, 1994.

Jenkins, Joseph. *The Humanure Handbook.* White River Junction, VT: Chelsea Green Publishing, 1999.

Kitcher, P. *The Lives to Come.* New York: Penguin, 1997.

Kuhse, H. and P. Singer, eds. *Bioethics: An Anthology.* Oxford, UK: Blackwell, 2006.

Kurzweil, Raymond. *The Singularity Is Near: When Humans Transcend Biology.* New York: Viking Press, 2005.

Lechner, Norbert. *Heating, Cooling, Lighting: Sustainable Design Methods for Architects.* Hoboken, NJ: Wiley, 2009.

Lehmann, J. and S. Joseph. *Biochar for Environmental Management.* London: Earthscan, 2009.

Lucena, Juan. *Defending the Nation: Policymaking in Science and Engineering Education from Sputnik to the War Against Terrorism.* Lanham, MD: University Press of America, 2005.

Mann, Charles. *1491: New Revelations of the Americas Before Columbus.* New York: Vintage, 2006.

Marcuse, Herbert. *One-Dimensional Man.* Boston, MA: Beacon Press, 1964.

Mazria, Edward. *The Passive Solar Energy Book*. New York: Rodale Press, 1979.

McDonough, William, and Michael Braungart. *Cradle to Cradle: Remaking the Way We Make Things*. New York: North Point Press, 2002.

McHarg, Ian L. *Design With Nature*. Hoboken, NJ: Wiley, 1992.

Milani, Brian. *Designing the Green Economy: The Postindustrial Alternative to Corporate Globalization*. Lanham, MD: Rowman and Littlefield, 2000.

Nussbaum, M. and C. Sunstein, eds. *Clones and Clones*. New York: W.W. Norton, 1998.

Ramlow, Bob and Benjamin Nusz. *Solar Water Heating: A Comprehensive Guide to Solar Water and Space Heating Systems*. Gabriola Island, British Columbia, Canada: New Society Publishers, 2006.

Reiss, M. and R. Straughan. *Improving Nature?* Cambridge, UK: Cambridge University Press, 1996.

Rogers, G. and Y. Mayhew. *Thermodynamics: Work and Heat Transfer*. Harlow, UK: Longman Scientific, 1996.

Ruse, M and C. Pynes, eds. *The Stem Cell Controversy*. New York: Prometheus Books, 2003.

Sandel, M. *The Case Against Perfection*. Cambridge, MA: Harvard University Press, 2007.

Shapin, Steven. *The Scientific Revolution*. Chicago: University of Chicago Press, 1996.

Shiva, V. *Biopiracy: The Plunder of Nature and Knowledge*. Cambridge, MA: South End Press, 1999.

Singer, P. and H. Kuhse. *Should the Baby Live?* Oxford, UK: Oxford University Press, 1985.

Singer, P. and A. M. Viens, eds. *The Cambridge Textbook of Bioethics*. Cambridge, UK: Cambridge University Press, 2008.

Steinbock, B. *The Oxford Handbook of Bioethics*. Oxford, UK: Oxford University Press, 2009.

Stitt, Fred A. *Ecological Design Handbook: Sustainable Strategies for Architecture, Landscape Architecture, Interior Design, and Planning*. New York: McGraw-Hill, 1999.

Van Der Ryn, Sim and Stuart Cowan. *Ecological Design*. Washington, DC: Island Press, 2007.

Williams, Daniel Edward. *Sustainable Design: Ecology, Architecture, and Planning*. Hoboken, NJ: Wiley, 2007.

Williams, P. *Waste Treatment and Disposal*. Hoboken, NJ: Wiley, 1998.

Winner, Langdon. *Autonomous Technology: Technics out of Control as a Theme in Political Thought*. Cambridge, MA: MIT Press, 1977.

Wisnioski, Matthew. *Engineers for Change: America's Culture Wars and the Making of New Meaning in Technology*. Cambridge, MA: MIT Press, 2010.

Yeang, Ken. *Ecodesign. A Manual for Ecological Design*. Hoboken, NJ: Wiley, 2008.

Zehner, Ozzie. *Coming Clean: The Dirty Truth About Clean Energy and the Real Future of Environmentalism*. Lincoln: University of Nebraska Press, 2011.

Zerzan, John, ed. *Against Civilization: Readings and Reflections*. Port Townsend, WA: Feral House, 2005.

Journals

Annual Review of Energy and the Environment

Bioresource Technology

Colorado Journal of International Environmental Law and Policy

Corporate Social Responsibility and Environment Management

Developing World Bioethics

Engineering Studies
Environmental and Resource Economics
Environmental Communication: A Journal of Nature and Culture
Environmental Quality Management

Harvard Environmental Law Review
History and Technology

Medicine, Health Care and Philosophy

Nature Photonics

Science, Technology and Human Value
Social Forces
Social Studies of Science
Stanford Environmental Law Journal

Technology and Culture

Water Science and Technology: Water Supply

Websites

Biochar Fund
  www.biocharfund.org

Carbon Footprint Calculator
  www.epa.gov/climatechange/emissions/ind_calculator.html

Carbon Tax Center
  www.carbontax.org

CorpWatch
  www.corpwatch.org

Global Footprint Network
  www.footprintnetwork.org

Green Electronic Council
  www.epeat.net

The Greener Blog
  www.greenenergytechnology.org

Greenhouse Gas Protocol
  www.ghgprotocol.org

Green Technology: Strategy and Leadership for Clean and Sustainable Communities
  www.green-technology.org

The Greenwashing Index
  www.greenwashingindex.com

Industry and Technology: EU Ecolabel
  ec.europa.eu/environment/ecolabel/index_en.htm

International Biochar Initiative
  www.biochar-international.org

Occupational Safety and Health Administration
  www.osha.gov

Scientific American: Green Technology
  www.scientificamerican.com/topic.cfm?id=green-technology

United Nations Development Programme
  www.un.org/en/development

United Nations 2015 Millennium Development Goals
  www.un.org/millenniumgoals

U.S. Department of Energy: Energy Efficiency and Renewable Energy
  www.eere.energy.gov

Yale Environment 360: Opinion, Analysis, Reporting and Debate
  e360.yale.edu

Green Technology Appendix

Biofuels Digest

www.biofuelsdigest.com

This is the website published by Ascension Publishing and edited by Jim Lane that, according to the Internet analysis services Alexa and Quantcast, is the most widely read biofuels daily in the world with readers in over 200 countries. It collects information from many sources and also publishes new material by the editor with a strong emphasis on industry news. The scope is international and includes opinion pieces, columns, news, financial analysis, policy information, research, and demand-side information about consumers and fleets. There is also a data section that includes studies and reports (many from outside sources such as the General Accounting Office and the Oak Ridge National Laboratory, and many of which are downloadable) on topics such as the impact of biofuels on the green jobs economy and the technical feasibility of using biomass to replace 30 percent of the current U.S. petroleum consumption. There is also a jobs section and information about the stocks that make up the Biofuels Digest Index

GreenBuilding: Improved Energy Efficiency for Non-Residential Buildings

www.eu-greenbuilding.org

This is the website of the GreenBuilding Programme (GBP) begun by the European Commission in 2004 to promote voluntary and cost-effective integration of renewable energy and improved energy efficiency in nonresidential buildings in Europe. The website includes basic information about green buildings and organizes technical and policy information as well as links to relevant outside organizations in categories or "Key Messages": sustainable summer comfort, heating, combined heat and power, solar hot water and heating, air conditioning, lighting, office equipment, and benchmarking. It also includes announcements on upcoming GreenBuilding events, contacts in different countries, a link to subscribe to the GreenBuilding quarterly newsletter, information about becoming a GreenBuilding Partner or Endorser, and an index of best practice documents related to sustainable building.

Green Chemistry

www.epa.gov/gcc

This is the website maintained by the U.S. Environmental Protection Agency (EPA), which provides basic information about green chemistry (sustainable chemistry) and EPA activities to promote green chemistry. The website also includes information about EPA programs and partnerships such as the Presidential Green Chemistry Challenge competition, the Joseph Breen Memorial Fellowship in Green Chemistry, the Kenneth G. Hancock memorial Award in Green Chemistry, cooperative activities with the American Chemical Society and the American Chemical Society Green Chemistry Institute. Information about grants and fellowships in green chemistry includes the Technology for a Sustainable Environment program, the EPA's Small Business Innovation Research Program, various National Science Foundation programs, and Technology Vision 2020. One section of the website is devoted to the P2 (Pollution Prevention) Recognition Project that honors companies for developing technologies and innovative chemistry, which aids in preventing pollution, including information about past winners and their projects. The website includes brief descriptions of and links to educational materials for green chemistry produced by other organizations, including Beyond Benign (for K–12 educators), Greener Education Materials for Chemists (including lab exercises, lecture materials, and multimedia content), the Green Chemistry Education Network, and the University of Scranton.

National Geographic Society: Energy

environment.nationalgeographic.com/environment/energy

This the website created by the National Geographic Society that presents information for the general reader on various topics related to the environment and energy. A particular feature of this website is the inclusion of high-quality photography and graphics, video, and/or interactive interfaces for many of the topics covered. Information is organized into general information about energy plus major sections on biofuels, fuel cells, hydropower, solar energy, wind power, and geothermal energy. Many other topics are also covered, including energy conservation, freshwater and energy, wind power, nuclear power, and greenhouse gases. A 2010 Greendex Map of the World provides interactive information on how different countries stack up on consumer behaviors, knowledge, and attitudes related to conservation. The website also includes buying guides for different types of products and features such as quizzes and calculators (for instance, to calculate your water footprint) to help users judge their own environmental knowledge and behavior.

Renewable & Alternative Fuels

www.eia.doe.gov/fuelrenewable.html

This is the website created and maintained by the U.S. Energy Information Administration that contains basic information about renewable energy (including estimates of current use in comparison to total energy use in the United States), reports on different forms of renewable energy, and data and statistics about its use. Data is categorized as U.S. or international and is organized into categories, including total capacity, generation and consumption; biomass; geothermal; hydro; solar; wind; alternative transportation fuel and alternative fueled vehicles; and ethanol. A kids' page offers more basic information and definitions about different types of renewable energy and how much they are used in the

United States along with games and activities, tips for saving energy, and a teacher's section including several energy conversion calculators and a glossary of terms. Numerous reports and analyses are available on the website, and historical summary data is available going back to 1949 (in some cases) on topics such as renewable energy production and consumption by primary energy source, estimated number of alternative fuel vehicles in use, and shipments of solar thermal collectors and photovoltaic cell and module shipments.

Society for Sustainable Mobility

www.osgv.org/about-society-sustainable-mobility

This is the website of a nonprofit organization established in 2005 that seeks to combine open design (including open source, public license technology) with sustainable technologies and sound business strategy to create a transportation alternative to conventional automobiles. Although the Society for Sustainability Mobility (SSM) engages in other sustainability activities, they are best known for developing the KernelTM Crossover, a hybrid-electric vehicle developed by an international team of engineers and that is expected to go into production in 2011. The website includes information about the open design process (analogous to open-source software, all design and test data is available for free on the Internet), general and technical data about the KernelTM Crossover (which is projected to operate at higher than 100 miles to the gallon-equivalent fuel economy and achieve a top speed of 125 miles per hour), a FAQ section about open design, electric cars, participating in the design process and SSM, and a blog that collects information about electric vehicles and other sustainability issues.

Sarah Boslaugh
Washington University in St. Louis

Index

Article entries and their page numbers are in **bold**.

industrial development in, 282
landfill gas produced in, 66
maglev trains in, 296
nuclear weapons in, 30
renewable energy policies in, 101
technological innovation in, 435
total resource recovery systems in, 449
university-industrial complex in, 432–433
urine diverting toilets in, 449
waste separation scheme in, 439
waste-to-energy incineration plants in, 442
wind turbines in, 458, 459
Giddens, Anthony, 149, 345, 389–390
Gill, William N., 306
Gingrey, Phil, 212
GIS. *See* **Geographic information systems (GIS)**
GIT. *See* **Information technology** (green information technology [GIT])
Globalization
anti-globalization, anti-Luddism and, 292
ecological modernization and, 151
engineering studies and, 160
global north *vs.* global south countries and, 216
green metrics of, 219–220
information technology and, 257
intellectual property rights and, 271
IT e-waste and, 257
negative impacts of, corporate social responsibility and, 219–220
of pesticides, 139
science and technology policy and, 354–355
Science and Technology Studies and, 357
technology and social change and, 412–414
See also Global warming
Global Reporting Initiative (GRI), 219, 220, 221
Global warming, 69, 197–201. *See also* **Carbon capture technology;** Climate change; Greenhouse gases (GHGs)
GM *See* Genetically modified (GM) biotechnology
Google, 237, 259, 266
Google Maps, 204
Graetzel, M., 127
Grätzel, Michael, 379
Greece, 70
Green Building Council. *See* U.S. Green Building Council
Green building materials, 206–210
air quality and, 209
building site selection and, 207

characteristics of, 207–208
costs, costs savings of, 207
decision-making and, 207
design techniques of, 208
Earthships and, 141–143
energy legislation and, 207
environmentally friendly approach of, 206, 207
examples of, 209–210
goals of, 206
government support of, 206–207
history of, 206
LEED standards and, 206
lighting techniques, 208
maintenance issues and, 209
passive solar designs, 208
renewable energy sources and, 208
salvaging from, 206
waste from, 206
water issues and, 209
white roofs, 208, 454–455
See also **Appliances, energy efficient; Green roofing; LEED standards; Sustainable design; Zero-energy building** (ZEB)
Green chemistry, 210–213
atom economy focus of, 211–212
awards in, 212, 213
cyclo addition reaction and, 212
government support for, 212
Green Chemistry Research and Development Act bill and, 212, 224–225
hydrogenation chemical reaction and, 211
industrial partners of, 212
journals and institutions of, 212
metathesis reaction and, 211–212
origins of, 210–211
pollution prevention focus of, 211–212
Presidential Green Chemistry Challenge Awards and, 212
principles of, 211
synthetic chemistry and, 211
Green electronics. *See* **Eco-electronics**
Green Electronics Council (GEC), 153–154, 259
Green Electronics Product Survey (Greenpeace), 147
Green engineering, 226–227
Green Grid, 255, 257–258
Greenhouse gases (GHGs)
Architecture 2030 Challenge and, 395
from biochar production, 53
carbon finance and, 81–85
desalination and, 132